Kwo Young (Ed.)

Nickel Metal Hydride Batteries

This book is a reprint of the Special Issue that appeared in the online, open access journal, *Batteries* (ISSN 2313-0105) from 2015–2016, available at:

http://www.mdpi.com/journal/batteries/special_issues/ni-mh-batteries

Guest Editor
Kwo Young
Department of Chemical Engineering and Material Science
Wayne State University
USA

Editorial Office
MDPI AG
St. Alban-Anlage 66
Basel, Switzerland

Publisher
Shu-Kun Lin

Managing Editor
Jing Su

1. Edition 2016

MDPI • Basel • Beijing • Wuhan • Barcelona • Belgrade

ISBN 978-3-03842-302-7 (Hbk)
ISBN 978-3-03842-303-4 (electronic)

Table of Contents

Chapter 2: Metal Hydride Alloys

Chapter 3: Electrolyte

Chapter 4: Analytic Methodology

List of Contributors

Leonid A. Bendersky Material Measurement Laboratory, National Institute of Standards and Technology, Gaithersburg, MD 20899, USA.

Shiuan Chang Department of Mechanical Engineering, Wayne State University, Detroit, MI 48202, USA.

Benjamin Chao BASF/Battery Materials-Ovonic, 2983 Waterview Drive, Rochester Hills, MI 48309, USA.

Cristian Fierro BASF/Battery Materials–Ovonic, 2983 Waterview Drive, Rochester Hills, MI 48309, USA.

Baoquan Huang BASF/Battery Materials-Ovonic, 2983 Waterview Drive, Rochester Hills, MI 48309, USA.

John Koch BASF/Battery Materials-Ovonic, 2983 Waterview Drive, Rochester Hills, MI 48309, USA.

Yi Liu Department of Chemistry, Wayne State University, Detroit, MI 48201, USA.

Tiejun Meng BASF/Battery Materials-Ovonic, 2983 Waterview Drive, Rochester Hills, MI 48309, USA.

Dhanashree Moghe Department of Chemical Engineering and Materials Science, Michigan State University, East Lansing, MI 48824, USA.

Negar Mosavati Department of Chemical Engineering and Materials Science, Wayne State University, Detroit, MI 48202, USA.

Jean Nei BASF/Battery Materials-Ovonic, 2983 Waterview Drive, Rochester Hills, MI 48309, USA.

K. Y. Simon Ng Department of Chemical Engineering and Materials Science, Wayne State University, Detroit, MI 48202, USA.

Taihei Ouchi BASF/Battery Materials-Ovonic, 2983 Waterview Drive, Rochester Hills, MI 48309, USA.

Damian Rotarov BASF/Battery Materials-Ovonic, 2983 Waterview Drive, Rochester Hills, MI 48309, USA.

Hao-Ting Shen BASF/Battery Materials-Ovonic, 2983 Waterview Drive, Rochester Hills, MI 48309, USA; Material Measurement Laboratory, National Institute of Standards and Technology, Gaithersburg, MD 20899, USA.

Lixin Wang BASF/Battery Materials-Ovonic, 2983 Waterview Drive, Rochester Hills, MI 48309, USA.

Diana F. Wong BASF/Battery Materials-Ovonic, 2983 Waterview Drive, Rochester Hills, MI 48309, USA; Department of Chemical Engineering and Materials Science, Wayne State University, Detroit, MI 48202, USA.

Xin Wu Department of Mechanical Engineering, Wayne State University, Detroit, MI 48202, USA.

Suli Yan Department of Chemical Engineering and Materials Science, Wayne State University, Detroit, MI 48202, USA.

Shigekazu Yasuoka FDK Corporation, 307-2 Koyagimachi, Takasaki 370-0071, Gunma, Japan.

Kwo-Hsiung Young Department of Chemical Engineering and Materials Science, Wayne State University, Detroit, MI 48202, USA; BASF/Battery Materials—Ovonic, 2983 Waterview Drive, Rochester Hills, MI 48309, USA.

About the Guest Editor

Kwo-hsiung Young is the Chief Scientist in the BASF-Ovonic located in Rochester Hills, Michigan, USA. He graduated from National Taiwan University, Republic of China, in 1982 with a BS in Electrical Engineering and graduated from Princeton University, New Jersey, in 1989, with a Ph.D. also in the Electrical Engineering Department. He has been in the research field of Ni/MH batteries for more than 25 years. He is one of the key inventors who aim to increase the power of NiMH battery technology, and successfully implemented the technology in EV and HEV with support from United State Advanced Battery Consortium (USABC). He has 38 US Patents in Ni/MH battery technology which form the basis to the licenses for battery manufactures. Dr. Young also serves as research professor at Wayne State University, Michigan, where he supervises Ph.D. students in electrochemical materials research. In recent years, he has published over 100 technical papers in the field of metal hydrides for electrochemical applications.

Preface to "Nickel Metal Hydride Batteries"

Nickel/metal hydride (Ni/MH) batteries are presently used extensively in hybrid electric vehicles (HEVs). More than 10 million HEVs based on NiMH batteries have been manufactured and driven, and NiMH battery chemistry is expected to continue dominating the HEV market with its proven abuse tolerance, wide operating-temperature range, and durable service life. With the main goal of achieving higher gravimetric energy density while maintaining safety and robustness advantages, continuous efforts in improving the performances of NiMH batteries are very much needed in order to explore their possible use in other applications, such as battery-powered electric vehicles, the stationary market, and more. Meanwhile, with the inherited high volumetric energy density, the NiMH battery may have a chance to return to application in portable electronic devices. In this Special Issue of *Batteries*, review papers, current research, and future projection in the materials, fabrication methods, cell integration and development, performance evaluation, failure analysis, and other subjects related to NiMH batteries are included.

Kwo Young
Guest Editor

Research in Nickel/Metal Hydride Batteries 2016

Kwo-Hsiung Young

Abstract: Nineteen papers focusing on recent research investigations in the field of nickel/metal hydride (Ni/MH) batteries have been selected for this Special Issue of Batteries. These papers summarize the joint efforts in Ni/MH battery research from BASF, Wayne State University, the National Institute of Standards and Technology, Michigan State University, and FDK during 2015–2016 through reviews of basic operational concepts, previous academic publications, issued US Patent and filed Japan Patent Applications, descriptions of current research results in advanced components and cell constructions, and projections of future works.

Reprinted from *Batteries*. Cite as: Young, K.-H. Research in Nickel/Metal Hydride Batteries 2016. *Batteries* **2016**, *2*, 31.

1. Introduction

Nickel/metal hydride (Ni/MH) rechargeable batteries are one of the important power sources for various consumer types of mobile applications, stationary energy storage, and, most distinctively, transportation usages. In the consumer market, more than one billion cylindrical cells are built annually to replace highly toxic Ni–Cd batteries and throw-away primary alkaline batteries with Ni/MH batteries of the same nominal voltage (1.2 V) and higher energy [1,2]. In stationary applications, Ni/MH batteries offer a wide operation/storage temperature range, a high energy density, and a very long service life [3–5]. For propulsion applications, more than 10 million hybrid electrical vehicles currently on the road are powered by Ni/MH batteries [6]. New applications in start–stop types of micro-hybrid electrical vehicles [7], temporary energy storage for train braking [8], ferries [9], and buses [10] are on the horizon.

Although Ni/MH batteries have an excellent track record for high abuse-tolerance and endurable service life, these batteries suffer from a relatively low gravitational energy density when comparing to rival Li-ion batteries [11]. The demand for higher mileage between charges limits the future perspectives of Ni/MH batteries in pure battery-powered electrical vehicles. In order to preempt the gap in energy density, ongoing research activities in Ni/MH are currently being conducted in the US, China, Japan, and Europe [12]. In this Special Issue of the journal Batteries, nineteen papers from research within the USA have been collected to reflect the current status of research and development in the area of Ni/MH batteries.

2. Contributions

The selected papers presented in this Special Issue are highlighted in this section. They are mainly from the research work conducted under a United States Department of Energy (DOE)–Advanced Research Projects Agency–Energy granted program (DE-AR0000386) and can be divided into four general categories: reviews on overall programs and Patents in the area (four papers); metal hydride (MH) alloys used as negative electrode active materials in Ni/MH batteries (eight papers); electrolyte composition and additives (two papers); and uses of analytic tools to investigate the nature and failure modes of components in Ni/MH batteries (five papers).

2.1. Reviews in Related Work

In this area, a single paper has been devoted to reviewing the major accomplishments of the Robust Affordable Next Generation Energy Storage System (RANGE) program funded by the DOE [13]; two papers are reviews of Patents, specifically those granted in the US [14] and applied in Japan [15]; and one review of the field of failure analysis of Ni/MH batteries [16]. In the RANGE program, new anodes, cathodes, and electrolytes—together with a new pouch type of cell assembly—were developed to boost the gravitational energy density of Ni/MH batteries to 145 Wh·kg^{-1} on the cell level. The combination of an advanced Si-anode with an extremely high potential capacity (up to 4000 mAh·g^{-1}) and an ionic liquid electrolyte has led to a new era of Ni/MH battery development [13]. In the paper reviewing US Patents on the subject of Ni/MH batteries, 350 US Patents were studied, beginning with active materials, to electrode fabrication, cell assembly, system integration, application, and finally recovery and recycling [14]. This paper also gives a brief introduction to the major components used in Ni/MH batteries. Another paper reviewing Japanese Patent Application takes a different approach. Instead of by subject manner, these Patent Applications were categorized by the filing company/institute [15]. Applications from nine top Ni/MH battery manufacturers, five major component suppliers, and three research institutes (all based in Japan) are included, with special emphasis on the evolution of melting/casting apparatuses, fabrication of paste electrode, and cell construction. The last review paper focuses on studies of failure modes and degradation mechanisms of Ni/MH batteries [16]. The paper first gives a brief introduction to the structure of Ni/MH batteries and the common experimental methods used in failure analysis. It then describes the capacity loss mechanism under various conditions (temperature, rate, and storage duration), and finally, presents methods for improving the cycle stability using six approaches: improvement to cell design, negative and positive electrodes, separator, electrolyte, and other components.

2.2. Metal Hydride Alloys

MH alloys are the active component in the negative electrodes of Ni/MH batteries and are capable of reversibly storing hydrogen in an electrochemical environment [17]. MH alloys with suitable metal-bond strengths for room-temperature electrochemical application can be categorized as solid-solution and pseudo-binary inter-metallic alloys, specifically A_3B, A_2B, AB, AB_2, AB_3, A_2B_7, and AB_5, where A is one or a combination of rare earth, alkaline earth, and light transition metal elements (Ti and Zr) and B is from the group of transition metals (mainly Ni) [18]. Comparisons of the general properties [19] and the high-rate potentials [20] of these alloy systems are available. Out of the eight available alloy systems, four are discussed in this Special Issue and are summarized as follows. Modifications of the A-site atom in body-centered-cubic (bcc) solid-solution alloys increases the storage capacity [21]. The effects of the incorporation of Mg or Ce in the Laves phase-based AB_2 MH alloys are discussed in [22,23], respectively. Formula optimization [24] and A-site substitution [25] in a series of Laves phase-related bcc alloys leads to a MH alloy suitable for electrical vehicle applications (P37 in [13]). TiNi-based AB MH alloys were investigated due to their low raw material costs and because they are free of rare earth elements [19]. Density function theory has also been employed to study the solubility of two ZrNi-based intermetallic alloys [26]. Last but not least, a new concept of using nickel hydroxide as the anode for Ni–Ni batteries is discussed [27]. In addition to the eight papers focusing on MH alloys, the failure mechanisms of a series of Co-substituted A_2B_7 superlattice alloys is discussed [28], and initial research activities focused on an Mg-based AB MH alloy can be found in the paper discussing the contribution of various hydroxides in the electrolyte [29].

2.3. Electrolyte

Part of the high-rate charge/discharge capabilities in Ni/MH batteries originates from the use of highly conductive alkaline electrolytes (30–35 wt% KOH). However, the highly corrosive nature of these electrolytes limits the choice of MH alloys. For example, extremely low cycle stabilities have been reported with Mg- [30,31] and V-containing [32] MH alloys. Therefore, studies focused on balancing corrosion and conductivity in the electrolytes were conducted through the choice of hydroxides [29] and salt additives [33]. In addition, the use of ionic liquid to replace alkaline solution as electrolyte was shown to be effective in reducing corrosion, which allowed attempts to develop high-capacity Si-anodes [13].

2.4. Analytic Methodology

Many analytic tools have been applied during the research and development of Ni/MH batteries. While analytical methods for MH alloy research can be found in one article [34], those involved in the failure analysis are summarized in a paper in this Special Issue [16]. In this Special Issue, the many uses for transmission electron microscopy (TEM) [35], electron backscatter diffraction (EBSD) [36,37], and X-ray energy dispersive spectroscopy (EDS) elemental mapping in a scanning electron microscope (SEM) [28], X-ray diffraction (XRD), and newly developed electrochemical pressure–concentration–temperature (PCT) measurements [38] were demonstrated to be effective for investigations into the microstructures and various mechanisms in electrochemistry and hydrogen gas–solid interactions. TEM results for a Si-doped AB_2 MH alloy [35] revealed a highly catalytic surface/interface microstructure which accounts for the superior low-temperature performance of the alloy and varies greatly from the conventional nano-Ni clusters embedded in surface oxide model [39,40]. The alignment in the crystallographic orientations of the constituent phases revealed by the EBSD technique [36,37] confirm the cleanness of the interface, which is therefore capable of generating synergetic effects and boosting the electrochemical performance of the multi-phase MH alloy systems [25]. A study comparing gaseous phase PCT and electrochemical PCT further distinguishes the synergetic effects in both environments [38]. The last paper exhibits a combination of analytic tools—including inductively coupled plasma, XRD, SEM, and EDS—to study the failure mechanism of AB_5 and A_2B_7 MH alloys after cycling at high temperature in a sealed-cell configuration [28].

3. Conclusions

The joint research efforts from BASF-Ovonic and their collaborators (2015–2016) are highlighted here through nineteen papers focused on the area of Ni/MH batteries in this Special Issue of Batteries. It has been demonstrated that achieving equalization of the energy density in the pack level between Ni/MH and rival Li-ion batteries is possible through the use of advanced components obtained from these studies. Future research will be focused on high-capacity Si-anodes, choice of high-voltages, multi-electron transfer cathodes, and implementation of the pouch cell design with the use of ionic liquid as the electrolyte.

Acknowledgments: The Guest Editor (Kwo-Hsiung Young) thanks both the colleagues who made impressive and important contributions to the articles and the editorial team at the publisher MDPI for rending precious guidance. Kwo-Hsiung Young is also obligated to RoseFigura Jordan at the Rockefeller University for refinement in his writing skill.

Conflicts of Interest: The author declares no conflict of interest.

References

1. Teraoka, H. NiMH Stationary Energy Storage—Extreme Temperature and Long Life Developments. In Proceedings of the 33rd International Battery Seminar & Exhibit, Fort Lauderdale, FL, USA, 21–24 March 2016.

2. HighPower International. The Current Status and Future Trend of Domestic and International Market of Ni/MH Batteries. 2014. Available online: http://cbea.com/u/cms/www/201406/06163842rc0l.pdf (accessed on 8 September 2016). (In Chinese)

3. Zelinsky, M.; Koch, J.; Fetcenko, M. *Heat Tolerant NiMH Batteries for Stationary Power*; Ovonic Battery Company: Rochester Hill, MI, USA, 2010.

4. Zelinsky, M.; Koch, J. Batteries and Heat—A Recipe for Success? Available online: www.battcon.com/Papers Final2013/16-Mike%20Zelinsky%20-%20Batteries%20and%20Heat.pdf (accessed on 8 September 2016).

5. Zelinsky, M.; Koch, J. Market Advancement of NiMH Batteries for Stationary Applications. Available online: www.battcon.com/PapersFinal2016/Zelinsky%20paper%202016.pdf (accessed on 8 September 2016).

6. Wikipedia. Hybrid Electric Vehicle. Available online: https://en.wikipedia.org/wiki/Hybrid_electric_vehicle (accessed on 8 September 2016).

7. Panasonic. Headquarters News—Panasonic's 12V Ni-MH Energy Recovery Systems in New Idle-Stop Minicars from Nissan and Mitsubishi. 2014. Available online: http://news.panasonic.com/global/press/data/2014/02/en140213-3/en140213-3.html (accessed on 8 September 2016).

8. Kawasaki Heavy Industry. Battery Power System (BPS) for Railways. Available online: http://global.kawasaki.com/en/energy/solutions/battery_energy/applications/bps.html (accessed on 8 September 2016).

9. Green City Ferries. MOVITZ—The World's First Supercharged Electrical Ferry. Available online: http://www.greencityferries.com/boatfleet/movitz/ (accessed on 8 September 2016).

10. Zibo Guoli New Power Source Technology Co., Ltd. Available online: http://www.glxdy.com (accessed on 8 September 2016).

11. Young, K.; Wang, C.; Wang, L.Y.; Strunz, K. Electrical Vehicle Battery Technologies. In *Electric Vehicle Integration into Modern Power Network*; Garcia-Valle, R., Lopes, J.A.P., Eds.; Springer: New York, NY, USA, 2013.

12. Yartys, V.A. Ti-Zr Based AB_2 Alloys for High Power Metal Hydride Batteries. In Proceedings of the 15th International Symposium on Metal-Hydrogen System, Interlaken, Switzerland, 7–12 August 2016.

13. Young, K.; Ng, K.Y.S.; Bendersky, L.A. A technical report of the Robust Affordable Next Generation Energy Storage System-BASF program. *Batteries* **2016**, *2*.

14. Chang, S.; Young, K.; Nei, J.; Fierro, C. Reviews on the U.S. Patents regarding nickel/metal hydride batteries. *Batteries* **2016**, *2*.

15. Ouchi, T.; Young, K.; Moghe, D. Reviews on the Japanese Patent Applications regarding nickel/metal hydride batteries. *Batteries* **2016**, *2*.

16. Young, K.; Yasuoka, S. Capacity degradation mechanisms in nickel/metal hydride batteries. *Batteries* **2016**, *2*.

17. Young, K. Stoichiometry in Inter-Metallic Compounds for Hydrogen Storage Applications. In *Stoichiometry and Materials Science: When Numbers Matter*; Innocenti, A., Kamarulzaman, N., Eds.; InTech: Rijeka, Croatia, 2012.

18. Young, K. Electrochemical Applications of Metal Hydrides. In *Compendium of Hydrogen Energy*; Barbir, F., Basile, A., Veziroğlu, T.N., Eds.; Woodhead Publishing Ltd.: Cambridge, UK, 2016; Volume 3, pp. 289–304.

19. Nei, J.; Young, K. Gaseous phase and electrochemical hydrogen storage properties of $Ti_{50}Zr_1Ni_{44}X_5$ (X = Ni, Cr, Mn, Fe, Co, or Cu) for nickel metal hydride battery applications. *Batteries* **2016**, *2*.

20. Young, K.; Nei, J. The current status of hydrogen storage alloy development for electrochemical applications. *Materials* **2013**, *6*, 4574–4608.

21. Young, K.; Ouchi, T.; Huang, B.; Nei, J. Structure, hydrogen storage, and electrochemical properties of body-centered-cubic $Ti_{40}V_{30}Cr_{15}Mn_{13}X_2$ alloys (X = B, Si, Mn, Ni, Zr, Nb, Mo, and La). *Batteries* **2015**, *1*, 74–90.

22. Chang, S.; Young, K.; Ouchi, T.; Meng, T.; Nei, J.; Wu, X. Studies on incorporation of Mg in Zr-based AB_2 metal hydride alloys. *Batteries* **2016**, *2*.

23. Young, K.; Ouchi, T.; Nei, J.; Moghe, D. The importance of rare-earth additions in Zr-based AB_2 metal hydride alloys. *Batteries* **2016**, *2*.

24. Young, K.; Wong, D.F.; Nei, J. Effects of vanadium/nickel contents in Laves phase-related body-centered-cubic solid solution metal hydride alloys. *Batteries* **2015**, *1*, 34–53.

25. Young, K.; Ouchi, T.; Meng, T.; Wong, D.F. Studies on the synergetic effects in multi-phase metal hydride alloys. *Batteries* **2016**, *2*.

26. Wong, D.F.; Young, K.; Ouchi, T.; Ng, K.Y.S. First-principles point defect models for Zr_7Ni_{10} and Zr_2Ni_7 phases. *Batteries* **2016**, *2*.

27. Wang, L.; Young, K.; Shen, H. New type of alkaline rechargeable battery—Ni-Ni battery. *Batteries* **2016**, *2*.

28. Meng, T.; Young, K.; Koch, J.; Ouchi, T.; Yasuoka, S. Failure mechanisms of nickel/metal hydride batteries with cobalt-substituted superlattice hydrogen-absorbing alloy anodes at 50 °C. *Batteries* **2016**, *2*.

29. Nei, J.; Young, K.; Rotarov, D. Studies on MgNi-based metal hydride electrode with aqueous electrolytes composed of various hydroxides. *Batteries* **2016**, *2*.

30. Mu, D.; Hatano, Y.; Abe, T.; Watanabe, K. Degradation kinetics of discharge capacity for amorphous Mg-Ni electrode. *J. Alloys Compd.* **2002**, *334*, 232–237.

31. Liu, J.; Jiao, L.; Yuan, H.; Wang, Y.; Liu, Q. Effect of discharge cut off voltage on cycle life of MgNi-based electrode for rechargeable Ni-MH batteries. *J. Alloys Compd.* **2005**, *403*, 270–274.

32. Yu, X.B.; Wu, Z.; Xia, B.J.; Xu, N.X. A Ti-V-based bcc phase alloy for use as metal hydride electrode with high discharge capacity. *J. Chem. Phys.* **2004**, *121*, 987–990.

33. Yan, S.; Young, K.; Ng, K.Y.S. Effects of salt additives to the KOH electrolyte used in Ni/MH batteries. *Batteries* **2015**, *1*, 54–73.

34. Young, K. Metal Hydride. In *Reference Module in Chemistry, Molecular Sciences and Chemical Engineering*; Reedijk, J., Ed.; Elsevier: Waltham, MA, USA, 2013.

35. Young, K.; Chao, B.; Nei, J. Microstructures of the activated Si-containing AB_2 metal hydride alloy surface by transmission electron microscope. *Batteries* **2016**, *2*.

36. Liu, Y.; Young, K. Microstructure investigation on metal hydride alloys by electron backscatter diffraction technique. *Batteries* **2016**, *2*.

37. Shen, H.-T.; Young, K.-H.; Meng, T.; Bendersky, L.A. Clean grain boundary found in C14/body-center-cubic multi-phase metal hydride alloys. *Batteries* **2016**, *2*.

38. Mosavati, N.; Young, K.; Meng, T.; Ng, K.Y.S. Electrochemical open-circuit voltage and pressure- concentration-temperature isotherm comparison for metal hydride alloys. *Batteries* **2016**, *2*.

39. Young, K.; Huang, B.; Regmi, R.K.; Lawes, G.; Liu, Y. Comparisons of metallic clusters imbedded in the surface oxide of AB_2, AB_5, and A_2B_7 alloys. *J. Alloys Compd.* **2010**, *506*, 831–840.

40. Young, K.; Chao, B.; Pawlik, D.; Shen, H. Transmission electron microscope studies in the surface oxide on the La-containing AB_2 metal hydride alloy. *J. Alloys Compd.* **2016**, *672*, 356–365.

Chapter 1:
Reviews

A Technical Report of the Robust Affordable Next Generation Energy Storage System-BASF Program

Kwo-hsiung Young, K. Y. Simon Ng and Leonid A. Bendersky

Abstract: The goal of the Robust Affordable Next Generation Energy Storage System (RANGE)-BASF program is to provide an alternative solution for the energy storage media that powers electric vehicles other than the existing Li-ion battery. With the use of a rare-earth-free metal hydride (MH) as the active negative electrode material, together with a core-shell type alpha-beta nickel hydroxide as the active positive electrode and a sealed pouch design, an energy density of 145 $Wh \cdot kg^{-1}$ and cost model of \$120 kWh^{-1} are shown to be feasible. Combined with the proven safety record and cycle stability, we have demonstrated the feasibility of using a Ni-MH battery in EV applications.

Reprinted from *Batteries*. Cite as: Young, K.-h.; Ng, K.Y.S.; Bendersky, L.A. A Technical Report of the Robust Affordable Next Generation Energy Storage System-BASF Program. *Batteries* **2016**, *2*, 2.

1. Introduction

On 21 August 2013, the US Department of Energy Advanced Research Projects Agency-Energy (ARPA-E) announced an award of \$3,873,537 to BASF under the Robust Affordable Next Generation Energy Storage System (RANGE) program [1]. The program was designed for a two-year period with BASF being the primary award recipient with partners from Wayne State University (WSU), the National Institute of Standards and Technology (NIST), and Strategic Analysis, Inc. (SAI). Dr. Ping Liu was the program director and stated in his review article that his goal for the RANGE program was to develop low-cost and/or high-specific-energy storage redox chemistries [2].

In 1999, Ni/MH batteries using an AB_2 metal hydride (MH) alloy were used to power EV1, which was the first commercialized EV made by General Motors Auto Co. With a Ni/MH pack of 26.4 kWh onboard, EV1 has a drive range of 160 miles between charges (which may take up to 8 h) [3]. The specific energy of Ni/MH battery used in EV1 was only about 52 $Wh \cdot kg^{-1}$. Since then, 15 years of continuous development in the Ni/MH technology has substantially improved both specific energy [4] and cycle stability [5]. Ni/MH battery has a relatively high energy density $(Wh \cdot L^{-1})$ but a rather low specific energy $(Wh \cdot kg^{-1})$ when compared to the rival Li-ion battery because of its high density of active materials in the anode (nickel-based alloys *versus* graphite). MgNi-based MH alloy is one of the candidates to replace the current rare-earth-metal-based AB_5 and A_2B_7 MH alloys because of the low cost, low density, and high hydrogen storage capacity [6]. However, it also suffers from severe oxidation in the conventional 30 wt% KOH electrolyte (AB (MgNi) Old in Figure 1).

Therefore, in order to implement the high-capacity MgNi-based MH alloy in the EV application, a thorough study of the electrolyte is needed. The tasks outlined before the start of the program are illustrated in Figure 2.

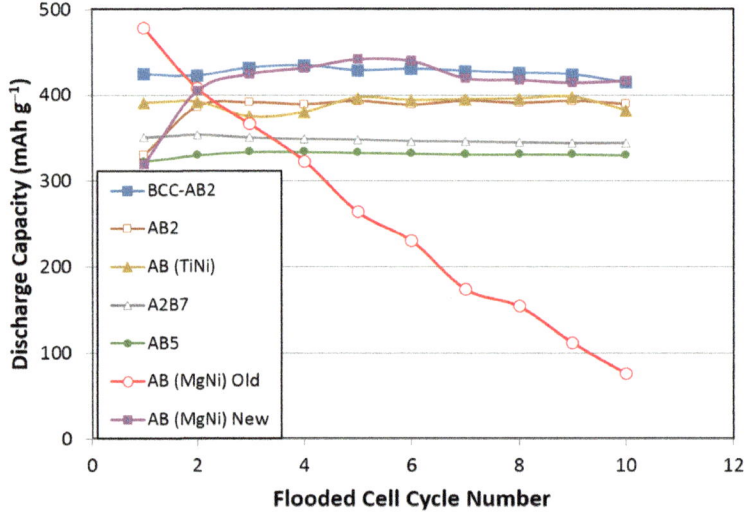

Figure 1. Electrochemical discharge capacity obtained in the flooded cell configuration with a very low discharge current density (4 mA·g^{-1}) for six commonly used metal hydride (MH) alloys used as active materials in the negative electrode. The mischmetal-based AB_5 alloy is the most used negative electrode material in today's commercial Ni/MH batteries. The AB (MgNi) Old was measured with a conventional 6M KOH, while the AB (MgNi) New was measured with a 1 M KOH + 1M NaOH electrolyte.

The scope of Task 1 focuses on the development of high-capacity MH alloys, which includes MgNi, BCC, and Si, as shown in Figure 3.

Task 2 is to develop an alternative electrolyte system to reduce the corrosive nature and possibly to expand the redox reaction voltage window (Figure 4). Task 3 is to develop high-capacity cathode active materials for the conventional alkaline electrolyte and the newly developed organic solvent/ionic-liquid-based electrolyte. Task 4 is to integrate these advanced materials in a 100-Ah cell developed for EV application (C/3 charge and discharge rates). Finally, Task 5 is designed to facilitate the commercialization of the Ni/MH battery for EV application. Three risk mitigation

(RM) plans were included: RM-A was to investigate carbon-based (carbon nanotube, graphite, and graphene) anode material; RM-B was to use perovskite oxide as the solid electrolyte; and RM-C was to develop non-hydroxide-based cathode material for the organic electrolyte. As for the contributions from different partners, NIST will support the microstructural analysis in Tasks 1 and 3, WSU will play an important role in Tasks 2 and 3, and SAI will focus on Task 5.

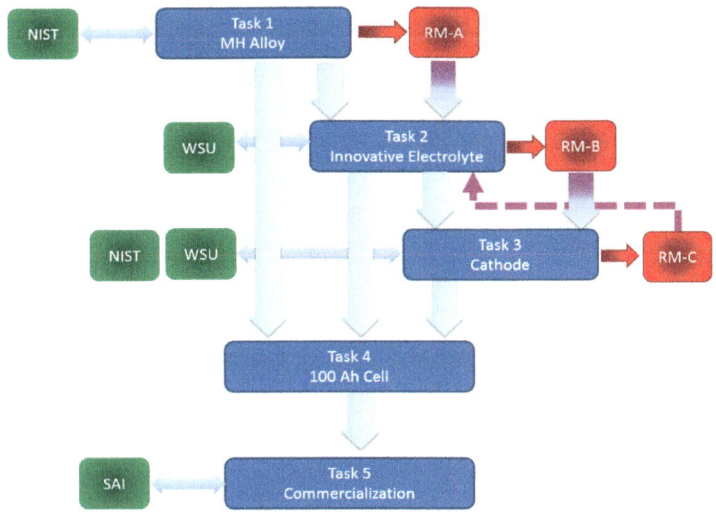

Figure 2. Schematic of tasks (in blue boxes), risk mitigations (RM, in red boxes), and interactions with partners (in green boxes) for the Robust Affordable Next Generation Energy Storage System (RANGE)-BASF program.

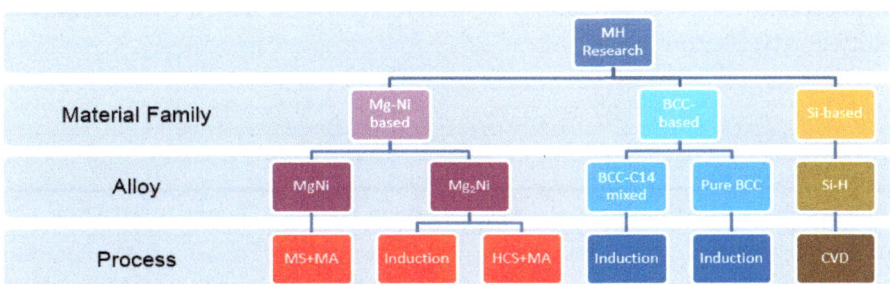

Figure 3. Scope of MH alloy selections in Task 1. MS: melt spin; MA: mechanical alloying; HCS: hydrogen combustion synthesis; and CVD: chemical vapor deposition.

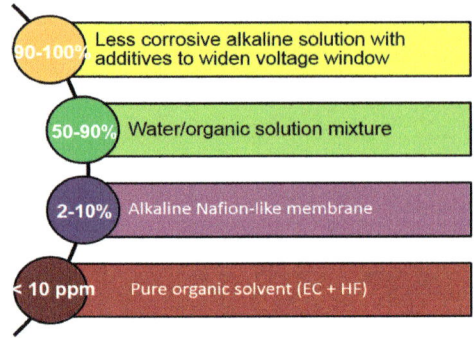

Figure 4. Four main directions for the studies of alternative electrolytes. Percentages in the circles represent the water concentration.

2. Technical Achievements

2.1. Metal Hydride Alloy in Negative Electrode

In the development of negative electrode (anode), five MH systems, MgNi, Mg_2Ni, BCC-C14, pure BCC, and Si, were investigated according to the original plan (Figure 3). Results of capacities are summarized in Table 1. MgNi (AR3: $Mg_{52}Ni_{39}Co_3Mn_6$) prepared by a melt-spinning and mechanical alloying (MA) process showed a good low-rate capacity but failed when the discharge current increased to 100 mA·g^{-1}. We tried five crystalline Mg_2Ni alloys doped with Cr, Co, Mn, and Si and the electrochemical results were disappointing. The BCC-C14 alloys went through a sequence of composition refinements [7–11] and the final champion, P37 with a composition of $Ti_{14.5}Zr_{1.7}V_{46.6}Cr_{11.9}Mn_{6.5}Co_{1.5}Ni_{16.9}Al_{0.4}$, showed the highest capacity under a 100 mA·g^{-1} discharge current, with a comparable cycle stability to AB_5 (Figure 5).

Table 1. Electrochemical discharge capacities (mAh·g^{-1}) for MH alloys developed in the RANGE-BASF program. AR3: $Mg_{52}Ni_{39}Co_3Mn_6$.

Alloy system	Alloy No.	Capacity @8 mA·g^{-1}	Capacity @100 mA·g^{-1}
MgNi	AR3	501	67
Mg_2Ni	$Mg_{66}Ni_{31}Mn_3$	18	147
BCC-C14	P12	460	259
BCC-C14	P37	414	400
Pure BCC	P08	408	340
Si	CVD a-Si:H	>3635	>3635

The transmission electron microscope (TEM) study on a complicated multi-phase BCC-C14 MH alloy (P7) shows clean boundaries between Laves C14,

Ti$_x$Ni$_y$ (B2 structure) and BCC structures (Figure 6). This clean phase boundary facilitates the hydrogen diffusion through the bulk of the alloy.

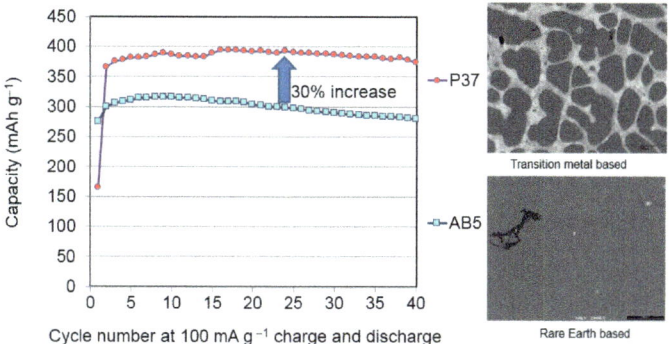

Figure 5. (**Left**) Comparison of capacities between P37 (BCC-C14 mixed phased) and AB5 (conventional) MH alloys in the first 40 cycles; and scanning electron microscopy (SEM) micrographs of the (**upper right**) P37 and (**lower right**) AB5 corners.

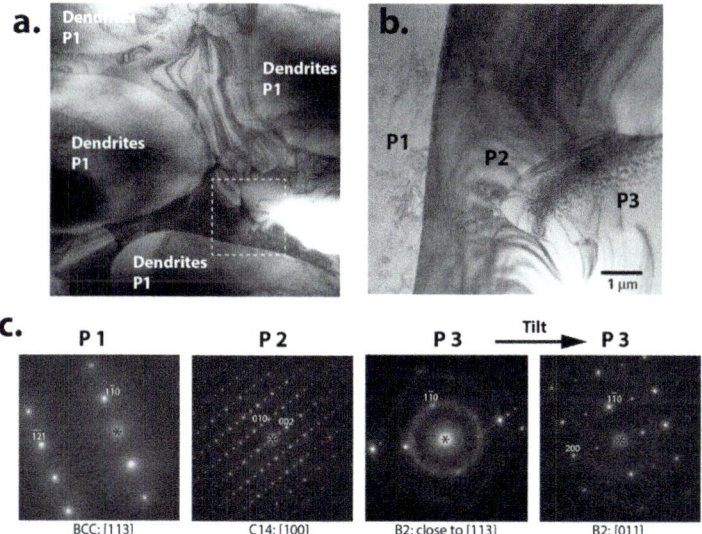

Figure 6. Transmission electron microscope (TEM) study of BCC-C14 alloy (P7): (**a**) TEM micrograph showing dendrites of phase 1 (P1) and an interdenritic region consisting of two phases shown in enlargement (**b**) as P2 and P3; (**c**) selected area electron diffraction patterns from different phases in (**b**) show that P1 is disordered BCC, P2—Laves C14, and P3—ordered BCC (proved to be B2 after tilting to [011] zone axis.

15

The further electron backscattering diffraction (EBSD) study of the same sample reveals the crystallographic orientation dependences among various constituent phases (Figure 7), which is more evidence that shows clean grain boundaries among various phases in the alloy.

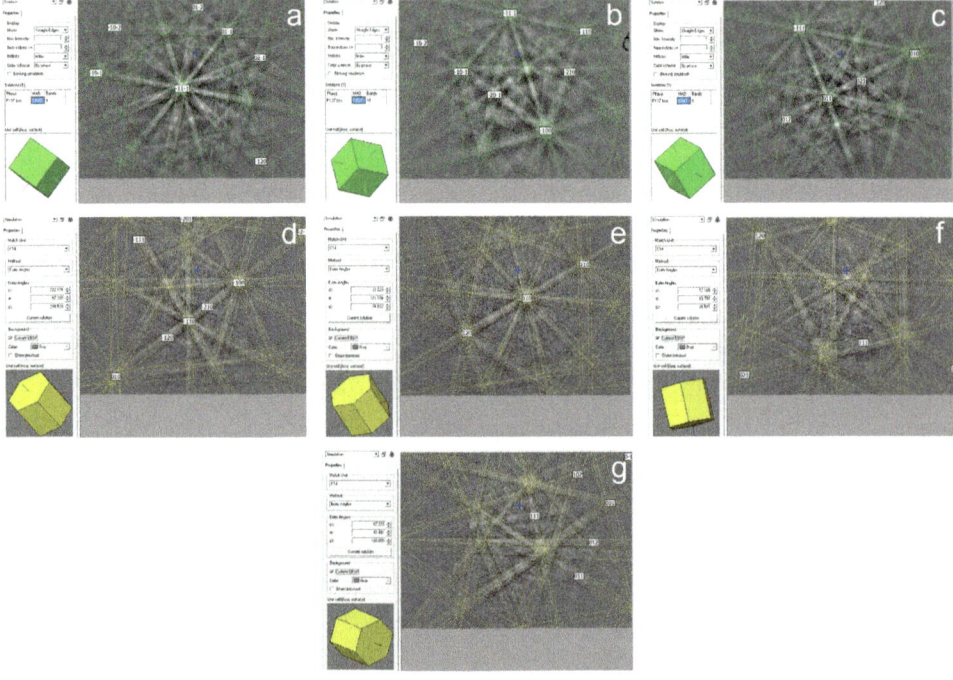

Figure 7. (**a–c**) Three BCC and (**d–g**) four C14 electron backscattering diffraction (EBSD) patterns collected from various grains on a typical BCC-C14 alloy (P7) with the computer generated matching orientation information. A few alignments in crystallographic orientation can be identified. For example, the <100>$_{BCC}$ in Figure 7c is aligned with the <0001>$_{C14}$ in both Figure 7d,e.

In the pure BCC MH alloy, the electrochemical discharge capacities increased from 408 mAh·g^{-1} [12] to 639 mAh·g^{-1} with optimizations in composition, process, and the electrolyte combination [13]. Finally, the Si-based thin film work showed the highest capacity of 3635 mAh·g^{-1} [14]. Both chemical vapor deposition (CVD) and physical vapor deposition (PVD) were employed to grow amorphous Si thin-films on the substrate (Figure 8). The high capacity and cycle stability of the Si-electrode can be seen in Figure 9.

Besides those five MH systems in Figure 3, we also studied other alternative negative electrode materials and received some unexpected results, such as

370-mAh·g^{-1} discharge capacity from perovskite oxide and 55-mAh·g^{-1} discharge capacity from an electrode containing NiCoMn(OH)$_6$.

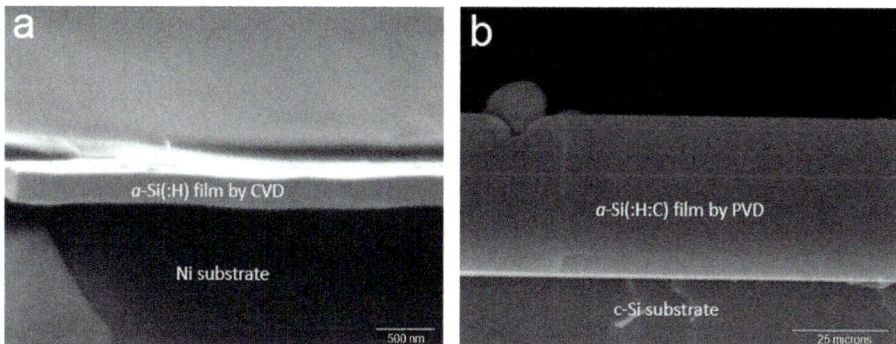

Figure 8. Cross-section SEM micrographs of (**a**) an *a*-Si:H thin film prepared by CVD and (**b**) an *a*-Si:H:C thin-film prepared by physical vapor deposition (PVD).

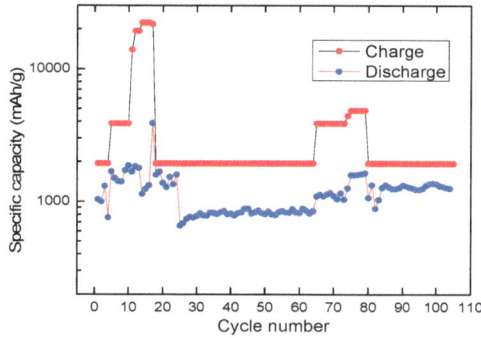

Figure 9. Charge-in and charge-out as functions of cycle number for an *a*-Si:H thin-film electrode prepared by CVD.

2.2. Development in Electrolyte

The electrolyte research in RANGE-BASF started from the modifications to the currently used 30 wt% KOH aqueous solution with different hydroxides and various salt additives to reduce the corrosion and then expanded to ionic liquid (IL) for the expansion of voltage window, which proved difficult with the current alkaline system. Three other directions highlighted in the original plan (Figure 4) faced different challenges and only exhibited marginal improvement over the currently used electrolyte.

Results from different hydroxides can be summarized by plotting the degradation in the first 15 cycles (normalized to that observed in 30 wt% KOH) in the MgNi alloy *versus* the conductivity in Figure 10. In general, we found the corrosion capability in various hydroxides in this sequence: LiOH > RbOH > KOH > NaOH > CsOH > TEAOH (tetraethylammonium hydroxide) [15]. The corrosion rate is closely related to the chemical activity of the hydroxide (as indicated by its oxidation potential, for example). TEAOH opens the window to other organic basic chemicals, which are currently under investigation. When a less corrosive electrolyte (1 M KOH + 1 M NaOH) was used, the cycle stability of MgNi alloy improved substantially as seen from Figure 1 (pink curve).

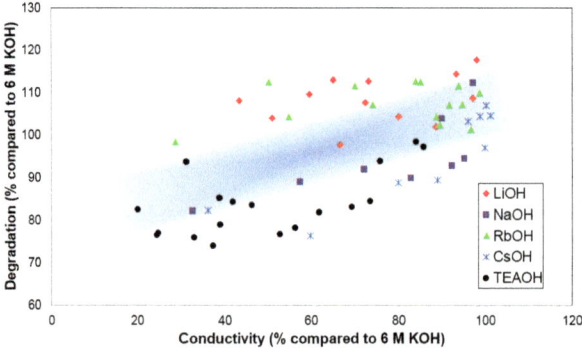

Figure 10. The degradation and conductivity of MgNi electrodes in various hydroxide as replacement for the 6 M KOH conventional electrolyte. The shaded area highlights the general trend of data points.

More than 50 soluble salts were tested as additives to the 6 M KOH electrolyte, and their performance results are summarized in Figure 11. The degradation rate was obtained from the first 15 cycles. The initial capacity and degradation of the MH alloy correlate well with the reduction potential of the alakine cations and radii of the halogen anions. A synergistic effect between KOH and some oxyacid salt additives was observed and greatly influenced by the nature of the salt additives [16]. The decrease in the degradation of the electrodes and the increase in the discharge capacity can be sttributed to two sources: a solid film formed on the MH alloy surface and a faster ionic conduction in the modified electrolyte [16]. In the end, we identified $NaC_2H_3O_2$, $KC_2H_3O_2$, K_2CO_3, Rb_2CO_3, Cs_2CO_3, K_3PO_4, Na_2WO_4, Rb_2SO_4, Cs_2SO_4, NaF, KF, and KBr to be responsible for increasing the corrosion resistance of the MgNi alloys [16].

The reason why Ni/MH has a lower specific energy compared to Li-ion battery is mainly because of the relatively low voltage (1.3 V *versus* 3.7 V) limited by the electrolysis voltage of water. The tandard electrode potential and discharge specific capacity of these two batteries are compared in Table 2. In order to make a direct comparison, standard hydrogen electrode (SHE) is used as the reference electrode, which is 0.1 V lower and 3.0 V higher than the commonly used Hg/HgO and Li-metal

(Li/Li$^+$) reference electrodes, respectively. On the cathode side, the Li-ion battery uses the +3/+4 oxidation state change in transition metals, which is 0.2 V higher than the +2/+3 oxidation state change in Ni/MH battery. The oxygen gas evolution limits the selection of cathode materials in Ni/MH battery. Likewise, the selection of anode materials for the Ni/MH battery is limited by the hydrogen gas evolution (-0.83 V *versus* SHE). The difference in specific capacities between Li-ion and Ni/MH batteries is much smaller compared to the difference in cell voltage. Therefore, increasing the cell voltage is one of the primary goals for the electrolyte development in RANGE-BASF. However, in the study with the conventional 30 wt% KOH electrolyte, only the addition of the anode materials collected from a commercial Ni-Zn battery showed a 0.2 V expansion on the anode side. Other Zn- and Pb-containing additives showed no effect in expanding the voltage window of the KOH electrolyte, which pushed us to pursue the IL.

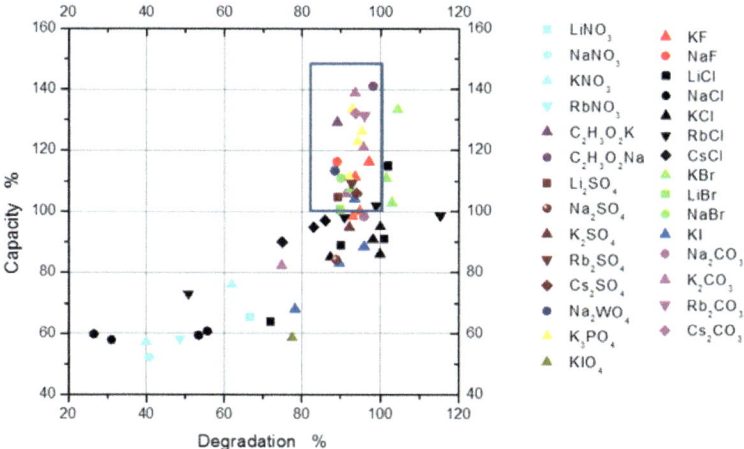

Figure 11. Discharge capacity and degradation of MgNi electrodes in 6 M KOH electrolyte with different salt additives. Salts in the blue rectangle increase the capacity and reduce the corrosion.

ILs are liquids consisting exclusively of cations and anions and exhibit superior physicochemical properties such as low melting points, high thermal and electrochemical stability, high ionic conductivity, negligible vapor pressure, and non-inflammability, which make them ideal candidates for electrochemical applications. After screening a few combinations of ILs and anhydrous acids, we found that the mixture of aprotic [EMIM][Ac] (1-ethyl-3-methylimidazolium acetate, structure shown in Figure 12) and glacial acetic acid was a suitable candidate to replace the conventional electrolyte (30 wt% KOH in H$_2$O) [17]. Acid is needed in aprotic IL to supply the protons to carry the charge.

19

Table 2. Comparison of conventional cathode and anode used in Li-ion and Ni/MH batteries. SHE: standard hydrogen electrode.

Electrode	Potential (*vs.* SHE)	Capacity (mAh· g^{-1})
Cathode in Li-ion	0.7 V	230
Cathode in current Ni/MH	0.5 V	280
Cathode in future Ni/MH (Mn)	1.5 V	1116
Anode in Li-ion	−3.0 V	330
Anode in current Ni/MH	−0.8 V	320
Anode in future Ni/MH (Si)	−1.0 V	3635

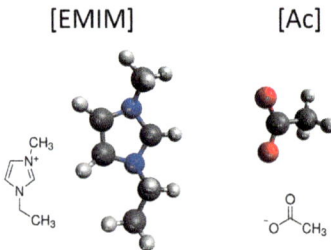

Figure 12. Molecular structure of 1-ethyl-3-methylimidazolium acetate ([EMIM][Ac]) that has been used extensively in the RANGE-BASF program.

A cyclic voltammogram (CV) of an AB$_5$ electrode in 2 M acetic acid in [EMIM][Ac] shows a 0.6 V separation between the hydride reduction and hydrogen gas evolution peaks (Figure 13a). Using [EMIM][Ac]/Ac mixtures electrolyte, we are able to study the Si-anode in this non-aqueous system since *a*-Si is highly reactive with the KOH electrolyte [18].

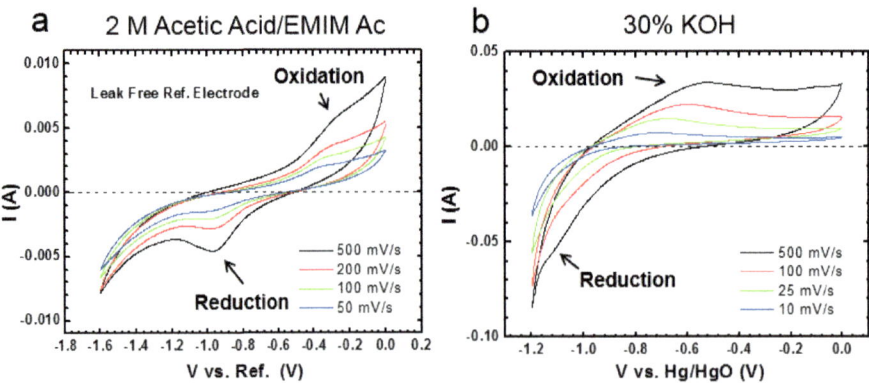

Figure 13. Cyclic voltammetry (CV) graph of an AB$_5$ electrode in: (**a**) an ionic liquid (IL) (2 M acetic acid in [EMIM][Ac]); and (**b**) aqueous electrolytes (30 wt% KOH).

Other development efforts in the electrolyte area include adding a Nafion coating on the negative electrode to reduce corrosion, and replacing aqueous electrolyte with solid polymer membrane (300 mAh·g^{-1}, Figure 14a,b) or gel (320 mAh·g^{-1}, Figure 12c), or non-aqueous 1 M triflic acid in PC:DMC (propylene carbonate:dimethyl carbonate) = 1:1 (50 mAh·g^{-1}).

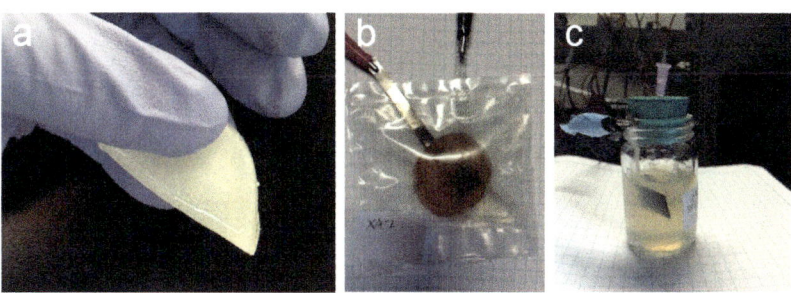

Figure 14. (**a,b**) A flexible PEO-PVA-KOH-H$_2$O gel film; and (**c**) PVA-KOH gel electrolyte. PEO: polyethylene oxide; and PVA: polyvinel alkahol.

2.3. Development in Positive Electrode

The development work on the positive electrode (cathode) started from the expansion of the oxidation swing window. Because of the limit from the oxygen gas evolution competition, the redox reaction for Ni(OH)$_2$ is conventionally between +2 and +3 oxidation states (Figure 15).

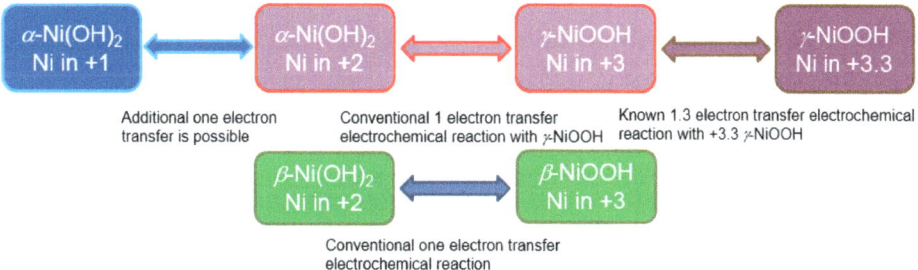

Figure 15. Changes in the oxidation states in Ni with different chemical reaction. Reactants in green boxes are used in the conventional Ni-based rechargeable batteries, while those in pink boxes are avoided to prevent electrode swelling/disintegration. The one in the blue box is the key element in the high-capacity positive electrode study in the RANGE-BASF program.

The Pourbaix diagram of Ni shows the +3/+4 redox reaction happens at about 0.2 V above the oxygen gas evolution potential (Figure 16). The γ-NiOOH is known to have a higher oxidation state (+3.3 or higher) but is still allowed in the aqueous solution [19]. However, transformation from β-Ni(OH)$_2$ into α-Ni(OH)$_2$ requires insertion of a water layer between the NiO$_2$ planes, which causes a swelling/disintegration of the positive electrode and is avoided by the addition of Zn or Cd [20].

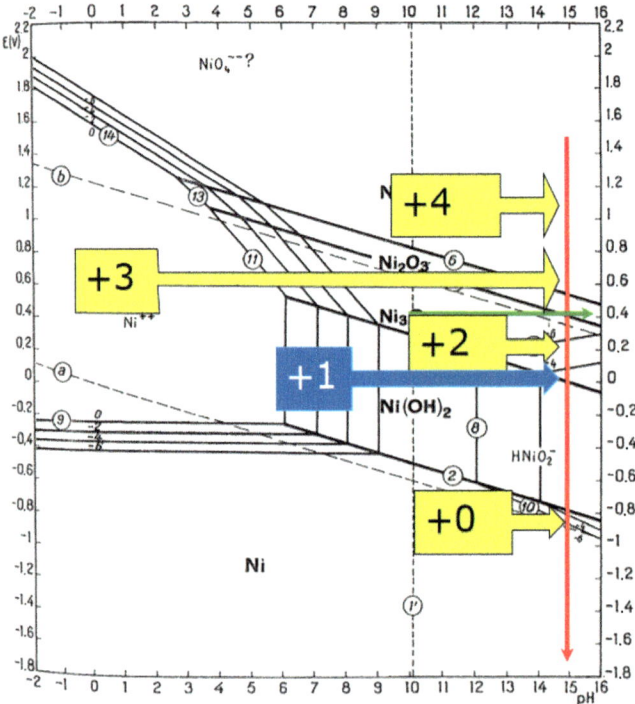

Figure 16. Pourbaix diagram showing different oxidation states of Ni at different voltages *vs.* standard hydrogen electrode (SHE). The red line corresponds to the pH value of the 30% KOH electrolyte. Oxidation states in yellow boxes are well-known, while the one in the blue box is the focus of the RANGE-BASF study.

In the RANGE-BASF program, we developed a continuous process to fabricate core-shell α-β Ni(OH)$_2$ (WM12) with a 50% increase in specific capacity and good cycle stability (Figure 17).

The X-ray energy dispersive spectroscopy (EDS)/scanning electron microscopy (SEM) study shows a higher Al (an α-phase promoter) content at the surface (Figure 18).

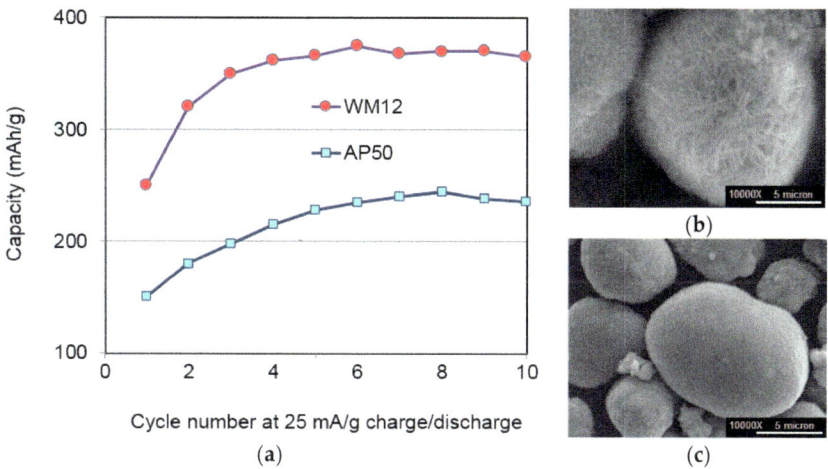

Figure 17. (**a**) Capacity comparison between core-shell α-β WM12 and conventional β-Ni(OH)$_2$ AP50 spherical particles. SEM micrographs of two materials: (**b**) WM12 and (**c**) AP50. Instead of a relative smooth surface of AP50, WM12 shows a very rough surface with a high reaction surface area.

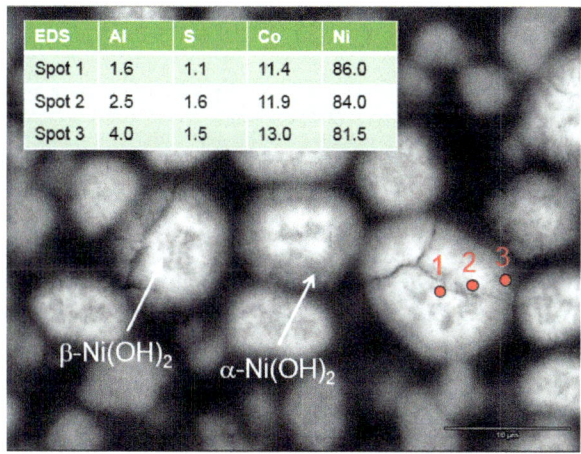

Figure 18. Cross-section SEM micrograph of the continuous process prepared WM12 Ni(OH)$_2$, showing that the core (β)-shell (α) structure has different Al-contents, as shown in the X-ray energy dispersive spectroscopy (EDS) results in the inset.

A TEM micrograph from the cross-section of a core-shell WM12 Ni(OH)$_2$ is shown in Figure 19a. The corresponding electron yields across the diffraction resembling the X-ray diffraction (XRD) patterns are shown in Figure 19b to show the microstructure of β-core and α-shell. The as-prepared material is purely β-phase Ni(OH)₂. During activation, the core with less Al remains the β-phase and the shell turns into α-phase with a higher Al-content. The surface is very fluffy, which allows α-phase to grow without disintegration. Besides the continuous process, WSU also developed batch processes to fabricate α-Ni(OH)₂ (Table 3).

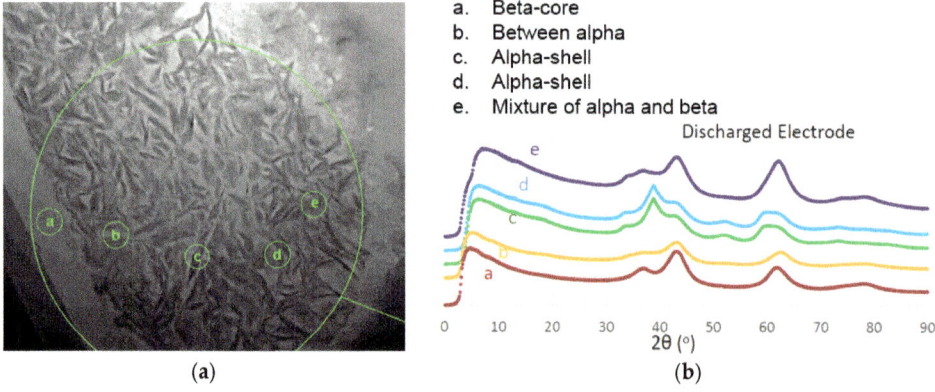

(a) (b)

Figure 19. (**a**) TEM micrograph of a cross-section of core-shell WM12 Ni(OH)₂; and (**b**) the corresponding electron yield across diffraction patterns.

Table 3. Design matrix and results for batch process prepare Ni(OH)₂ performed in Wayne State University (WSU).

Parameter	Ni-1	Ni-2	Ni-3	Ni-4
Preparation	Single-step precipitation	Multi-step precipitation	Homogeneous precipitation	Homogeneous precipitation
Precipitants	KOH	NaOH, Na₂CO₃, NH₄OH, NH₄Cl	Urea, Tween-20	Urea
Composition	100% Ni	86% Ni, 14% Al	86% Ni, 14% Al	86% Ni, 14% Al
Structure	β-Ni(OH)₂	α-Ni(OH)₂	α-Ni(OH)₂	α-Ni(OH)₂
Activation cycles	6	5	12	11
Maximum capacity (mAh·g⁻¹)	260	346	305	310

Since the material is α-phase before cycling, it does not go through the β-α transition. Therefore, the integrity of the Ni(OH)₂ particle is preserved (Figure 20), and the capacity is very stable (Figure 21). XRD analysis verifies the α-phase structure in both the pristine and cycled materials (Figure 22).

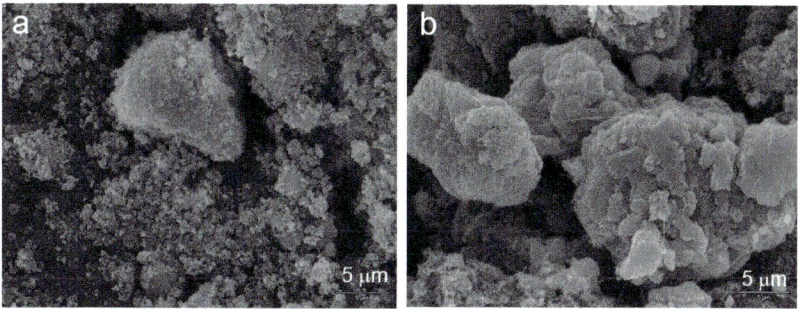

Figure 20. SEM micrographs of (**a**) pristine and (**b**) cycled α-Ni(OH)$_2$ prepared by a batch co-precipitation process.

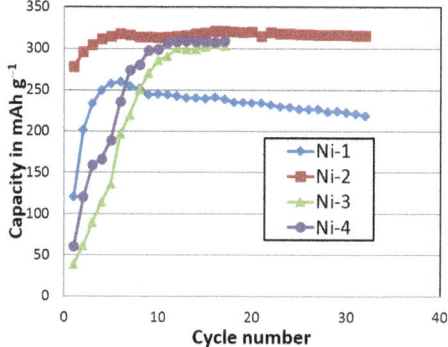

Figure 21. Cycle stability of one β-Ni(OH)$_2$ (Ni-1) and three α-Ni(OH)$_2$ (Ni-2, Ni-3, and Ni-4) samples prepared by a batch co-precipitation process.

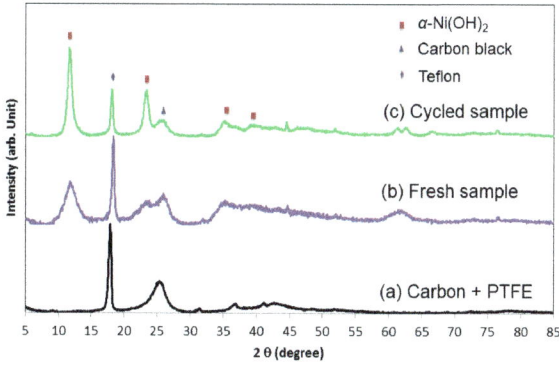

Figure 22. X-ray diffraction (XRD) of (**a**) the additives, (**b**) a pristine, and (**c**) a cycled a α-Ni(OH)$_2$ (Ni-2) prepared by a batch co-precipitation process. From the decrease in the width of the diffraction pattern, the crystallite of α-Ni(OH)$_2$ is found to increase with electrochemical cyclings.

In this program, we also investigated the possibility of further reducing α-Ni(OH)$_2$ into +1 or even lower oxidation state to maximize the range of redox reaction (Figure 16). Discharge into lower potential shows an additional plateau at about 0.8 V *versus* standard AB$_5$ negative electrode, which may be related to the +1 oxidation state of Ni, as demonstrated in the literature [21]. The combined number of electron-transfer per Ni atom may exceed 1 when properly activated (Table 4).

Table 4. Discharge specific capacity (Q) and the number of electrons transferred during the redox reaction per Ni atom (N_e) for the first 11 cycles for the Ni(OH)$_2$ samples developed in the RANGE-BASF program. $Q_{Control}$, Q_{S1}, Q_{S2} and Q_{S3} are the discharge specific capacities of the control sample (Ni$_{0.91}$Co$_{0.045}$ Zn$_{0.045}$(OH)$_2$), Sample 1 (Ni$_{0.94}$Co$_{0.06}$(OH)$_2$), Sample 2 (Ni$_{0.85}$Co$_{0.05}$Al$_{0.10}$(OH)$_2$) and Sample 3 (Ni$_{0.69}$Co$_{0.05}$Zn$_{0.06}$Al$_{0.20}$(OH)$_2$) for a discharge current density of 25 mA·g^{-1}, respectively.

Cycle number	1	2	3	4	5	6	7	8	9	10	11
$Q_{Control}$ (mAh·g^{-1})	150	180	198	215	228	235	240	243	240	244	238
N_e	0.59	0.71	0.78	0.85	0.90	0.93	0.95	0.96	0.95	0.97	0.94
Q_{S1} (mAh·g^{-1})	95	119	141	139	148	149	163	143	245	423	343
N_e	0.36	0.45	0.54	0.53	0.56	0.57	0.62	0.55	0.93	1.61	1.31
Q_{S2} (mAh·g^{-1})	172	216	238	234	243	270	335	635	599	463	259
N_e	0.69	0.86	0.95	0.94	0.97	1.08	1.34	2.54	2.39	1.85	1.03
Q_{S3} (mAh·g^{-1})	152	175	190	223	379	530	426	365	329	296	236
N_e	0.71	0.82	0.89	1.04	1.77	2.48	1.99	1.70	1.54	1.38	1.10

2.4. Cell Assembly

Three types of test fixtures were used in the RANGE-BASF program. The first one is a conventional open-air flooded half-cell plastic bag with two or three terminals sandwiched between two pieces of acrylic plates (Figure 23a). The second is a sealed three-terminal Swagelok T-type cell (Figure 23b). The third port (top) can be connected to a reference electrode, a pressure relief value, or a pressure transducer. The third one is a sealed pouch-type full-cell that uses the same materials as those used in the Li-ion pouch cell (Figure 23c).

The Ni/MH pouch cell has the advantage of easy fabrication and high specific energy density [20]. The development of the pouch cell took about a month, and a cell with a specific energy of 127 Wh·kg^{-1} was developed in the end with a conventional AB$_5$ MH alloy and newly developed WM12 active materials (Figure 24).

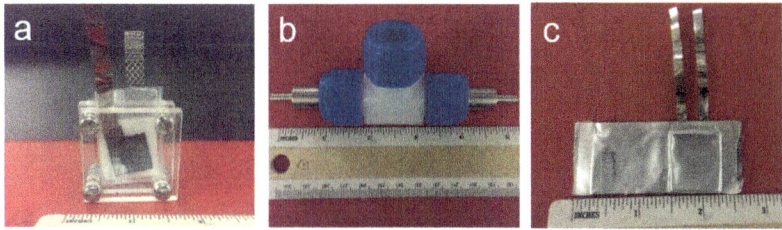

Figure 23. Pictures of a: (**a**) flooded half-cell; (**b**) swagelok sealed half-cell; and (**c**) sealed pouch cell.

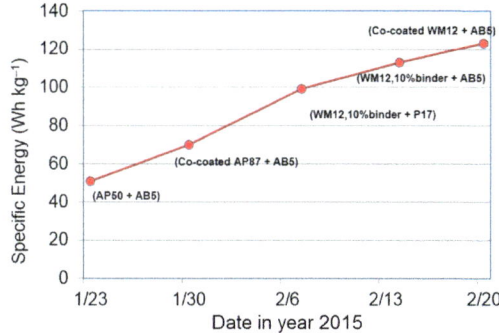

Figure 24. Development timeline of sealed pouch cell showing improvement in the specific energy density from 1 January to 20 February 2015 in BASF-Ovonic.

After applying the newly developed P37 MH alloy, we expect a specific energy of 145 Wh·kg^{-1} can be realized in a 100-Ah pouch-type Ni/MH battery (Table 5).

Table 5. Design of a 100 Ah pouch-type Ni/MH battery with the cell dimension of $20 \times 12 \times 1.8$ cm^3 and an N/P ratio of 1.2. Total cell weight is 826 g and the projected gravimetric and volumetric energy densities are 145 Wh·kg^{-1} and 278 Wh·L^{-1}, respectively.

Cell component	Active weight (g)	Additives weight (g)	Substrates weight (g)	Total weight (g)	Area (cm^2)
Positive electrode	303	34	69	406	1060
Negative electrode	316	0	23	339	1060
Separator	-	-	-	13	-
Ni-tab	-	-	-	4	-
Electrolyte	-	-	-	52	-
Pouch (Al foil)	-	-	-	12	-

3. Conclusions

Throughout this RANGE-BASF program, we discovered a lot of new materials and processes and came up with new directions for future development of the Ni/MH battery, which resulted in 22 US Patent Applications. In our opinion, the advancement of the Ni/MH battery in this one-and-a-half-year period exceeds what had been previously accomplished. The next generation of Ni/MH based on the P37, WM12, and pouch design should generate a battery with higher specific energy on a pack-level that considers savings in battery- and thermal- management systems (Figure 25). We will continue to apply these results to the product, and further the research of ILs and novel cathode material areas, which will further improve the specific energy and power of Ni/MH to a level competitive with the Li-S battery (Figure 25).

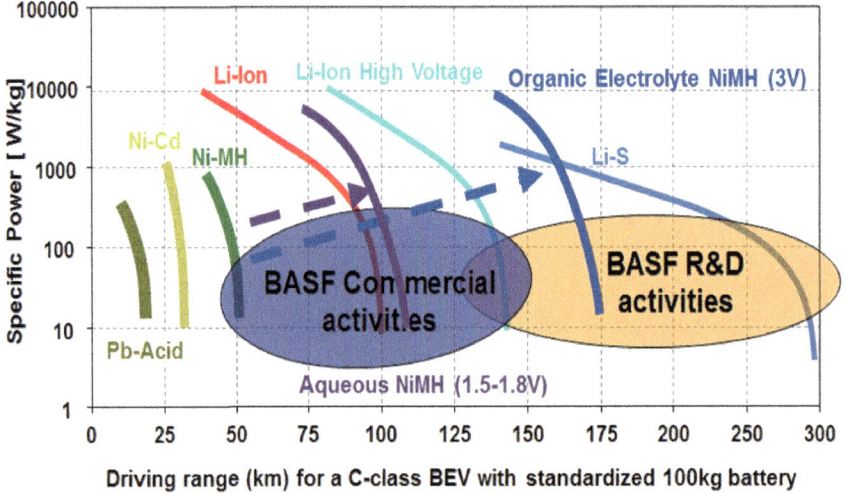

Figure 25. Current and future development goals of the next-generation Ni/MH batteries. While the aqueous NiMH uses advance materials developed in the RANGE-BASF program, such as WM12 and P37, the organic electrolyte NiMH will employ IL, the Si-based anode, and the Mn-based cathode.

Acknowledgments: This work was financially supported by Advanced Research Projects Agency-Energy (ARPA-E) under the Robust Affordable Next Generation Energy Storage Systems (RANGE) program (DE-AR0000386). The authors would like to thank the following individuals for their help: Simon Ng, Shuli Yan, Venkateswara Chitturi, Shiuan Chang, Marissa Kerrigan (Wayne State University), Michael A. Fetcenko, Benjamin Reichman, Michael Zelinsky, William Mays, Benjamin Chao, Jean Nei, Baoquan Huang, Tiejun Meng, Lixin Wang, Jun Im, Diana F. Wong, Cristian Fierro, Dennis Corrigan, Taihei Ouchi, David Beglau, Jonathan Tao, Paul Gasiorowski, David Pawlik, Allen Chan, Ryan J. Blankenship, Su Cronogue, Chaolan Hu, Rose Bertolini, Theodore Jurczak (BASF-Ovonic), and Haoting Shen (NIST).

Author Contributions: Kwo-hsiung Young, K. Y. Simon Ng, and Leonid A. Bendersky led research teams for RANGE program in BASF-Ovonic, Wayne State University, and Nation Institute of Standards and Technology, respectively.

Conflicts of Interest: The authors declare no conflicts of interest.

References

1. High Performance NiMH Alloy for Next-Generation Batteries. Available online: http://arpa-e.energy.gov/ ?q=slick-sheet-project/rare-earth-free-ev-batteries (accessed on 10 November 2015).
2. Liu, P.; Ross, R.; Newman, A. Long-range, low-cost electric vehicles enabled by robust energy storage. *MRS Energy Sustain.* **2015**, *2*.
3. General Motors EV1. Available online: https://en.wikipedia.org/wiki/General_Motors_EV1 (accessed on 12 November 2015).
4. Overview: All Eneloop Batteries 2005–2015. Available online: http://budgetlightforum.com/node/7336 (accessed on 12 November 2015).
5. Kai, T.; Ishida, J.; Yasuoka, S.; Takeno, K. The Effect of Nickel-Metal Hydride Battery's Characteristics with Structure of the Alloy. In Proceedings of the 54th Battery Symposium in Japan, Osaka, Japan, 7–9 October 2012; p. 210.
6. Young, K.; Nei, J. The current status of hydrogen storage alloy development for electrochemical applications. *Materials* **2013**, *6*, 4574–4608.
7. Young, K.; Nei, J.; Wong, D.F.; Wang, L. Structural, hydrogen storage, and electrochemical properties of Laves phase-related body-centered-cubic solid solution metal hydride alloys. *Int. J. Hydrog. Energy* **2014**, *39*, 21489–21499.
8. Young, K.; Wong, D.F.; Wang, L. Effect of Ti/Cr content on the microstructures and hydrogen properties of Laves phase-related body-centered-cubic solid solution alloys. *J. Alloys Compd.* **2015**, *622*, 885–893.
9. Young, K.; Wong, D.F.; Wang, L. Annealing effect on Laves phase-related body-centered-cubic solid solution alloys. *J. Alloys Compd.* **2016**, *654*, 216–225.
10. Young, K.; Ouchi, T.; Nei, J.; Meng, T. Effect of Cr, Zr, V, Mn, Fe, and Co to the hydride properties of Laves phase-related body-centered-cubic solid solution alloys. *J. Power Sources* **2015**, *281*, 164–172.
11. Young, K.; Wong, D.F.; Nei, J. Effects of vanadium/nickel contents in Laves phase-related body-centered-cubic solid solution metal hydride alloys. *Batteries* **2015**, *1*, 34–53.
12. Young, K.; Ouchi, T.; Huang, B.; Nei, J. Structure, hydrogen storage, and electrochemical properties of body-centered-cubic $Ti_{40}V_{30}Cr_{15}Mn_{13}X_2$ alloys (X = B, Si, Mn, Ni, Zr, Nb, Mo, and La). *Batteries* **2015**, *1*, 74–90.
13. Young, K.; Ouchi, T.; Meng, T.; Nei, J. Studies in molybdenum/manganese content in the duel body-centered-cubic phased metal hydride alloys. Unpublished work. **2016**.
14. Meng, T.; Young, K.; Beglau, D.; Yan, S.; Zeng, P.; Cheng, M.M. Hydrogenated amorphous silicon thin film anode for proton conducting batteries. *J. Power Sources* **2016**, *302*, 31–38.
15. Yan, S.; Young, K.; Nei, J.; Rotarov, D.; Ng, K.Y.S. Studies in various hydroxides used as electrolyte in Ni/MH batteries. *Batteries* **2016**, *2*, to be submitted for publication.

16. Yan, S.; Young, K.; Ng, K.Y.S. Effects of salt additives to the KOH electrolyte used in Ni/MH batteries. *Batteries* **2015**, *1*, 54–73.

17. Meng, T.; Young, K.; Wong, D.F.; Nei, J. Ionic liquid-based non-aqueous electrolyte for nickel metal hydride batteries. Unpublished work. **2016**.

18. Kiani, A.; Venkatakrishnan, K.; Tan, B. Maskless lithography using patterned amorphous silicon layer induced by femtosecond laser irradiation. *NSTI Nanotech* **2010**, *2*, 276–279.

19. Corrigan, D.A.; Bendert, R.M. Effect of coprecipitated metal ions on the electrochemistry of nickel hydroxide thin films: Cylic voltammetry in 1 M KOH. *J. Electrochem. Soc.* **1989**, *136*, 723–728.

20. Young, K.; Yasuoka, S. Capacity degradation mechanisms in nickel/metal hydride batteries. *Batteries* **2016**, *2*.

21. Bonneviot, L.; Cai, F.X.; Che, M.; Kermarec, M.; Legendre, O.; Lepetit, C.; Olivier, D. Preparation of mono- or zerovalent nickel by single or successive one-electron-transfer steps in the photoreduction of silica-supported nickel catalysts. *J. Phys. Chem.* **1987**, *91*, 5912–5921.

Reviews on the U.S. Patents Regarding Nickel/Metal Hydride Batteries

Shiuan Chang, Kwo-hsiung Young, Jean Nei and Cristian Fierro

Abstract: U.S. patents filed on the topic of nickel/metal hydride (Ni/MH) batteries have been reviewed, starting from active materials, to electrode fabrication, cell assembly, multi-cell construction, system integration, application, and finally recovering and recycling. In each category, a general description about the principle and direction of development is given. Both the metal hydride (MH) alloy and nickel hydroxide as active materials in negative and positive electrodes, respectively, are reviewed extensively. Both thermal and battery management systems (BMSs) are also discussed.

Reprinted from *Batteries*. Cite as: Chang, S.; Young, K-h.; Nei, J.; Fierro, C. Reviews on the U.S. Patents Regarding Nickel/Metal Hydride Batteries. *Batteries* **2016**, *2*, 2.

1. Introduction

The nickel/metal hydride (Ni/MH) battery is an essential electrochemical device for consumer, propulsion, and stationary energy storages. Since its commercial debut in the late 1980s, many researchers have worked diligently in the Ni/MH battery field. Their contributions were publicized through two routes: academic publications and patent applications. While there are several key reviews available on the former [1–16], there is not one on the latter. Therefore, we have organized and summarized the patents related to Ni/MH battery technology in the current review and its companion—Reviews on the Japanese Patents Regarding Nickel/Metal Hydride Batteries [17].

2. Results

As shown in Figure 1, the current review is divided into six categories in the chronological order of technology development, which are active materials → electrode fabrication → cell assembly → multi-cell construction → system integration → application. Because of the extremely large number of U.S. patents on these subjects, we have selected and presented the most representative patents historically and technologically in the current review. More related U.S. patent documents can be found on each U.S. patent presented here.

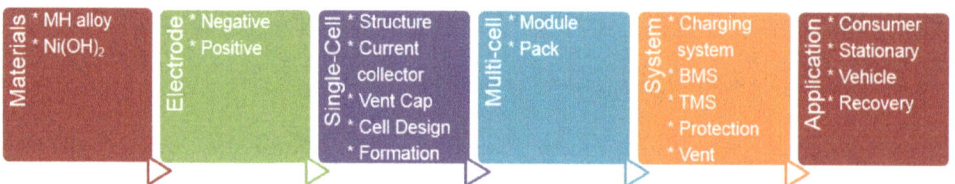

Figure 1. Content of this paper. MH: metal hydride.

2.1. Active Materials

The half-cell electrochemical reaction for the negative electrode is:

$$M + H_2O + e^- \leftrightarrows MH + OH^- \text{ (forward : charge, reverse : discharge)} \quad (1)$$

where M is a hydrogen storage alloy capable of storing hydrogen reversibly and MH is the corresponding metal hydride (MH). During charge, applied voltage splits water into protons and hydroxide ions. Driven by the voltage and diffusion caused by the concentration difference, protons enter into the bulk of the alloy and then meet with electrons from the current collector, which is attached to a power supply. During discharge, protons at the alloy surface recombine with hydroxide ions in the electrolyte. Electrons are injected into the circuit load through the current collector in order to maintain charge neutrality in the negative electrode.

The counter reaction for the negative electrode is:

$$Ni(OH)_2 + OH^- \leftrightarrows NiOOH + H_2O + e^- \text{ (forward : charge, reverse : discharge)} \quad (2)$$

where $Ni(OH)_2$ is the active material going through +2/+3 oxidation state change during the electrochemical reactions. During charge, protons deprived of $Ni(OH)_2$ move to the surface of the positive electrode because of the applied voltage and recombine with hydroxide ions in the electrolyte. In order to maintain charge neutrality, electrons are injected into the power supply through the current collector. During discharge, water at the surface of the positive electrode splits into protons and hydroxide ions. Protons then enter into NiOOH and neutralize with electrons through the load to complete the circuit with the negative electrode. Shown in the net reaction Equation (3), charge/discharge reaction does not change the overall hydroxide concentration/pH value, which is different from the working principle of Ni/Cd battery Equation (4).

$$M + Ni(OH)_2 \leftrightarrows MH + NiOOH \text{ (forward : charge, reverse : discharge)} \quad (3)$$

$$Cd(OH)_2 + 2Ni(OH)_2 \leftrightarrows Cd + 2NiOOH + 2H_2O \text{ (forward : charge, reverse : discharge)} \quad (4)$$

Both the inventions and further developmental works of MH alloy and $Ni(OH)_2$ are presented in the next two sections.

2.1.1. Metal Hydride Alloys in the Negative Electrode

Klaus Beccu contributed to the earliest patent using MH alloy as an active material in the negative electrode to construct a Ni/MH battery in 1970 [18]. The alloy used had a composition of $Ti_{85}Ni_{10}Cu_3V_2$ (in wt%), which was later expanded to any hydride of the third, fourth, or fifth group of transition metal (for example, Ti) [19]. Two types of MH alloys were used in the first commercialized Ni/MH batteries introduced in 1989. While Japanese companies such as Matsushita [20–22] (later changed its name to Panasonic), Toshiba [23], Sanyo [24,25], and Yuasa [26] used the misch metal-based (mixtures of light rare earth elements such as La, Ce, Pr, and Nd) AB_5 MH alloy, the Ovonic Battery Company (OBC, Troy, MI, USA) chose the transitional metal-based AB_2 MH alloy as the active material in the negative electrode. General properties of these two alloy families are listed in Table 1. The AB_5 MH alloy eventually dominated the market due to the substantially lower price of misch metal since 2000.

Table 1. Property comparison between the AB_2 and AB_5 MH alloy families. FCC, 1C/1C, and DOD denote face-centered-cubic, 1C charge and discharge rates, and depth of discharge, respectively.

Property	AB_2 MH alloy	AB_5 MH alloy
Basic crystal structure	Hexagonal C14 and FCC C15	Hexagonal CaCu$_5$
Composition example	$Ti_{12}Zr_{21}Ni_{38}V_{10}Cr_5Mn_{12}Co_{1.5}Al_{0.5}$	$La_{10}Ce_5Pr_{0.5}Nd_{1.5}Ni_{60}Co_{12}Mn_6Al_5$
Discharge capacity	$340–440\ mAh\cdot g^{-1}$	$300–330\ mAh\cdot g^{-1}$
High-rate dischargeability	Acceptable	Excellent
Cycle life (1C/1C, 100% DOD)	800	1200
Self-discharge	Acceptable (V-free alloy)	Acceptable
High-temperature storage	Acceptable due to leach-out	Bad due to surface passivation
Activation	Modest	Easy
Low temperature	Good	Modest
Fabrication method	Casting, hydriding, and grinding	Casting and grinding
Raw material cost	V is expensive	Volatile due to rare earth price fluctuation
Low-cost version	V and Co-free alloy	Pr, Nd, and Co-free alloy

Discovery of the hydrogen storage capability of the rare earth-based AB_5 intermetallic alloy was accidental [27]. Guegen [28] filed the first patent using $LaNi_5$ modified with Ti, Ca, Ba, Cr, and/or Cu with improved capacity as negative electrode material in 1978. In 1984, Willems *et al.* [29] filed a patent adding Co and/or Cu in $LaNi_5$ to reduce the lattice expansion during hydride process and consequently extending the cycle life. Based on the idea of utilizing misch metal in AB_5 for gaseous phase hydrogen storage originated by Osumi and his coworkers in 1979 [30] and 1983 [31], Kanda and Sato from Toshiba [32] filed a U.S. patent based on Mn, Al-modified misch metal-based AB_5 MH alloy for Ni/MH battery in 1986. Subsequently, the Japanese companies filed many patents optimizing the AB_5 formula (examples can be found in [17,33–42]) and finalized on using Co, Al, Mn, and Ni as the B-site elements. OBC also filed a patent for adding Zr or Si to extend the cycle life of Cu-containing AB_5 MH alloy [43].

In the field of AB_2 MH alloy, Gamo and his coworkers [44] from Matsushita filed a patent about its gaseous phase hydrogen storage application in 1979, and Sapru and her coworkers [45] at OBC filed the first patent for battery application in 1985. Since then, many patents have been filed on the AB_2 MH alloys for electrochemical applications by OBC [46–55], Matsushita [56–58], and SAFT [59].

Other MH alloy families, including body-centered-cubic (BCC) solid solution [60–62], BCC-AB_2 composite [63,64], Mg, Ni-based alloy [65–68], TiNi [69], and composite [70,71], were also proposed as negative electrode active materials; however, they were not commercialized. The only MH alloy that successfully succeeded AB_2 and AB_5 in commercialization is the A_2B_7 superlattice alloy [72–76]. Development works of the A_2B_7 MH alloy done in academia can be found in a previous review article [10]. Several non-conventional MH alloy fabrication processes, such as centrifugal casting [77], double-roller casting [78], melt-spin [34,79], gas atomization [80], strip casting [81,82], and a chemical process using misch metal oxide and/or hydroxide as raw material [83], have been patented as well, but only strip casting is used for mass production besides the conventional induction melting.

Besides patents on a specific alloy family, there are several patents that contribute across the board and independently of alloy selection. Sapru *et al.* [84] filed a patent about the advantage of increasing the degree of disorder in a multi-component MH alloy; as a result, the electrochemical performance is improved. Such material is disordered and characterized as a composite of amorphous, microcrystalline, and polycrystalline components lacking long-range compositional order and with three or more phases [85]. Another group of patents not confined to a specific alloy family addresses the structure designs of surface oxide, which is a catalyst for the electrochemical reaction [52,86–88] (Figure 2).

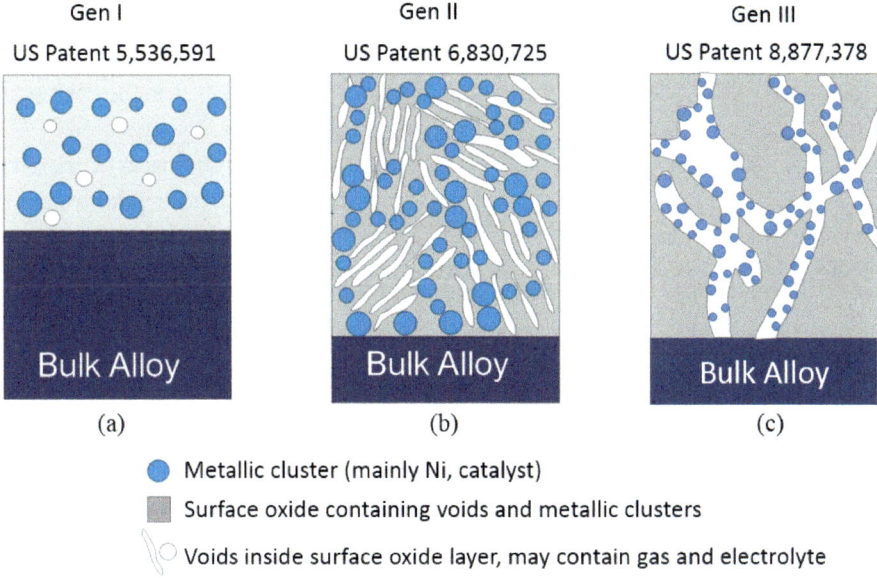

Gen I	Gen II	Gen III
US Patent 5,536,591	US Patent 6,830,725	US Patent 8,877,378

| (a) | (b) | (c) |

● Metallic cluster (mainly Ni, catalyst)

▦ Surface oxide containing voids and metallic clusters

○ Voids inside surface oxide layer, may contain gas and electrolyte

Figure 2. Schematic diagrams of the evolution in MH alloy surface oxide from (**a**) metallic clusters and voids randomly distributed in the surface oxide to (**b**) metallic clusters and channels randomly distributed in the surface oxide and (**c**) metallic clusters only on the surface of channels.

2.1.2. Nickel Hydroxide in the Positive Electrode

Soon after the Swedish inventor Waldemar Jungner filed a number of Swedish and German Patents on the concepts of Ni/Cd and Ni/Fe batteries in 1899 [89], a famous American inventor, Thomas A. Edison, filed several U.S. patents on Ni/Fe and Ni/Cd rechargeable batteries utilizing nickel hydroxide as active material in the positive electrode in 1901 [90,91]. The same chemistry, $Ni(OH)_2$, has been used as a positive electrode active material for the entire alkaline battery family because of its high electrochemical reversibility, appropriate voltage (the highest among several options but lower than the oxygen gas evolution potential), and low cost. The original positive electrode used in Ni/MH batteries was inherited from those in Ni/Cd and Ni/Fe batteries [92–95]. The electrode is manufactured by first pasting filamentary Ni onto a substrate (such as perforated foil) and then "sintered" under a high temperature annealing furnace in a nitrogen/hydrogen atmosphere, where binders used in the pasting process are burned away to leave a conductive skeleton of nickel with a typical average pore size of about 30 μm [96]. Next, a chemical impregnation process is carried out to produce the active $Ni(OH)_2$ by consecutively dipping the electrode between the $Ni(NO_3)_2$ and NaOH baths [97,98]. This kind of $Ni(OH)_2$ electrode is the sintered type. A new pasted type of positive electrode using spherical Ni(OH)$_2$ particles together with the CoO additive (for forming a CoOOH conductive network on the particle surface) was first patented by Oshitani and his coworkers in Yuasa in 1989 [26,99], and soon became the mainstream for Ni/MH battery due to its high volumetric and gravimetric energy densities. The spherical particle is originally produced by a two-step batch chemical precipitation process, which consists of ammonia complex formation and precipitation in two different reactors. Later on, Fierro and his coworkers at OBC [100] developed a continuous stirring single reactor process, and the smaller particles near the bottom of the reactor and the larger ones are separated by the centrifugal force from a bottom blade, leaving the container with an overflow (Figure 3). This method has been widely adopted in today's modern manufacturing. The produced spherical Ni(OH)$_2$ powder is then mixed with binders and pasted onto the Ni foam substrate to form the pasted type of positive electrode. The main advantage of using the pasted type of positive electrode is the increase in gravimetric energy density; however, rate-capability, cycle stability, and cost are sacrificed.

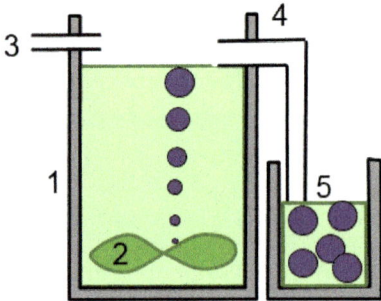

Figure 3. Schematic diagram of a continuous stirring tank reactor used to co-precipitate hydroxides of metals with similar solubilities, such as Co, Ni, and Zn. 1: Stainless-steel vessel; 2: mixer blade; 3: raw material inlet; 4: product overflow; and 5: product container.

There are many patents modifying the basic Ni(OH)$_2$ for performance improvement, such as increasing utilization rate by chemical modification through co-precipitation [25,101–111], promoting higher electron transfer per Ni atom by introducing the gamma phase with an oxidation state between +3 and +4 [112–117], alternating the microstructure by various process changes [118–120], and adding a conductive coating on the Ni(OH)$_2$ core [121]. Moreover, OBC filed four patents regarding the incorporation of fine Ni fiber in the co-precipitation reactor for the purpose of increasing the native conductivity of Ni(OH)$_2$ [122–125]. The crystalline size is an important indicator for the degree of disorder (density of stacking fault along the c-axis) of Ni(OH)$_2$ and affects the electrochemical performance, and an evaluation method was patented by OBC [126]. In order to increase the conductivity or resistance to poisoning, a core-shell type of structure obtained by sequential precipitation or other coating methods was invented [108,124,125,127–129].

2.2. Electrode Fabrication

There are two electrodes in a typical Ni/MH battery, namely the negative and positive electrodes, and they are made of MH alloy and nickel hydroxide, respectively. During service (discharge), MH in the negative electrode is oxidized into metal, and therefore the negative electrode is also called an anode. Furthermore, the positive electrode can be termed a cathode since NiOOH is reduced to Ni(OH)$_2$ during discharge.

Patents regarding electrode fabrications of both the anode and cathode, including component, substrate, and construction, are discussed in the following sections.

2.2.1. Negative Electrode (Anode)

The negative electrode in a commercial Ni/MH battery can be fabricated by any of the following methods: dry compaction, wet paste method, or dry paste method (Figure 4).

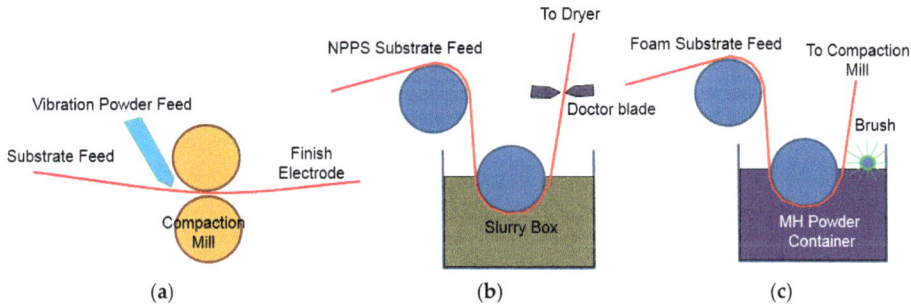

Figure 4. Schematic diagrams of three commonly used nickel/metal hydride (Ni/MH) battery negative electrode fabrication methods: (**a**) dry compaction; (**b**) wet paste method; and (**c**) dry paste method.

OBC filed a patent for its dry compaction manufacturing process in 1989 [130]. While the wet paste method was inherited from the conventional alkaline battery technology, negative electrode fabrication by dry pasting onto the Ni foam through brushing has gained popularity in just the last few years [131]. The pros and cons of each method are summarized in Table 2.

Table 2. Comparisons among three commonly used Ni/MH battery negative electrode fabrication methods. Both the dry compaction and dry paste method can be processed without a binder, and therefore the power density is higher from both cases compared to that from the wet paste method.

Methods	Pros	Cons
Dry compaction	• Higher power density • Lower cost and simple process	• Difficult to control the thickness precisely
Wet paste method	• Longer cycle life • Lower cost	• Lower power density • Complicated process
Dry paste method	• Simple process • Higher power density	• Higher cost

Expanded Ni substrate is typically used for the dry compaction method to reduce the burr that happens with the use of Ni mesh substrate. Ni-plated perforated steel (NPPS) plate [132] and Ni foam are commonly used for the wet paste and dry paste processes, respectively. Other types of substrates, such as porous Ni [133], porous Cu [134], welded Cu [135], cellular metal [136], and conductive ceramic [137], were also patented. Substrate-less electrode fabrication was also proposed, where the active material is first mixed with a conductive metal and subsequently forms the electrode without a substrate [138].

Additives such as carbon and metal powder can be added in the negative electrode to increase the conductivity [139]. Other additives such as rare earth oxides and GeO_2 can be used to reduce the oxidation of the negative electrode [140,141]. Moreover, binders are required for the wet paste method, and many candidates were patented in the past [142–144]. Polyvinyl alcohol is the most popular binder material used in the negative electrode of Ni/MH battery.

In order to remove the passive oxide formed during powder fabrication and form a thin, porous, and electrochemically active oxide on the surface of MH alloy, activation is needed and can be executed at the powder or electrode level with acid [145] or alkaline [146] or in the sealed cell during formation. Furthermore, performing a controllable oxidation on MH alloy before fabricating it into electrode is preferred for the purpose of reducing the native oxide thickness [147,148]. Also, oxygen gas evolution during fast charge was found to be reduced by increasing the final surface roughness [149]. A surface fluorination process was invented for performance enhancement as well [150]. In terms of construction, an auxiliary negative electrode was added to reduce the cell pressure of button cell [151].

2.2.2. Positive Electrode (Cathode)

The most common type of positive electrode in today's Ni/MH battery is the pasted type. Spherical $Ni(OH)_2$ co-precipitated with one or more additional metal hydroxides (Co and Zn being the most common) is mixed with additives, binder, and solvent (water, methanol, or ethylene glycol) to form a paste. The paste is then applied to the Ni-foam substrate, calendared, and dried [152]. A continuous positive electrode fabrication method was patented by OBC in 1994 [97]. The active material in the tab area can be removed or avoided by water wash or ultrasonic cleaning, using a pre-taped foam substrate [153] or reduction by hot hydrogen [154]. Besides the commonly used Ni foam patented by Inco in 1990 [155], metal foil [156], metal mesh [157], Ni mesh/fabric/metal fiber composite [158], and Ni-plated porous Cu/Cu alloy skeleton [159] were also proposed as substrates. Both non-conducting [160] and conducting binders [161] can be used. Since the intrinsic conductivity of spherical particle is poor, a CoOOH conductive network on the surface has been indispensable since the beginning of spherical $Ni(OH)_2$ application [99], an idea originated from its use in the sintered electrode [162]. Conversion from Co^{2+} to Co^{3+} is usually completed in the sealed cell during the formation process. Pre-oxidation of Co at the electrode level was also patented [163]. Modifications and improvements on the basic Co-compound additives were proposed [23,102,127,164–169]. Other additives such as Ca compound [170], Mn compound [171,172], Al compound [173], carbon [174], and silicate [175] were patented as well. For high temperature application, oxides of rare earth elements (Y_2O_3 and Yb_2O_3 being the most common) are added [176–178]. In today's commercial Ni/MH battery, Y_2O_3 is a must additive in the positive electrode. Besides the pasting method, an electrodeless plating technique was proposed for applying the active material onto the substrate, but it is not commercialized [179,180].

2.3. Cell Assembly

Several key areas such as components (negative and positive electrodes, current collector, venting cap, case, electrolyte, and separator), construction type, design of capacity ratio, and formation process need to be considered in order to complete the cell assembly. The following sections summarize the patents concerning the aforementioned items.

2.3.1. Cell Construction

There are five conventional designs of Ni/MH batteries, and a comparison among them can be found in Table 3. Moreover, a new pouch type of Ni/MH battery providing high energy density was recently proposed [181]. The cylindrical type of Ni/MH battery is the most common and mainly used to replace the disposable primary battery. While the typical cylindrical cell is made by winding the electrodes/separator spirally, a patent from Matsushita used the technique of layering hollow cylindrical electrodes in the form of concentric circles to increase the mutual facing area between the negative and positive electrodes [182]. Examples of prismatic Ni/MH batteries can be found in patents filed by Energy Conversion Devices Inc. (ECD) [183,184] and Toshiba [185–187] and show the most scalability up to 1 kWh. The typical prismatic cells are composed of a stack of electrodes with tabs welded together. Toyota filed a patent using a single piece of folded electrode stack (one separator and multiple positive and negative electrode) in a rectangular cell [188], similar to the design of GIGACELL developed by Kawasaki Heavy Industry [189]. The case and electrode assembly may not be in the same shape. For example, optimizing the amount of electrolyte by placing a spirally wound electrode assembly in a rectangular casing was proposed [190]. A flat wafer cell, similar in design to the coin cell [191], was patented with the use of conductive carbon-filled polymeric outer layers as electrode contacts [192,193].

Table 3. Comparison among five common types of Ni/MH battery. SS: stainless steel. EV: electric vehicle; and HEV: hybrid electric vehicle.

Type	Description	Pros	Cons
Consumer	• Cylindrical • SS case	• Mass production • Compatible with other alkaline batteries	• Size limitation
Stick	• Prismatic • SS case	• Mass production • Easy for integration	• Higher cost • Lower energy density
EV	• Prismatic • SS case	• Large format (100 Ah)	• Hand assembly • High cost
HEV	• Prismatic • Plastic case	• High-volume mass production	• Lower pressure rating • Poor heat transfer
Coin	• Button-shaped • SS case	• High-volume mass production • Compact for medical application	• No vent • Cannot be scaled up

While only separators are placed between the adjacent negative and positive electrodes in the typical Ni/MH cell construction, a special "bipolar" design (originated from its use for a lead-acid battery [194]) also uses a metal sheet as substrate for both electrodes (one side as negative electrode substrate and the other side as positive electrode substrate) (Figure 5). This design substantially reduces the contact resistance between adjacent cells [195–198]. Placing a hydrophobic layer between two adjacent negative electrodes to reduce the hydrogen pressure buildup in the cell is another special arrangement [199]. A monoblock of multi-cell assembly was designed to reduce the contact resistance and improve heat dissipation by integrating flow channels within, which allows the gas or liquid coolant to flow between adjacent cells [200]. A lightweight design where the electrode stacks are connected together in a honeycomb structure was also proposed [201].

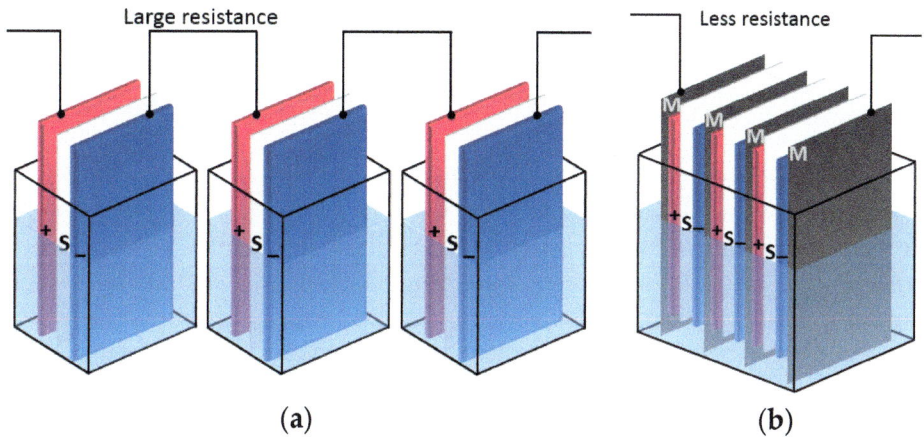

(a)   (b)

Figure 5. Schematic diagrams of (a) regular three cells in series; and (b) a bipolar design with three internal cells. +, −, S, and M denote the positive electrode, negative electrode, separator, and metal sheet, respectively.

2.3.2. Current Collector

Electrodes are connected to the terminals located outside of the cell casing through current collectors. Conventional consumer-type of Ni/MH battery (for application of up to 0.5C rate) uses a single Ni tab (or Ni-plated stainless steel (SS) tab) to collect the current from each electrode. The negative tab is stapled onto the negative electrode, and the positive tab is welded to the part of positive electrode without active material (tab area cleaned by ultrasonic cleaning or water jet). Cu was also proposed as the tab material [135]. Generally, the tab only connects to one side of the electrode: in a cylindrical cell, the positive tab is on the top and electrically connected to the safety valve, and the negative tab is on the bottom and grounded to the case; in a prismatic cell, both the positive and negative tabs are on the top. However, in a patent filed by Park, tabs on both sides of the electrode were proposed to reduce the impedance [202]. Usually, the tab is directly connected to the substrate, but there is also a patent suggesting the technique of folding the end of the electrode, which provides a coupling surface for the connection to the tab or terminal [203]. Matsushita [204–206], Sanyo [207], Yuasa [208], and Johnson Controls [209] patented special current collector designs for high-power cylindrical cells (for applications in power tool and hybrid electric vehicle (HEV)). Both laser welding [210] and electron beam welding [211] are popular techniques for welding multiple pieces of tabs together.

2.3.3. Other Components

Other components essential to the construction of Ni/MH battery include a safety vent, case, separator, and electrolyte. Most types of Ni/MH batteries require a safety vent to manage the extra pressure buildup during high-rate charge and also near the end of service life except a coin cell because of its special medical application requirement. The original safety vent designs were inherited from those available for the Ni/Cd battery. Through the years, companies such as ECD [212], Moltec [213], Sanyo [214], Toyota [215,216], and Johnson Controls [217–219] proposed a variety of improvements in the safety vent design. While the pressure limit in a cylindrical cell is around 350–400 psi, in a prismatic cell it is reduced to 100–120 psi because of the easy deformation of the case. Panasonic and Toyota together patented the idea of using a hydrogen-permeable case to release the overpressure within of the cell [220]. Another special case design with multiple ridges on the outside was also developed to improve the cooling performance and mechanical strength of the case [221].

Several different non-woven fabrics were proposed as separator materials previously [222–229], and the most commonly used is the grafted polypropylene/ polyethylene type. Recently, a sulfonated separator [230] has become very popular for low self-discharge application since it can trap the nitrogen-containing compound in the cell, which is the main cause of the shutting effect that results in self-discharge [231]. A ceramic, polymer, or composite separator capable of conducting alkali ions was also proposed by Ceramatec [232,233], which can be used to eliminate the cross-contamination of ions leached out from the negative electrode onto the separator and positive electrode. Metal-organic framework, coordination polymer, or covalent-organic framework can serve as separator as well [234]. While most of the electrolyte works were performed in academia [235,236], there are several patents in the area of electrolytes for the Ni/MH battery. For example, Toshiba filed a patent for adding W- and Na-containing compounds to the electrolyte [237]. Certain separators can also serve as electrolytes, and they can be categorized into the gel/polymer and solid types. Electrolyte leakage can be reduced by the use of the gel/polymer type electrolyte. Motorola filed a patent with a KOH-dispersed polyvinyl alcohol or polyvinyl acetate polymer electrolyte [238]. Furthermore, Matsushita filed a patent on a gel electrolyte composed of a water absorbent polymer, a water repellent, and an aqueous alkaline solution [239]. In the solid-state battery field, silicon nitride [240], lithium nitride [241], halide [241], oxysalt [241], phosphorus sulfide [241], and oxide with a perovskite structure [242] were suggested. Among these, perovskite is the most promising. $ABO_{3-\delta}$ perovskite, where A is a rare earth or alkaline earth element with a large radius and B is an element with a smaller radius, is capable of proton conduction. The structure of such an oxide, if prepared under proper conditions (e.g., acceptor-doped), is prone to a large density of oxygen vacancies. When it reacts with water, oxygen from water occupies the vacancy, and the remaining two protons are then attached to two separate oxygen ions, which creates a proton-conducting path. The proton conductivity depends on the vacancy density and also the distance between neighboring oxygen ions.

2.3.4. Cell Design

In a typical Ni/MH cell, an oversized negative electrode is needed to create an overcharge reservoir (OCR) and an overdischarge reservoir (ODR) (Figure 6) [243,244]. OCR is created to absorb the hydrogen gas produced during overcharge and also to provide a recombination center to react with the oxygen gas evolved from the positive electrode. ODR is created to prevent oxidation of the negative electrode during overdischarge (cell reversal). Typical negative-to-positive ratios (N/P) are around 1.2 for a high-capacity cell, 1.4–1.6 for a consumer cell, and 1.8–2.2 for a high-power cell [245]. Methodology for the N/P design can be found in several patents [245–247].

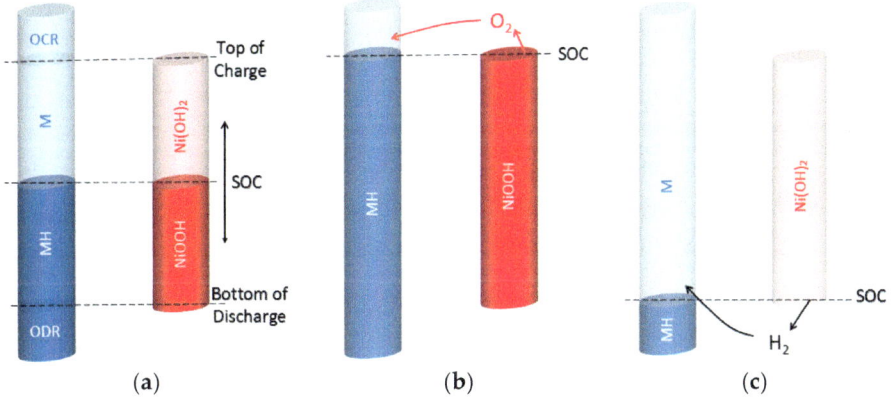

Figure 6. Schematic diagrams demonstrating the positive-limit cell design in Ni/MH battery: (**a**) during regular operation (operating within the limit); (**b**) in the state of overcharge, where the oxygen gas evolved from the positive electrode recombines with the excess hydrogen stored in the negative electrode (OCR prevents hydrogen evolution from the negative electrode); and (**c**) in the state of overdischarge, where the hydrogen gas evolved from the positive electrode is stored in the negative electrode (ODR prevents oxygen gas evolution from the negative electrode). SOC: state of charge.

ODR does not exist when the battery is first built. Later on, ODR starts to increase due to oxidations of MH alloy (Equation (5)), $Co(OH)_2$ (Equation (6)), separator, and other metal components in the cell.

$$3M + H_2O \rightarrow MO + 2M + H_2 \rightarrow MO + 2MH \text{ (assuming metal in } +1 \text{ oxidation state)} \quad (5)$$

$$Co(OH)_2 + M \rightarrow CoOOH + MH \quad (6)$$

With further cycling, ODR continues to grow and suppresses OCR, and finally the cell vents without the protection from OCR. Therefore, it is important to introduce a "pre-charge" to the positive electrode in order to compensate for the increase in ODR during formation when a low N/P is used [248,249].

2.3.5. Formation Process

After the cell is closed, a series of formation/activation is performed. The process is composed of two parts: a thermal activation and an electrochemical formation (Figure 7). Thermal activation dissolves the Co-compound additives in the positive electrode and also dissolves the native oxide formed on the surface of MH alloy during fabrication. Electrochemical formation applies several charge/discharge cycles to crack the MH alloy and create new and electrochemically active surfaces, and it also converts the dissolved Co into a CoOOH conductive network on the surface of $Ni(OH)_2$. Various formation schemes have been patented [250–254], but they may need to be optimized to prevent cell venting. Also, it was shown that keeping a low state of charge (SOC) during battery routine operation helps ensure a long cycle life [251].

2.4. Multi-Cell Construction

The nominal voltage of a single Ni/MH cell is 1.2 V. In order to increase the voltage, several cells have to be connected in series to form a multi-cell module. For consumer use, a module composed of 4 (4.8 V) or 10 (12 V) cells is common. For the propulsion application, multiple modules are assembled together to form a pack. For example, the 1.3 kWh braking energy storage system for the Toyota 2010 HEV Prius (Generation IV) is composed of 168 cells in 28 modules, and each cell is a 6.5 Ah Ni/MH prismatic cell [255]. Patents regarding the module and pack constructions are reviewed in the following sections.

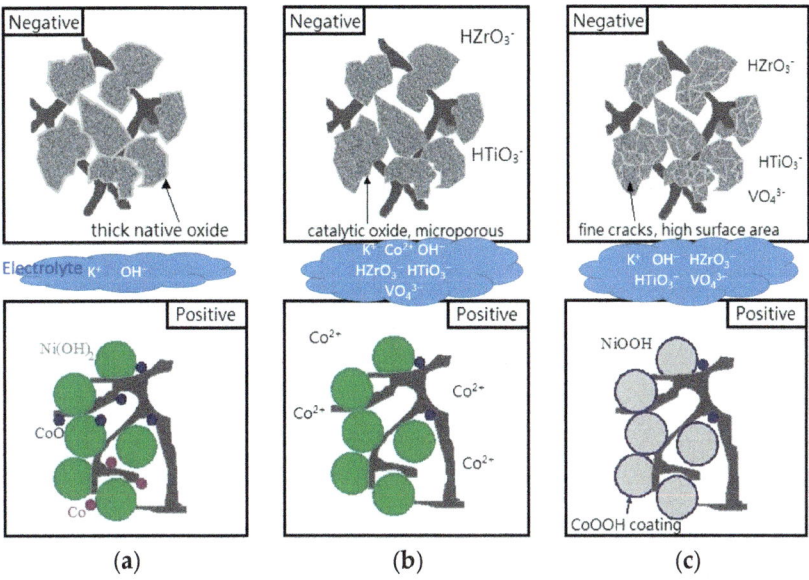

Figure 7. Schematic diagrams showing the formation process of Ni/MH battery: (**a**) before formation, where the MH alloy surface is covered with a thick native oxide; (**b**) after thermal activation, where a new catalytic oxide with metallic Ni embedded within is formed on the MH alloy surface, and bi-valent Co is dissolved; and (**c**) after electrochemical formation, where new surfaces are formed due to MH alloy cracking, and tri-valent Co conductive network is formed on the spherical particle surface in the positive electrode.

2.4.1. Module Assembly

A module can be made of a series of reparably individual cells with metal [184, 256,257] or plastic cases [258], or it can be built with cells isolated by compartment dividers in a single case [259]. Inter-cell conduction is the subject of a number of patents [260,261]. Special module designs for cooling performance improvement include strategic placement of thermally conductive case material and negative electrodes [183] and adding coolant flow channels [262], heat exchanger [263,264], or cooling fins [265,266] between adjacent cells. A compressible two-plate spacer was designed to be placed between adjacent cells for module swelling reduction during operation [267]. For smaller applications in cellphones [268], portable power tools [269], and personal health devices [270], a multi-cell module is the final pack integrated with other system components.

2.4.2. Pack Assembly

When assembling several battery modules into a single pack, both the electrical interconnection between modules and integrity of the connectors are very important. While the former can be ensured by a dedicated welding methodology [271], the latter can be secured by strategically filling the gaps between modules with packaging material [272]. Thermal management of Ni/MH battery pack is also critical to achieve a long service life, which can be done by using either forced air [273,274] or a liquid coolant [275–277]. Moreover, the use of a heat sink is helpful to increase the thermal dissipation rate of the battery pack [278].

2.5. System Integration

For application of a larger energy storage system (for example, in EV), many sub-systems need to be integrated, such as a charging system, a battery management system (BMS), a protection system, and a thermal management system (TMS), to ensure a smooth and safe operation. Patents on these sub-systems are discussed in the following sections.

2.5.1. Charging System

For consumer application, a charger is a separate unit from the battery. In the early days, a different charger was needed for the Ni/MH battery but with a very similar design principle. Later charger hardware improvements made charging both Ni/Cd and Ni/MH batteries with the same charger possible. Numerous patents regarding the portable charger hardware design are available [279–294]. Most of the charging algorithms are composed of several different cut-off schemes, such as ΔT (temperature change), $-\Delta V$ (negative voltage change), dV/dt (voltage change with time), dT/dt (temperature change with time), maximum input, and V_t (terminal voltage) [295–320]. End-of-charge can also be detected by electronic circuit alone [321]. In order to fast charge, special hardware [322] and algorithms [323,324] are required. For application in EV, the charging unit is integrated into BMS and is a part of the system.

2.5.2. Battery Management System

When a large number of cells are used together, BMS is indispensable [325]. Even in a portable power tool, the SOC estimation of Ni/MH battery is necessary [326]. Due to its high abuse tolerance, BMS for Ni/MH battery system is much simpler than that for a Li-ion battery system. For example, voltage monitoring and SOC recording at the cell level in Ni/MH battery system are not necessary. Therefore, for Ni/MH battery systems used in vehicles, the main functions of BMS are monitoring charge/discharge [327–332] and voltage [333–336], estimating SOC [337–344] and state-of-health (SOH) [345–351], communicating with the on-board computer system, and handling emergencies. In order to obtain an accurate estimation, several methodologies, such as integral amp-hour [352–354], self-discharge calculation [355], calibration from voltage variation [356,357], and impedance measurement [358,359] are used in combination.

2.5.3. Protection System

Originated from the overcharge protection circuit invented for Ni/Cd battery [360–364], a similar design was patented for Ni/MH batteries [365,366]. Over-temperature protection [367] and under-voltage [368] circuitries for Ni/MH battery were previously patented. The design of putting a capacitor in parallel with the battery was also invented to offer an emergency power source [369–371]. The addition of a high-voltage connector was suggested for electrical shock prevention [372]. Finally, a venting system for automobiles equipped with Ni/MH battery system was proposed to release hydrogen from a strategic location [219].

2.5.4. Thermal Management System

Thermal dissipation for Ni/MH battery is performed either at the cell [184,256,257] or module level [189,262,263,265,266,373–376]. Compared to TMS for Li-ion battery pack, a much simpler TMS is needed for Ni/MH battery pack [373]. For example, the Ni/MH battery system equipped in Toyota HEV Prius requires only a three-speed 12 V DC blower [377], and its motor speed is determined by both the take-in air temperature and cell temperature. Furthermore, Toyota filed a patent regarding the design of cooling control in a battery pack [378].

2.6. Others

As the last part of this review, U.S. patents on the application and recycling of Ni/MH battery are presented.

2.6.1. Application

Key applications for Ni/MH battery are summarized in Table 4 [11]. For a consumer battery, a hybrid Ni/MH battery/capacitor device was patented by Motorola to handle pulse power communication [379,380]. Patents filed by OBC [381] and Delphi [382] described the use of Ni/MH batteries in HEV. Applications in uninterrupted power supply (UPS) [318,383] and grid [384–386] were also patented. A window construction combining Ni/MH battery and solar cell was proposed [387].

Table 4. Key applications for Ni/MH battery in the market. UPS: uninterrupted power supply.

Application	Examples	Main Competitor	Advantages over Competitor
Consumer	AA, AAA, cordless phone, shaver, power tool	Ni/Cd	• Higher energy density • Environmentally safe
Propulsion	EV, HEV	Li-ion	• Long service life • Excellent abuse tolerance
UPS	Cellphone communication hub, vending machine	Lead-acid	• Higher energy density • Longer service life • Environmental safe
Grid	Solar and wind energy storage, voltage equalizer	Lead-acid	• Higher energy density • Longer service life • Environmental safe • Higher instant charge acceptance

2.6.2. Revivification and Recycling

In order to reuse the battery pack, a refreshment method was proposed to replace or re-inject electrolyte into cells with bad performance [388]. At the stage where recovery is impossible, the Ni/MH battery is ready for recycling. Because of the high flammability of the negative electrode, rupturing the battery under anaerobic conditions and flooding the battery interior with CO_2 in an anaerobic chamber is recommended [389]. Recycling can be performed at the component level by separating the MH alloy (AB_2 [390] and AB_5 [391]) and binder [392] or at the individual element level by aiming at rare-earth metals [393–397], Zr [395], Fe [396], Al [396], Co [396,397] and Ni [397,398].

3. Conclusions

U.S. patents on the Ni/MH battery have been reviewed and categorized. With close to 400 U.S. patents cited, this report summarizes the efforts to improve the performance and expand the applicability of Ni/MH batteries undertaken by scientists and engineers working in various companies; these can be used to study the principle and fabrication process systematically. By learning what has been patented previously, we can continue to explore uncharted territory in the science and engineering of Ni/MH batteries.

Acknowledgments: Shiuan Chang collected and organized the information. Kwo-hsiung Young and Cristian Fierro validated the information in the area of negative and positive electrodes, respectably. Jean Nei prepared the manuscript.

Author Contributions: All the authors made a significant contribution to the writing of this manuscript.

Conflicts of Interest: The authors declare no conflict of interest.

Abbreviations

Ni/MH	Nickel/metal hydride
MH	Metal hydride
OBC	Ovonic Battery Company
FCC	Face-centered-cubic
1C/1C	1C charge and discharge rates
DOD	Depth of discharge
BCC	Body-centered-cubic
NPPS	Ni-plated perforated steel
ECD	Energy conversion devices
SS	Stainless steel
EV	Electric vehicle
HEV	Hybrid electric vehicle
OCR	Overcharge reservoir
ODR	Overdischarge reservoir
N/P	Negative-to-positive ratio
SOC	State of charge
BMS	Battery management system
TMS	Thermal management system
ΔT	Temperature change
$-\Delta V$	Negative voltage change
dV/dt	Voltage change with time
dT/dt	Temperature change with time
V_t	Terminal voltage
SOH	State of health
UPS	Uninterrupted power supply

References

1. Sakai, T. Secondary Battery with Hydrogen Storage Alloys. In *Hydrogen Storage Alloys—Fundamentals and Frontier Technology*; Tamura, H., Ed.; NTS Inc.: Tokyo, Japan, 1998; p. 411.
2. Osumi, Y. *Suiso Kyuzou Goukin*; Agune Co. Ltd.: Tokyo, Japan, 1999; pp. 473–507.
3. Kleperis, J.; Wójcik, G.; Czerwinski, A.; Skowronski, J.; Kopczyk, M.; Beltowska-Brzezinska, M. Electrochemical behavior of metal hydride. *J. Solid State Electrochem.* **2001**, *5*, 229–249.

4.   Feng, F.; Geng, M.; Northwood, D.O. Electrochemical behaviour of intermetallic-based metal hydrides used in Ni/metal hydride (MH) batteries: A review. *Int. J. Hydrog. Energy* **2001**, *26*, 725–734.

5.   Hong, K. The development of hydrogen storage electrode alloys for nickel hydride batteries. *J. Power Sources* **2001**, *96*, 85–89.

6.   Cuevas, F.; Joubert, J.M.; Latroche, M.; Percheron-Guégan, A. Intermetallic compounds as negative electrodes of Ni/MH batteries. *Appl. Phys. A* **2001**, *72*, 225–238.

7.   Petrii, O.A.; Levin, E.E. Hydrogen-accumulating materials in electrochemical systems. *Russ. J. Gen. Chem.* **2007**, *77*, 790–796.

8.   Zhao, X.; Ma, L. Recent progress in hydrogen storage alloys for nickel/metal hydride secondary batteries. *Int. J. Hydrog. Energy* **2009**, *34*, 4788–4796.

9.   Liu, Y.; Pan, H.; Gao, M.; Wang, Q. Advanced hydrogen storage alloys for Ni/MH rechargeable batteries. *J. Mater. Chem.* **2011**, *21*, 4743–4755.

10.   Young, K.; Nei, J. The current status of hydrogen storage alloy development for electrochemical applications. *Materials* **2013**, *6*, 4574–4608.

11.   Young, K. Electrochemical Applications of Metal Hydrides. In *Compendium of Hydrogen Energy*; Barbir, F., Basile, A., Veziroğlu, T.N., Eds.; Woodhead Publishing Ltd.: Cambridge, UK, 2015; Volume 3, pp. 289–304.

12.   Oliva, P.; Leonardi, J.; Laurent, J.F.; Delmas, C.; Braconnier, J.J.; Figlarz, M.; Fievet, F.; Guibert, A.C. Review of the structure and the electrochemistry of nickel hydroxide and oxy-hydroxides. *J. Power Sources* **1982**, *8*, 229–255.

13.   Barton, R.T.; Mitchell, P.J.; Hampson, N.A. The electrotechnology of the nickel positive electrode: A review of the recent literature. *Surf. Coat. Technol.* **1986**, *28*, 1–9.

14.   Yuan, A.; Zhang, J.; Cao, C. Research progress of nickel hydroxide electrodes. *Chin. J. Power Sources* **2001**, *25*, 53–59.

15.   Vidotti, M.; Torresi, R.; Torresi, S.I. Nickel hydroxide modified electrodes: A review study concerning its structural and electrochemical properties aiming the application in electrocatalysis, electrochromism and secondary batteries. *Quim. Nova* **2010**, *33*, 2176–2186.

16.   Hall, D.S.; Lockwood, D.J.; Bock, C.; McDougall, B.R. Nickel hydroxides and related materials: A review of their structures, synthesis and properties. *Proceed. R. Soc. A Math. Phys. Eng. Sci.* **2015**, *471*.

17.   Ouchi, T.; Young, K.; Moghe, D. Reviews on the Japanese patent applications regarding nickel/metal hydride batteries. *Batteries* **2016**, *2*. .

18.   Beccu, K. Electrical Accumulator with a Metal Hydride Serving as the Cathodic Reactive Material Arranged in Suspension in the Electrolyte. U.S. Patent 3,520,728, 14 July 1970.

19.   Beccu, K. Accumulator Electrode with Capacity for Strong Hydrogen and Method of Manufacturing Said Electrode. U.S. Patent 3,669,745, 13 June 1972.

20.   Ikoma, M.; Kawano, H.; Matsumoto, I.; Yanagihara, N. Sealed Storage Battery and Method for Making Its Electrode. U.S. Patent 4,837,119, 6 June 1989.

21.   Ikoma, M.; Kawano, H.; Takahashi, O.; Matsumoto, I.; Ikeyama, M. Alkaline Storage Battery Using Hydrogen Absorbing Alloy. U.S. Patent 4,925,748, 15 May 1990.

22. Ikoma, M.; Kawano, H.; Ito, Y.; Matsumoto, I. Sealed Type Nickel-Hydride Battery and Production Process Thereof. U.S. Patent 4,935,318, 19 June 1990.

23. Hasebe, H.; Takeno, K.; Mitsuyasu, K.; Sato, Y.; Takahashi, H.; Hayashida, H.; Sawatari, I.; Ishiwa, K.; Hata, K.; Yoshida, K.; *et al.* Nickel-Metal Hydride Secondary Cell. U.S. Patent 5,032,475, 16 July 1991.

24. Yamano, M.; Sakai, T.; Furukawa, N.; Murakami, S.; Matsumoto, T. Metal/Hydrogen Alkaline Storage Battery. U.S. Patent 4,636,445, 13 January 1987.

25. Nakahori, S.; Honda, H. Method of Manufacturing Nickel Hydroxide Electrode for Alkaline Storage Cell. U.S. Patent 4,844,948, 4 July 1989.

26. Oshitani, M.; Hasegawa, K.; Yufu, H. Alkaline Battery with a Nickel Electrode. U.S. Patent 4,985,318, 15 January 1991.

27. Willems, J.J.G.; Buschow, K.H.J. From permanent magnets to rechargeable hydride electrodes. *J. Less Comm. Met.* **1987**, *129*, 13–30.

28. Born Guegen, A.P.; Achard, J.C.; Loriers, J.; Bonnemay, M.; Bronoël, G.; Sarradin, J.; Schlapbach, L. Electrode Materials Based on Lanthanum and Nickel, and Electrochemical Uses of Such Materials. U.S. Patent 4,107,405, 15 August 1978.

29. Willems, J.J.G.S.A.; van Beek, J.R.G.C.M.; Buschow, K.H.J. Electrochemical Cell Comprising Stable Hydride-Forming Material. U.S. Patent 4,487,817, 11 December 1984.

30. Osumi, Y.; Suzuki, H.; Kato, A.; Nakane, M.; Miyake, Y. Alloy for Occlusion of Hydrogen. U.S. Patent 4,147,536, 3 April 1979.

31. Osumi, Y.; Suzuki, H.; Kato, A.; Oguro, K.; Nakane, N. Alloy for Occlusion of Hydrogen. U.S. Patent 4,396,576, 2 August 1983.

32. Kanda, M.; Sato, Y. Hermetically Sealed Metallic Oxide-Hydrogen Battery Using Hydrogen Storage Alloy. U.S. Patent 4,605,603, 12 August 1986.

33. Mori, H.; Hasegawa, K.; Watada, M.; Oshitani, M. Metal Hydride Electrode. U.S. Patent 5,393,616, 28 February 1995.

34. Hasebe, H.; Inada, S.; Isozaki, Y.; Inaba, T.; Sawa, T.; Horie, H.; Yagi, N.; Shizu, H.; Kanazawa, Y. Hydrogen-absorbing Alloy for Battery, Method of Manufacturing the Same, and Secondary Nickel-Metal Hydride Battery. U.S. Patent 5,843,372, 1 December 1998.

35. Hirosawa, T.; Ikemachi, T. Hydrogen Storage Alloy Electrode and Manufacturing Method of the Same. U.S. Patent 5,766,792, 16 June 1998.

36. Hazama, T. Hydrogen Absorbing Electrode for Use in Nickel-Metal Hydride Secondary Batteries. U.S. Patent 5,284,619, 8 February 1994.

37. Lichtenberg, F. Alloys for Use as Active Material for the Negative Electrode of an Alkaline, Rechargeable Nickel-Metal Hydride Battery, and Process for Its Production. U.S. Patent 5,738,958, 14 April 1998.

38. Ohyama, H.; Nakatsuji, K.; Kikuyama, S.; Dansui, Y. Negative Electrode Material for Nickel-Metal Hydride Battery and Treatment Method Thereof, and Nickel-Metal Hydride Battery. U.S. Patent 8,202,650, 19 June 2012.

39. Kojima, Y.; Ikeda, H.; Furukawa, S.; Sugiyama, K.; Kobayashi, N. Hydrogen-Absorbing Alloy and Electrode for Nickel-Metal Hydride Secondary Batteries. U.S. Patent 8,377,374, 19 February 2013.

40. Young, K.; Fetcenko, M.A.; Huang, B.; Chao, B. NdNi5 Alloys for Hydrogen Storage and Ni-MH Batteries. U.S. Patent 8,968,644, 3 March 2015.

41. Ito, S.; Sakamoto, H. Nickel-Metal Hydride Battery. U.S. Patent 8,119,284, 21 February 2012.

42. Lv, H. High-Temperature Ni-MH Battery and a Method for Making the Same. U.S. Patent 6,689,514, 10 February 2004.

43. Fetcenko, M.A.; Young, K.; Ovshinsky, S.R.; Ouchi, T. Hydrogen Storage Alloys Having Improved Cycle Life and low Temperature Operating Characteristics. U.S. Patent 7,393,500, 1 July 2008.

44. Gamo, T.; Moriwaki, Y.; Yamashita, T.; Fukuda, M. Method of Making a Hydrogen Storage Alloy and Product. U.S. Patent 4,144,103, 13 March 1979.

45. Sapru, K.; Hong, K.; Fetcenko, M.A.; Venkatesan, S. Hydrogen Storage Materials and Methods of Sizing and Preparing the Same for Electrochemical Applications. U.S. Patent 4,551,400, 5 November 1985.

46. Venkatesan, S.; Reichman, B.; Fetcenko, M.A. Enhanced Charge Retention Electrochemical Hydrogen Storage Alloys and an Enhanced Charge Retention Electrochemical Cell. U.S. Patent 4,728,586, 1 March 1988.

47. Fetcenko, M.A.; Ovshinsky, S.R. Catalytic Hydrogen Storage Electrode Materials for Use in Electrochemical Cells and Electrochemical Cells Incorporating the Materials. U.S. Patent 5,104,617, 14 April 1992.

48. Fetcenko, M.A.; Ovshinsky, S.R. Metastable Hydrogen Storage Alloy Material. U.S. Patent 5,135,589, 4 August 1992.

49. Fetcenko, M.A.; Ovshinsky, S.R.; Kajita, K. Electrode Alloy Having Decreased Hydrogen Overpressure and/or Low Self-Discharge. U.S. Patent 5,238,756, 24 August 1993.

50. Fetcenko, M.A.; Ovshinsky, S.R. Metal Hydride Cells Having Improved Cycle Life and Charge Retention. U.S. Patent 5,330,861, 19 July 1994.

51. Ovshinsky, S.R.; Fetcenko, M.A. Electrochemical Hydrogen Storage Alloys and Batteries fabricated from These Alloys Having Significantly Improved Capacity. U.S. Patent 5,407,761, 18 April 1995.

52. Fetcenko, M.A.; Ovshinsky, S.R.; Chao, B.S.; Reichman, B. Electrochemical Hydrogen Storage Alloys for Nickel Metal Hydride Batteries. U.S. Patent 5,536,591, 16 July 1996.

53. Ovshinsky, S.R.; Fetcenko, M.A.; Im, J.S.; Young, K.; Chao, B.S.; Reichman, B. Hydrogen Storage Materials Having a High Density of Non-Conventional Useable Hydrogen Storing Sites. U.S. Patent 5,840,440, 24 November 1998.

54. Fetcenko, M.A.; Young, K.; Ovshinsky, S.R.; Reichman, B.; Koch, J.; Mays, W. Modified Electrochemical Hydrogen Storage Alloy Having Increased Capacity, Rate Capability and Catalytic Activity. U.S. Patent 6,270,719, 7 August 2001.

55. Fetcenko, M.A.; Young, K.; Fierro, C. Hydrogen Storage Battery and Methods for Making. U.S. Patent 6,593,024, 15 July 2003.

56. Seri, H.; Yamamura, Y.; Tsuji, Y.; Moriwaki, Y.; Iwaki, T. Hydrogen Storage Alloy Electrodes. U.S. Patent 5,468,309, 21 November 1995.

57. Tsuji, Y.; Yamamoto, O.; Yamamura, Y.; Seri, H.; Toyoguchi, Y. Hydrogen Storage Alloy and Electrode Therefrom. U.S. Patent 5,753,054, 19 May 1998.

58. Izumi, Y.; Moriwaki, Y.; Yamashita, K.; Tokuhiro, T. Nickel-Metal Hydride Storage Battery and Alloy for Configuring Negative Electrode of the Same. U.S. Patent 5,962,156, 5 October 1999.

59. Knosp, B.; Bouet, J.; Jordy, C.; Mimoun, M.; Gicquel, D. Hydridable Material for the Negative Electrode in a Nickel-Metal Hydride Storage Battery. U.S. Patent 5,626,987, 6 May 1997.

60. Iba, H.; Akiba, E. Hydrogen-Absorbing Alloy. U.S. Patent 5,968,291, 19 October 1999.

61. Iba, H.; Akiba, E. Hydrogen-Absorbing Alloy and Process for Preparing the Same. U.S. Patent 6,153,032, 28 November 2000.

62. Nichimura, K.; Yano, M.; Nakamura, H.; Hamasaki, K.; Tokuda, M.; Itoh, Y. Nickel-Metal Hydride Storage Battery. U.S. Patent 6,632,567, 14 October 2003.

63. Tsukahara, M.; Takahashi, K.; Mishima, T.; Isomura, A.; Sakai, T.; Miyamura, H.; Uehara, I. Hydrogen-Occluding Alloy and Hydrogen-Occluding Alloy Electrode. U.S. Patent 5,690,799, 25 November 1997.

64. Tsukahara, M.; Takahashi, K.; Mishima, T.; Isomura, A.; Sakai, T.; Miyamura, H.; Uehara, I. Hydrogen-Occluding Alloy and Hydrogen-Occluding Alloy Electrode. U.S. Patent 5,776,626, 7 July 1998.

65. Ovishinsky, S.R.; Fetcenko, M.A. Electrochemical Hydrogen Storage Alloys and Batteries Fabricated from Mg Containing Base Alloy. U.S. Patent 5,506,069 A, 9 April 1996.

66. Ovishinsky, S.R.; Fetcenko, M.A.; Reichman, B.; Young, K.; Chao, B.; Im, J. Electrochemical Hydrogen Storage Alloys and Batteries Fabricated from Mg Containing Base Alloys. U.S. Patent 5,616,432 A, 1 April 1997.

67. Fetcenki, M.A.; Young, K.; Ouchi, T.; Reinhout, M.; Ovshinsky, S.R. Mg-Ni Hydrogen Storage Composite Having High Storage Capacity and Excellent Room Temperature Kinetics. U.S. Patent 7,211,541, 1 May 2007.

68. Ovshinsky, S.R.; Fetcenko, M.A.; Im, J.; Chao, B.; Reichman, B.; Young, K. Electrochemical Hydrogen Storage Alloys and Batteries Containing Heterogeneous Powder Particles. U.S. Patent 5,554,456, 10 September 1996.

69. Ting, J.; Habel, U.; Peretti, M.W.; Eisen, W.B.; Young, R.; Chao, B.; Huang, B. Electrochemical Hydrogen Storage Alloys for Nickel Metal Hydride Batteries, Fuel Cells and Methods of Manufacturing Same. U.S. Patent 6,500,583, 31 December 2002.

70. Komori, K.; Matsuda, H.; Toyoguchi, Y. Active Material for Hydrogen Storage Alloy Electrode and Method for Producing the Same. U.S. Patent 6,444,361, 3 September 2002.

71. Arvidsson, J.; Carlström, R.; Hallén, H.; Löfgren, S. Powder Composition and Process for the Preparation Thereof. U.S. Patent 6,245,165, 12 June 2001.

72. Hayashida, H.; Kitayama, H.; Yamamoto, M.; Inada, S.; Sakai, I.; Kono, T.; Yoshida, H.; Inaba, T.; Kanda, M. Nickel-Hydrogen Secondary Battery. U.S. Patent 6,200,705, 13 March 2001.

73. Kakeya, T.; Kanemoto, M.; Kuzuhara, M.; Ozaki, T.; Watada, M.; Sakai, T. Hydrogen Storage Alloy, Hydrogen Storage Alloy Electrode, Secondary Battery, and Method for Producing Hydrogen Storage Alloy. U.S. Patent 8,343,660, 1 January 2013.

74. Ozaki, T.; Sakai, T.; Kanemoto, M.; Kakeya, T.; Kuzuhara, M.; Watada, M. Hydrogen Absorbing Alloy, Production Method Thereof, and Secondary Battery. U.S. Patent 8,343,659, 1 January 2013.

75. Kono, T.; Sakai, I.; Yoshida, H.; Inaba, T.; Yamamoto, M.; Takeno, S. Hydrogen-Absorbing Alloy, Electrode and Secondary Battery. U.S. Patent 6,214,492, 10 April 2001.

76. Hayashida, H.; Yamamoto, M.; Kitayama, H.; Inada, S.; Sakai, I.; Kono, T.; Yoshida, H.; Inaba, T.; Kanda, M. Nickel-Hydrogen Secondary Battery. U.S. Patent 6,248,475, 19 June 2001.

77. Hasebe, H.; Sori, N.; Arai, T. Nickel-Metal Hydride Secondary Cell, and Method of Manufacturing the Same, Hydrogen Absorbing Alloy Particles for Cell, Method of Manufacturing the Same. U.S. Patent 5,219,678, 15 June 1993.

78. Inaba, T.; Sawa, T.; Inada, S.; Kawashima, F.; Sato, N.; Sakamoto, T.; Okamura, M.; Arai, T.; Hashimoto, K. Hydrogen-Absorbing Alloy for Battery and Secondary Nickel-Metal Hydride Battery. U.S. Patent 5,753,386, 19 May 1998.

79. Ovshinsky, S.R.; Young, R.T.; Chao, B. Hydrogen Storage Alloys and Methods and Improved Nickel Metal Hydride Electrodes and Batteries Using Same. U.S. Patent 6,210,498, 3 April 2001.

80. Chen, Y.; Cai, Y.; Yu, D.; Dai, X. Method and an Equipment for Producing Rapid Condensation Hydrogen Storage Alloy Powder. U.S. Patent 6,174,345, 16 January 2001.

81. Yamamoto, K.; Miyake, Y.; Okada, C.; Kitazume, N. Method for Production of Rare Earth Metal-Nickel Hydrogen Occlusive Alloy Ingot. U.S. Patent 5,680,896, 28 October 1997.

82. Sun, B.; Chen, X. Vacuum Induction Melting and Strip Casting Equipment for Rare Earth Permanent Magnetic Alloy. U.S. Patent 20140352909 A1, 4 December 2014.

83. Shinya, N.; Sugahara, H.; Ishii, M. Process for Producing Hydrogen Absorbing Alloy Powder and Hydrogen Absorbing Alloy Electrode. U.S. Patent 6,284,066, 4 September 2001.

84. Sapru, K.; Reichman, B.; Reger, A.; Ovshinsky, S.R. Rechargeable Battery and Electrode Used Therein. U.S. Patent 4,623,597, 18 November 1986.

85. Ovshinsky, S.R.; Fetcenko, M.A. Electrochemical Hydrogen Storage Alloys and Batteries Fabricated These Alloys Having Significantly Improved Performance Characteristics. U.S. Patent 5,277,999, 11 January 1994.

86. Fetcenko, M.A.; Young, K.; Ovshinsky, S.R.; Reichman, B.; Koch, J.; Mays, W. Modified Electrochemical Hydrogen Storage Alloy Having Increased Capacity, Rate Capability and Catalytic Activity. U.S. Patent 6,740,448, 25 May 2004.

87. Fetcenko, M.A.; Ovshinsky, S.R.; Young, K.; Reichman, B.; Ouchi, T.; Koch, J.; Mays, W. Hydrogen Storage Alloys Having a High Porosity Surface Layer. U.S. Patent 6,830,725, 14 December 2004.

88. Young, K.; Reichman, B.; Fetcenko, M.A. Metal Hydride Alloy with Catalyst Particles and Channels. U.S. Patent 8,877,378, 4 November 2014.

89. Wikipedia. Nickel-Iron Battery. Available online: https://en.wikipedia.org/wiki/Nickel-iron_battery (accessed on 9 February 2016).

90. Edison, T.A. Reversible Galvanic Battery. U.S. Patent 678722 A, 16 July 1901.

91. Edison, T.A. Reversible Galvanic Battery. U.S. Patent 692507, 4 February 1902.
92. Gutjahr, M.A.; Schmid, R.; Beccu, K.D. Method of Manufacturing Positive Nickel Hydroxide Electrodes. U.S. Patent 3,877,987 A, 15 April 1975.
93. Gutjahr, M.A.; Schmid, R.; Beccu, K.D. Method of Manufacturing Positive Nickel Hydroxide Electrodes. U.S. Patent 3,926,671 A, 16 December 1975.
94. Fritts, D.H.; Leonard, J.F. Method of Nickel Electrode Production. U.S. Patent 4,399,005, 16 August 1983.
95. Fritts, D.H.; Leonard, J.F. Method for Fabricating Battery Plaque and Low Shear Nickel Electrode. U.S. Patent 4,628,593, 16 December 1986.
96. Fetcenko, M.; Koch, J. Nickel-Metal Hydride Batteries. In *Linden's Handbook of Batteries*, 4th ed.; Reddy, T.B., Ed.; McGraw-Hill: New York, NY, USA, 2011; p. 229.
97. Ovshinsky, S.R.; Fetcenko, M.A.; Venkatesan, S.; Holland, A. Compositionally and Structurally Disordered Multiphase Nickel Hydroxide Positive Electrode for Alkaline Rechargeable Electrochemical Cells. U.S. Patent 5,344,728, 6 September 1994.
98. Ovshinsky, S.R.; Fetcenko, M.A.; Venkatesan, S.; Holland, A. Compositionally and Structurally Disordered Multiphase Nickel Hydroxide Positive Electrode Containing Modifiers. U.S. Patent 5,948,564, 7 September 1999.
99. Oshitani, M.; Yufu, H. Nickel Electrode for Alkaline Battery and Battery Using Said Nickel Electrode. U.S. Patent 4,844,999, 4 July 1989.
100. Fierro, C.; Fetcenko, M.A.; Young, K.; Ovshinsky, S.R.; Sommers, B.; Harrison, C. Nickel Hydroxide Positive Electrode Material Exhibiting Improved Conductivity and Engineered Activation Energy. U.S. Patent 6,228,535, 8 May 2001.
101. Watada, M.; Oshitani, M.; Onishi, M. Nickel Electrode for Alkaline Battery. U.S. Patent 5,366,831, 22 November 1994.
102. Ohta, K.; Matsuda, H.; Ikoma, M.; Morishita, N.; Toyoguchi, Y. Nickel Positive Electrode for Alkaline Storage Battery and Sealed Nickel-Hydrogen Storage Battery Using Nickel Positive Electrode. U.S. Patent 5,571,636, 5 November 1996.
103. Yamawaki, A.; Nakahori, S.; Tadokoro, M.; Hamamatsu, T.; Baba, Y. Nickel Electrode Active Material; a Nickel Electrode and a Nickel Alkali Storage Cell Using Such Nickel Electrode Active Material; and Production Methods of Such Material, Electrode, and Cell. U.S. Patent 5,629,111, 13 May 1997.
104. Ovshinsky, S.R.; Fetcenko, M.A.; Venkatesan, S.; Holland, A. Compositionally and Structurally Disordered Multiphase Nickel Hydroxide Positive Electrode for Alkaline Rechargeable Electrochemical Cells. U.S. Patent 5,637,423, 10 June 1997.
105. Ikoma, M.; Akutsu, N.; Enokido, M.; Yoshii, F.; Kaiya, H.; Tsuda, S. Nickel Hydroxide Active Material Powder and Nickel Positive Electrode and Alkali Storage Battery Using Them. U.S. Patent 5,700,596, 23 December 1997.
106. Matsuda, H.; Okada, Y.; Ohta, K.; Toyoguchi, Y. Active Material and Positive Electrode for Alkaline Storage Battery. U.S. Patent 5,773,169, 30 June 1998.
107. Joo, K.; Choi, J.; Choi, K.; Kim, G.; Lee, S. Active Material for Nickel Electrode and Nickel Electrode Having the Same. U.S. Patent 5,789,113, 4 August 1998.

108. Corrigan, D.; Fierro, C.; Martin, F.J.; Ovshinsky, S.R.; Xu, L. Nickel Battery Electrode Having Multiple Composition Nickel Hydroxide Active Materials. U.S. Patent 5,861,225, 19 January 1999.

109. Okada, Y.; Ohta, K.; Matsuda, H.; Toyoguchi, Y. Nickel Hydroxide Positive Electrode Active Material Having a Surface Layer Containing a Solid Solution Nickel Hydroxide With Manganese Incorporated Therein. U.S. Patent 6,066,416, 23 May 2000.

110. Dansui, Y.; Kato, F.; Suzuki, K.; Yuasa, K. Nickel/Metal Hydride Storage Battery. U.S. Patent 6,074,785, 13 June 2000.

111. Fierro, C.; Fetcenko, M.A.; Young, K.; Ovshinsky, S.R.; Sommers, B.; Harrison, C. Nickel Hydroxide Positive Electrode Material Exhibiting Improved Conductivity and Engineered Activation Energy. U.S. Patent 6,447,953, 10 September 2002.

112. Jackovitz, J.F.; Pantier, E.A. Alkali Slurry Ozonation to Produce a High Capacity Nickel Battery Material. U.S. Patent 4,481,128, 6 November 1984.

113. Ovshinsky, S.R.; Corrigan, D.; Venkatesan, S.; Young, R.; Fierro, C.; Fetcenko, M.A. Chemically and Compositionally Modified Solid Solution Disordered Multiphase Nickel Hydroxide Positive Electrode for Alkaline Rechargeable Electrochemical Cells. U.S. Patent 5,348,822, 20 September 1994.

114. Ovshinsky, S.R.; Young, R.T. Nickel Metal Hydride Battery Containing a Modified Disordered Multiphase Nickel Aluminum Based Positive Electrode. U.S. Patent 5,567,549, 22 October 1996.

115. Young, R.; Ovshinsky, S.R.; Xu, L. Beta to Gamma Phase Cycleable Electrochemically Active Nickel Hydroxide Material. U.S. Patent 5,905,003, 18 May 1999.

116. Ovshinsky, S.R.; Young, R.T.; Xu, L.; Kumar, S. Active Nickel Hydroxide Material Having Controlled Water Content. U.S. Patent 6,019,955, 1 February 2000.

117. Singh, D.B. Gamma NiOOH Nickel Electrodes. U.S. Patent 6,020,088, 1 February 2000.

118. Aladjov, B. Battery-Grade Nickel Hydroxide and Method for Its Preparation. U.S. Patent 5,788,943, 4 August 1998.

119. Olbrich, A.; Meese-Marktscheffel, J.; Stoller, V.; Erb, M.; Albrecht, S.; Gille, G.; Maikowske, G.; Schrumpf, F.; Schmoll, J.; Jahn, M. Nickel Hydroxide and Method for Producing Same. U.S. Patent 7,563,431, 21 July 2009.

120. Kato, F.; Tanigawa, F.; Dansui, Y.; Yuasa, K. Positive Electrode Material for Alkaline Storage Battery, Method of Producing the Same, and Alkaline Storage Battery Using the Same. U.S. Patent 6,083,642, 4 July 2000.

121. Junichi, I.; Yuri, K.; Tetsushi, M.; Toyoshi, I. Nickel Hydroxide Particles Having an α- or β-Cobalt Hydroxide Coating Layer for Use in Alkali Batteries and a Process for Producing the Nickel Hydroxide. U.S. Patent 6,040,007, 21 March 2000.

122. Fetcenko, M.A.; Fierro, C.; Ovshinsky, S.R.; Sommers, B.; Reichman, B.; Young, K.; Mays, W. Composite Positive Electrode Material and Method for Making Same. U.S. Patent 6,177,213, 23 January 2001.

123. Fetcenko, M.A.; Fierro, C.; Ovshinsky, S.R.; Sommers, B.; Reichman, B.; Young, K.; Mays, W. Composite Positive Electrode Material and Method for Making Same. U.S. Patent 6,348,285, 19 February 2002.

124. Fetcenko, M.A.; Fierro, C.; Ovshinsky, S.R.; Sommers, B.; Reichman, B.; Young, K.; Mays, W. Composite Positive Electrode Material and Method for Making Same. U.S. Patent 6,548,209, 15 April 2003.

125. Fetcenko, M.A.; Fierro, C.; Ovshinsky, S.R.; Sommers, B.; Reichman, B.; Young, K.; Mays, W. Composite Positive Electrode Material and Method for Making Same. U.S. Patent 6,569,566, 27 May 2003.

126. Fierro, C.; Fetcenko, M.A.; Young, K.; Ovshinsky, S.R.; Sommers, B.; Harrison, C. Nickel Hydroxide Electrode Material With Improved Microstructure and Method for Making the Same. U.S. Patent 7,294,434, 13 November 2007.

127. Ovshinsky, S.R.; Fetcenko, M.A.; Fierro, C.; Gifford, P.R.; Corrigan, D.A.; Benson, P.; Martin, F.J. Enhanced Nickel Hydroxide Positive Electrode Materials for Alkaline Rechargeable Electrochemical Cells. U.S. Patent 5,523,182, 4 June 1996.

128. Fierro, C.; Fetcenko, M.A.; Ovshinsky, S.R.; Corrigan, D.A.; Sommers, B.; Zallen, A. Nickel Hydroxide Electrode Material and Method for Making the Same. U.S. Patent 6,416,903, 9 July 2002.

129. Tokuda, M.; Ogasawara, T.; Yano, M.; Fujitani, S. Nickel Hydroxide Electrode for Alkaline Storage Battery and Alkaline Storage Battery. U.S. Patent 6,835,498, 28 December 2004.

130. Wolff, M.; Nuss, M.A.; Fetcenko, M.A.; Lijoi, A.L. Method for the Continuous Fabrication of Hydrogen Storage Alloy Negative Electrodes. U.S. Patent 4,820,481, 11 April 1989.

131. Wang, Y.; Li, W.; Sun, L.; Wang, S.; Wang, W.; Li, C.; Wang, J. Electrode of Charging Battery and Method and Equipment for Making Electrode. CN Patent 1,085,896, 29 May 2002.

132. Hasebe, H.; Takeno, K.; Sato, Y.; Takahashi, H.; Hayashida, H.; Mitsuyasu, K.; Swatari, I. Nickel-Metal Hydride Secondary Cell. U.S. Patent 5,053,292, 1 October 1991.

133. Yun, K.S.; Cho, B.W.; Cho, W.I.; Paik, C.H. Fabrication Method for Paste-Type Metal Hydride Electrode. U.S. Patent 5,682,592, 28 October 1997.

134. Venkatesan, S.; Reichman, B.; Ovshinsky, S.R.; Prasad, B.; Corrigan, D.A. High Power Nickel-Metal Hydride Batteries and High Power Electrodes for Use Therein. U.S. Patent 5,856,047, 5 January 1999.

135. Reichman, B.; Venkatesan, S.; Ovshinsky, S.R.; Fetcenko, M.A. Nickel-Metal Hydride Batteries Having High Power Electrodes and Low-Resistance Electrode Connections. U.S. Patent 5,851,698, 22 December 1998.

136. Towsley, F.E. Method for Making Electrodes for Nickel-Metal Hydride Batteries. U.S. Patent 6,881,234, 19 April 2005.

137. James, D.; Allison, D.B., II; Kelley, J.J.; Doe, J.B. Electrical Energy Devices. U.S. Patent 6,451,485, 17 September 2002.

138. Moriwaki, Y.; Karanaka, S.; Iwasaki, M.; Yamasaki, Y.; Maeda, A. Hydrogen Storage Alloy Electrode, Battery Including the Same and Method for Producing the Both. U.S. Patent 6,610,445, 26 August 2003.

139. Nakayama, S. Negative Electrode and Nickel-Metal Hydride Storage Battery Using the Same. U.S. Patent 7,534,529, 19 May 2009.

140. Magari, Y.; Tanaka, T.; Akita, H.; Shinyama, K.; Funahashi, A.; Nohma, T. Nickel-Metal Hydride Storage Battery. U.S. Patent 6,924,062, 2 August 2005.

141. Akita, H.; Tanaka, T.; Magari, Y.; Shinyama, K.; Funahashi, A.; Nohma, T. Nickel-Metal Hydride Storage Battery. U.S. Patent 6,926,998, 9 August 2005.

142. Frye, B.; Pensabene, S.; Puglisi, V. Electrode Structure for Nickel Metal Hydride Cells. U.S. Patent 5,478,594, 26 December 1995.

143. Kawase, H.; Morishita, S.; Towata, S.; Suzuki, K.; Abe, K. Sealed. U.S. Patent 5,948,563, 7 September 1999.

144. Matsuura, Y.; Maeda, R.; Shinyama, K.; Tanaka, T.; Nohma, T.; Yonezu, I. Hydrogen Absorbing Alloy Electrode and Nickel-Metal Hydride Battery. U.S. Patent 6,475,671, 5 November 2002.

145. Kuribayashi, Y.; Sugahara, H.; Ishii, M.; Shima, S. Hydrogen Absorbing Alloy Powder and Electrodes Formed of the Hydrogen Absorbing Alloy Powder. U.S. Patent 6,235,130, 22 May 2001.

146. Ovshinsky, S.R.; Aladjov, B.; Venkatesan, S.; Dhar, S.K.; Hopper, T.; Fok, K. Method of Activating Hydrogen Storage Alloy Electrode. U.S. Patent 6,605,375, 12 August 2003.

147. Young, K.; Fetcenko, M.A.; Ovshinsky, S.R. Hydrogen Storage Powder and Process for Preparing the Same. U.S. Patent 6,461,766, 8 October 2002.

148. Noréus, D.; Zhou, Y. Method for Improving the Properties of Alloy Powders for NiMH Batteries. U.S. Patent 7,056,397, 6 June 2006.

149. Okawa, T.; Murakami, T.; Aoki, K.; Usui, H. Nickel Metal Hydride Rechargeable Battery and Method for Manufacturing Negative Electrode Thereof. U.S. Patent 8,389,159, 5 March 2013.

150. Matsuura, Y.; Nogami, M.; Maeda, R.; Shinyama, K.; Yonezu, I.; Nishio, K. Metal Hydride Alkaline Storage Cell and Manufacturing Method Thereof. U.S. Patent 6,852,447, 8 February 2005.

151. Köhler, U.; Chen, G.; Lindner, J. Gastight, Sealed Metal Oxide/metal Hydride Storage Battery. U.S. Patent 5,639,569, 17 June 1997.

152. Ferrando, W. Suspension Method of Impregnating Active Material into Composite Nickel Plaque. U.S. Patent 4,574,096, 4 March 1986.

153. Amiel, O.; Belkhir, I.; Freluche, J.; Pineau, N.; Dupuy, C.; Babin, S. Non-Sintered Electrode With Three-Dimensional Support for a Secondary Electrochemical Cell Having an Alkaline Electrolyte. U.S. Patent 6,656,640, 2 December 2003.

154. Pensabene, S.F.; Royalty, R.L. Electrode Having a Conductive Contact Area and Method of Making the Same. U.S. Patent 5,196,281, 23 March 1993.

155. Babjak, J.; Ettel, V.A.; Paserin, V. Method of Forming Nickel Foam. U.S. Patent 4,957,543, 18 September 1990.

156. Kawano, H.; Hayashi, T.; Matsumoto, I. Non-Sintered Type Nickel Electrode. U.S. Patent 5,824,435, 20 October 1998.

157. Ohnishi, M.; Watada, M.; Oshitani, M. Nickel Electrode and Alkaline Battery Using the Same. U.S. Patent 5,200,282, 6 April 1993.

158. Kobayashi, T. Nickel Electrode Plate for an Alkaline Storage Battery. U.S. Patent 5,677,088, 14 October 1997.

159. Harada, K.; Ishii, M.; Watanabe, K.; Yamanaka, S. Process for Preparing Metallic Porous Body, Electrode Substrate for Battery and Process for Preparing the Same. U.S. Patent 5,640,669, 17 June 1997.

160. Venkatesan, S.; Aladjov, B.; Fok, K.; Hopper, T.; Ovshinsky, S.R. Nickel Hydroxide Paste with Molasses Binder. U.S. Patent 6,818,348, 16 November 2004.

161. Ovshinsky, S.R.; Aladjov, B.; Tekkanat, B.; Venkatesan, S.; Dhar, S.K. Active Electrode Composition With Conductive Polymeric Binder. U.S. Patent 7,238,446, 3 July 2007.

162. Lee, W.W. Cobalt Treatment of Nickel Composite Electrode Surfaces. U.S. Patent 4,595,463, 17 June 1986.

163. Geng, M.; Phillips, J.; Mohanta, S. Nickel Hydroxide Electrode for Rechargeable Batteries. U.S. Patent 8,048,566, 1 November 2011.

164. Hyashi, K.; Tomioka, K.; Morishita, N.; Ikeyama, M.; Ikoma, M. Nickel Positive Electrode for Alkaline Rechargeable Batteries and Nickel Metal Hydride Cells. U.S. Patent 5,968,684, 19 October 1999.

165. Seyama, Y.; Sasaki, H.; Murata, T. Positive Active Material for Alkaline Battery and Electrode Using the Same. U.S. Patent 6,251,538, 26 June 2001.

166. Maeda, T.; Shinyama, K.; Matsuura, Y.; Nogami, M.; Yonezu, I.; Nishio, K. Nickel-Hydrogen Storage Battery. U.S. Patent 6,472,101, 29 October 2002.

167. Miyamoto, K.; Bando, N. Active Material for Positive Electrode for Alkaline Secondary Cell and Method for Producing the Same, and Alkaline Secondary Cell Using the Active Material for Positive Electrode and Method for Producing the Same. U.S. Patent 6,528,209, 4 March 2003.

168. Seyama, Y.; Sasaki, H.; Murata, T. Positive Active Material for Alkaline Battery and Electrode Using the Same. U.S. Patent 6,558,842, 6 May 2003.

169. Sakamoto, H.; Ohkawa, K.; Yuasa, S. Alkali Battery Positive Electrode Active Material, Alkali Battery Positive Electrode, Alkali Battery, and Method for Manufacturing Alkali Battery Positive Electrode Active Material. Patent WO2006011430 A1, 2 February 2006.

170. Pensabene, S.F.; Puglisi, V.J. Positive Nickel Electrode for Nickel Metal Hydride Cells. U.S. Patent 5,466,546, 14 November 1995.

171. Nakayama, S.; Yuasa, K.; Kaiya, H. Nickel Positive Electrode Plate and Alkaline Storage Battery. U.S. Patent 6,803,148, 12 October 2004.

172. Nakayama, S.; Yuasa, K.; Kaiya, H. Nickel Positive Electrode Plate and Alkaline Storage Battery. U.S. Patent 7,364,818, 29 April 2008.

173. Bauerlein, P. Ni/Metal Hydride Secondary Element. U.S. Patent 6,881,519, 19 April 2005.

174. Ovshinsky, S.R.; Corrigon, D.A.; Benson, P.; Fierro, C.A. Nickel Metal Hydride Battery Containing a Modified Disordered Multiphase Nickel Hydroxide Positive Electrode. U.S. Patent 5,569,563, 29 October 1996.

175. Dansui, Y.; Suzuki, T.; Kasahara, H.; Yao, T. Nickel-Metal Hydride Secondary Battery Comprising a Compound Silicate. U.S. Patent 6,461,767, 8 October 2002.

176. Maeda, A.; Kimiya, H.; Moriwaki, Y.; Matsumoto, I. Alkaline Storage Battery. U.S. Patent 6,338,917, 15 January 2002.

177. Dansui, Y.; Suzuki, T.; Kasahara, H. Method of Preparing a Nickel Positive Electrode Active Material. U.S. Patent 7,147,676, 12 December 2006.

178. Hayashi, K.; Tomioka, K.; Morishita, N.; Ikoma, M. Nickel Positive Electrode and Alkaline Storage Battery Using the Same. U.S. Patent 6,027,834, 22 February 2000.

179. Klein, M. Method for Preparing Conductive Electrochemically Active Material. U.S. Patent 5,585,142, 17 December 1996.

180. Klein, M. Method for Fabricating a Battery Electrode. U.S. Patent 5,611,823, 18 March 1997.

181. Young, K.; Ng, K.Y.S.; Bendersky, L.A. A technical report of the robust affordable next generation energy storage system-BASF program. *Batteries* **2016**, *2*.

182. Yoshinaka, T.; Yamamoto, S.; Inagaki, T.; Nakamura, Y.; Tanigawa, F.; Kaiya, H.; Takeshima, H. Sealed Cylindrical Nickel-Metal Hydride Storage Battery. U.S. Patent 6,805,995, 19 October 2004.

183. Ovshinsky, S.R.; Fetcenko, M.A.; Holland, A.; Dean, K.; Fillmore, D. Optimized Cell Pack for Large Sealed Nickel-Metal Hydride Batteries. U.S. Patent 5,558,950, 24 September 1996.

184. Ovshinsky, S.R.; Corrigan, D.A.; Venkatesan, S.; Dhar, S.K.; Holland, A.; Fillmore, D.; Higley, L.; Gow, P.; Himmler, R.; Karditsas, N.; *et al.* Mechanical and Thermal Improvements in Metal Hydride Batteries, Battery Modules and Battery Packs. U.S. Patent 5,879,831, 9 March 1999.

185. Kozawa, H.; Kojima, K.; Ono, T.; Yanagawa, H.; Kitazume, H.; Taguchi, K. Rectangular Nickel-Metal Hydride Secondary Cell. U.S. Patent 5,372,897, 13 December 1994.

186. Kozawa, H.; Kojima, K.; Ono, T.; Yanagawa, H.; Kitazume, H.; Taguchi, K. Method of Making a Rectangular Nickel-Metal Hydride Secondary Cell. U.S. Patent 5,490,867, 13 February 1996.

187. Kozawa, H.; Kojima, K.; Ono, T.; Yanagawa, H.; Kitazume, H.; Taguchi, K. Method of Manufacturing a Nickel-Metal Hydride Secondary Cell. U.S. Patent 5,537,733, 23 July 1996.

188. Yuasa, S.; Morishita, N.; Taniguchi, A.; Ikoma, M. Rectangular Alkaline Storage Battery and Battery Module and Battery Pack Using the Same. U.S. Patent 20020022179 A1, 21 February 2002.

189. Kawasaki Heavy Industries. Bipolar 3D Design Is the Secret to the High Capacity and Rapid Charge/Discharge of the GIGACELL. Available online: https://global.kawasaki.com/en/energy/solutions/ battery_energy/about_gigacell/ structure.html (accessed on 9 February 2016).

190. Kato, F.; Tanigawa, F.; Yuasa, K. Nickel-Metal Hydride Storage Battery and Assembly of the Same. U.S. Patent 6,958,200, 25 October 2005.

191. Sasaki, E.W.; Owens, B.B.; Passerini, S. Rechargeable Battery. U.S. Patent 6,265,100, 24 July 2001.

192. Klein, M. Method of Making Electrodes for Bipolar Electrochemical Battery. U.S. Patent 5,478,363, 26 December 1995.

193. Klein, M. Bipolar Electrochemical Battery of Stacked Wafer Cells. U.S. Patent 5,393,617, 28 February 1995.

194. Mrotek, E.N.; Kao, W. Bipolar Battery and Method of Making Same. U.S. Patent 5,688,615, 18 November 1997.

195. Fredriksson, L.; Puester, N. Bipolar Battery and Biplate Assembly. U.S. Patent 7,097,937, 29 August 2006.

196. Fredriksson, L.; Puester, N.H. Bipolar Battery and a Method for Manufacturing a Bipolar Battery. U.S. Patent 7,258,949, 21 August 2007.

197. Fredriksson, L.; Puester, N. Bipolar Battery and a Biplate Assembly. U.S. Patent 7,767,337, 3 August 2010.

198. Puester, N.H.; Hock, D.; Fredriksson, L. Method for Manufacturing a Bipolar Battery with a Gasket. U.S. Patent 8,470,469, 25 June 2013.

199. Benczur-Uermoessy, G. Gastight Prismatic Nickel-Metal Hydride Cell. U.S. Patent 7,205,065, 17 April 2007.

200. Corrigan, D.A.; Gow, P.; Higley, L.R.; Muller, M.D.; Osgood, A.; Ovshinsky, S.R.; Payne, J.; Puttaiah, R. Monoblock Battery Assembly. U.S. Patent 6,255,015, 3 July 2001.

201. Lyman, P.C. Battery. U.S. Patent 5,567,544, 22 October 1996.

202. Park, Y. Secondary Battery. U.S. Patent 2013/0149579 A1, 13 June 2013.

203. Fuhr, J.D.; Dougherty, T.J.; Dinkelman, J.P. Battery Cell. U.S. Patent 8,609,278, 17 December 2013.

204. Hirano, F. Cylindrical Storage Battery. U.S. Patent 6,703,158, 9 March 2004.

205. Hashimoto, T. Negative Electrode Current Collector, Negative Electrode Using the Same, and Non-Aqueous Electrolytic Secondary Cell. U.S. Patent 7,150,942, 19 December 2006.

206. Nakamaru, H. Current Collector. U.S. Patent D522,965, 13 June 2006.

207. Kometani, S.; Nagase, T.; Masuda, Y.; Asanuma, H.; Enishi, E.; Kondo, T. Collector Used for an Alkali Storage Battery. U.S. Patent 6,979,514, 27 December 2005.

208. Mori, H.; Sakamoto, K.; Bandou, T.; Okabe, K. Nickel Metal-Hydride Battery. U.S. Patent 7,867,655, 11 January 2011.

209. Fuhr, J.D.; Bowen, G.K.; Dinkleman, J.P.; Dougherty, T.J.; Tsutsui, W.; Bonin, C. Current Collector for An Electromechanical Cell. U.S. Patent 8,679,678, 25 March 2014.

210. Shinohara, W.; Yamamoto, Y.; Hosokawa, H.; Yamauchi, Y. Method of Manufacturing Sealed Battery and Sealed Battery. U.S. Patent 6,843,811, 18 January 2005.

211. Ling, P.T.P.; Ng, S.O.A. Prismatic Battery Cells, Batteries with Prismatic Battery Cells and Methods of Making Same. U.S. Patent 20060040176 A1, 23 February 2006.

212. Dean, K.; Holland, A.; Ovshinsky, H.C.; Fetcenko, M.; Venkatesan, S.; Dhar, S. Hydrogen Containment Cover Assembly for Sealing the Cell Can of a Rechargeable Electrochemical Hydrogen Storage Cell. U.S. Patent 5,171,647, 15 December 1992.

213. Pate, P.E. Electrochemical Cell Safety Vent. U.S. Patent 6,080,505, 27 June 2000.

214. Sugita, N. Sealed Battery with Less Electrolyte Leakage. U.S. Patent 6,838,207, 4 January 2005.

215. Komori, K.; Matsuura, T.; Hamada, S.; Eto, T.; Nakamura, Y. Nickel Metal Hydride Storage Battery With a Safety Valve for Relieving Excess Gas Pressure in the Battery When the Safety Valve Is Open, the Safety Valve Having a Hydrogen-Permeable Value Member for Allowing Hydrogen-Gas Leakage Therethrough When the Safety Valve Is Closed. U.S. Patent 7,758,994, 20 July 2010.

216. Hamada, S.; Matsuura, T.; Eto, T.; Miyamoto, H. Nickel-Metal Hydride Storage Battery. U.S. Patent 7,807,282, 5 October 2010.

217. Fuhr, J.D.; Trester, D.; Houchin-Miller, G.; Bonin, C.; Pacheco, A.C. Battery Module with Sealed Vent Chamber. U.S. Patent 8,999,538, 7 April 2015.

218. Fuhr, J.D.; Tsutsui, W.; Houchin-Miller, G.P.; Bonin, C.M. Vent for Electrochemical Cell. U.S. Patent 8,945,740, 3 February 2015.

219. Tyler, M.; Fuhr, J. Device for Aiding in the Fracture of a Vent of an Electrochemical Cell. U.S. Patent 9,105,902, 11 August 2015.

220. Komori, K.; Ogata, Y.; Adachi, A.; Maekawa, K.; Fujioka, N. Nickel Metal Hydride Storage Battery. U.S. Patent 7,452,629, 18 November 2008.

221. Hamada, S.; Eto, T. Sealed Rechargeable Battery and Battery Module. U.S. Patent 7,393,611, 1 July 2008.

222. Kung, J.K. Battery Separator for Nickel/Metal Hydride Batteries. U.S. Patent 5,298,348, 29 March 1994.

223. Senyarich, S.; Viaud, P. Storage Cell Having an Alkaline Electrolyte, in Particular a Storage Cell of Nickel-Cadmium or Nickel Metal Hydride. U.S. Patent 5,939,222, 17 August 1999.

224. Whear, J.K.; Yaritz, J.G. Nonwoven Separator for a Nickel-Metal Hydride Battery. U.S. Patent 6,537,696, 25 March 2003.

225. Tsukiashi, M.; Teraoka, H.; Hata, K.; Tajima, M. Battery Separator and Manufacturing Method Thereof, and Alkali Secondary Battery Having the Separator Incorporated Therein. U.S. Patent 6,623,809, 23 September 2003.

226. Harada, Y.; Tanaka, T.; Magari, Y.; Shinyama, K.; Nohma, T.; Yonezu, I. Separator for Nickel-Metal Hydride Storage Battery and Nickel-Metal Hydride Storage Battery. U.S. Patent 7,052,800, 30 May 2006.

227. Harada, Y.; Tanaka, T.; Shinyama, K.; Nohma, T.; Yonezu, I. Separator for Alkaline Secondary Battery, Method for Preparing the Same Alkaline Secondary Battery. U.S. Patent 6,790,562, 14 September 2004.

228. Tsuji, Y.; Nakai, H.; Muraoka, Y. Nickel-Metal Hydride Storage Battery. U.S. Patent 7,435,511, 14 October 2008.

229. Lim, H.S.; Arora, P. Separator Media for Electrochemical Cells. U.S. Patent 20140134498 A1, 15 May 2014.

230. Sato, N.; Komori, K.; Morishita, N. Separator for Alkaline Storage Battery and Alkaline Storage Battery Using the Same. U.S. Patent 6,933,079, 23 August 2005.

231. Young, K.; Yasuoka, S. Capacity degradation mechanisms in nickel/metal hydride batteries. *Batteries* **2016**, 2.

232. Joshi, A.V.; Gordon, J.H.; Bhavaraju, S. Nickel-Metal Hydride Battery Using Alkali Ion Conducting Separator. U.S. Patent 8,012,621, 6 September 2011.

233. Joshi, A.V.; Gordon, J.H.; Bhavaraju, S. Nickel-Metal Hydride/Hydrogen Hybrid Battery Using Alkali Ion Conducting Separator. U.S. Patent 9,209,445, 8 December 2015.

234. Alkordi, M.H.; Eddaoudi, M. Electrode Separator. U.S. Patent 20130280611 A1, 24 October 2013.

235. Yan, S.; Young, K.; Ng, K.Y.S. Effects of salt additives to the KOH electrolyte used in Ni/MH batteries. *Batteries* **2015**, *1*, 54–73.

236. Young, K.; Nei, J.; Rotarov, D. Studies in various hydroxides used as electrolyte in Ni/MH batteries. *Batteries* **2016**, to be submitted for publication.

237. Kitayama, H.; Hayashida, H.; Yamamoto, M.; Bando, N.; Miyamoto, K.; Suzuki, H. Nickel-Metal Hydride Secondary Battery. U.S. Patent 6,399,247, 4 June 2002.

238. Anani, A.A.; Reichert, V.R.; Massaroni, K.M. Metal Hydride Electrochemical Cell Having a Polymer Electrolyte. U.S. Patent 5,541,019, 30 July 1996.

239. Iwakura, C.; Furukawa, N.; Izumi, Y.; Moriwaki, Y. Nickel-Metal Hydride Storage Battery and Production Method Thereof. U.S. Patent 7,022,434, 4 April 2006.

240. Ovshinsky, S.R.; Young, R. Solid State Battery Using a Hydrogenated Silicon Nitride Electrolyte. U.S. Patent 5,552,242, 3 September 1996.

241. Nakamura, Y. Power Storage Device. U.S. Patent 7,803,486, 28 September 2010.

242. Young, K.; Fetcenko, M.A. Low Cost, High Power, High Energy Density, Solid-State, Bipolar Metal Hydride Batteries. U.S. Patent 8,974,948, 10 March 2015.

243. Van Deutekom, H.J.H. Rechargeable Electrochemical Cell. U.S. Patent 4,214,043, 22 July 1980.

244. Van Deutekom, H.J.H. Rechargeable Electrochemical Cell. U.S. Patent 4,312,928, 26 January 1982.

245. Ikezoe, M. Sealed Nickel-Metal Hydride Storage Cells and Hybrid Electric Having the Storage Cells. U.S. Patent 7,353,894, 8 April 2008.

246. Takee, M.; Tadokoro, M.; Ise, T.; Yamawaki, A. Nickel-Metal Hydride Storage Cell Having a High Capacity and an Excellent Cycle Characteristic and Manufacturing. U.S. Patent 6,368,748, 9 April 2002.

247. Berlureau, T.; Liska, J. Sealed Nickel-Metal Hydride Storage Cell. U.S. Patent 6,444,349, 3 September 2002.

248. Onishi, M.; Fukunaga, H.; Isogai, M.; Nagai, R. Nickel Metal-Hydride Cell. U.S. Patent 6,593,031, 15 July 2003.

249. Young, K.; Fierro, C.; Reichman, B.; Fetcenko, M.A.; Koch, J.; Zallen, A. Nickel Metal Hydride Battery Design. U.S. Patent 7,261,970, 28 August 2007.

250. Hasebe, H.; Tsuruta, S.; Yoshida, H.; Yamamoto, M.; Kanno, K.; Ishitsuka, K.; Komiyama, K.; Oppata, H. Alkaline Secondary Battery Manufacturing Method, Alkaline Secondary Battery Positive Electrode, Alkaline Secondary Battery, and a Method of Manufacturing an Initially Charged Alkaline Secondary Battery. U.S. Patent 5,708,349, 13 January 1998.

251. Nanamoto, K.; Umehara, Y. Method for Manufacturing Nickel-Metal-Hydride Battery. U.S. Patent 5,814,108, 29 September 1998.

252. Singh, D.B. Method of Forming CoOOH and NiOOH in a NiMH Electrochemical Cell and An Electrochemical Cell Formed Thereby. U.S. Patent 6,270,535, 7 August 2001.

253. Onishi, M.; Tomioka, K.; Fujioka, N.; Ikoma, M. Method for Producing a Nickel Metal-Hydride Storage Battery. U.S. Patent 6,669,742, 30 December 2003.

254. Morishita, N.; Ito, S.; Seri, H. Nickel-Metal Hydride Rechargeable Battery. U.S. Patent 7,560,188, 14 July 2009.

255. Toyota Prius Battery. Available online: http://www.toyotapriusbattery.com (accessed on 9 February 2016).

256. Ovshinsky, S.R.; Corrigan, D.A.; Venkatesan, S.; Dhar, S.K.; Holland, A.; Fillmore, D.; Higley, L.; Gow, P.; Himmler, R.; Karditsas, N.; *et al.* Mechanical and Thermal Improvements in Metal Hydride Batteries, Battery Modules and Battery Packs. U.S. Patent 6,372,377, 16 April 2002.

257. Ovshinsky, S.R.; Corrigan, D.A.; Venkatesan, S.; Dhar, S.K.; Holland, A.; Fillmore, D.; Higley, L.; Gow, P.; Himmler, R.; Karditsas, N.; *et al.* Mechanical and Thermal Improvements in Metal Hydride Batteries, Battery Modules and Battery Packs. U.S. Patent 6,878,485, 12 April 2005.

258. Morishita, N.; Hamada, S.; Matsuda, H.; Ikoma, M. Cell and Module Battery of Sealed Nickel-Metal Hydride Storage. U.S. Patent 5,747,186, 5 May 1998.

259. Hamada, S.; Kasahara, H.; Ito, S.; Ando, H.; Eto, T. Nickel-Metal Hydride Secondary Battery Module and Secondary Battery Module Manufacturing Method. U.S. Patent 8,785,014, 22 July 2014.

260. Hamada, S.; Inoue, H.; Fujioka, N.; Ikoma, M. Battery Module. U.S. Patent 6,551,741, 22 April 2003.

261. Muis, P. Interconnectors for a Battery Assembly. U.S. Patent 8,541,130, 24 September 2013.

262. Yang, J.H. Battery Module of High Cooling Efficiency. U.S. Patent 7,955,726, 7 June 2011.

263. Houchin-Miller, G.P.; Pacheco, A.C.; Wiegmann, M.; Joswig, R.; Hoh, M.; Balk, M.; Obasih, K. Battery System with Heat Exchanger. U.S. Patent 8,603,660, 10 December 2013.

264. Houchin-Miller, G.P.; Pacheco, A.C.; Wiegmann, M.; Joswig, R.; Hoh, M.; Balk, M.; Obasih, K. Battery System with Heat Exchanger. U.S. Patent 9,225,045, 29 December 2015.

265. Gadawski, T.J.; Payne, J. Battery Module and Method for Cooling the Battery Module. U.S. Patent 8,399,118 B2, 19 March 2013.

266. Koetting, W.; Payne, J. Battery Module and Method for Cooling the Battery Module. U.S. Patent 8,399,119 B2, 19 March 2013.

267. Shinyashiki, Y.; Fujiwara, M.; Maeda, H.; Funahashi, A. Method for Compressing Individual Cells in Battery Module. U.S. Patent 8,309,247, 13 November 2012.

268. Chamberlain, C.P.; Lamb, D.; Lauby, W.J.; Voorheis, H.T. Compact, Shock Resistant Battery Pack. U.S. Patent 5,466,545, 14 November 1995.

269. Wheeler, D.K.; Moores, R.G., Jr.; Walter, R.T. Battery Pack for Cordless Device. U.S. Patent 7,273,676, 25 September 2007.

270. Adams, P.; Ellis, R.; Newton, J. Battery Pack Control Circuit for Handheld Battery Operated Device. U.S. Patent 8,548,633, 1 October 2013.

271. Pyo, K.R. Battery Pack and Manufacturing Method for the Same. U.S. Patent 8,835,045, 16 September 2014.

272. Shih, C.; Chang, C.; Yeh, S.; Liu, L.; Wu, S. Battery Pack. U.S. Patent 20130171492 A1, 4 July 2013.

273. Mita, Y. Battery Module and Temperature-Controlling Apparatus for Battery. U.S. Patent 5,456,994, 10 October 1995.

274. Inui, K.; Nakanishi, T. Battery Pack. U.S. Patent 6,504,342, 7 January 2003.

275. Ogata, Y.; Hamada, S. Battery Pack Cooling Structure. U.S. Patent 6,709,783, 23 March 2004.

276. Inui, K.; Etoh, T. Fluid-Cooled Battery Pack System. U.S. Patent 6,953,638, 11 October 2005.

277. Ovshinsky, S.R.; Corrigan, D.A.; Venkatesan, S.; Dhar, S.K.; Holland, A.; Fillmore, D.; Higley, L.; Gow, P.; Himmler, R.; Karditsas, N.; et al. Mechanical and Thermal Improvements in Metal Hydride Batteries, Battery Modules and Battery Packs. U.S. Patent 7,217,473, 15 May 2007.

278. Fuhr, J.D.; Houchin-Miller, G.P.; Swoyer, J.L.; Pacheco, A.C.; Zhang, X.; Obashih, K.M. Prismatic Electrochemical Cell. U.S. Patent 20130216872 A1, 22 August 2013.

279. Yuen, T.K. Battery Charging Circuit for Charging NiMH and NiCd Batteries. U.S. Patent 5,489,836, 6 February 1996.

280. Feldstein, R.S. NiCd/NiMH Battery Charger. U.S. Patent 5,523,668, 4 June 1996.

281. McClure, M.S. Battery Monitoring and Charging Control Unit. U.S. Patent 5,563,496, 8 October 1996.

282. Seong, H.; Im, S. Battery Charging Circuit with Charging Rate Control. U.S. Patent 5,698,963, 16 December 1997.

283. Im, S.; Seoung, H.; Choi, B. Fast-Charging Device for Rechargeable Batteries. U.S. Patent 5,717,311, 10 February 1998.

284. Cheon, K. Dual Battery Charging Device for Charging Nickel Metal-Hydride and Lithium-Ion Batteries. U.S. Patent 5,744,937, 28 April 1998.

285. Gaza, B.S. Microcontrolled Battery Charger. U.S. Patent 5,764,030, 9 June 1998.

286. Yeon, S. Charge Mode Control in a Battery Charger. U.S. Patent 5,821,736, 13 October 1998.

287. Gaza, B.S. Microcontrolled Battery Charger. U.S. Patent 5,998,966, 7 December 1999.

288. Yuen, T.K. Battery Charger. U.S. Patent 6,208,148, 27 March 2001.

289. Tsenter, B. Battery Charger and Method of Charging Nickel Based Batteries. U.S. Patent 6,313,605, 6 November 2001.

290. Gignac, R.G. Temperature/Voltage Controlled Battery Charging Circuit. U.S. Patent 6,707,273, 16 March 2004.

291. Santana, G.L., Jr. Battery, Battery Charger, Electrical System and Method of Charging a Battery. U.S. Patent 6,924,620, 2 August 2005.

292. Morgan, R. Rechargeable Batteries. U.S. Patent 7,459,882, 2 December 2008.

293. Ishikawa, Y.; Okabayashi, H.; Sakai, M.; Kobayashi, M. Charging Apparatus. U.S. Patent 20130285600 A1, 31 October 2013.

294. Guang, H.T.; Hua, L.W.; Wentink, R.F. Multiple Cell Battery Charger Configured With a Parallel Topology. U.S. Patent 8,860,372, 14 October 2014.

295. Geodken, T.J. Battery Charger Having Variable-Magnitude Charging Current Source. U.S. Patent 5,166,596, 24 November 1992.

296. Dias, D.R.; Mumper, E.W. Battery Charger Systems and Methods. U.S. Patent 5,488,284, 30 January 1996.

297. Kamke, J.E. Apparatus and Method for Maintaining the Charge of a Battery. U.S. Patent 5,493,198, 20 February 1996.

298. Hanselmann, D.; Mayer, B.; Nutz, K.; Weller, S. Charging Method for Storage Batteries. U.S. Patent 5,537,023, 16 July 1996.

299. Takeda, T. Control Method and Control Apparatus for Secondary Battery Charging in Constant Current Charging Method. U.S. Patent 5,627,451, 6 May 1997.

300. Sage, G.E. Pulse-Charge Battery Charger. U.S. Patent 5,633,574, 27 May 1997.

301. Im, S.; Seoung, H.; Choi, B. Method and Apparatus for Detecting a Full-Charge Condition While Charging a Battery. U.S. Patent 5,691,624, 25 November 1997.

302. Podrazhansky, Y.M.; Tsenter, B. Control and Termination of a Battery Charging Process. U.S. Patent 5,694,023, 2 December 1997.

303. Kuno, H. Battery Charger and Method for Completing Charging at Designated Time. U.S. Patent 5,736,834, 7 April 1998.

304. Lane, R.W. Method and Apparatus for Charging Batteries. U.S. Patent 5,739,672, 14 April 1998.

305. Wieczorek, R. Method for Charging a Rechargeable Battery. U.S. Patent 5,773,956, 30 June 1998.

306. Kwan, H.; Chen, Y.; Chu, C.; Chen, S.; Wang, C. Automatic Battery Charging System Using Lowest Charge Current Detection. U.S. Patent 5,844,398, 1 December 1998.

307. Paryani, A.; Sando, Y. Refreshing Charge Control Method and Apparatus to Extend the Life of Batteries. U.S. Patent 6,011,380, 4 January 2000.

308. Freiman, J.F. Temperature Compensated Voltage Limited Fast Charge of Nickel Cadmium and Nickel Metal Hydride Battery Packs. U.S. Patent 6,020,722, 1 February 2000.

309. Hardie, J.O. Battery Charger. U.S. Patent 6,211,655, 3 April 2001.

310. Pavlovic, V.S. Method and Apparatus for Charging Batteries Utilizing Heterogeneous Reaction Kinetics. U.S. Patent 6,495,992, 17 December 2002.

311. Patino, J.; Fiske, J.D. Battery Charging Algorithm. U.S. Patent 6,731,096, 4 May 2004.

312. Patino, J. Method and System for Battery State of Charge Estimation by Using Measured Changes in Voltage. U.S. Patent 7,098,666, 29 August 2006.

313. Hoff, C.M.; Nelson, J.E. Controlling Re-Charge of a Nickel Metal-Hydride (NiMH) or Nickel Cadmium (NiCd) Battery. U.S. Patent 7,129,676 B2, 31 October 2006.

314. Meng, C.; Daou, Y. Systems and Methods for Temperature-Dependent Battery Charging. U.S. Patent 7,615,969, 10 November 2009.

315. Kung, S.; Shen, Y. Method for Charging Portable Electronic Device. U.S. Patent 7,772,807, 10 August 2010.

316. Wong, K.P.; Ng, S.O.A.; Yeo, C.W. Method and System for Determining the SOC of a Rechargeable Battery. U.S. Patent 7,888,911, 15 February 2011.

317. Kaplan, J.; Furlan, J.L.W.; Schaefer, P.R.; Bacho, E.V.; Chu, J.Y.H.; Lin, R.; Inman, R.T.; Orner, W. System and Method for Charging Rechargeable Batteries in a Digital Camera. U.S. Patent 8,259,221, 4 September 2012.

318. Hoff, C.M.; Nelson, J.E. Method and System for Charging a NiMH or NiCd Battery. U.S. Patent 7,405,538, 29 July 2008.

319. Esnard-Domerego, D.; Walley, J.; Wang, L. Sink Current Adaptation Based on Power Supply Detection. U.S. Patent 8,729,867, 20 May 2014.

320. Yuen, T.K. Battery Charger. U.S. Patent 6,225,789, 1 May 2001.

321. Uchida, Z. Quick Charge Control Apparatus and Control Method Thereof. U.S. Patent 5,200,690, 6 April 1993.

322. Yau, K.W.; Li, Y.C.; Bai, L. Intelligent Serial Battery Charger. U.S. Patent 7,557,538, 7 July 2009.

323. Nicolai, J. Method for the Fast Charging of a Battery and Integrated Circuit for the Implementation of This Method. U.S. Patent 5,612,607, 18 March 1997.

324. Fernandez, J.M.; Coapstick, R.S. Method for Ultra-Rapidly Charging a Rechargeable Battery Using Multi-Mode Regulation in a Vehicular Recharging System. U.S. Patent 5,986,430, 16 November 1999.

325. Young, K.; Wang, C.; Wang, L.Y.; Strunz, K. Electric Vehicle Battery Technologies. In *Electric Vehicle Integration into Modern Power Networks*; Garcia-Valle, R., Lopes, J.A.P., Eds.; Springer Science, Business Media: New York, NY, USA, 2013; pp. 15–56.

326. Johnson, T.W.; Rosenbecker, J.J.; Meyer, G.D.; Zeiler, J.M.; Glasgow, K.L.; Zick, J.A.; Brozek, J.M.; Scheucher, K.F. Method and System for Battery Protection. U.S. Patent 9,112,248, 18 August 2015.

327. Carrier, D.A.; Seman, A.E., Jr.; Howard, G.S.; Brotto, D.C.; Trinh, D.T.; Watts, F.S.; Choksi, S.S.; Zhang, Q.J. Battery Pack for Cordless Power Tools. U.S. Patent 7,728,553, 1 June 2010.

328. Kim, S.; Song, J. Battery Management System, Method of Controlling the Same, and Energy Storage System Including the Battery Management System. U.S. Patent 8,806,240, 12 August 2014.

329. Duncan, W.D.; Hyde, R.A.; Kare, J.T. Management of a Remote Electric Vehicle Traction Battery System. U.S. Patent 20140351107 A1, 27 November 2014.

330. Lin, X.; Stefanopoulou, A.; Anderson, R.D.; Li, Y. Detection of Imbalance across Multiple Battery Cells Measured by the Same Voltage Sensor. U.S. Patent 20140361743 A1, 11 December 2014.

331. Duncan, W.D.; Hyde, R.A.; Kare, J.T. Managed Electric Vehicle Traction Battery System. U.S. Patent 9,002,537, 7 April 2015.

332. Phouc, D.V.; Wieczorek, R.; Zeising, E.; Hruska, L.W.; Taylor, A.H.; Friel, D.D.; Hull, M.P. Battery Pack Having a Processor Controlled Battery Operating System. U.S. Patent 5,691,621 A, 25 November 1997.

333. Townsley, D.B.; Blanc, J.J. Circuit Offering Sequential Discharge and Simultaneous Charge for a Multiple Battery System and Method for Charging Multiple Batteries. U.S. Patent 5,666,006, 9 September 1997.

334. Kadouchi, E.; Watanabe, Y.; Kinoshita, M.; Ito, N.; Takata, K. Monitoring Apparatus for a Series Assembly of Battery Modules. U.S. Patent 6,020,717, 1 February 2000.

335. Hess, R.L.; Cooper, P.R.; Interiano, A.; Freiman, J.F. Battery Charge Monitor and Fuel Gauge. U.S. Patent 5,315,228, 24 May 1994.

336. Wood, S.J.; Maubert, E.; Veglio, O. Cell Diagnostic System and Method. U.S. Patent 8,788,225, 22 July 2014.

337. Rathmann, R. Battery Pack and a Method for Monitoring Remaining Capacity of a Battery Pack. U.S. Patent 5,955,869, 21 September 1999.

338. Ying, R.Y. State of Charge Algorithm for a Battery. U.S. Patent 6,356,083, 12 March 2002.

339. Kikuchi, Y. Battery Control System. U.S. Patent 6,600,293, 29 July 2003.

340. Bockelmann, T.R.; Hope, M.E.; Zou, Z.; Kang, X. Battery Control System for Hybrid Vehicle and Method for Controlling a Hybrid Vehicle Battery. U.S. Patent 7,489,101, 10 February 2009.

341. Melichar, R.J. Battery State of Charge Voltage Hysteresis Estimator. U.S. Patent 7,570,024, 4 August 2009.

342. Fassnacht, J. Method for Regulating the State of Charge of an Energy Accumulator in a Vehicle Having a Hybrid Drive Unit. U.S. Patent 7,934,573, 3 May 2011.

343. Izumi, J. Upper-Limit of State-of-Charge Estimating Device and Upper-Limit of State-of-Charge Estimating Method. U.S. Patent 8,912,761, 16 December 2014.

344. Takahashi, K.; Nishi, Y.; Tomura, S.; Takemoto, T.; Haga, N.; Fuchimoto, T.; Sugimoto, T. Control System of Vehicle. U.S. Patent 8,498,766, 30 July 2013.

345. Galbraith, R.E.; Gisi, J.M.; Norgaard, S.P.; Reetz, D.D.; Ziebarth, D.J. Method and Apparatus for Estimating the Service Life of a Battery. U.S. Patent 6,191,556, 20 February 2001.

346. Galbraith, R.E.; Gisi, J.M.; Norgaard, S.P.; Reetz, D.D.; Ziebarth, D.J. Method and Apparatus for Estimating the Service Life of a Battery. U.S. Patent 6,271,647, 7 August 2001.

347. Juncker, C.; Christensen, F.K. Battery Life Estimation. U.S. Patent 6,538,449, 25 March 2003.

348. Singh, P.; Fennie, C., Jr.; Reisner, D.E. Method and System for Determining State-of-Health of a Nickel-Metal Hydride Battery Using an Intelligent System. U.S. Patent 7,051,008, 23 May 2006.

349. Staton, K.L. Detecting an End of Life for a Battery Using a Difference between an Unloaded Battery Voltage and a Loaded Battery Voltage. U.S. Patent 8,410,783, 2 April 2013.

350. Center, M.B. Method for Estimating Battery Life in a Hybrid Powertrain. U.S. Patent 8,204,702, 19 June 2012.

351. Tsenter, B.J.; James, J.E. State of Health Recognition of Secondary Batteries. U.S. Patent 7,605,591, 20 October 2009.

352. Interiano, A.; Hess, R.L.; Cooper, P.R.; Freiman, J.F. Battery Charge Monitor to Determine Fast Charge Termination. U.S. Patent 5,200,689, 6 April 1993.

353. Van Phuoc, D.; Wieczorek, R.; Zeising, E.; Hruska, L.W.; Taylor, A.H.; Friel, D.D.; Hull, M.P. Battery Pack Having a Processor Controlled Battery Operating System. U.S. Patent 5,652,502 A, 29 July 1997.

354. Tate, E.D., Jr.; Verbrugge, M.W.; Sarbacker, S.D. State of Charge Prediction Method and Apparatus for a Battery. U.S. Patent 6,441,586, 27 August 2002.

355. Yao, T.; Kasahara, H.; Suzuki, T.; Masui, M.; Konishi, H. Method for Managing Back-up Power Source. U.S. Patent 6,097,176, 1 August 2000.

356. Vebrugge, M.W.; Tate, E.D., Jr.; Sarbacker, S.D.; Koch, B.J. Quasi-Adaptive Method for Determining a Battery's State of Charge. U.S. Patent 6,359,419 B1, 19 March 2002.

357. Furukawa, T. Method for Estimating State of Charge of a Rechargeable Battery. U.S. Patent 8,000,915, 16 August 2011.

358. Lee, J.; Lee, H.; Lee, J.; Lee, H.; Kim, D. Method for a Measuring Residual Capacity of a Ni/MH Cell. U.S. Patent 5,701,078, 23 December 1997.

359. Kinoshita, T.; Miyazaki, H.; Emori, A.; Okoshi, T.; Hirasawa, T. Battery System, Battery Monitoring Method and Apparatus. U.S. Patent 7,173,397, 6 February 2007.

360. Shinohara, S.; Takano, N. Battery Charger. U.S. Patent 5,444,353, 22 August 1995.

361. Garrett, S.M.; Fernandez, J.M.; Patino, J. Ultrafast Rechargeable Battery Pack and Method of Charging Same. U.S. Patent 5,576,612, 19 November 1996.

362. Van Phuoc, D.; Wieczorek, R.; Zeising, E.; Hruska, L.W.; Hull, M.P.; Taylor, A.H.; Friel, D.D. Battery Pack Having a Processor Controlled Battery Operating System. U.S. Patent 5,633,573 A, 27 May 1997.

363. Tibbs, B.L. Recharging Method and Temperature-Responsive Overcharge Protection Circuit for a Rechargeable Battery Pack Having Two Terminals. U.S. Patent 5,708,350, 13 January 1998.

364. McGrath, F.D.; Kellogg, N.D. Rechargeable Battery Having Overcharge Protection Circuit and Method of Charging Rechargeable Battery. U.S. Patent 5,939,865, 17 August 1999.

365. Saeki, M.; Ozawa, H.; Tsukuni, T.; Takeda, Y. Protection Circuit and Battery Unit. U.S. Patent 6,051,955, 18 April 2000.

366. Narita, I. Over Discharge Protection Circuit for a Rechargeable Battery. U.S. Patent 5,729,061, 17 March 1998.

367. Gamboa, P. Method and Apparatus for Charging and Discharging a Rechargeable Battery. U.S. Patent 7,541,781, 2 June 2009.

368. Johnson, T.W.; Rosenbecker, J.J.; Meyer, G.D.; Zeiler, J.M.; Glasgow, K.L.; Zick, J.Z.; Brozek, J.M.; Scheucher, K.F. Method and System for Battery Protection. U.S. Patent 7,589,500, 15 September 2009.

369. Van Phuoc, D.; Wieczorek, R.; Zeising, E.; Hruska, L.W.; Taylor, A.H.; Friel, D.D.; Hull, M.P. Battery Pack Having a Processor Controlled Battery Operating System. U.S. Patent 5,646,508 A, 8 July 1997.

370. Glonner, H.; Strobl, W.; Rump, R.; Mallog, J. Energy Storage System. U.S. Patent 7,591,331, 22 September 2009.

371. Reis, A.; Whitmer, D.K.; Muthu, M.E.J.; Korutla, S.K. Integrated Energy Storage Unit. U.S. Patent 8,481,203, 9 July 2013.

372. Garascia, M.P.; Arena, A.P.; Wood, S.J. Battery System. U.S. Patent 8,235,732, 7 August 2012.

373. Zhu, D.; Treharne, W.D. System and Method for Thermal Management of a Vehicle Power Source. U.S. Patent 7,683,582, 23 March 2010.

374. Fuhr, J.D.; Obasih, K.; De Keuster, R.M.; Mack, R.; Houchin-Miller, G.P.; Patel, D. Battery Module Having a Cell Tray with Thermal Management Features. U.S. Patent 20140234687 A1, 21 August 2014.

375. Hamada, S. Battery Pack. U.S. Patent 7,189,474, 13 March 2007.

376. Wood, S.J.; Tiedemann, W.H. Battery System Including a Device Configured to Route Effluent Away from Battery Modules within the Battery System. U.S. Patent 7,846,572, 7 December 2010.

377. Zolot, M.; Pesaran, A.A.; Mihalic, M. *Thermal Evaluation of Toyota Prius Battery Pack*; National Renewable Energy Laboratory: Golden, CO, USA, 2002.

378. Murata, T. Power Source Apparatus. U.S. Patent 7,974,095, 5 July 2011.

379. Thomas, G.; Moré, G. Hybrid Energy Storage System. U.S. Patent 5,587,250, 24 December 1996.

380. Thomas, G.; Moré, G. Hybrid Energy Storage System. U.S. Patent 5,849,426, 15 December 1998.

381. Kruger, D.D.; Young, G.; Moore, S.W. Method and Apparatus for Providing and Storing Power in a Vehicle. U.S. Patent 6,577,099, 10 June 2003.

382. Ovshinsky, S.R.; Stempel, R.C. Very Low Emission Hybrid Electric Vehicle Incorporating an Integrated Propulsion System Including a Hydrogen Powered Internal Combustion Engine and a High Power Ni-MH Battery Pack. U.S. Patent 6,565,836, 20 May 2003.

383. Lee, E. Energy Storage System and Method of Controlling the Same. U.S. Patent 9,041,354, 26 May 2015.

384. Park, J.; Choi, J. Energy Storage System and Controlling Method of the Same. U.S. Patent 20130169064 A1, 4 July 2013.

385. Yoo, H. Method of Measuring Voltage of Battery Pack and Energy Storage System Including the Battery Pack. U.S. Patent 2014/0035365 A1, 6 February 2014.

386. Jung, Y. Battery Pack, Controlling Method of the Same, and Energy Storage System Including the Battery Pack. U.S. Patent 20140078632 A1, 20 March 2014.

387. Pietrangelo, N.J.; Moran, T.J. Window Construction Combining NiMH Technology and Solar Power. U.S. Patent 20080236654 A1, 2 October 2008.

388. Komori, K.; Fujioka, N.; Kimura, T.; Yamashita, H.; Takahashi, Y. Method for Recycling Battery Pack. U.S. Patent 6,936,371, 30 August 2005.

389. Sloop, S.E. Recycling Batteries Having Basic Electrolytes. U.S. Patent 20130302223 A1, 14 November 2013.

390. Smith, W.N.; Swoffer, S. Process for the Recovery of Metals from Used Nickel/metal Hydride Batteries. U.S. Patent 8,246,717 B1, 21 August 2012.

391. Smith, W.N.; Swoffer, S. Process for the Recovery of AB5 Alloy from Used Nickel/metal Hydride Batteries. U.S. Patent 8,696,788, 15 April 2014.

392. Prickett, O.G.; Czajkowski, R.; Citta, N.C.; Klein, M.R.; Huston, E.L.; Galbraith, P.W. Reclamation of Active Material From Metal Hydride Electrochemical Cells. U.S. Patent 6,180,278, 30 January 2001.

393. Smith, W.N.; Swoffer, S. Process for the Recovery of Metals from Used Nickel/metal Hydride Batteries. U.S. Patent 8,252,085 B1, 28 August 2012.

394. Jacobson, A.J.; Samarasekere, P. Methods and Systems for Recovering Rare Earth Elements. U.S. Patent 20140311294 A1, 23 October 2014.

395. Burlingame, N.H.; Burlingame, S. Method for Recycling of Rare Earth and Zirconium Oxide Materials. U.S. Patent 8,940,256, 27 January 2015.

396. Kleinsorgen, K.; KöHler, U.; Bouvier, A.; FöLzer, A. Process for the Recovery of Metals from Used Nickel/metal Hydride Storage Batteries. U.S. Patent 5,858,061, 12 January 1999.

397. Kikuta, N.; Asano, S.; Takano, M. Method for Separating Nickel and Cobalt from Active Material Contained in Spent Nickel-Metal Hydride Battery. U.S. Patent 8,888,892, 18 November 2014.

398. Kikuta, N.; Asano, S.; Takano, M. Method for Producing Nickel-Containing Acid Solution. U.S. Patent 8,974,754, 10 March 2015.

Reviews on the Japanese Patent Applications Regarding Nickel/Metal Hydride Batteries

Taihei Ouchi, Kwo-Hsiung Young and Dhanashree Moghe

Abstract: The Japanese Patent Applications filed on the topic of nickel/metal hydride (Ni/MH) batteries have been reviewed. Patent applications filed by the top nine battery manufacturers (Matsushita, Sanyo, Hitachi Maxell, Yuasa, Toshiba, FDK, Furukawa, Japan Storage, and Shin-kobe), five component suppliers (Tanaka, Mitsui, Santoku, Japan Metals & Chemicals Co. (JMC), and Shin-Etsu), and three research institutes (Industrial Research Institute (ISI), Agency of Industrial Science and Technology (AIST), and Toyota R & D) were chosen as the main subjects for this review, based on their production volume and contribution to the field. By reviewing these patent applications, we can have a clear picture of the technology development in the Japanese battery industry. These patent applications also provide insights, know-how, and future directions for engineers and scientists working in the rechargeable battery field.

Reprinted from *Batteries*. Cite as: Ouchi, T.; Young, K.-H.; Moghe, D. Reviews on the Japanese Patent Applications Regarding Nickel/Metal Hydride Batteries. *Batteries* **2016**, *2*, 21.

1. Introduction

Since the debut of commercial nickel/metal hydride (Ni/MH) batteries in the late 1980's, over one thousand Japanese Patents have been granted. Many Japanese companies have worked very diligently in the past and some are still very active in the new product and market development. As a result, the Ni/MH battery market aims to maintain a smooth growth into the future (Figure 1), especially in the hybrid electric vehicle (HEV) and stationary markets. It was predicted by Yano Research Institute (Tokyo, Japan) that the use of Ni/MH in HEVs will increase four-fold from 2014 to 2020 [1] (Figure 2). Although some key inventions conceived in Japan have also been filed to the U.S. Patent and Trademark Office (for a review, see [2]), many of them have not. In order to facilitate the research and development world outside Japan, a review of the Japanese Patent Applications on the subject of Ni/MH batteries is presented here. The summary of the Japanese Patent Application in English is available online by inputting the H-number (before 2000) or Year-number (since 2000) in the *"A: publication of patent application"* row [3]. Machine translations of the claims and specification into English are available by selecting *"Detail"* in the corresponding webpage.

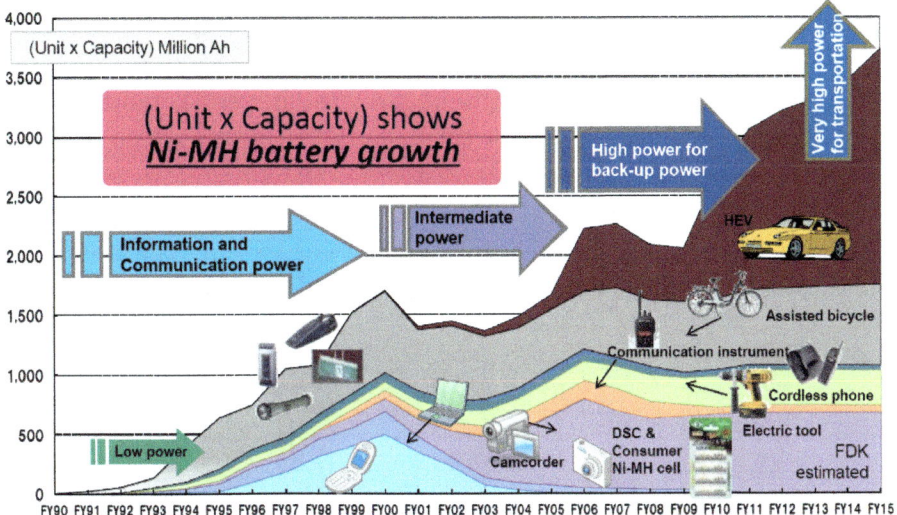

Figure 1. The nickel/metal hydride (Ni/MH) market prediction by FDK. While the demand for Ni/MH in hybrid electric vehicle (HEV) is increasing, uses in other areas remain stable. Courtesy of FDK Corp. DSC stands for digital still camera. The power requirement for Ni/MH batteries started slowly and grows strongly into transportation applications (HEV, start-stop vehicle, and trains) as indicated by the arrows.

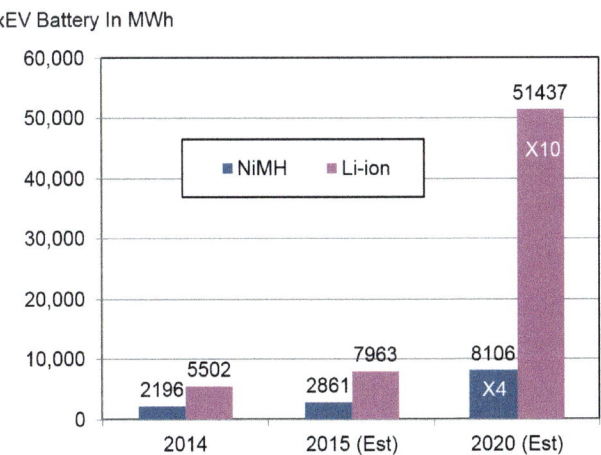

Figure 2. Prediction of Ni/MH and Li-ion batteries in the xEV, including electric vehicles, plug-in HEV, and HEV. Data from Yano's Report [1].

73

2. Results

In this review, we have separated patent applications filed in Japan from battery manufacturers, component supplying companies, and independent research institutes. The reason for selecting the patent application, instead of the patent itself, is to increase the scope of coverage. For each company, we picked the most influential patent applications, based on the length of effective days, and summarized their contents here. The background of these companies is briefly introduced in Table 1.

2.1. Battery Manufacturers

Nine Ni/MH manufacturers were chosen for this section. Examples of their consumer products are shown in Figure 3.

Figure 3. Ni/MH consumer products from major Japanese manufacturers—from left to right: Matsushita (Panasonic), Sanyo, Maxell, Yuasa, Toshiba, FDK, Furukawa, GS-Saft, and Shin-kobe. All are cylindrical, except the stick-type (gum) made by GS-Saft for portable digital assistant and cell phone use.

Table 1. Brief introduction to the Japanese companies included in this review. The number of consolidated employees includes those in the parent company who are not directly involved in the battery business. JMC: Japan Metals & Chemicals; MH: metal hydride; and AIST: Agency of Industrial Science and Technology.

Company Name	Established Date	Main Business	Trademark	Employee	Remark
Matsushita (Panasonic)	1918	Home appliances and electronics	Panasonic	Consolidated: 254,084	-
SANYO Electric	1947	Home appliances and electronics	SANYO	Merge into Panasonic in December 2009	Ni/MH division was sold to FDK in January 2010
Hitachi Maxell	1960	Energy, industrial materials and electronic appliance/consumer products	maxell	Consolidated: 4053	-
Yuasa	1918	Storage batteries and other products	YUASA GS YUASA	Consolidated: 14,506 as GS Yuasa	Managed integration through the establishment of a holding company (GS Yuasa Corporation) in April 2004.
Toshiba Battery	1954	Storage batteries and other products	TOSHIBA	Merge into Toshiba Home Appliances, and dissolved in 2009	Ni/MH div. was sold to SANYO in 2000
FDK	1950	Various kinds of batteries, rechargeable batteries, battery devices, electronic components, and devices	FDK	4917 consolidated basis	Acquired Sanyo Energy Twicell Co., Ltd. and Sanyo Energy Tottori Co., Ltd. in 2000
Furukawa Battery	1950	Various kinds of batteries, rechargeable batteries, battery devices, electronic components, and devices	FB FURUKAWA BATTERY	Consolidated: 1999	-
Japan Storage Battery	1917	Storage batteries and other products	GS GS YUASA	Consolidated: 14,506 as GS Yuasa	Managed integration through the establishment of a holding company (GS Yuasa Corporation) in April 2004.

Table 1. *Cont.*

Company Name	Established Date	Main Business	Trademark	Employee	Remark
Shin-Kobe Electric Machinery	1969 (consolidation)	Various kinds of batteries, rechargeable batteries, battery devices, electronic components and devices, and polymer science materials	新神戸電機株式会社	Merge into Hitachi Chemical, and dissolved in Jan. 2016	Subsidiary of Hitachi Chemical in March 2012
Tanaka Chemical	1957	Positive electrode materials for rechargeable batteries, catalyst materials, inorganic chemical products	Tanaka Chemical Corporation	180	-
Mitsui Kinzoku	1950	Manufacturing and sales of functional materials, electronic materials, and automotive parts/components, etc.	MITSUI KINZOKU	Consolidated: 10,804	-
Santoku	1949	Rare earths	三徳	229	-
JMC	1917	Ferroalloy, MH and non-ferrous metals, ferrite and ceramic products, geothermal energy consulting service, and electric power generation	JMC Japan Metals & Chemicals Co., Ltd.	Consolidated: 921	-
Shin-Etsu Chemical	1926	PVC/Chlor-Alkali, silicones, specialty chemicals, electronics & functional material, and diversified	Shin-Etsu	Consolidated: 18,276	-
AIST	1952	AIST takes initiative as a leading research institute for innovation while placing its major emphasis on research that offers practical benefits to the world	AIST	2929	Reformed in 2001 (National Institute of Advanced Industrial Science and Technology)

2.1.1. Matsushita

Matsushita Electric Industrial (Kadoma, Osaka, Japan) was one of the first companies to sell commercial Ni/MH batteries. In 1996, Matsushita and Toyota formed a joint venture, Panasonic EV Energy Co. (PEVE, now Primearth EV Co., Okasaki, Shizuoka, Japan), to manufacture Ni/MH batteries for propulsion application. In 2008, its name was changed to "Panasonic Corporation" and this company continued to manufacture Ni/MH consumer batteries in Wuxi, Jiangsu, China. The Ni/MH batteries, under the tradename Evolta, reached a cycle life of 1600 by sharing the positive electrode developed for their primary cell with the same trade name [4], which was the world record of the longest life for Ni/MH batteries in 2008. In 2009, Panasonic acquired Sanyo and sold its major stocks of PEVE to Toyota. The consumer battery manufacturing section was sold to FDK and the cylindrical cell manufacturing facility in Shonan, Kanagawa, Japan was sold to the Chinese Corun Company (Changsha, Hunan, China). Today, Panasonic continues to operate Sanyo's facility in Awajishima, making high-power cylindrical cell for start-stop vehicles. In the melt-and-cast preparation of MH alloys, Matsushita filed two patent applications: one on water atomization (Figure 4, [5]) and the other one on a double-roller casting mechanism (Figure 5, [6]).

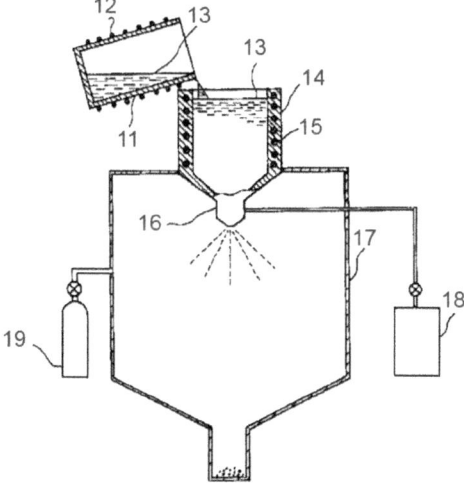

Figure 4. Schematic diagram of a water atomization system patented by Matsushita. Numbered parts are 11: induction melting furnace; 12: coil; 13: MH alloy; 14: holding furnace; 15: coil; 16: injection nozzle; 17: collection tank; 18: high pressure water pump; and 19: nitrogen cylinder. Reproduced from [5].

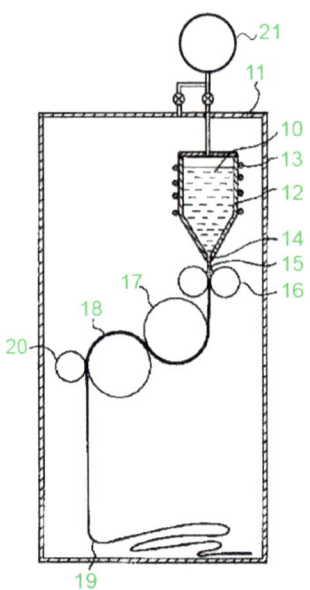

Figure 5. Schematic diagram of a double-roller casting system patented by Matsushita. Numbered parts are 10: MH alloy; 11: vacuum chamber; 12: induction furnace; 13: coil; 14: nozzle; 15: melt; 16: high temperature double-roller; 17, 18: water-cooled rollers; 19: alloy film; 20: pressing roller; and 21: Ar gas cylinder. Reproduced from [6].

A few patent applications filed between 1996 and 1998 were selected to demonstrate the coverage Matsushita acquired in the area of Ni/MH batteries: an AB_5 metal hydride (MH) alloy containing a small amount of Fe, Cr, or Cu to lower the raw material cost [7], a Zr-based AB_2 MH alloy with improved cycle life [8], a negative electrode of high surface area with a carbon content between 350 ppm and 1000 ppm [9], a positive electrode with spherical particles with an average diameter of less than 1.7 μm and the percentage of the particles smaller than 1.0 μm of below 20% [10], the effect of the addition of compounds from Y, Er, and Yb in the positive electrode to improve the high temperature performance [11], addition of ions from rare earth elements in the electrolyte to suppress the oxidation and capacity degradation of the negative electrode [12], $Co(OH)_2$ coating on the $Ni(OH)_2$ spherical particle surface to increase the cycle stability by increasing the electrolyte holding power [13], a design of negative-to-positive ratio from 1.5 to 2.0 for the uninterrupted power supplier application [14], and a surface treatment method to improve the uniformity in capacity degradation during cycling (Figure 6, from [15]) and its application in high-rate batteries [16].

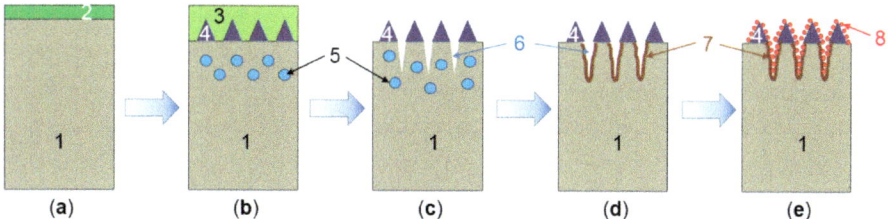

Figure 6. A schematic diagram of a surface preparation method for a negative electrode to pre-age the MH alloy in order to improve the uniformity in cycle degradation in the cell. From left to right, showing the surface microstructure (**a**) after a wet-grinding process; (**b**) after an alkaline treatment; (**c**) after an acid treatment; (**d**) after a de-hydrogenation process; and (**e**) after the final hydroxide ion modification. In the diagram, the components indicated by numbers are 1: bulk ingot; 2: hydroxide and incomplete oxide layer; 3: hydroxide from the rare earth element layer; 4: metallic Ni; 5: hydrogen atoms; 6: pore; 7: Ni-rich layer; and 8: hydroxide ions. This surface treatment has become the standard procedure for Ni/MH batteries used in the propulsion industry. Reproduced from [15].

2.1.2. Sanyo

Sanyo Electric Co. (Osaka, Japan) was the other company selling the first-generation commercial Ni/MH batteries [17]. It began manufacturing Ni/MH in 1990. In 2000, Sanyo purchased Toshiba's Ni/MH manufacturing facility (Takasaki, Gumma, Japan) and became the largest producer of Ni/MH batteries in the world. In 2005, Sanyo introduced a low self-discharge Ni/MH battery, an eneloop, with less than 15% capacity loss during the first year of storage [18]. The battery is based on the A_2B_7 superlattice Mg and rare earth elements containing MH alloys [19,20]. In 2009, Sanyo was acquired by Panasonic. In 2014, its Ni/MH manufacturing facility in Suzhou, Jiangsu, China was closed and the production line was moved back to Japan, where it continued to produce Ni/MH batteries under the operation of FDK.

The representative patent applications filed by Sanyo include the following concepts: addition of fluoro-resin powder on the negative electrode surface to increase the gas recombination rate during overcharge [21], addition of polytetrafluoroethylene (PTFE) to the negative electrode for increasing its mechanical strength [22], a coating of the co-precipitated hydroxide from Zn, Cd, Mg, or Ca on the Ni(OH)$_2$ surface to prevent capacity loss due to over-discharge [23], a melt-spin fabrication method for preparing MH alloy with improved high-rate dischargeability (HRD) at a low temperature [24], a requirement in the uniformity of MH alloy particle size for better cycle life [25], addition of 0.5–10 ppm Y into electrolyte [26], a β-CoOOH coating [27] to improve the utilization of positive electrode active material, addition of Mn to prevent leakage of the electrolyte [28], a partially reduced metal between the MH alloy particles to improve the conductivity of the negative electrode [29], a special substrate made of metallic form with a hollow structured skeleton to prevent the breakage of substrate during winding [30], and a Li-containing Co coating to the $Ni(OH)_2$ spherical particles to improve the utilization of the positive electrode active material [31]. After merging into Panasonic, most of the Sanyo's patents belong to the new owner, except the last two, which were acquired by FDK.

2.1.3. Hitachi Maxell

Hitachi Maxell Ltd. (Tokyo, Japan) is a consumer electronics manufacturer in Japan. In the early days, Hitachi Maxell teamed with Ovonic Battery Company (Troy, MI, USA) and manufactured Ni/MH batteries based on the AB_2 MH alloys. In contrast to the previous two companies, Maxell's patent applications are more related to the fabrication at the cell level, instead of raw materials or electrodes. Some examples are: addition of a PTFE layer on the surface of MH alloy to increase the conductivity and mechanical strength of the negative electrode [32], a design of a safety-vent to prevent rupture of the battery [33], a disordered nickel hydroxide co-precipitated with 1.5–2.8 wt% Zn to increase the capacity [34], addition of corrosion preventives in the electrolyte, such as K_2SiO_3, $K_4Si_3O_8$, K_4SiO_4, K_6SiO_6, K_3PO_4, K_4PO_7, K_2CrO_4 [35], Mo, W, and Cr ions [36] to improve cycle stability, a requirement in the uniformity of spherical particle size in the positive electrode [37], a method of assembling two different chemistries in one hybrid container [38], an electrolyte filling amount of 1.4 cc· $(Ah)^{-1}$ [39], a Zn-Co layer on the $Ni(OH)_2$ spherical particle to increase the utilization and lower the raw material cost [40], and an Ni-rich secondary phase in the AB_5 MH alloy to improve low-temperature HRD [41,42].

2.1.4. Yuasa

Yuasa Battery Co. (Takatsuki, Osaka, Japan) was the first automobile battery company in Japan (1920). Yuasa was also the first company to develop spherical $Ni(OH)_2$ for the high-density positive electrode used in Ni/MH batteries. Oshitani and Yufu [43] disclosed a β-$Co(OH)_2$-coated $Ni(OH)_2$ in his patent application filed in 1988 , which is still used in today's Ni/MH batteries (Figure 7). In 1997, Yuasa formed a joint venture with Taiwanese Delta Electronics (Taipei, Taiwan) in Tianjin, China to manufacture Ni/MH batteries, which was productive until a recent shutdown in 2015 due to an accidental explosion. In 2004, Yuasa merged with Japan Storage Battery Co. and formed GS Yuasa Co. The company has a product similar to FDK's eneloop—the eNiTIME. It comes pre-charged and can be used right out of the box [44].

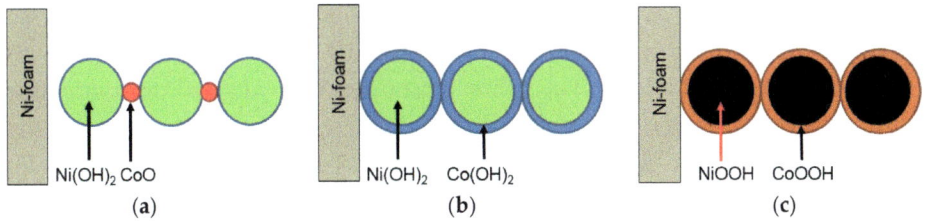

Figure 7. Schematic diagrams showing the formation of a conductive network in the positive electrode: (**a**) as prepared with CoO additives; (**b**) after the dissolution and re-deposition of Co(OH)$_2$; and (**c**) after the first charge into CoOOH. Reproduced from [43].

Since 1980, Yuasa has filed 327 patent applications on alkaline batteries [45] that cover both the positive and negative electrodes. We have picked the following as the most representative patents with an emphasis on: a bi-polar design electrode design [46], an ex-situ oxidation applied to the Co-conductive network of the positive electrode [47], a dual polyolefin/vinyl monomer composite used as a separator to reduce the self-discharge [48], a formation algorithm to increase the protection against overcharge and over-discharge [49], addition of Y and Yb compounds in the positive electrode to reduce the corrosion in the negative electrode [50], addition of a sulfonic group to the benzene nucleus forming a separator material with a lower self-discharge [51], a positive electrode with Co metal and/or Co oxide and at least one rare earth compound, alkaline earth metal compound and zinc oxide [52], an negative electrode additive made from Ce metal or a complex of Ce metal and another metal with a BET surface area >50 m$^2 \cdot$g^{-1} [53] or fine Ni powder (<2 µm) [54] to improve the performance of the battery, and MH alloy containing a rare earth element with an atomic number >62 (Eu, Gd, Tb, Dy, Ho, Er, Tm, Yb, and Lu) [55].

2.1.5. Toshiba

Toshiba Corp. (Tokyo, Japan) is a heavily diversified conglomerate corporation in electronics, nuclear, and electrical appliances. It was one of the earliest Ni/MH manufacturers. Toshiba, together with Duracell (Bethel, CT, USA) and Varta Battery (Elleangen, Germany, now bankrupted), formed a joint venture, 3C Alliance LLP, in 1996 and achieved annual sales of 100 million cells. Toshiba sold its Ni/MH operation to Sanyo in 2000. Examples of patent applications from Toshiba include: a battery module with space filled with resin to improve the temperature uniformity during charge [56], a negative electrode with polymer binder containing PTEE of an average molecular weight of 200,000–1,000,000 [57], an Mg-containing (>10 at%) MH alloy with some unique X-ray diffraction (XRD) intensity ratio characteristics [58], Mg-based AB$_2$ and A$_2$B MH alloys as negative electrode active materials [59], a high energy density cell with Mn content between 0.6 wt% and 2.6 wt% [60], a method to increase the loading of a negative electrode [61], a method to suppress the self-discharge in the early stage of cycle life [62], improvement in the low-temperature performance [63], Mg-containing MH alloys with layered crystal structure [64], and an MH alloy containing Mg, a rare earth element, and alkaline metals [65]. The last six patents now belong to Yuasa.

In 2000, right before transferring the Ni/MH business to Sanyo, scientists at Toshiba filed a series of patent applications regarding newly discovered superlattice alloys, trying to summarize the laboratory findings up to that day, without significant conclusions. These patent applications may be considered immature compared to today's technology, which has been fine-tuned by Sanyo and FDK, but they are good examples of how a large battery company performed research, starting from ground level. They include patent applications about engineering the Mg content within an MH alloy particle [66], improvement in low-temperature performance by increasing the surface metallic Ni content [67], improvement in capacity and cycle life by controlling the lattice parameter c in the crystal structure [68], a rapid cooling method to prevent loss of Mg [69], and an optimized composition range for capacity and cycle life (Figure 8, data from [70]).

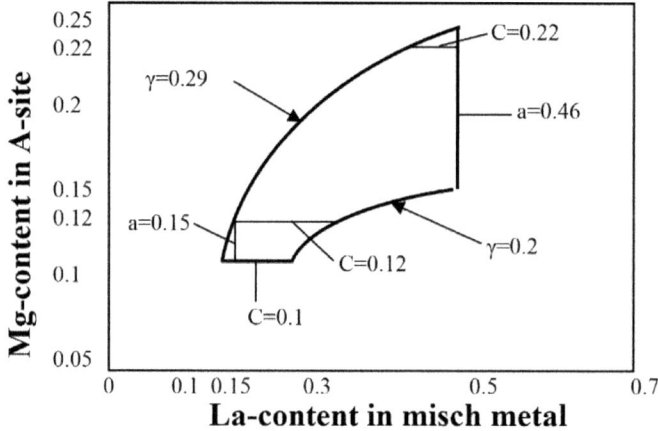

Figure 8. A composition map of superlattice MH alloys showing range optimized by capacity and cycle life [70]. The x-axis is the La content in the misch metal (La, Ce, Pr, and Nd) represented by "a" in the chart. The y-axis is the Mg content in the A-site atoms (Mg, La, Ce, Pr, and Nd) represented by "C" in the chart. In the chart, γ is a quantity defined by $C + 0.025/a$.

2.1.6. FDK

FDK Corp. (original name: Fuji Denki Kagaku) was originally an electronic company making primary Li batteries and other electronic components. In 2010, it acquired the Twicell Division (Ni/MH battery branch making eneloop batteries) of Sanyo and continues to operate as the world's largest producer of Ni/MH batteries in Takasaki, Japan. One example, showing the continuity of research from one company to the next company, is shown in Figures 9 and 10. Since its debut, the superlattice alloy was targeted to replace the AB_5 MH alloy with the goal of increasing the capacity. However, a corrosion problem came with the Mg component and led to a short cycle life. Therefore, a ten-year campaign to find the most suitable oxidation inhibitor for superlattice alloys started in Toshiba, then Sanyo, finally accomplished by FDK.

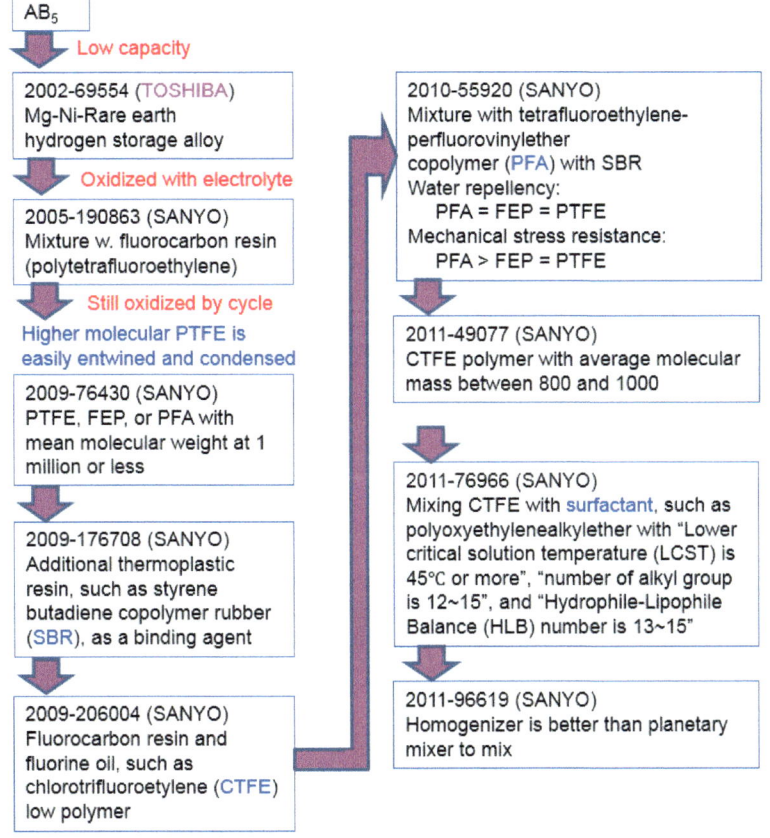

Figure 9. Early Japanese Patent Applications regarding the use of newly developed superlattice MH alloys.

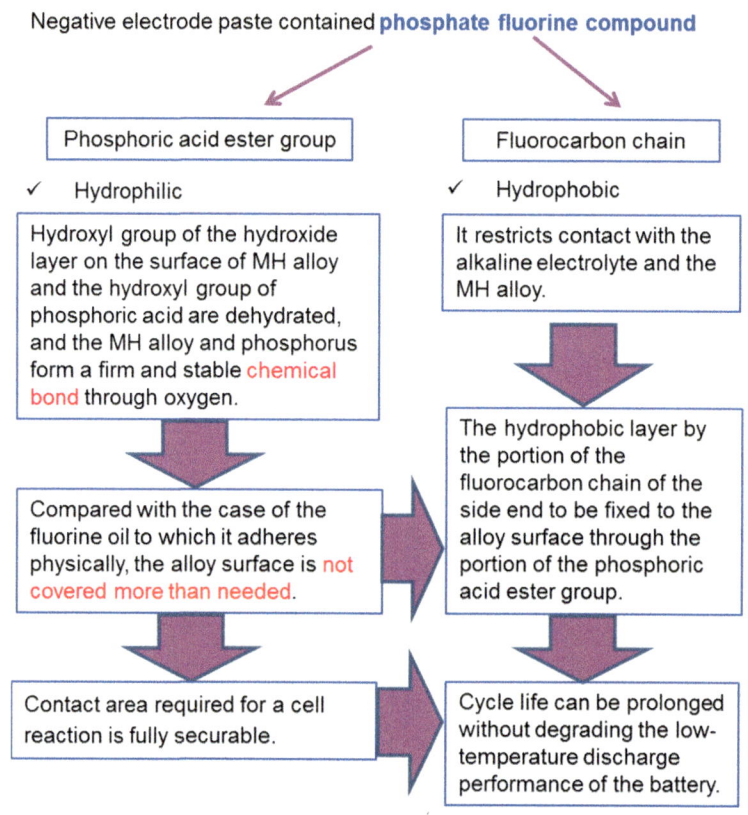

Figure 10. In a continuation of previous efforts (Figure 9), FDK filed a patent application to finalize the design of additives in negative electrode paste to extend the cycle life of cells, using superlattice alloys to 5000+ [71].

FDK, being the only remaining domestic consumer Ni/MH battery producer, filed patent applications that were centered on: an LHLHHH type of stacking sequence along the c-axis for a superlattice alloy where L and H are A_2B_4 and AB_5 layers, respectively [72], larger MH alloy particles used in the outside rim of a cylindrical cell to improve its cycle stability [73], a positive electrode using nickel hydroxide from co-precipitation of Mg to reduce the voltage lowering during storage [74], a Ca-containing [75] and an Sm-rich [76], Zr-containing [77] superlattice alloy to increase both the capacity and the cycle life, an electrolyte recipe composed of KOH, NaOH, and LiOH to optimize cycle life and low-temperature characteristics [78], an NaOH-dominant electrolyte for improving both cycle life and self-discharge performances [79], a TeO_2-containing electrolyte to improve low-temperature discharge characteristics [80], the additives in the negative electrode containing CaF_2, CaS, and/or $CaCl_2$ [81], and AlF_3 [82] to improve alloy performance, application of MH alloy particle of different sizes onto each side of the electrode to prevent short-circuit [83], an NaF-containing electrolyte to improve the self-discharge characteristics of the cell [84], a low-La (<25% in A-site element) superlattice alloy to have improved voltage standing during long-period storage [85], and a multi-layer construction for a negative electrode with different binder contents to improve the capacity degradation and short-circuit of the cell [86].

2.1.7. Furukawa

Furukawa Battery Co. (Yokohama, Kanagawa, Japan) is a company that manufactures mainly lead-acid and alkaline batteries. It constructed a Ni/MH plant in Iwaki, Fukushima, Japan in 1993. In 1997, it was one of seven Ni/MH battery manufacturers in Japan (the others were Matsushita, Sanyo, Toshiba, Yuasa, GS-Saft, and Hitachi-Maxell [87]). Patent applications filed by Furukawa include: a 5%–15% porosity in the negative electrode assembly to suppress the pressure build-up during cycling [88], a sponge nickel sheet as the positive electrode substrate and a punched nickel sheet (30%–44% open space) as the negative electrode substrate to increase the energy density of the cell [89], a formation process including both electrical activation and thermal aging processes to ensure high HRD [90], a heat treatment (100–250 °C) of the MH alloy powder before electrode assembly to suppress the pressure increase during overcharge [91], a mixture of $Ni(OH)_2$, CoO, and ZnO as the active material in a positive electrode to improve cycle life, utilization, HRD, and temperature range of the cell [92], and an optimized amount of electrolyte filling, 85%–97% of the available space in the cell to lower the cell pressure during charge [93].

2.1.8. Japan Storage

The GS-Japan Storage (Tokyo, Japan) Company was the first lead-acid manufacturer in Japan (1895). It formed a joint venture with the French Saft (Bagnolet, France), GS-Saft, producing Ni-Cd, and later Ni/MH and Li-ion batteries in 1986. The company later merged into Yuasa in 2004. Some of their patent applications include: the addition of a metal layer next to a synthetic-resin-made container to suppress the self-discharge [94], a thermoplastic synthetic-resin-made relief valve with a melting point higher than 270 °C to improve the safety of the cell [95], addition of an Mn compound to the positive electrode to reduce the hydrogen reaction on the positive electrode surface and suppress the self-discharge of the cell [96], an activation scheme composed of electrical formation and heat storage to improve the HRD [97], and an increase in the negative electrode area to suppress pressure during charge [98].

2.1.9. Shin-Kobe

Shin-kobe Electric Machinery Co. (Tokyo, Japan) is a manufacturer of lead-acid, Ni/MH, and Li-ion batteries. Their Ni/MH product was sold through Hitachi with the trade name CellAce. In 2013, their business was merged into Hitachi Chemical (Tokyo, Japan). Shin-kobe's patent applications include: addition of conductive fibers in the negative electrode to reduce its electric resistivity [99], adding hydroxyl methyl cellulose as a ductile binder in the negative electrode to improve cell performance [100], a substrate fabrication method for mixing MH and Ni powder on the surface of polyurethane (PU) foam and removing the PU in a heated reduction atmosphere [101], addition of ethylene-vinyl acetate-acrylate copolymer, acrylate-styrene-alkyl acrylate copolymer, methylcellose, and nickel powder in the slurry for the negative electrode to improve the gas absorption ability [102], and addition of $CaCO_3$ to reduce the oxygen evolution in the negative electrode during over-discharge (cell reversal) [103].

2.1.10. Other Battery Manufacturers

Other than those consumer battery manufacturers introduced above, there are companies focused on other markets. For example, PEVE currently produces the highest Ah amount of batteries, supporting the HEV business of Toyota. All their inventions are related to battery pack assembly, battery management systems (BMS), and testing. Examples include: a failure sensing mechanism [104], a cooling controlling method [105], a current detection method [106], an electric current leakage detecting device [107], a voltage measuring method [108] in the battery pack, a state-of-charge (SOC) estimation method [109,110], a testing method for relay contact welding [111], BMS for vehicles [112], and an activation algorithm [113]. Kawasaki Heavy Industry (Kyoto, Japan) developed a GIGACELL high-power Ni/MH battery for stationary and locomotive applications [114]. They have filed patent applications to cover: 3D batteries [115], addition of conductive fibers, carbon particles, Ni foil, Ni-plates fiber, etc. to increase the conductivity and connection of the negative electrode in Ni/MH batteries [116], use of fibrous MH alloys in the negative electrode to improve charge and discharge rate capability [117], a fold design for electrode stacking (the basis of GIGACELL) [118–120], design of a battery charger for railroad applications [121], an SOC estimation algorithm [122], a pressure-regulating device [123], a battery module [124], and a battery system [125]. Honda Denki (Tokyo, Japan) developed a flooded type of long-lasting Ni/MH battery (Table 2), but they did not file any patent application for technology originating from the flooded Ni-Fe battery [126]. Varta Battery filed a Japanese patent application about adding Mn-oxide in the positive electrode [127]. Saft, a French company, also filed a Japanese patent application on a C14 Laves phase alloy for negative electrodes [128]. Samsung Display Devices (Yongin-Gun, Korea) filed a Japanese patent application about adding an insulating layer at the edge of one electrode to prevent short-circuit [129]. Ovonic Battery Company also has a few Japanese patent applications covering disordered AB$_2$ [130–134], MgNi-based [135–137] MH alloys, a disordered positive electrode [138–140], and a mono-block design [141]. Its parent company, Energy Conversion Devices Inc. (Troy, MI, USA) also filed a patent application regarding a dry compaction method for making a negative electrode belt [142].

Table 2. Characteristics of the flooded type Ni/MH batteries made by Honda Denki. Information from their brochure.

	Type	MHS	MHM
Main components	Positive plate	Oxy nickel hydroxide (pocket type)	
	Negative plane	Metal hydride (pocket type)	
	Separator	Synthetic resin	
	Cell container	Synthetic resin (translucent)	
	Electrolyte	Potassium hydroxide	
	Specific gravity (20 °C)	1.21	
Capacity (Ah)	Range	20–1000	
	Discharge rate (h)	5	
Voltage (V/cell)	Nominal	1.2	
	Floating charge	1.39	1.38
Recommended back-up time		More than 60 min	More than 30 min
Expected life		More than 15 years	

2.2. Component Suppliers

2.2.1. Tanaka Chemical Co.

Tanaka Chemical Co. (Fukui, Japan) is the main producer of spherical Ni(OH)$_2$ that can act as positive electrode active materials. In the early days, Tanaka monopolized the spherical Ni(OH)$_2$ market, but has now been replaced by many Chinese manufacturers. In 1996, Tanaka filed five patent applications covering: surface coating of α-Co(OH)$_2$ [143] or β-Co(OH)$_2$ [144] and conversion into γ-CoOOH with a high conductivity on the surface of Ni(OH)$_2$, addition of salts from sulfuric acid, nitric acid, boric acid, and phosphoric acid to increase the solubility of Co(OH)$_2$ in the electrolyte [145], a co-precipitation of hydroxide with Ni, Co, Zn, Y (or Cd) as the positive electrode active material [146], and an increase in the adhesion of Co coating to the Ni(OH)$_2$ spherical particles to prevent peeling [147]. In 1998, Tanaka filed another three patent applications covering: an auxiliary conductive binder addition into the positive electrode paste with α-Co(OH)$_2$ [148], an additional coating of Ni or Ni alloy on the Co(OH)$_2$ coating on the Ni(OH)$_2$ particles [149], and an oxidation process for $Co(OH)_2$ coating in an oxidation environment with heat to save the cost [150].

2.2.2. Mitsui Metal

Mitsui Bussan Metals Co. (Tokyo, Japan) was the major AB_5 MH alloy producer that has used the conventional induction melting and casting method since the debut of Ni/MH battery. It was the major AB_5 MH alloy supplier to Matsushita. Patent applications from Mitsui are all about AB_5 MH alloys which include: a columnar structure-dominate (>80%) of AB_5 ingot which facilitates the early pulverization [151], a wet crushing and grinding process for AB_5 ingots [152], addition of Mo or Cr in the AB_5 formula to increase the capacity of the alloy [153], partial replacement of rare earth elements with Ca to reduce the cost [154], quick quench plus annealing to make alloys with heterogeneous strain to improve the cycle life [155], addition of Cu and B in the alloy formula to improve HRD of the cell [156], a low-cost Cu-containing AB_5 MH alloy with the lattice parameter c between 4.062 Å and 4.069 Å [157], an over-stoichiometric MH alloy with B/A between 0.05 and 5.25 and a lattice parameter c greater than 4.049 Å [158], and a low-Co Cu (Fe)-containing AB_5 MH alloy with the lattice parameter c greater than 4.042 Å [159].

2.2.3. Santoku Metal

Santoku Metal Industry Co. (Kobe, Hyogo, Japan) manufactures both MH alloy and NdFeB (for permanent magnetic applications) based on rare-earth elements using a strip-casting method. In 2000, the company name was changed to Santoku Corporation. Currently, the main production facility for making MH alloys was moved to its joint venture in China, Baotou Santoku Battery Materials Co. (Baotou, Inner Mongolia, China). Santoku's patent applications are also all about the development of MH alloys, such as: a method of making rare-earth-containing hydrogen storage alloy (filed in 1983) [160], an asymmetric alloy microstructure in the network segregation of AB_5 MH alloy [161], a storage method for rare-earth-containing alloys [162], a rapid quench with cooling rate between 1000 °C·s^{-1} and 10,000 °C·s^{-1} [163], and a strip-casting method of making rare-earth-containing alloy (Figure 11) [164].

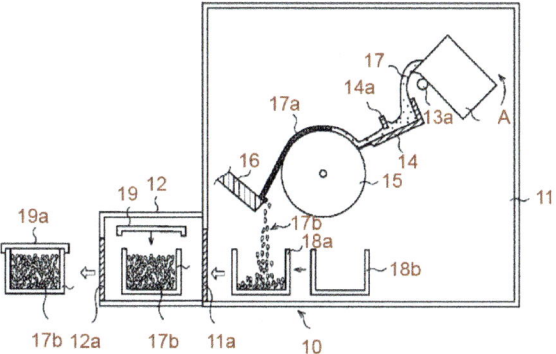

Figure 11. Schematic diagram of a strip-casting system patented by Santoku and widely adapted in China for production of Nd-B-Fe magnetic materials and AB_5 MH alloys. The numbers indicate the following: 10: manufacturing system; 11: the first chamber; 12: the second chamber; 11a and 12a: shutter valve; 13: induction furnace; 14: tundish; 15: rotating roll; 16: alloy crushing board; 17a: ribbons; 17b: crushed powder; 18, 18a and 18b: containers; 19 and 19a: top cover of a water-cooled container. Reproduced from [164].

2.2.4. Japan Metals & Chemicals Co.

In the early days, Japan Metals & Chemicals Co. (JMC, Tokyo, Japan) worked closely with the battery manufacturer Toshiba. With the ownership of Takasaki Plant transferring from Toshiba to Sanyo and finally to FDK, JMC remained the largest MH alloys supplier in Japan. With a proprietary fabrication method, JMC is one of the few MH alloy suppliers that can produce A_2B_7-related MH alloy in a full production scale (the other one is Santoku), which is a key component used in the eneloop low self-discharge consumer and GIGACELL stationary batteries. Besides battery materials, JMC also serves as the pioneering and continuing contributor in the gaseous hydrogen storage application of MH alloys. The first patent application filed by JMC in 1980 on MH alloys is about a fabrication method of TiFe binary alloy [165]. Later patent applications extended to the quaternary alloy system Ti-Fe-Zr-(Nb or Mn) [166]. The first AB_5 MH alloy patent application filed in 1986 by JMC was regarding the addition of Co, Al, and Sn to reduce the hysteresis in the pressure–concentration–temperature absorption and desorption isotherms [167]. In 1987, JMC also patented a precipitation method of making $Ni(OH)_2$ [168]. JMC filed a La-rich AB_5 MH alloy composition with Ni, Co, Mn, and Al in 1993 [169]. Porous 3D-foam substrates made from Ni were proposed as substrates for both the positive and negative electrodes [170]. A patent application about a recovery methodology for Ni/MH battery scrap was filed in 1994 [171]. Other patent applications filed by JMC include: the use of Ag-plated Ni-foam as substrate to improve corrosion resistance [172], an Mg-Y MH alloy [173], an Fe-containing over-stoichiometry AB_5 MH alloy [174], a gas atomization method for acid or alkaline solutions [175], a pre-oxidation by CO_2 for a freshly ground MH alloy [176], an $MgNi_2$-based MH alloy [177], an acid etch plus alkaline solution activation method for MH alloys [178], a low-cost Ni-plated Fe alloy substrate [179], additives of Li composite for a spinel composite in a negative electrode to improve the cycle stability [180], an activation in alkaline solution with a complexing agent to reduce the pulverization rate [181], a two-phase AB_5 MH alloys to reduce the pulverization rate [182], addition of Sn into the AB_5 alloy formula and a lattice constant ratio c/a between 0.8055 and 0.8070 [183], an AB_5 with a Mg-rich secondary phase to maintain the HRD even with reduction in the Co content to save the cost [184], a multi-phase superlattice alloy with >90% of $PuNi_3$, Ce_2Ni_7, and Pr_5Co_{19} phases and also an annealing condition of 1000 °C for 10 h to reduce the abundance of AB_5 phase to 3% [185], a Ti-, Zr-, and Hf- containing (0.2–1 at% in A-site atoms) AB_5 MH alloy to increase capacity and cycle stability [186], an AB_2 phase containing a superlattice MH alloy for better capacity, cycle life, and HRD [187], and a low-cost multi-phase MH alloy with $CaCu_5$ (40–90 wt%), Ce_5Co_{19} (5–39 wt%), and Pr_5Co_{19} (3–20 wt%) phases [188].

2.2.5. Shin-Etsu Chemical

Shin-Etsu Chemical (Tokyo, Japan) is a chemical company that has developed the various binder materials used in the paste electrode for Ni/MH batteries. It filed a few patent applications for: a main binder material using hydroxyalkyl cellulose in the negative electrode [189], an organic binder with Sm-containing MH alloy [190], a negative electrode slurry with binders of hydroxyalkyl alkyl cellulose, alkyl cellulose, carboxymethyl cellulose, and water-soluble natural polysaccharide, and at least one selected from the group consisting of polyhydric alcohols having two or more hydroxyl groups in one molecule and water-soluble polyoles which are polyalkanol amines [191], a negative electrode slurry containing hydroxyalkyl alkyl cellulose, alkyl cellulose, carboxymetyl cellulose and water-soluble natural polysaccharides, and a denatured polyvinyl alcohol (PVA) [192]. Shin-Etsu is also an MH alloy supplier in Japan and filed an application for an MH alloy fabrication method without the need for annealing (Figure 12, [193]).

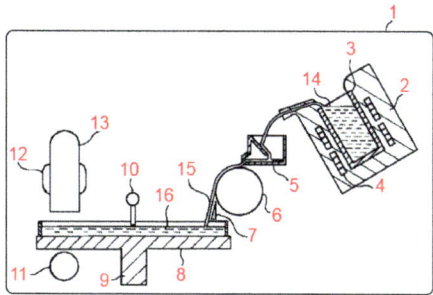

Figure 12. Schematic diagram of a strip-casting with the water-cooled pancake-mode system patented by Shin-Etsu. The numbers indicate 1: vacuum chamber; 2: induction furnace; 3: crucible; 4: radio frequency coil; 5: tundish; 6: cooling roll; 7: scrapper; 8: cooling table; 9: rotation shaft; 10: leveling arm; 11: gas inlet; 12: heat exchange fan; 13: fan duct; 14: melt; 15: alloy ribbons and flakes; and 16: alloy ribbons and flakes on a cooling table. Reproduced from [193].

2.2.6. Other Suppliers

As $Ni(OH)_2$ suppliers, Ise Chemicals Corp. (Tokyo, Japan) has filed patent applications about a high density spherical particle from the co-precipitation of Ni, Zn, and Co [194,195] and a Co-compound-coated $Ni(OH)_2$ [196] for use in the positive electrode. Sumitomo Chemical Co. (Tokyo, Japan) has also filed a patent application about high tap density $Ni(OH)_2$ spherical particles [197] and another one about a sintered substrate for Ni-based positive electrodes [198]. Furukawa Electric (Tokyo, Japan) filed a patent application regarding the use of a gas atomization technique to make MH alloys (Figure 13, [199]). Chuo Denki Kogyo Co. (Myoko, Niigata, Japan) is an MH alloy supplier who filed patent applications involving a pre-oxidation environment including partial pressure of oxygen and some organic compounds during grinding [200], an acid etch process to facilitate the formation of an Ni-rich phase on the alloy surface [201], a hyper-stoichiometric AB_5 MH alloy with an improved high-rate and low-temperature performances [202], an acid etch in mineral acid solution with Co, Ni, Fe, salt of rare earth elements or boric acid and other organic compounds [203]. Mitsubishi Metal Corp. (later changed name to Mitsubishi Materials Corp, Tokyo, Japan) filed patent applications about the use of AB_2 [204–206], Mn-free [207] and S-, P-containing AB_5 [208], and AB_5/A_2B_7 mixed phase [209] MH alloys as the negative electrode active materials. Daido Steel (Nagoya, Aichi, Japan) filed three AB_2 MH alloy fabrication method patent applications [210–212]. TDK Corp. (Tokyo, Japan) filed two patent applications about using V-based body-centered-cubic (bcc) MH alloys as negative electrode materials [213,214] and one about arc melting together with melt-spinning to produce bcc-type MH alloys, which are very sensitive to the oxygen content from the oxide crucible (Figure 14, [215]). Japan Steel Works Ltd. (Tokyo, Japan) filed a patent application about an alkaline treatment for activation [216], a method using Mg-alloy [217], Nb-containing AB_2 [218,219], Fe-containing [220], Mn-containing [221] and Mg-free [222] hyper-stoichiometric A_2B_7 superlattice alloys for negative electrodes, and a cold crucible melting apparatus (Figure 15, [223]). Showa Denko (Tokyo, Japan) also patented a centrifugal casting method for making MH alloys (Figure 16, [224]). Sumitomo Metal Mining Co. (Tokyo, Japan), a major $Ni(OH)_2$ supplier, filed patent applications regarding porous metal substrates [225], recycling from scrap of Ni/MH batteries [226,227], and a highly flexible metal form with resin fiber [228]. Katayama Special Industries, Ltd. (Osaka, Japan) is a supplier of a very flexible substrate for negative electrodes from power metallurgical processes [229]. The company also filed a patent application to cover the areas of a reinforced can structure with a hard layer to decrease the contact resistance [230] and a wet-paste electrode fabrication method [231]. A special tri-layer with a punched resin plate in the middle was patented by Toyoda Automatic Loom Works Ltd. (now Toyota Industries Co., Kariya-shi, Aichi, Japan) to maintain electrolyte folding power [232]. Toyo Ink SC Holding Co. (Tokyo, Japan) filed a patent application of ink mixture for foil-shape Ni/MH batteries [233]. Japan Vilene Co. (Tokyo, Japan) is a major sulfonated separator supplier and they filed patent applications to cover a PVA sheet laminated with nonwoven fabric [234], a separator made of polyolefin fiber graphed with polypropylene and polyethylene [235], and a fabrication method for a separator with higher acid resistance [236]. A mechanical alloying apparatus using an attritor to prepare MH alloys was patented by Nasu Denki Tekko Co. (Tokyo, Japan, Figure 17 [237].)

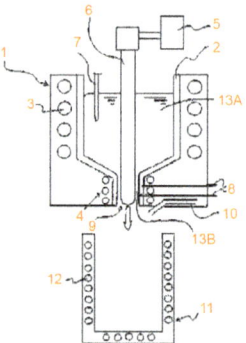

Figure 13. Schematic diagram of a gas atomization technique patented by Furukawa Electric with numbered parts, 1: melting furnace; 2: inner wall of crucible; 3: upper induction coil; 4: lower induction coil; 5: motor; 6: stopper rod; 7 and 8: thermocouples; 9: slurry out; 10: gas nozzle; 11: water-cooled mold; 12: water pipe; 13A: melt; and 13B: partially solidified liquid. Reproduced from [199].

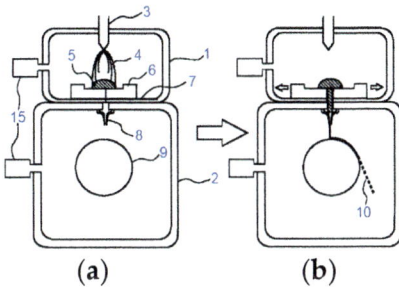

(a)   **(b)**

Figure 14. Schematic diagram of an arc melt and melt-spin preparation method for a body-centered-cubic (bcc) alloy by TDK Corp. with numbered parts, 1: melting chamber; 2: casting chamber; 3: arc electrode; 4: arc; 5: molten metal; 6: hearth; 7: bottom plate; 8: nozzle; 9: rotary roll; 10: ribbon; 15: inert atmosphere providers during (**a**) melting and (**b**) casting. Reproduced from [215].

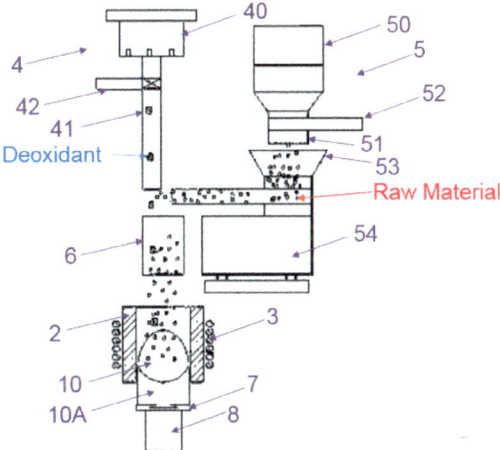

Figure 15. Schematic diagram of a magnetic levitation (skull) melting method patented by Japan Steel Works. Raw material and deoxidant were fed into the top of the furnace. Parts labeled by numbers are; 2: smelter; 3: induction coil; 4: deoxidant dispatcher; 40: container; 41: fall tube; 42: gate valve; 5: raw material feeder; 50: container; 51: fall tube; 52: gate valve; 53: hopper; 54: vibration feeder; 6: mixing box; 7: bottom of smelter; 8: drawing mechanism; 10: molten metal; and 10A: lower part of molten metal. Reproduced from [223].

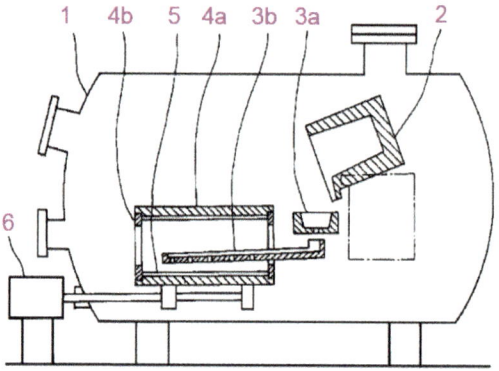

Figure 16. Schematic diagram of a centrifugal casting method patented by Showa Denko. Parts labeled by numbers are 1: vacuum chamber; 2: induction furnace; 3a: primary stationary tundish; 3b: secondary shuttling tundish; 4a: rotary cylindrical mold; 4b: end plate; 5: ingot; and 6: rotation drive. Reproduced from [224].

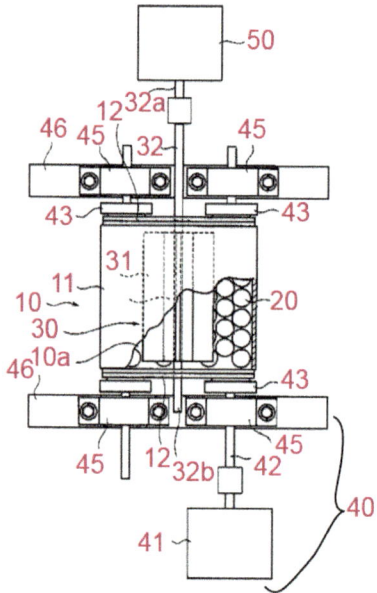

Figure 17. Schematic diagram of mechanical alloying by Nasu Denki Tekko Co. with numbered parts, 10: container; 11: drum section; 12: lateral section; 20: balls; 30: rotary body; 31: blade; 32: rotation axis; 40: slewing mechanism; 41: motor; 42: auxiliary roller; 43: transmission shaft; 45: holder; 46: base plate; and 50: motor. Reproduced from [237].

2.3. Research Institutes

2.3.1. Industrial Research Institute

The Japan Government Industrial Research Institute (ISI, Osaka, Japan) was established in 1918 and later merged into the National Institute of Advanced Industrial Science and Technology (AIST) in 2001. ISI made two remarkable contributions to the development of Ni/MH technologies. In the early days, research works conducted by Osumi and his team [238–241] on the hydrogen storage properties of the mish-metal based AB_5 MH alloys paved the road for Ni/MH battery commercialization. In 1993, a research team led by Kadir et al. [242] unveiled the potential of Mg-Ni-based superlattice alloys, which are the base of today's eneloop products. Examples of patent applications filed by ISI are: MH alloys with a Ni-rich surface made by electrical formation before the cells are assembled [243], a multi-phase MH alloy made from a Ti-V bcc solid solution and a TiNi phase [244], a multi-phase MH alloy made from AB_5 main phase and an AB–AB_3 minor phase, where A is mainly Ti or Zr and B is mainly Ni [245], addition of a gas recombination electrode made from Pd and Pt to reduce the oxygen evolution potential [246], and an AB_3 MH alloy containing rare earth and alkaline earth elements [247].

2.3.2. National Institute of Advanced Industrial Science and Technology

AIST (Tsukuba, Ibaraki, Japan) was reorganized in 2001 as an incorporated administrative agency integrating 15 research institutes (including ISI) in Japan. Today, AIST still continues its support for the hydrogen storage and battery industries. AIST filed a patent application with Honda Motor using $TiFe_{1-x}Pd_x$ as a low-cost alternative negative electrode active material [248]. AIST also filed a patent application with JMC regarding a Laves phase based AB_2 MH alloy for use as a negative electrode active material [249]. In addition, AIST filed a patent application with Mitsubishi Kasei Corp. (Tokyo, Japan) about the use of a hypo-stoichiometric AB_5 MH alloy [250]. Together with Hitachi Maxell and Yuasa, AIST filed two patent applications about Mg-containing multi-phase MH alloys [251] and rectangular Ni/MH batteries from wound electrodes [252]. With Yuasa, AIST filed two patent applications for using Pr_5Co_{19}-structured [253] and layer-structured, along c-direction [254], MH alloys in the negative electrode. AIST also filed two patent applications with Asahi Kasei Corp. (Tokyo, Japan) about using a separator with a porous polyolefin surface with a pore size between 0.01 μm and 1 μm [255] and a pore size smaller than those in the positive and negative electrodes [256]. AIST and Sumitomo Electric Industries filed a patent application regarding the use of a Ni-coated non-woven fabric as the substrate to increase the loading of the negative electrode active materials [257]. AIST and Kawasaki Heavy Industry filed two patent applications about the use of a Ni-coated carbon fiber as the substrate and a final coating of an ion permeable polymer layer on top of the negative electrode [258] and a Co-free superlattice MH alloy [259]. Patent applications filed solely by AIST itself demonstrate the use of Mg-containing MH alloy in the Ni/MH battery [260] and a porous foil substrate made from metal powder or fiber [261].

2.3.3. Toyota Central R & D Lab

Toyota Central Research & Development Laboratories, Inc. was established in 1960 in Nagakute-Shi, Aichi, Japan. It has approximately 1000 researchers working on resource conservation, energy conservation, environmental preservation, and safety. This company has been very active in the research of hydrogen storage alloys and fuel cell applications. Their patent applications in the field of Ni/MH batteries include: a surface treatment involving HF and NH_4F mixed solution [262] and stream [263] to activate MH alloys, a porous metallic film on the substrate to suppress the deformation of the electrode [264], a high tap density positive material [265], a surface to reduce hydrogen evolution during over-charge [266], a conductive powder in the negative electrode to improve the HRD [267], a positive electrode suitable for high-temperature applications [268], a method of suppressing or preventing memory effects by adding an anion exchange membrane to the anode [269].

2.3.4. Other Research Institutes in Japan

Japan Aerospace Exploration Agency (JAXA, Tsukuba, Ibaraki, Japan) has described a hydrophobic metal sponge layer inside a button cell as a gas recombination center and the patent belongs to Toshiba [270]. Institute of Energy Engineering Inc. (Tokyo, Japan) offered a solution to cool down large-format batteries by inserting coolant-flowing channels inside the battery [271]. Imura Materials Development Institute (Kariya, Aichi, Japan) filed patent applications regarding fabrication of Mg-Cu composites [272] and V–(Ta, Nb)(Cr, Mn) [273] materials for hydrogen storage applications, and V-based bcc [274], Ti–V solid solution bcc [275], bcc–TiNi multi-phase [276] alloys for battery applications. Jointly with AIST, Imura also filed a patent application about a deoxidation process for V-raw materials [277]. The International Center for Environmental Technology Transfer (ICETT, Yokkaichi, Mie, Japan) was established in 1990 to coordinate efforts from industry, academia, and the government to solve pollution issues. It filed two patent applications with GS Yuasa for the use of LiOH as an additive in the electrolyte to prolong the cycle life and improve the HRD performance [278] and a current-collector design to reduce internal resistivity [279].

2.4. Analysis

From the Japanese Patent Applications reviewed above, it is clear that scientists and engineers in Japan contributed a large amount of efforts in the composition fine-tuning, manufacturing process improvement, and special applications. Their research covers from raw materials, active ingredients, additives, electrodes, cells, modules, to the entire battery-powered systems. These patented approaches may not be adopted in today's commercial products, but they remain as alternative choices when the demand or market situation changes. It is, thus, important to keep track of these possible solutions to the past or current problems. Compared to the Japanese Patent Applications, those filed in the US tend to be more fundamental and less specific. For example, the methodologies to improve ultra-low-temperature performance of the Ni-MH battery can be found in US Patent 7,344,677 describing a surface oxide structure allowing electrolyte transportation at low-temperature [280] and a Japanese Patent Application filing the use of TeO_2 in the electrolyte [80]. The US Patent is less material-specific and provides a general direction for the improvement. Therefore, our suggestion is that, while the Japanese Patent Applications provide more direct solutions to engineers working in the field, the US Patents (and applications) are better guides for scientists looking for new ideas and approaches.

3. Conclusions

We have listed the key Japanese patent applications from the major Ni/MH battery makers and component suppliers in Japan, covering a variety of topics, including the composition and process design of the active materials and additives, electrode fabrication process, cell assembly (Figure 18), and battery system integration. In addition, we also use two examples to demonstrate the continuity and the competition of the research in areas of additives to the negative electrode associated with the superlattice alloys and the apparatus of MH alloy fabrication using different methods. Compared to the US Patents, the Japanese patent applications usually describe a specific composition formula range or detailed description of the design with narrower scope and less elaboration. However, they are described in full detail and are easier to interpret than the US Patents.

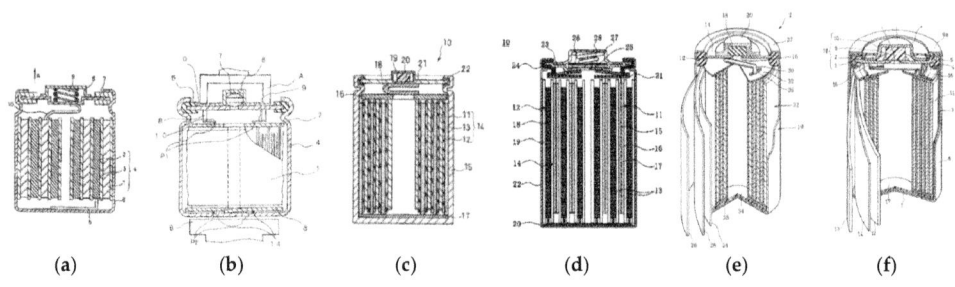

| (a) | (b) | (c) | (d) | (e) | (f) |

Figure 18. Schematic diagrams of (**a**) a single tap (10) design from Sanyo [21]; (**b**) a multiple welding spots (P) design from Yuasa [281]; (**c**) an addition of current-collect plate (16) design from Panasonic [282]; (**d**) a foldable end-plate design (21) from FDK [283]; (**e**) an insulating plate (32) design from FDK [77]; and (**f**) a positive electrode current collector (15) design from Panasonic [284].

Acknowledgments: Authors appreciate assistance from Su Cronogue and Tiejun Meng.

Author Contributions: Taihei Ouchi collected and analyzed the information. Kwo-hsiung Young prepared the manuscript and drawings. Dhanashree Moghe helped in information analysis and manuscript preparation.

Conflicts of Interest: The authors declare no conflict of interest.

Abbreviations

Ni/MH	Nickel/metal hydride
HEV	Hybrid electric vehicle
AIST	Agency of Industrial Science and Technology
PEVE	Panasonic EV Energy (now Primearth EV Energy)
MH	Metal hydride
PTFE	Polytetrafluoroethylene
HRD	High-rate dischargeability
XRD	X-ray diffractometer
PU	Polyurethane
BMS	Battery management system
JMC	Japan Metals & Chemicals Co.
PVA	Polyvinyl alcohol
ISI	Industrial Research Institute
bcc	Body-centered-cubic
JAXA	Japan Aerospace Exploration Agency
ICETT	International Center for Environmental Technology Transfer

References

1. *Global xEV Batteries Market: Key Research Findings 2014*; Yano Research Institute: Tokyo, Japan, 2015.
2. Chang, S.; Young, K.; Nei, J.; Fierro, C. Reviews on the U.S. Patents regarding nickel/metal hydride batteries. *Batteries* **2016**, *2*.
3. Webpage of Japan Platform for Patent Information. Available online: https://www4.j-platpat.inpit.go.jp/eng/tokujitsu/tkbs_en/TKBS_EN_GM101_Top.action (accessed on 24 June 2016).
4. Panasonic EVOLTA NiMH. Available online: http://www.panasonic.com/global/consumer/battery/panasonic_rechargeable.html (accessed on 9 April 2016).
5. Yamamoto, T.; Komori, K.; Suzuki, G.; Yamaguchi, S.; Kimura, T.; Ikoma, M.; Toyoguchi, Y. Production of Hydrogen Storage Alloy Powder. Jpn. Pat. Appl. H07-048602, 21 February 1995.
6. Yamamoto, T.; Tsuji, M.; Komori, K.; Suzuki, G.; Yamaguchi, S.; Toyoguchi, Y. Production of Hydrogen Storage Alloy Powder and Nickel-Hydrogen Battery. Jpn. Pat. Appl. H07-054016, 28 February 1995.
7. Furuike, K.; Ebihara, T.; Tanaka, A.; Yuasa, K.; Kaiya, H. Sealed Alkaline Battery. Jpn. Pat. Appl. H9-213319, 15 August 1997.
8. Izumi, Y.; Moriwaki, Y.; Yamashita, K.; Tokuhiro, T. Nickel-Hydrogen Storage Battery. Jpn. Pat. Appl. H09-231969, 5 September 1997.
9. Imamura, K.; Oomura, A.; Yuasa, K.; Kaiya, H. Hydrogen Storage Alloy Powder for Battery and Manufacture Thereof. Jpn. Pat. Appl. H09-245784, 19 September 1997.
10. Hayashi, S.; Tomioka, K.; Morishita, N.; Ikeyama, S.; Ikoma, M. Nickel Positive Electrode for Alkaline Storage Battery and Nickel Hydrogen Storage Battery Using This Electrode. Jpn. Pat. Appl. H10-064537, 6 March 1998.
11. Hayashi, S.; Tomioka, K.; Morishita, N.; Ikoma, M. Nickel Positive Electrode and Nickel-Hydrogen Storage Battery Using It. Jpn. Pat. Appl. H10-106556, 24 April 1998.
12. Yuasa, S.; Komori, K.; Matsuda, H.; Ikoma, M. Alkaline Storage Battery. Jpn. Pat. Appl. H10-106620, 24 April 1998.
13. Kasahara, H.; Yao, T.; Dansui, Y.; Konishi, H. Alkaline Storage Battery and Manufacture of Its Positive Electrode Active Material. Jpn. Pat. Appl. H11-185749, 9 July 1999.
14. Kasahara, H.; Yao, T.; Suzuki, T.; Masui, M.; Konishi, H. Nickel-Hydrogen Storage Battery for Backup Power Supply. Jpn. Pat. Appl. H11-329481, 30 November 1999.
15. Kikuyama, T.; Ebihara, T.; Miyahara, A.; Ou, S.; Yuasa, K. Hydrogen Storage Alloy for Battery, Its Manufacture and Alkaline Storage Battery Using It. Jpn. Pat. Appl. H11-339793, 10 December 1999.
16. Ebihara, T.; Kikuyama, T.; Miyahara, A.; Ou, S.; Yuasa, K. Alkaline Storage Battery, Hydrogen Storage Alloy Electrode and Its Manufacture. Jpn. Pat. Appl. 2000-090920, 31 March 2000.
17. Sakai, T. Secondary Battery with Hydrogen Storage Alloys. In *Hydrogen Storage Alloys—Fundamentals and Frontier Technology*; Tamura, H., Ed.; NTS Inc.: Tokyo, Japan, 1998; p. 411.

18. Wikipedia, The Free Encyclopedia. Eneloop. Available online: https://en.wikipedia. org/wiki/Eneloop (accessed on 9 April 2016).

19. Yasuoka, S.; Magari, Y.; Murata, T.; Tanaka, T.; Ishida, J.; Nakamura, H.; Nohma, T.; Kihara, M.; Baba, Y.; Teraoka, H. Development of high-capacity nickel-metal hydride batteries using superlattice hydrogen-absorbing alloys. *J. Power Sources* **2006**, *156*, 662–666.

20. Teraoka, H. Development of Low Self-Discharge Nickel-Metal Hydride Battery. Available online: http://www.scribd.com/doc/9704685/Teraoka-Article-En (accessed on 24 June 2016).

21. Takizawa, Y.; Ueda, T.; Kanekawa, I. Metal Hydride Storage Battery. Jpn. Pat. Appl. H05-242908, 21 September 1993.

22. Shiraiwa, S.; Ueda, T.; Takizawa, Y. Metal Hydride Battery. Jpn. Pat. Appl. H07-065825, 10 March 1995.

23. Yamawaki, A.; Nakahori, S.; Hamamatsu, T.; Baba, Y. Non-Sintered Nickel Electrode for Alkaline Storage Battery, Manufacture Thereof and Alkaline Storage Battery. Jpn. Pat. Appl. H08-148146, 7 June 1996.

24. Nogami, M.; Matsuura, Y.; Kimoto, M.; Higashiyama, N.; Tokuda, M.; Isono, T.; Yonezu, I.; Nishio, K. Hydrogen Storage Alloy Electrode for Alkaline Storage Battery. Jpn. Pat. Appl. H09-147903, 6 June 1997.

25. Ise, T. Metal Hydride Storage Battery. Jpn. Pat. Appl. H09-245782, 19 September 1997.

26. Niiyama, K.; Tokuda, M.; Satoguchi, K.; Yano, M.; Nogami, M.; Yonezu, I.; Nishio, K. Alkaline Storage Battery. Jpn. Pat. Appl. H10-021904, 23 January 1998.

27. Satoguchi, K.; Tokuda, M.; Niiyama, K.; Yano, M.; Nogami, M.; Yonezu, I.; Nishio, K. Manufacture of Unsintered Nickel Electrode for Alkaline Storage Battery. Jpn. Pat. Appl. H10-188970, 21 July 1998.

28. Suzuki, S.; Tokuda, M.; Kimoto, M.; Yano, M.; Fujitani, S.; Nishio, K. Sealed Alkaline Storage Battery. Jpn. Pat. Appl. H10-214621, 11 August 1998.

29. Matsuura, Y.; Nogami, M.; Maeda, R.; Niiyama, K.; Yonezu, I.; Nishio, K. Hydrogen Storage Alloy Electrode for Alkaline Storage Battery, and the Alkaline Storage Battery Using Thereof. Jpn. Pat. Appl. H11-031504, 2 February 1999.

30. Kamiyoshi, K.; Ozaki, K.; Otsuki, K. Alkaline Storage Battery. Jpn. Pat. Appl. H09-213342, 15 August 1997.

31. Niiyama, K.; Yano, M.; Maeda, R.; Nogami, M.; Yonezu, I.; Nishio, K. Non-Sintered Nickel Electrode for Alkaline Storage Battery. Jpn. Pat. Appl. H09-259878, 3 October 1997.

32. Uenae, K. Hydrogen Storage Alloy Electrode and Alkaline Secondary Cell Using It. Jpn. Pat. Appl. H05-159798, 25 June 1993.

33. Horiie, H.; Nagai, T. Cylindrical Alkaline Secondary Battery. Jpn. Pat. Appl. H05-226001, 3 September 1993.

34. Hattori, H.; Kido, H.; Ishida, O.; Nagai, T. Alkaline Secondary Battery. Jpn. Pat. Appl. H05-290841, 5 November 1993.

35. Edamoto, T.; Wada, S. Nickel Hydride Secondary Cell. Jpn. Pat. Appl. H08-050918, 20 February 1996.

36. Fukunaga, H.; Nagai, T. Hydride Secondary Battery. Jpn. Pat. Appl. H08-190931, 23 July 1996.

37. Fukunaga, H.; Tamakoshi, H.; Nagai, T.; Tateishi, S. Hydride Secondary Battery and Manufacture Thereof. Jpn. Pat. Appl. H10-040910, 13 February 1998.

38. Takai, M.; Fukunaga, H.; Nagai, T. Assembled Battery of Hydride Secondary Battery. Jpn. Pat. Appl. H10-208768, 7 August 1998.

39. Fukunaga, H.; Nagai, T. Closed Type Hydride Secondary Battery. Jpn. Pat. Appl. H11-149939, 2 June 1999.

40. Fukunaga, H.; Tamakoshi, H.; Fujimoto, Y.; Nagai, T. Hydride Secondary Battery and Its Manufacture. Jpn. Pat. Appl. 2000-149934, 30 May 2000.

41. Ono, H.; Tamakoshi, H.; Fukunaga, H.; Nagai, T. Alkaline Storage Battery. Jpn. Pat. Appl. 2000-353520, 19 December 2000.

42. Ono, H.; Tamakoshi, H.; Fukunaga, H.; Nagai, T. Alkaline Storage Battery. Jpn. Pat. Appl. 2000-353542, 19 December 2000.

43. Oshitani, M.; Yufu, H. Nickel Electrode for Alkaline Battery. Jpn. Pat. Appl. H01-272050, 31 October 1989.

44. GS Yuasa. ENiTIME: Sealed Nickel-Metal Hydride Rechargeable Battery. Available online: http://pdf.directindustry.com/pdf/gs-yuasa/enitime/12414-296249.html (accessed on 9 April 2016).

45. Tomita, M. Alkaline Battery. Jpn. Pat. Appl. S55-009354, 23 January 1980.

46. Watada, M.; Matsumura, Y.; Miyake, N.; Oshitani, M. Electrode for Alkaline Storage Battery and Alkaline Storage Battery Using This Electrode. Jpn. Pat. Appl. H06-314567, 8 November 1994.

47. Bougauchi, T.; Nakagawa, H.; Kishimoto, N.; Yamane, M. Nickel Electrode Plate and Manufacture Thereof and Alkaline Storage Battery Using It. Jpn. Pat. Appl. H07-114920, 2 May 1995.

48. Tani, A. Nickel-Hydrogen Storage Battery. Jpn. Pat. Appl. H08-115739, 7 May 1996.

49. Ito, T.; Harada, T.; Arahi, K.; Yufu, H. Forming Method of Sealed Alkaline Storage Battery. Jpn. Pat. Appl. H08-153543, 11 June 1996.

50. Furukawa, K.; Tanaka, T.; Onishi, M.; Oshitani, M. Alkaline Storage Battery. Jpn. Pat. Appl. H09-092279, 4 April 1997.

51. Tani, A.; Kurokuzuhara, M. Separator for Nickel-Hydrogen Storage Battery. Jpn. Pat. Appl. H10-031990, 3 February 1998.

52. Matsumura, Y.; Tanaka, T.; Kurokuzuhara, M.; Tani, A.; Watada, M.; Oshitani, M. Sealed Nickel-Hydrogen Storage Battery. Jpn. Pat. Appl. H10-074536, 17 March 1998.

53. Kanemoto, M.; Kodama, M.; Kurokuzuhara, M. Hydrogen-Storing Alloy Electrode and Nickel Hydrogen Battery Using Same. Jpn. Pat. Appl. 2001-283854, 12 October 2001.

54. Kanemoto, M.; Kodama, M.; Kurokuzuhara, M.; Watada, M. Hydrogen Storage Alloy Electrode. Jpn. Pat. Appl. 2002-042800, 8 February 2002.

55. Kanemoto, M.; Kurokuzuhara, M.; Kodama, M.; Sakamoto, K.; Watada, M. Hydrogen Occlusion Alloy Powder, Hydrogen Occlusion Alloy Electrode, and Nickel-Hydrogen Storage Battery Using the Same. Jpn. Pat. Appl. 2004-124132, 22 April 2004.

56. Hasebe, H.; Takeno, K.; Ikeda, K.; Sato, Y. Nickel Hydrogen Secondary Cell Module. Jpn. Pat. Appl. H05-036392, 12 February 1993.

57. Uchiyama, M.; Takeno, K. Alkaline Secondary Battery. Jpn. Pat. Appl. H07-211313, 11 August 1995.

58. Tsuruta, S.; Kono, R.; Kanda, M. Hydrogen Storage Alloy and Alkaline Secondary Battery. Jpn. Pat. Appl. H10-251782, 22 September 1998.

59. Kono, R.; Kanda, M. Hydrogen Storage Alloy, Cathode for Battery and Alkaline Secondary Battery. Jpn. Pat. Appl. H10-251791, 22 September 1998.

60. Mukai, K.; Ishizuka, S.; Takeno, K. Alkaline Secondary Battery. Jpn. Pat. Appl. H10-294106, 4 November 1998.

61. Hosobuchi, K.; Gama, M. Metallic Oxide Hydrogen Secondary Battery. Jpn. Pat. Appl. H07-326354, 12 December 1995.

62. Hayashida, H.; Kitayama, H.; Yamamoto, M.; Sakai, I.; Kono, R.; Yoshida, H.; Inaba, T.; Inada, S.; Kanda, M. Nickel-Hydrogen Secondary Battery. Jpn. Pat. Appl. H11-162460, 18 June 1999.

63. Kitayama, H.; Hayashida, H.; Yamamoto, M.; Sakai, I.; Kono, R.; Yoshida, H.; Inaba, T.; Inada, S.; Kanda, M. Nickel-Hydrogen Secondary Battery. Jpn. Pat. Appl. H11-162503, 18 June 1999.

64. Kono, R.; Sakai, I.; Yoshida, H.; Inaba, T.; Yamamoto, M.; Takeno, S. Hydrogen Storage Alloy and Secondary Battery. Jpn. Pat. Appl. 2000-073132, 7 March 2000.

65. Kono, R.; Sakai, I.; Yoshida, H.; Inaba, T.; Yamamoto, M. Hydrogen Storage Alloy and Secondary Battery. Jpn. Pat. Appl. 2000-265228, 26 September 2000.

66. Yoshida, H.; Yamamoto, M.; Sakai, I.; Inaba, T.; Takabayashi, J.; Irie, S.; Suzuki, H.; Takeno, K. Hydrogen Storage Alloy, Alkali Secondary Battery, Hybrid Car and Electric Vehicle. Jpn. Pat. Appl. 2002-069554, 8 March 2002.

67. Inaba, T.; Sakai, I.; Yoshida, H.; Takabayashi, J.; Yamamoto, M.; Suzuki, H.; Irie, S.; Takeno, K. Nickel Hydrogen Secondary Battery, Hybrid Car and Electric Vehicle. Jpn. Pat. Appl. 2002-083593, 22 March 2002.

68. Kawashima, F.; Sakamoto, T.; Arai, T. Hydrogen Storage Alloy and Nickel-Hydrogen Secondary Battery Using the Same. Jpn. Pat. Appl. 2002-105563, 10 April 2002.

69. Sakamoto, T.; Kawashima, F.; Arai, T. Hydrogen Storage Alloy, Its Production Method and Nickel-Hydrogen Secondary Battery Using the Same. Jpn. Pat. Appl. 2002-105564, 10 April 2002.

70. Sakai, I.; Inaba, T.; Yoshida, H.; Yamamoto, M.; Irie, S.; Suzuki, H.; Takeno, K. Hydrogen Storage Alloy, Secondary Battery, Hybrid Vehicle, and Electric Vehicle. Jpn. Pat. Appl. 2002-164045, 7 June 2002.

71. Saguchi, A.; Kihara, M.; Endo, T. Negative Electrode for Alkaline Secondary Battery, and Alkaline Secondary Battery Comprising the Negative Electrode. Jpn. Pat. Appl. 2012-134110, 12 July 2012.

72. Ishida, J.; Yasuoka, S.; Inui, H.; Kishida, K. Hydrogen Absorbing Alloy and Alkaline Storage Battery Manufactured Using the Hydrogen Absorbing Alloy. Jpn. Pat. Appl. 2012-174639, 10 September 2012.

73. Kihara, M.; Endo, T.; Sato, T.; Saguchi, A.; Wada, S.; Mugima, I.; Nakamura, T.; Asanuma, H.; Tamura, M. Negative Electrode for Nickel-Hydrogen Secondary Battery, and Nickel-Hydrogen Secondary Battery Using the Negative Electrode. Jpn. Pat. Appl. 2012-256522, 27 December 2012.

74. Kihara, M.; Takei, M.; Yamane, T. Nickel-Hydrogen Secondary Battery. Jpn. Pat. Appl. 2013-030345, 7 February 2013.

75. Ishida, J. Hydrogen Storage Alloy and Nickel-Hydrogen Secondary Battery Using the Same. Jpn. Pat. Appl. 2013-100585, 23 May 2013.

76. Ishida, J.; Yasuoka, S. Hydrogen Storage Alloy and Nickel-Hydrogen Secondary Battery Using the Same. Jpn. Pat. Appl. 2013-108105, 6 June 2013.

77. Ishida, J.; Kai, T. Hydrogen Storage Alloy and Nickel Hydride Secondary Battery Using the Hydrogen Storage Alloy. Jpn. Pat. Appl. 2014-145122, 14 August 2014.

78. Yamane, T.; Takei, M.; Imoto, Y.; Ito, T. Nickel Hydrogen Secondary Battery. Jpn. Pat. Appl. 2013-206867, 7 October 2013.

79. Kai, T.; Ishida, J. Nickel Hydrogen Secondary Battery and Negative Electrode for Nickel Hydrogen Secondary Battery. Jpn. Pat. Appl. 2014-026844, 6 February 2014.

80. Kai, T.; Ishida, J. Nickel Hydrogen Secondary Battery. Jpn. Pat. Appl. 2014-146557, 14 August 2014.

81. Kihara, M.; Saguchi, A.; Takei, M.; Ito, T.; Imoto, Y. Nickel Hydrogen Secondary Battery. Jpn. Pat. Appl. 2014-089879, 15 May 2014.

82. Ishida, J.; Kai, T. Negative Electrode for Nickel-Hydrogen Secondary Battery, and Nickel-Hydrogen Secondary Battery Using the Same. Jpn. Pat. Appl. 2014-207086, 30 October 2014.

83. Sato, T. Nickel Hydrogen Storage Battery. Jpn. Pat. Appl. 2015-008107, 15 January 2015.

84. Kai, T.; Ishida, J. Nickel Hydrogen Secondary Battery. Jpn. Pat. Appl. 2015-103497, 4 June 2015.

85. Ishida, J.; Kai, T. Nickel Hydrogen Secondary Battery. Jpn. Pat. Appl. 2015-195108, 5 November 2015.

86. Takasu, D. Negative Electrode for Nickel Hydrogen Secondary Battery and Nickel Hydrogen Secondary Battery Using This Negative Electrode. Jpn. Pat. Appl. 2015-201334, 12 November 2015.

87. Ushara, I.; Sakai, T.; Ishikawa, H. The state of research and development for applications of metal hydrides in Japan. *J. Alloys Compd.* **1997**, *253–254*, 635–641.

88. Furukawa, A. Negative Electrode for Nickel-Hydrogen Secondary Battery. Jpn. Pat. Appl. H06-231758, 19 August 1994.

89. Furukawa, A. Nickel-Hydrogen Secondary Battery. Jpn. Pat. Appl. H06-231761, 19 August 1994.

90. Furukawa, A. Manufacture of Sealed Nickel-Hydrogen Secondary Battery. Jpn. Pat. Appl. H06-251800, 9 September 1994.

91. Furukawa, A. Production of Powdery Hydrogen Occluding Alloy for Negative Electrode of Nickel-Hydrogen Secondary Battery and Production of Negative Electrode for Nickel-Hydrogen Secondary Battery. Jpn. Pat. Appl. H06-279980, 4 October 1994.

92. Furukawa, A. Nickel-Hydrogen Secondary Battery. Jpn. Pat. Appl. H06-283195, 7 October 1994.

93. Furukawa, A. Sealed Nickel-Hydrogen Secondary Battery. Jpn. Pat. Appl. H06-283196, 7 October 1994.

94. Murata, T. Sealed Battery. Jpn. Pat. Appl. H05-021045, 29 January 1993.

95. Murata, T. Enclosed Type Nickel/Metal Hydride Storage Battery. Jpn. Pat. Appl. H05-041204, 19 February 1993.

96. Murata, T. Nickel-Metal Hydride Storage Battery. Jpn. Pat. Appl. H05-121073, 18 May 1993.

97. Shichimoto, K. Manufacture of Nickel Metal Hydride Cell. Jpn. Pat. Appl. H08-050919, 20 February 1996.

98. Nanamoto, K. Sealed Alkaline Storage Battery. Jpn. Pat. Appl. H09-199162, 31 July 1997.

99. Nomura, Y.; Ogura, T.; Kobayashi, K.; Tsuda, T. Hydrogen Storage Alloy Electrode and Manufacture Thereof. Jpn. Pat. Appl. H05-003029, 8 January 1993.

100. Konuki, T.; Kobayashi, K. Hydrogen Storage Alloy Electrode and Manufacture Thereof. Jpn. Pat. Appl. H06-163043, 10 June 1994.

101. Yamaguchi, T.; Watanabe, K.; Yoshida, M.; Kamigata, Y. Hydrogen Storage Electrode and Manufacture Thereof. Jpn. Pat. Appl. H08-050898, 20 February 1996.

102. Kobayashi, K.; Tamagawa, T. Hydrogen Storage Alloy Electrode and Nickel-Hydrogen Storage Battery. Jpn. Pat. Appl. H08-264175, 11 October 1996.

103. Minoura, S.; Kobayashi, K.; Ogura, T. Hydrogen Storage Alloy Electrode for Alkaline Storage Battery. Jpn. Pat. Appl. H10-092422, 10 April 1998.

104. Minamiura, K. Method and Device for Sensing Failure in Battery Pack System. Jpn. Pat. Appl. 2003-142165, 16 May 2003.

105. Minamiura, K. Controlling Method and Device for Cooling of Battery. Jpn. Pat. Appl. 2003-142166, 16 May 2003.

106. Yudahira, H. Battery Power Unit and Its Current Detecting Method. Jpn. Pat. Appl. 2003-168488, 13 June 2003.

107. Yudahira, H. Electric Leakage Detecting Device. Jpn. Pat. Appl. 2003-194870, 9 July 2003.

108. Yudahira, H. Voltage Measurement Device and Method, as well as Battery Pack System. Jpn. Pat. Appl. 2003-197273, 11 July 2003.

109. Murakami, T. Estimating Method of Polarized Voltage of Secondary Battery, Estimating Method and Device of Residual Capacity of Secondary Battery, as well as Battery Pack System. Jpn. Pat. Appl. 2003-197275, 11 July 2003.

110. Ueda, T.; Morishita, N.; Okawa, K.; Nakao, Y. Remaining Capacity Arithmetic Unit and Remaining Capacity Computing Method of Secondary Battery. Jpn. Pat. Appl. 2004-361313, 24 December 2004.

111. Morimoto, N. Testing Method for Relay Contact Welding in Battery Power Supply. Jpn. Pat. Appl. 2003-209907, 25 July 2003.

112. Nakanishi, T.; Torii, Y. Controller for Motorized Vehicle, Motorized Vehicle Equipped with It. Jpn. Pat. Appl. 2004-048937, 12 February 2004.

113. Katakura, Y.; Kajikawa, T. Manufacturing Method of Nickel Metal Hydride Storage Battery. Jpn. Pat. Appl. 2010-153261, 8 July 2010.

114. Kawasaki Heavy Industries. Battery Energy Storage System-GIGACELL. Available online: https://global.kawasaki.com/en/energy/solutions/battery_energy/ (accessed on 9 April 2016).

115. Tsutsumi, K. Three-Dimensional Battery. Jpn. Pat. Appl. 2002-141101, 17 May 2002.

116. Tsutsumi, K.; Nishimura, K.; Mitsuta, S. Active Material for Battery and Its Manufacturing Method. Jpn. Pat. Appl. 2003-197187, 11 July 2003.

117. Tsutsumi, K.; Nishimura, K.; Mitsuta, S. Electrode Using Fibrous Hydrogen Storage Alloy, Battery Using Fibrous Hydrogen Storage Alloy and Electric Double Layer Capacitor. Jpn. Pat. Appl. 2004-022332, 22 January 2004.

118. Tsutsumi, K.; Nishimura, K. Battery. Jpn. Pat. Appl. 2008-041522, 21 February 2008.

119. Nishimura, K.; Tsutsumi, K. Square Battery. Jpn. Pat. Appl. 2010-129360, 10 June 2010.

120. Tsutsumi, K.; Nakoji, M.; Origuchi, T.; Ogawa, S.; Nakayama, N. Alkaline Storage Battery. Jpn. Pat. Appl. 2010-177071, 12 August 2010.

121. Ishida, T.; Sugiyama, S. Charger for Battery in Railroad Vehicle. Jpn. Pat. Appl. 2008-172857, 24 July 2008.

122. Koyano, K.; Miyamoto, Y.; Hayashi, M.; Sawai, T.; Yoshiyama, E.; Tsutsumi, K. Charged State Estimation Method and Device for Secondary Battery. Jpn. Pat. Appl. 2009-244057, 22 October 2009.

123. Nishimura, K.; Tsutsumi, K. Pressure Regulating Device of Battery. Jpn. Pat. Appl. 2009-301888, 24 December 2009.

124. Nishimura, K.; Takagaki, K.; Ugawa, K.; Ide, T. Battery Module. Jpn. Pat. Appl. 2013-168216, 29 August 2013.

125. Nishimura, K. Battery System. Jpn. Pat. Appl. 2013-143295, 22 July 2013.

126. Eguro, T.; Hashiguchi, J.; Koga, Y. Chemical Forming Method for High Energy Type Fe-Ni Battery. Jpn. Pat. Appl. H02-075168, 14 March 1990.

127. Lichtenberg, F.; Kleinsorgen, K.; Hofmann, G. Nickel/Metal Hydride Secondary Battery. Jpn. Pat. Appl. H07-211317, 11 August 1995.

128. Knosp, B.; Bouet, J.; Jordy, C.; Mimoun, M.; Gicquel, D. Material Capable of Hydrogenation for Negative Electrode of Nickel-Metal Hydride Storage Battery. Jpn. Pat. Appl. H08-017435, 19 January 1996.

129. Kang, S. Nickel Metal Hydride Storage Battery and Manufacture Thereof. Jpn. Pat. Appl. H09-171818, 30 June 1997.

130. Fetcenko, M.; Ovshinsky, S.; Chao, B.; Reichman, B. Improved Electrochemical Hydrogen Storage Alloy for Nickel Hydride Metal Battery. Jpn. Pat. Appl. 2000-144288, 26 May 2000.

131. Ovshinsky, S.; Fetcenko, M.; Im, J.; Young, K.; Chao, B.; Reichman, B. Hydrogen Occluding Material Having Abnormal Site Capable of Occluding Hydrogen at High Density. Jpn. Pat. Appl. 2002-088430, 27 March 2002.

132. Fetcenko, M.; Ovshinsky, S.; Chao, B.; Reichman, B. Improved Electrochemical Hydrogen Storage Alloy for Nickel Hydride Metal Battery. Jpn. Pat. Appl. 2002-088436, 27 March 2002.

133. Ovshinsky, S.; Fetcenko, M.; Im, J.; Young, K.; Chao, B.; Reichman, B. Hydrogen Storage Material Having High Density of Non-Conventional Usable Hydrogen Storing Sites. Jpn. Pat. Appl. 2002-241874, 28 August 2002.

134. Fetcenko, M.; Young, K.; Ovshinsky, S.; Reichman, B.; Koch, J.; Mays, W. Modified Electrochemical Hydrogen Storage Alloy Having Increased Capacity, Rate Capability and Catalytic Activity. Jpn. Pat. Appl. 2006-183148, 13 July 2006.

135. Ovshinsky, S.; Fetcenko, M. Secondary Battery Fabricated of Electrochemical Hydrogen Storage Alloy and Mg-Contained Base Alloy. Jpn. Pat. Appl. 2003-217578, 31 July 2003.

136. Ovshinsky, S.; Fetcenko, M.; Reichman, B.; Young, K.; Chao, B.; Im, J. Electrochemical Hydrogen Storage Alloy and Battery Fabricated from Magnesium-Containing Base Alloy. Jpn. Pat. Appl. 2003-247038, 5 September 2003.

137. Fetcenko, M.; Young, K.; Ouchi, T.; Reinhout, M.; Ovshinsky, S. Mg-Ni Hydrogen Storage Composite Having High Storage Capacity and Excellent Room Temperature Kinetics. Jpn. Pat. Appl. 2007-119906, 17 May 2007.

138. Fetcenko, M.; Fierro, C.; Ovshinsky, S.; Sommers, B.; Reichman, B.; Young, K.; Mays, W. Composite Positive Electrode Material and Its Manufacturing Method. Jpn. Pat. Appl. 2004-214210, 29 July 2004.

139. Ovshinsky, S.; Fetcenko, M.; Fierro, C.; Gifford, P.; Corrigan, D.; Benson, P.; Martin, F. High Performance Nickel Hydroxide Positive Electrode Electrode Material for Alkaline Rechargeable Electrochemical Battery. Jpn. Pat. Appl. 2006-054190, 23 February 2006.

140. Fetcenko, M.; Fierro, C.; Ovshinsky, S.; Sommers, B.; Reichman, B.; Young, K.; Mays, W. Composite Positive Electrode Material and Method for Making Same. Jpn. Pat. Appl. 2010-282973, 16 December 2010.

141. Corrigan, D.; Gow, P.; Higley, L.; Muller, M.; Osgood, A.; Ovshinsky, S.; Payne, J.; Puttaiah, R. Monoblock Battery Assembly. Jpn. Pat. Appl. 2010-225591, 7 October 2010.

142. Wolff, M.; Nuss, M.; Fetcenko, M.; Lijoi, A. Continuous Manufacture of Hydrogen Storage Alloy Cathode. Jpn. Pat. Appl. H01-286255, 17 November 1989.

143. Imaizumi, J.; Kawasaki, Y.; Makino, T.; Iida, T. Alpha-Cobalt Hydroxide Layer-Coated Nickel Hydroxide for Alkaline Storage Battery and Manufacture Thereof. Jpn. Pat. Appl. H10-012236, 16 January 1998.

144. Imaizumi, J.; Kawasaki, Y.; Makino, T.; Iida, T. Beta-Cobalt Hydroxide Layer-Coated Nickel Hydroxide for Alkaline Storage Battery and Manufacture Thereof. Jpn. Pat. Appl. H10-012237, 16 January 1998.

145. Watada, M.; Oshitani, M.; Imaizumi, J.; Iida, T. Positive Electrode Active Material for Alkali Storage Battery, Its Manufacture and Positive Electrode for Alkali Storage Battery. Jpn. Pat. Appl. H10-027608, 27 January 1998.

146. Usui, T.; Makino, T.; Iida, T. Nickel Hydroxide for Alkaline Storage Battery and Manufacture Thereof. Jpn. Pat. Appl. H10-097856, 14 April 1998.

147. Watada, M.; Oshitani, M.; Imaizumi, J.; Iida, T. Positive Electrode Active Material for Alkaline Storage Battery and Manufacture Thereof, and Positive Electrode for Alkaline Storage Battery. Jpn. Pat. Appl. H10-188973, 21 July 1998.

148. Sakai, T.; Ishihara, K.; Imaizumi, J. Nickel Positive Electrode for Alkaline Secondary Battery. Jpn. Pat. Appl. 2000-077068, 14 March 2000.

149. Hosoe, A.; Imaizumi, J.; Iida, T. Nickel Positive Electrode Active Material for Alkaline Battery and Its Manufacture. Jpn. Pat. Appl. 2000-082463, 21 March 2000.

150. Tanaka, T.; Makino, T.; Iida, T. Manufacture of Positive Electrode Active Material for Alkaline Storage Battery. Jpn. Pat. Appl. 2000-268820, 29 September 2000.

151. Hirayama, S. Ingot for Hydrogen Storage Alloy Powder and Production of the Powder. Jpn. Pat. Appl. H04-358008, 11 December 1992.

152. Kuji, T.; Kitakado, M.; Yasuda, K.; Hanawa, K.; Nitta, S.; Dobashi, M. Method for Surface-Modifying Hydrogen Storage Alloy. Jpn. Pat. Appl. H07-316610, 5 December 1995.

153. Sasaki, M.; Hirayama, S.; Sumimoto, S. Hydrogen Storage Alloy and Its Production. Jpn. Pat. Appl. H09-031573, 4 February 1997.

154. Yasuda, K.; Nakayama, S. Hydrogen Storage Alloy. Jpn. Pat. Appl. H09-316573, 9 December 1997.

155. Sakaguchi, Y.; Nakayama, S.; Yasuda, K. Hydrogen Storage Alloy and Its Production. Jpn. Pat. Appl. H10-088261, 7 April 1998.

156. Sumimoto, S.; Sakai, M.; Uchiyama, A.; Hirayama, S.; Yasuda, K.; Nakayama, S.; Ebihara, T.; Yuasa, K. Hydrogen Storage Alloy and Electrode for Nickel-Hydrogen Battery Using It. Jpn. Pat. Appl. H10-152739, 9 June 1998.

157. Yasuda, K.; Hirayama, S.; Sakai, M.; Uchiyama, A.; Sakaguchi, Y.; Nakayama, S. Hydrogen Storage Alloy and Its Production. Jpn. Pat. Appl. H11-152533, 8 June 1999.

158. Yasuda, K.; Sakaguchi, Y.; Uchiyama, A.; Mukai, D.; Kikukawa, S. Hydrogen Storage Alloy and Its Manufacture. Jpn. Pat. Appl. 2000-234133, 29 August 2000.

159. Yasuda, K.; Sakaguchi, Y.; Kikukawa, S. Hydrogen Storage Alloy and Its Production Method. Jpn. Pat. Appl. 2001-348636, 18 December 2001.

160. Toma, H. Manufacture of Hydrogen Occluding Alloy of Rare Earth Metal-Nickel System. Jpn. Pat. Appl. S59-140301, 11 August 1984.

161. Yamamoto, K.; Miyake, Y.; Okada, T.; Kitatsume, N. Rare Earth-Nickel Hydrogen Storage Alloy Ingot and Its Production. Jpn. Pat. Appl. H05-320792, 3 December 1993.

162. Nishigaki, N.; Kitatsume, N.; Okada, T. Method for Preserving Rare Earth-Transition Metal Alloy. Jpn. Pat. Appl. H06-264102, 20 September 1994.

163. Kaneko, A. Rare Earth Metal-Nickel Based Hydrogen Storage Alloy and Its Production, and Cathode for Nickel-Hydrogen Secondary Battery. Jpn. Pat. Appl. H09-025529, 28 January 1997.

164. Yamamoto, K.; Okada, T. Production System of Alloy Containing Rare Earth Metal. Jpn. Pat. Appl. H09-155507, 17 June 1997.

165. Oota, T.; Yamaguchi, M.; Gashiyuu, S.; Noda, K.; Oku, K.; Konno, H.; Sasai, K. Hydrogen Occluding Alloy and Its Manufacture. Jpn. Pat. Appl. S56-136957, 26 October 1981.

166. Noda, K.; Oku, K.; Konno, H.; Sasai, K.; Onoe, K.; Kashiyuu, S. Material for Storage of Hydrogen. Jpn. Pat. Appl. S57-075138, 11 May 1982.

167. Sasai, O.; Hayamizu, N.; Uotani, S. Hydrogen Storage Material. Jpn. Pat. Appl. S63-047345, 29 February 1988.

168. Uemura, M.; Yamamoto, I.; Tsubata, A.; Hayashi, T. Production of Nickel Hydroxide. Jpn. Pat. Appl. S64-042330, 14 February 1989.

169. Tomioka, H.; Matsubara, Y.; Hayamizu, N. Hydrogen Storage Alloy for Nickel/Hydrogen Battery. Jpn. Pat. Appl. H07-029570, 31 January 1995.

170. Uehara, H.; Ishikawa, H.; Nishida, J.; Sakuma, T.; Saito, N. Electrode for Alkaline Secondary Battery. Jpn. Pat. Appl. H07-073885, 17 March 1995.

171. Sasai, K.; Hayamizu, N.; Nakamura, M.; Kenmochi, Y.; Honda, J. Method for Recovering Valuable Material from Scrapped Nickel Hydrogen Occluding Alloy Secondary Battery. Jpn. Pat. Appl. H08-020825, 23 January 1996.

172. Kamisaka, K.; Nakayama, Y.; Nishida, J.; Igarashi, K.; Sakuma, T. Electrode Material of Silver Plated Nickel Based Porous Metal and Its Production. Jpn. Pat. Appl. H08-337894, 24 December 1996.

173. Takahashi, S.; Osawa, M.; Shimizu, H. Production of Magnesium-Yttrium Hydrogen Storage Alloy. Jpn. Pat. Appl. H09-125172, 13 May 1997.

174. Saito, N.; Takahashi, M.; Sasai, K. Hydrogen Storage Alloy for Battery. Jpn. Pat. Appl. H09-298059, 18 November 1997.

175. Obuchi, I.; Haraikawa, N. Production of Powder-Type Hydrogen-Storing Alloy. Jpn. Pat. Appl. H09-316505, 9 December 1997.

176. Sanoki, H.; Sugimoto, T.; Kudo, K. Method for Stabilizing Hydrogen Storage Alloy. Jpn. Pat. Appl. H10-195503, 28 July 1998.

177. Terashita, N.; Takahashi, M.; Kobayashi, K.; Sasai, K. Production of Amorphous Magnesium-Nickel Base Hydrogen Storage Alloy. Jpn. Pat. Appl. H11-269572, 5 October 1999.

178. Osawa, M.; Muromachi, N. Surface Treatment of Hydrogen Storage Alloy. Jpn. Pat. Appl. H11-335867, 7 December 1999.

179. Sakai, T.; Ikenaga, M.; Nishino, H.; Nishida, J. Alkaline Secondary Battery Electrode Substrate. Jpn. Pat. Appl. 2000-040516, 8 February 2000.

180. Yoshikawa, T.; Osawa, M.; Muromachi, N.; Endo, T.; Ogura, H. Negative Electrode for Secondary Battery. Jpn. Pat. Appl. 2000-090919, 31 March 2000.

181. Kobayashi, K.; Ogura, H.; Osawa, M.; Muromachi, N.; Harada, R.; Kimura, M.; Toyoshima, H. Hydrogen Storage Alloy Negative Electrode. Jpn. Pat. Appl. 2000-182608, 30 June 2000.

182. Saito, N.; Sugimoto, T.; Haneda, T.; Osawa, M.; Yoshikawa, T. Hydrogen Storage Alloy. Jpn. Pat. Appl. 2001-279354, 10 October 2001.

183. Saito, N.; Sugimoto, T.; Haneda, T.; Osawa, M.; Yoshikawa, T. Hydrogen Storage Alloy for Secondary Battery. Jpn. Pat. Appl. 2001-279355, 10 October 2001.

184. Terashita, N.; Ito, N.; Osawa, M.; Takahashi, S.; Tsunokake, S.; Hamura, K. Hydrogen Storage Alloy, Its Manufacturing Method and Nickel Hydrogen Secondary Battery. Jpn. Pat. Appl. 2007-056309, 8 March 2007.

185. Osawa, M.; Ito, N.; Terashita, N.; Takahashi, S.; Tsunokake, S. Hydrogen Storage Alloy and Nickel-Hydride Secondary Battery. Jpn. Pat. Appl. 2007-291474, 8 November 2007.

186. Kobayashi, K.; Takahashi, S.; Hayamizu, N. Secondary Battery Hydrogen Storage Alloy. Jpn. Pat. Appl. 2008-210809, 11 September 2008.

187. Terashita, N.; Osawa, M.; Kudo, K.; Soma, Y.; Tsunokake, S. Hydrogen Storage Alloy, and Nickel Hydrogen Secondary Battery. Jpn. Pat. Appl. 2012-067357, 5 April 2012.

188. Sakai, T.; Saito, M.; Mukai, T.; Tsunokake, S.; Osawa, M. Hydrogen-Storage Alloy, Hydrogen-Storage Alloy Electrode, and Nickel-Hydrogen Secondary Battery. Jpn. Pat. Appl. 2012-102343, 31 May 2012.

189. Fukui, I.; Kuribayashi, Y. Hydrogen Storage Alloy Electrode and Manufacture Thereof. Jpn. Pat. Appl. H08-088001, 2 April 1996.

190. Shintani, H.; Sugahara, Y. Hydrogen Storage Alloy-Containing Composition and Electrode Using It. Jpn. Pat. Appl. H08-329934, 13 December 1996.

191. Fukui, I. Manufacture of Hydrogen Storage Alloy Electrode. Jpn. Pat. Appl. H09-171820, 30 June 1997.

192. Fukui, I. Hydrogen Storage Alloy Electrode and Its Manufacture. Jpn. Pat. Appl. H11-329437, 30 November 1999.

193. Kasashima, M.; Hashimoto, T.; Minowa, T. Apparatus for Producing Hydrogen-Storage Alloy and Production Thereof. Jpn. Pat. Appl. 2000-158098, 13 June 2000.

194. Komata, N.; Hibi, K.; Ohashi, S.; Aizawa, A.; Yaginuma, T. Production of Nickel-Containing Hydroxide. Jpn. Pat. Appl. H11-130441, 18 May 1999.

195. Hibi, K.; Wakai, E.; Oki, T.; Ban, S.; Furushima, K.; Shioda, M. Method for Producing Spherical High Density Nickel Hydroxide and Spherical High Density Nickel Hydroxide Powder for Anode Active Substance of Alkaline Secondary Battery. Jpn. Pat. Appl. 2003-002665, 8 January 2003.

196. Hibi, K.; Wakai, E.; Oki, T.; Furushima, K.; Ono, M.; Shioda, M. Nickel Positive Electrode Active Material for Alkaline Secondary Battery and Manufacturing Method of Cobalt Compound-Coated Nickel Hydroxide Particle. Jpn. Pat. Appl. 2003-157840, 30 May 2003.

197. Tsuji, S.; Son, H.; Uchida, Y. Spherical Nickel Hydroxide Powder and Method for Producing Same. Jpn. Pat. Appl. 2006-151795, 15 June 2006.

198. Matsubara, N.; Tanaka, M.; Ishikawa, S.; Shimizu, T.; Kitamura, N. Method for Manufacturing Nickel Sintered Substrate for Alkaline Storage Battery Positive Electrode and Nickel Sintered Substrate for Alkaline Storage Battery Positive Electrode. Jpn. Pat. Appl. 2015-198061, 9 November 2015.

199. Shibuya, S.; Hideno, A. Porous Hydrogen Storage Alloy. Jpn. Pat. Appl. H01-309937, 14 December 1989.

200. Kojima, Y.; Yamamoto, K.; Furukawa, S. Method of Producing Hydrogen Storage Alloy Powder. Jpn. Pat. Appl. 2004-169125, 17 June 2004.

201. Ishikawa, R.; Miyashita, T.; Yugamidani, M. Production of Hydrogen Occluding Alloy. Jpn. Pat. Appl. H06-088150, 29 March 1994.

202. Kojima, Y.; Ikeda, H.; Furukawa, S.; Sugiyama, K.; Kobayashi, N. Hydrogen-Storage Alloy, and Electrode for Nickel-Hydrogen Battery. Jpn. Pat. Appl. 2008-184660, 14 August 2008.

203. Hatakeyama, S.; Kojima, Y.; Miyashita, T.; Furukawa, S.; Sugiyama, K.; Tsuchiya, E.; Yabe, T. Manufacture of Hydrogen Storage Alloy Powder, and Ni-Hydrogen Battery. Jpn. Pat. Appl. H10-021907, 23 January 1998.

204. Doi, H.; Yabuki, T. Hydrogen Storage Ni-Base Alloy and Closed Type Ni-Hydrogen Storage Battery. Jpn. Pat. Appl. H02-111836, 24 April 1990.

205. Doi, H.; Yabuki, T. Hydrogen Storage Ni-Based Alloy and Closed Ni-Hydrogen Battery. Jpn. Pat. Appl. H02-194140, 31 July 1990.

206. Doi, H.; Yabuki, T. Hydrogen Storage Ni-Zr Series Alloy and Closed-Type Ni-Hydrogen Storage Battery. Jpn. Pat. Appl. H02-263944, 26 October 1990.

207. Nishikawa, S.; Takeshita, T. Hydrogen Storage Alloy Excellent in Corrosion Resistance and Negative Electrode for Secondary Battery Using It. Jpn. Pat. Appl. H05-156395, 22 June 1993.

208. Kita, K.; Sugawara, K.; Wada, M.; Murai, T.; Isobe, T. Hydrogen Storage Alloy Enabling High Rate Discharge of Battery. Jpn. Pat. Appl. 2000-345261, 12 December 2000.

209. Kita, K.; Sugawara, K.; Wada, M.; Murai, T.; Isobe, T. Hydrogen Storage Alloy. Jpn. Pat. Appl. 2000-073131, 7 March 2000.

210. Okochi, T.; Omori, H.; Kanekawa, A. Production of Hydrogen Storage Alloy. Jpn. Pat. Appl. H07-188799, 25 July 1995.

211. Nagase, I.; Shimizu, T.; Matsuyama, M. Production of Hydrogen Storage Alloy. Jpn. Pat. Appl. H07-305123, 21 November 1995.

212. Nagase, I.; Nishinakagawa, T.; Kimura, Y. Production of Hydrogen Storage Alloy Powder. Jpn. Pat. Appl. 2000-073101, 7 March 2000.

213. Matsukawa, A.; Odakawa, Y.; Fukuno, A.; Yamashita, S. Hydrogen Storage Alloy, Its Manufacture, and Secondary Battery. Jpn. Pat. Appl. 2000-265234, 26 September 2000.

214. Matsukawa, A.; Odakawa, Y.; Fukuno, A.; Yamashita, S. Hydrogen Storage Alloy, Its Manufacture, and Secondary Battery. Jpn. Pat. Appl. 2000-265235, 26 September 2000.

215. Sano, T.; Yamashita, S.; Cho, T.; Okada, M. Alloy Manufacturing Apparatus and Manufacturing Method of Hydrogen Storage Alloy. Jpn. Pat. Appl. 2002-331336, 19 November 2002.

216. Miyaki, T.; Terao, K.; Takahashi, T.; Kabutomori, T.; Wakizaka, Y. Manufacture of Hydrogen Storage Alloy Electrode. Jpn. Pat. Appl. H09-213317, 15 August 1997.

217. Miyaki, T.; Kabutomori, T.; Wakizaka, Y.; Morozumi, S.; Minegishi, T. Manufacture of Hydrogen Storage Alloy Electrode. Jpn. Pat. Appl. H08-203510, 9 August 1996.

218. Miyaki, T.; Kabutomori, T. Hydrogen Storage Alloy Electrode. Jpn. Pat. Appl. H09-027321, 28 January 1997.

219. Kabutomori, T.; Miyaki, T.; Terao, K. Electrode of Hydrogen Storage Alloy. Jpn. Pat. Appl. H09-213320, 15 August 1997.

220. Miyaki, T.; Aoki, K.; Goto, T.; Ito, H. Hydrogen Storage Alloy, Producing Method Therefor and Hydrogen Storage Alloy Electrode Made of Same Alloy. Jpn. Pat. Appl. 2001-192756, 17 July 2001.

221. Miyaki, T.; Aoki, K.; Goto, T.; Ito, H. Hydrogen Storage Alloy, Producing Method Therefor and Hydrogen Storage Alloy Electrode Made of Same Alloy. Jpn. Pat. Appl. 2001-192758, 17 July 2001.

222. Miyaki, T.; Aoki, K.; Goto, T.; Ito, H. Hydrogen Storage Alloy, Producing Method Therefor and Hydrogen Storage Alloy Electrode Made of Same Alloy. Jpn. Pat. Appl. 2001-192757, 17 July 2001.

223. Aoki, K.; Muro, M.; Kakihara, H. Method of Producing Hydrogen Storage Alloy and Cold Crucible Melting Apparatus. Jpn. Pat. Appl. 2010-242145, 28 October 2010.

224. Hirose, Y.; Sasaki, S.; Hasegawa, H.; Hosono, U. Manufacture of Hydrogen Storage Alloy. Jpn. Pat. Appl. H09-180716, 11 July 1997.

225. Nishi, T.; Oka, Y. Metal Porous Body, Its Manufacture, and Battery Electrode Plate Using the Same. Jpn. Pat. Appl. H09-153365, 10 June 1997.

226. Fukui, A.; Imamura, M. Method for Recovering Valuable Metal from Scrap of Nickel Metal Hydride Secondary Battery. Jpn. Pat. Appl. 2003-041326, 13 February 2003.

227. Toki, N.; Kudo, T.; Asano, S. Method of Recovering Metal from Used Nickel-Metal Hydride Battery. Jpn. Pat. Appl. 2010-174366, 12 August 2010.

228. Harada, K.; Kato, M.; Saito, H.; Tsuchida, H.; Omura, T. Current Collector and Electrode Base Plate for Battery and Their Manufacturing Method. Jpn. Pat. Appl. 2006-310261, 9 November 2006.

229. Sugikawa, H. Production of Metallic Sheet and Metallic Sheet Produced by This Method. Jpn. Pat. Appl. H09-287006, 4 November 1997.

230. Sugikawa, H. Battery Can and Forming Material Thereof. Jpn. Pat. Appl. H05-021044, 29 January 1993.

231. Sugikawa, H. Method for Manufacturing Electrode Plate for Battery, Electrode Plate Manufactured by the Method, and Battery Provided with the Electrode Plate. Jpn. Pat. Appl. 2000-173603, 23 June 2000.

232. Kamiya, Y. Nickel-Metal Hydride Battery. Jpn. Pat. Appl. H08-227707, 3 September 1996.

233. Oshima, Y.; Kimura, K.; Nishimura, K.; Takasaki, T.; Ikeda, T. Mixture Ink for Forming Foil-Shape Collector Positive Electrode of Nickel-Metal Hydride Secondary Battery. Jpn. Pat. Appl. 2013-138001, 11 July 2013.

234. Tanaka, M.; Tokutake, N.; Kondo, Y.; Yamazaki, H.; Hirooka, M. Separator for Alkaline Battery and Alkaline Battery Using It. Jpn. Pat. Appl. H06-036753, 10 February 1994.

235. Hirooka, M.; Arimura, T.; Kawatsu, Y. Separator for Alkaline Battery. Jpn. Pat. Appl. H08-273650, 18 October 1996.

236. Sato, K.; Tanaka, M.; Hirooka, M. Separator for Alkaline Battery and Manufacture of the Separator. Jpn. Pat. Appl. 2000-106162, 11 April 2000.

237. Abe, M.; Tokuyama, E.; Okazaki, D.; Kokaji, T. Ball Mill Device, Method for Producing Hydrogen Storage Alloy Powder Using the Device and Hydrogen Storage Alloy Powder. Jpn. Pat. Appl. 2006-111909, 27 April 2006.

238. Osumi, Y.; Kato, A.; Suzuki, H.; Nakane, M.; Miyake, Y. Hydrogen absorption-desorption characteristics of mischmetal-nickel-aluminum alloys. *J. Less Comm. Met.* **1979**, *66*, 67–75.

239. Osumi, Y.; Suzuki, H.; Kato, A.; Oguro, K.; Nakane, M. Development of misch metal-nickel and titanium-cobalt hydrides for hydrogen storage. *J. Less Comm. Met.* **1980**, *74*, 271–277.

240. Osumi, Y.; Suzuki, H.; Kato, A.; Oruro, K. Hydrogen absorption-desorption characteristics of $MnNi_{5-x}Al(Mn)_{y-z}M_z$ and $MmNi_{5-x}AlMn_yM_2$ alloys (Mm = misch metal). *J. Less Comm. Met.* **1983**, *89*, 287–292.

241. Osumi, Y. *Suiso Kyuzou Goukin*; Agune Co. Ltd.: Tokyo, Japan, 1993.

242. Kadir, K.; Sakai, T.; Uehara, I. Synthesis and structure determination of a new series of hydrogen storage alloys; RMg_2Ni_9 (R = La, Cel, Pr, Nd, Sm and Gd) built from $MgNi_2$ Laves-type layers alternating with AB_5 layers. *J. Alloys Compd.* **1997**, *257*, 115–121.

243. Kuriyama, N.; Sakai, T.; Miyamura, H.; Uehara, H. Manufacture of Hydrogen Storage Electrode. Jpn. Pat. Appl. H06-283164, 7 October 1994.

244. Tsukahara, M.; Takahashi, K.; Mishima, T.; Isomura, A.; Sakai, T.; Miyamura, H.; Uehara, H. Hydrogen Occluding Alloy and Hydrogen Occluding Alloy Electrode. Jpn. Pat. Appl. H07-268513, 17 October 1995.

245. Sakai, T.; Madono, J.; Miyamura, H.; Uehara, H. Hydrogen Storage Alloy and Electrode of It. Jpn. Pat. Appl. H07-278708, 24 October 1995.

246. Sakai, T.; Iwaki, T. Manufacture of Alkaline Secondary Battery and Catalytic Electrode Body. Jpn. Pat. Appl. H07-282860, 27 October 1995.

247. Kariimu, K.; Sakai, T.; Takeshita, H.; Uehara, H. New Hydrogen Storage Alloy and Hydrogen Electrode Using the Alloy. Jpn. Pat. Appl. H11-217643, 10 August 1999.

248. Sakai, T.; Takeshita, H.; Uehara, H.; Yamashita, I. Alkaline Secondary Battery Negative Electrode and Alkaline Secondary Battery. Jpn. Pat. Appl. H11-307088, 5 November 1999.

249. Osawa, M.; Tsunokake, S.; Katsura, S.; Iwaki, T.; Sakai, T. Hydrogen Storage Alloy and Nickel-Hydrogen Battery. Jpn. Pat. Appl. 2015-113522, 22 June 2015.

250. Mishima, R.; Sekine, T.; Sakai, T.; Ishikawa, H.; Miyamura, H.; Uehara, H. Hydrogen Storage Alloy and Its Production. Jpn. Pat. Appl. H06-145851, 27 May 1994.

251. Sakai, T.; Fukunaga, H.; Tanaka, T. Hydrogen Storage Alloy, and Electrode Using the Same. Jpn. Pat. Appl. 2003-059784, 6 March 2003.

252. Sakai, T.; Fukunaga, H.; Matsumoto, N.; Tanaka, T. Rectangular Nickel-Hydrogen Battery. Jpn. Pat. Appl. 2005-235421, 2 September 2005.

253. Kanemoto, M.; Kakeya, T.; Kurokuzuhara, M.; Watada, M.; Ozaki, T.; Sakai, T. Hydrogen Storage Alloy and Nickel-Hydrogen Storage Battery. Jpn. Pat. Appl. 2008-163421, 17 July 2008.

254. Ozaki, T.; Kanemoto, M.; Kakeya, T.; Kurokuzuhara, M.; Watada, M.; Sakai, T. Nickel-Hydrogen Storage Battery. Jpn. Pat. Appl. 2009-163986, 23 July 2009.

255. Kondo, T.; Yamamizu, T.; Sakai, T.; Kuriyama, N. Nickel Hydride Secondary Battery. Jpn. Pat. Appl. 2002-063889, 28 February 2002.

256. Kondo, T.; Yamamizu, T.; Sakai, T. Nickel Hydrogen Secondary Battery. Jpn. Pat. Appl. 2002-157988, 31 May 2002.

257. Iwaki, T.; Yao, M.; Sakai, T.; Okuno, K.; Kato, M.; Boku, T. Hydrogen Absorbing Alloy Negative Electrode for Alkaline Battery. Jpn. Pat. Appl. 2008-117579, 22 May 2008.

258. Mukai, T.; Takasaki, T.; Sakai, T.; Iwaki, T.; Tsutsumi, K.; Nishimura, K. Alloy Negative Electrode for Fiber Battery. Jpn. Pat. Appl. 2010-160912, 22 July 2010.

259. Takasaki, T.; Nishimura, K.; Fukunaga, H.; Tsutsumi, K.; Saito, M.; Mikai, T.; Sakai, T. Cobalt-Free Alkaline Secondary Battery. Jpn. Pat. Appl. 2012-204177, 22 October 2012.

260. Hashimoto, H.; Son, M.; Abe, T. Hydrogen Storage Material, and Production Method Therefor. Jpn. Pat. Appl. 2004-204309, 22 July 2004.

261. Sakai, T.; Iwaki, T. Electrode for Secondary Battery, and the Secondary Battery Using the Same. Jpn. Pat. Appl. 2003-326738, 18 September 2003.

262. Towata, S.; Ito, K.; Kadoura, H. Surface Treatment of Hydrogen Occlusion Alloy Material, Activation Treatment of Hydrogen Occlusion Alloy Electrode, Activating Solution, and Hydrogen Occlusion Alloy Electrode Having Excellent Initial Activity. Jpn. Pat. Appl. H08-291391, 5 November 1996.

263. Towata, S.; Ito, K.; Yamakawa, S.; Abe, K.; Oya, Y.; Morishita, S.; Kawase, Y. Surface Treatment Method of Hydrogen Storage Alloy by Steam and Alloy Obtained Thereby. Jpn. Pat. Appl. H09-180715, 11 July 1997.

264. Yoshida, T.; Miyano, K.; Itou, T.; Towata, S.; Ito, K. Hydrogen Storage Alloy Unit and Manufacture Thereof. Jpn. Pat. Appl. H10-012227, 16 January 1998.

265. Morishita, S.; Kondo, Y.; Towata, S.; Abe, K.; Muta, M.; Kinoshita, K. Nickel Positive Electrode Active Material for Alkali Storage Battery. Jpn. Pat. Appl. H10-074514, 17 March 1998.

266. Kondo, Y.; Morishita, S.; Oya, Y.; Towata, S.; Muta, M.; Kinoshita, K. Stabilizing Method for Hydrogen Storage Material. Jpn. Pat. Appl. H10-130860, 19 May 1998.

267. Muta, M.; Kinoshita, K.; Morishita, S.; Towata, S. Conductive Powder for Hydrogen Storage Alloy Negative Electrode. Jpn. Pat. Appl. H11-111299, 23 April 1999.

268. Morishita, S.; Towata, S.; Imaizumi, J.; Usui, T. Nickel Hydroxide for Positive Electrode Active Material of Alkaline Secondary Battery, Alkaline Secondary Battery Using Same, Its Characteristics Evaluation Method and Manufacturing Method. Jpn. Pat. Appl. 2002-208400, 26 July 2002.

269. Kobayashi, T.; Kondo, Y.; Matsuo, H.; Sasaki, I.; Ito, Y.; Nozaki, H.; Nonaka, T.; Senoo, Y.; Ukiyou, Y.; Ito, M. Alkaline Storage Battery. Jpn. Pat. Appl. 2005-347089, 15 December 2005.

270. Yamawaki, K.; Kuwajima, S.; Suzuki, N.; Shirogami, T. Bipolar Metal-Hydrogen Secondary Battery. Jpn. Pat. Appl. H07-014618, 17 January 1995.

271. Tsutsumi, K. Laminate Battery and Laminate Battery System. Jpn. Pat. Appl. 2013-080698, 2 May 2013.

272. Ueda, K.; Tsukahara, M.; Kamiya, Y.; Kikuchi, S. Manufacturing Method for Mg-Cu Composite Material and Hydrogen Occluding Alloy. Jpn. Pat. Appl. 2005-178095, 7 July 2005.

273. Kamiya, Y.; Tsukahara, M. Hydrogen Storage Alloy and Hydrogen Storage Vessel. Jpn. Pat. Appl. 2006-028632, 2 February 2006.

274. Takahashi, K.; Tsukahara, M.; Mishima, T.; Isomura, A. Hydrogen Storage Alloy Electrode and Manufacture Thereof. Jpn. Pat. Appl. H08-236107, 13 September 1996.

275. Tsukahara, M.; Takahashi, K.; Mishima, T.; Isomura, A. Hydrogen Storage Alloy and Hydrogen Storage Alloy Electrode. Jpn. Pat. Appl. H08-269655, 15 October 1996.

276. Kamiya, Y.; Tsukahara, M.; Takahashi, K.; Isomura, A.; Sakai, T.; Takeshita, H. Hydrogen Storage Alloy, Hydrogen Storage Alloy Electrode and Production of Hydrogen Storage Alloy. Jpn. Pat. Appl. 2000-096179, 4 April 2000.

277. Tsukahara, M.; Kamiya, Y.; Takahashi, K.; Isomura, A.; Sakai, T.; Kuriyama, N.; Takeshita, H. Production of Hydrogen Storage Alloy, Alloy Thereof and Electrode Using the Alloy. Jpn. Pat. Appl. H11-106847, 20 April 1999.

278. Furukawa, K.; Sakamoto, K.; Mori, H.; Kishimoto, M.; Okabe, K. Sealed Nickel-Hydrogen Secondary Battery. Jpn. Pat. Appl. 2006-147327, 8 June 2006.

279. Furukawa, K.; Harada, Y. Battery. Jpn. Pat. Appl. 2007-234486, 13 September 2007.

280. Young, K.; Ouchi, T.; Fetcenko, M.A. Hydrogen Storage Alloys Having Improved Cycle Life and Low Temperature Operating Characteristics. U.S. Patent 7,344,677, 18 March 2008.

281. Sakamoto, K.; Bando, H.; Mori, H.; Okabe, K. Nickel Hydrogen Battery and Its Manufacturing Method. Jpn. Pat. Appl. 2007-012573, 18 January 2007.

282. Morimoto, K.; Saito, H.; Inaba, Y. Method of Manufacturing Nickel Metal Hydride Storage Battery. Jpn. Pat. Appl. 2010-010097, 14 January 2010.

283. Sakatani, T.; Sugui, H.; Ochi, M.; Kawase, R. Nickel Hydrogen Storage Battery and Battery System. Jpn. Pat. Appl. 2014-089896, 15 May 2014.

284. Sumiyama, S.; Shibuya, N.; Okabe, A. Nickel Hydrogen Storage Battery. Jpn. Pat. Appl. 2015-173058, 1 October 2015.

Capacity Degradation Mechanisms in Nickel/Metal Hydride Batteries

Kwo-hsiung Young and Shigekazu Yasuoka

Abstract: The consistency in capacity degradation in a multi-cell pack (>100 cells) is critical for ensuring long service life for propulsion applications. As the first step of optimizing a battery system design, academic publications regarding the capacity degradation mechanisms and possible solutions for cycled nickel/metal hydride (Ni/MH) rechargeable batteries under various usage conditions are reviewed. The commonly used analytic methods for determining the failure mode are also presented here. The most common failure mode of a Ni/MH battery is an increase in the cell impedance due to electrolyte dry-out that occurs from venting and active electrode material degradation/disintegration. This work provides a summary of effective methods to extend Ni/MH cell cycle life through negative electrode formula optimizations and binder selection, positive electrode additives and coatings, electrolyte optimization, cell design, and others. Methods of reviving and recycling used/spent batteries are also reviewed.

Reprinted from *Batteries*. Cite as: Young, K.-h.; Yasuoka, S. Capacity Degradation Mechanisms in Nickel/Metal Hydride Batteries. *Batteries* **2016**, *2*, 3.

1. Introduction

Nickel/metal hydride (Ni/MH) batteries are widely used in many energy storage applications. Cycle stability is one of the key criteria in judging the performance of rechargeable battery technology. The general observations regarding failed Ni/MH cells are summarized in Figure 1. In order to further investigate the mechanisms of capacity degradation and their relevant solutions to extend cycling under normal and abuse conditions, we have chosen to begin with a review of the significance of the Ni/MH battery in the overall battery market, its basic structure and chemistry, and the analytical tools used to study and characterize its performance, and to mainly focus on academic publications and reports. Patents detailing solutions for extending cycle life are reviewed in two separate articles [1,2].

1.1. Significance of Nickel/Metal Hydride Batteries

Ni/MH batteries using an alkaline KOH electrolyte have been commercialized for more than 25 years [3]. Because of its durability, abuse tolerance, compact size, and environmental friendliness, Ni/MH battery applications have steadily expanded from the traditional consumer market to include propulsion and telecommunications.

However, due to its relatively low gravimetric energy density compared to the rival Li-ion battery, the Ni/MH battery lost part of its market share in portable electronic devices, such as notebook computers, cell phones, and digital cameras. In the meantime, Ni/MH battery technology also invaded the primary alkaline battery market because of its voltage compatibility and low self-discharge [4–6], as well as the NiCd power tool market for its non-toxicity [7,8]. The success of Ni/MH in powering hybrid electric vehicles (HEV) developed by a handful of automobile manufacturers stems from its wide temperature range, abuse tolerance, superb cycle stability, high charge and discharge rate capabilities, and environmental friendliness [9]. One analyst even predicted a fourfold increase in Ni/MH battery sales for HEV and EV markets from 2014 to 2020 [10]. Although the current industries making pure battery-powered electric vehicles embrace Li-ion battery technology, a Ni/MH pouch cell developed under a five million dollar Advanced Research Projects Agency-Energy (ARPA-E) program has demonstrated a specific energy of 127 Wh·kg^{-1} at the cell level with an estimated target of 148 Wh·kg^{-1} [11]. With the recent breakthrough of high-energy Si-negative electrodes capable of storing 3635 mAh·g^{-1} (about ten times the current A_2B_7 alloy) [12], the future of Ni/MH in the EV application appears very bright. From the beginning of their competition, Ni/MH batteries have had higher volumetric energy density than Li-ion batteries, due to the high density active materials (rare earth metal (RE) and transition metal *versus* carbon-based products in the anode and nickel hydroxide *versus* lithiated transition metal oxides in the cathode). With the improvements in specific energy, a resurgence in Ni/MH batteries for applications that place a premium on space rather than weight, such as portable displays, wearable electronic devices, and medical devices, can be expected. As for large-scale high-power temporary energy storage applications, the GIGACELL, made by Kawasaki Heavy Industries, demonstrates superior performance using the Ni/MH chemistry [13,14]. In stationary applications, its excellent cycle stability and wide operating temperatures, combined with the low cost and easy manufacturability, have made Ni/MH the best choice [15]. The overall outlook for Ni/MH battery technology shows that it has tremendous potential in various energy storage applications following these new scientific discoveries and process improvements—a far cry from being the 25-year obsolete veteran in the battery business.

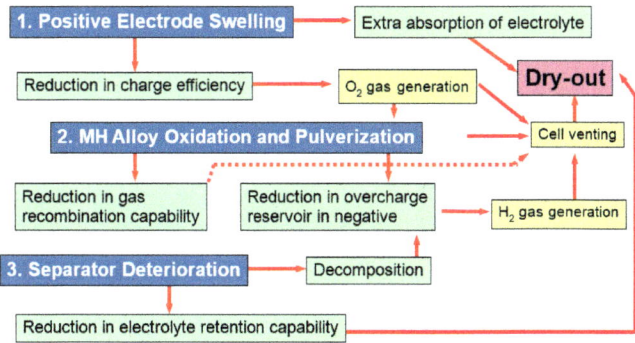

Figure 1. Schematic diagram of three key factors leading to the major failure mode of nickel/metal hydride (Ni/MH) cells—electrolyte dry-out.

1.2. Basic Structure of Nickel/Metal Hydride Battery

There are basically seven different types of Ni/MH batteries: cylindrical with metal cases, stick (bubble gum shape), prismatic with metal cases, prismatic with plastic cases, button cell, pouch cell [16], and flooded cell [17] (Figure 2). All but button cell and pouch cell have a safety valve installed to prevent explosions from gas build-up. A simple comparison between various construction types is shown in Table 1. They all share some common parts: positive electrode, negative electrode, separator, electrolyte, case, and safety valve (except button and pouch cells). The basic electrochemistry reactions for the positive electrode, negative electrode, and full cell are:

$$Ni(OH)_2 + OH^- \leftrightarrows NiOOH + H_2O + e^- \text{ (forward : charge, reverse : discharge)} \quad (1)$$

$$M + H_2O + e^- \leftrightarrows MH + OH^- \text{ (forward : charge, reverse : discharge)} \quad (2)$$

$$Ni(OH)_2 + M \leftrightarrows NiOOH + MH \text{ (forward : charge, reverse : discharge)} \quad (3)$$

where M is the hydrogen storage metal/alloy and MH is the hydride of metal M. During the charge process, bi-valent Ni is oxidized into the tri-valent state while metal M is reduced by the absorbed hydrogen atom. The most commonly used positive electrode in the current Ni/MH battery technology consists of active materials made of co-precipitated spherical hydroxides from Ni, Co, and Zn and some binders, pasted onto Ni-foam via a wet method. Recently, a dry application of spherical powder onto Ni-foam with no binder followed by immediate compaction has also been used to increase the energy and power density of the cell. In some high-temperature/high-rate applications, old sintered-type positive electrodes, based on fibrous Ni on stainless steel plate, are still in commission. Co-coating of the spherical particles and additives such as metallic Co and/or CoO and rare earth

117

element (RE) oxides in the positive electrode paste are also popular. A review of the synthesis and properties of $Ni(OH)_2$ was recently reported [18].

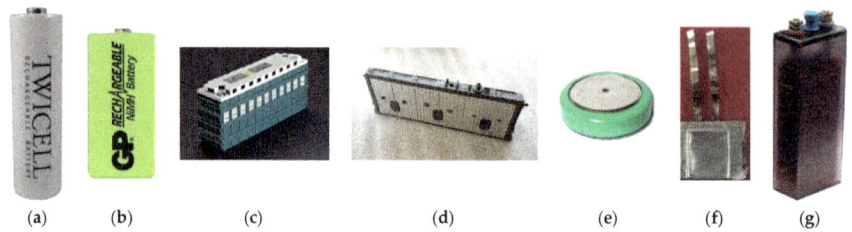

(a)   (b)   (c)   (d)   (e)   (f)   (g)

Figure 2. Ni/MH batteries in: (**a**) cylindrical; (**b**) stick; (**c**) metallic prismatic; (**d**) plastic prismatic; (**e**) button; (**f**) pouch; and (**g**) flooded configurations.

Table 1. Comparison of different types of Ni/MH battery packaging.

Shape	Case material	Sealed	Manufacturability	Cost	Energy density	Heat dissipation	Abuse tolerance
Cylindrical	Metal	Yes	Easy	Low	High	Easy	High
Stick	Metal	Yes	Medium	Low	High	Easy	High
Prismatic	Metal	Yes	Medium	High	Low	Easy	Med
Prismatic	Plastic	Yes	Medium	High	Low	Hard	Med
Button	Metal	Yes	Easy	Low	Low	Easy	Low
Pouch	Al foil	Yes	Easy	Low	Very high	Easy	Low
Cylindrical/prismatic	Plastic/flooded	No	Easy	Low	Low	Hard	High

The most common metal hydride (MH) alloy used in the negative electrode is a RE-based AB_5 alloy. A typical atomic composition is $La_{10.5}Ce_{4.3}Pr_{0.5}Nd_{1.4}Ni_{60.0}Co_{12.7}Mn_{5.9}Al_{4.7}$. Recently, RE- based A_2B_7 MH alloys have gained popularity in high-energy and low self-discharge consumer type applications [4–6]. A typical atomic composition of this type is $La_{6.7}Pr_{6.3}Nd_{6.3}Ni_{72.8}Al_{4.0}$. Recent progress in MH alloys for Ni/MH battery applications can be found in the following review article [19]. The negative electrode can be prepared by dry-compacting the MH powder directly onto a Ni-mesh, Cu-mesh, expanded Ni, Ni foam, or expanded Cu substrates without the use of a binder, or by wet-pasting a slurry with MH alloy, binder, and/or additives onto nickel plated perforated stainless steel (NPPS).

A 30% KOH solution is widely used as the electrolyte for Ni/MH batteries due to the balance of conductivity and freezing point temperature. Performance comparisons for other concentrations [20] and alkaline metal hydroxides [21] are available. A small amount of LiOH ($1.5 \text{ g} \cdot L^{-1}$), which has higher chemical reactivity, is added to boost low-temperature performance, while in high-temperature applications, part or all of the KOH is replaced by the less reactive (corrosive) NaOH to reduce corrosion. In the current standard mass production of Ni/MH cells, no other specific additive is added to the electrolyte.

Grafted polypropylene (PP)/polyethylene (PE) non-woven fabric is today's standard separator material, and an overview has been published by Kritzer and Cook [22]. While the regular separator is white in color, the sulfonated separator is brownish and offers benefits to low self-discharge due to its ability to trap redox shuttle substances, especially the nitrogen-containing compounds [23]. Both types of separators can be found in current NiMH batteries.

1.3. Experimental Methods Used in Failure Analysis

A few analytic tools are frequently used to identify the failure mode of a cycled Ni/MH battery [24–26]. Scanning electron microscope (SEM) with X-ray energy dispersive spectroscopy (EDS) capability is commonly used to examine the degree of pulverization, phase segregation, degree of oxidation, physical size changes, and trapping of particulates. The different features found between the secondary electron image and the backscattering electron image can be extrapolated into the changes in the average atomic weight of the area of interest. EDS mapping is especially useful in studying elemental distribution (for example, oxygen) in a relatively large area (10–100 μm scale). While gas chromatography (GC) is used to identify the gas composition in the cell, inductively coupled plasma (ICP) is used to examine the metallic composition of any solid (electrode, separator, tap, *etc.*) or liquid (remaining electrolyte and solution attained through Soxhlet extraction) content from the autopsy of a cycled cell. Titration is another method to determine the content of a specific element [27]. X-ray diffraction (XRD) is an important tool to study oxide formation, phase changes, and microstructure changes in both negative [25,26,28,29] and positive [30] electrodes.

Other tools are used less frequently in failure analysis. For example, transmission electron microscope (TEM) is sometimes used to study the microstructure and composition of the surface oxide from a cycled cell [31–33]. Magnetic susceptibility (MS) measurements can be used to monitor the evolution of the count and size of metallic Ni-clusters embedded in the surface oxide [26,34,35]. Both X-ray photoelectron spectroscopy (XPS) [36–40] and Auger electron spectroscopy (AES) [41] have been used to study the surface composition, with the former being able to identify the oxidation state. The acoustic emission (AE) technique has also been used to study the volume change and pulverization of the MH alloy [42,43]. Fourier transform infrared spectroscopy (FTIR) has been used to study the OH^- ligand in $Ni(OH)_2$ [44–47]. Raman spectroscopy (RS) is another optical measurement used to characterize the changes in the separator and positive electrode [37,46–48]. Electrochemical impedance spectroscopy (EIS) or AC impedance measurements are usually used to isolate components with different degrees of degradation [49–52]. Polarization curves [53–56] and cyclic voltammetry (CV) [57–60] are other electrochemical tools that can be used to study the evolution of electrode surface changes. Besides experiments with real batteries, empirical capacity degradation models have also been previously developed [61–63].

2. Capacity Degradation

Battery failure can be separated into two categories: accidental and long-term degradation. The former includes fire, electrical short-circuit, and physical damage. In Table 2, we have listed a few common symptoms and possible causes that originated the failure of the batteries. Long-term capacity loss under various test conditions is discussed in the remainder of this section.

Table 2. Common Ni/MH battery failure symptoms and possible causes.

Symptom	Reasons	Possible causes
Battery short-circuit	Direct conducting path between two electrodes developed	• Separator punch-through • Conducting debris from Cu-impurities • Deformation of electrode causing direct contact between taps
Battery open-circuit	Breakage of inside connection	• Electrode breakage due to expansion/distortion • Broken tap connection • Complete electrolyte dry-out
Battery abuse	Over-discharge and overcharge	• Unbalanced capacity in positive and negative electrode • Mismatched charger
Capacity decrease	Electrode degradation	• Pulverization/oxidation of MH alloys in negative electrode • Pulverization in spherical particle due to formation of γ-NiOOH phase • Decrease in the Co-conductive network in the positive electrode
Power decrease and impedance increase	Electrolyte dry-out	• Venting from improper cell-balance • Consumption due to oxidation
	Electrode degradation	• Reduction in electrode active materials • Increase of the surface oxide of negative electrode • Loss of co-conductive network in positive electrode
	Separator degradation	• Increase in fiber diameter • Reduction in pore volume • Impurity trapped internally • Decomposition
Overheat during charge	Micro-shorting	• Conductive debris accumulation in separator
White deposits	Electrolyte leak from venting	• Improper closing of the cell • Off-balance in the remaining electrode capacity • Deterioration of gas recombination capability at the surface of MH alloy • Heavily oxidized electrode and/or electrolyte • Failure in the safety vent

120

2.1. Capacity Loss During Normal Cycling at Room Temperature

There are two types of capacity loss during cycling: reversible and irreversible. The reversible part is also called self-discharge, which mainly occurs through six pathways: shuttling effects from nitrogen containing compounds [64], shuttling effects from soluble ions of multi-valence transition metals [65], micro-shorts [66] from conducting/semiconducting deposits trapped in the separator [67,68], hydrogen gas desorption from MH alloys [69–73], direct reaction between hydrogen gas and NiOOH [69,73,74], and CoOOH protective/conductive coating breakdown due to contamination from leached MH alloys [6]. Self-discharge accelerates with rises in the environmental temperature. There is basically no self-discharge at below −5 °C [75]. Before the low self-discharge Ni/MH battery was introduced (the Eneloop cell from Sanyo using a combination of improved MH alloy, separator, and positive active materials [6]), cells initially had a monthly 20%–30% capacity reduction, which was then improved to a monthly loss of 5%–10% at room temperature [76]. Modern low-self discharge Ni/MH consumer batteries have self-discharge rates of less than 20% per year [6]. An automatically triggered re-charging algorithm may be necessary for large-scale applications [77]. Common methods used to suppress self-discharge in Ni/MH batteries are summarized in Table 3. The irreversible capacity loss, which leads to failure of the battery, covers the majority of this review.

Table 3. Summary of common methods used to suppress self-discharge in Ni/MH batteries. The star system used in the effectiveness column in Tables 3–9 was meant to show the relative strength in each method to address the problem based on authors' own experience. Interested readers are encouraged to read the original article and to form their own opinions. PP: polypropylene; PTFE: polytetrafluoroethylene; and CMC: carboxymethyl cellulose.

Method	Direct impact	Environmental impact	Cost impact	Effectiveness	References
Use of a sulfonated separator	Removal of N-containing compounds	None	Modest	★★★★★	[22,78,79]
Use of an acrylic acid grafted PP separator	Reduction in Al- and Mn-debris formation in separator	None	None	★★★★	[80]
Removal of Co and Mn in A_2B_7 MH alloy	Reduction in debris formation in separator	None	None	★★★★★	[6,81]
Increase of the amount of electrolyte	Reduction in the hydrogen diffusion in electrolyte	None	None	★★★★	[82]
Removal of Cu-containing components	Reduction in micro-short	None	None	★★★★★	[83–85]
PTFE coating on positive electrode	Suppression of reaction between NiOOH and H_2	None	Negligible	★★★★	[86]

121

Table 3. *Cont.*

Method	Direct impact	Environmental impact	Cost impact	Effectiveness	References
CMC solution dipping	Suppression of oxygen evolution	None	Negligible	★★★★	[87]
Micro-encapsulation of Cu on MH alloy	Decrease in H_2 released from MH alloy	None	Modest	★★★	[88]
Ni-B alloy coating on MH alloy	Formation of a protection layer	None	Modest	★★★	[89]
Alkaline treatment of negative electrode	Reduction of leach-out of Mn and Al	None	Modest	★★★★	[90]
Addition of LiOH and NaOH into electrolyte	Reduction in electrolyte corrosion capabilities	None	None	★★★★	[75]
Addition of $Al_2(SO_4)_3$ into electrolyte	Reduction in MH alloy corrosion	None	Negligible	★★	[91]

Irreversible capacity losses under regular cycling conditions (temperature between 20 °C and 30 °C, rated below 2C with one or a combination of reasonable cut-off schemas during over-charge, such as those used in [26,29]) can be categorized into five main categories: degradation of negative electrode active material (MH alloy), degradation of positive electrode active material (spherical $Ni(OH)_2$), disintegration of the negative electrode, disintegration of the positive electrode, and venting of cells.

Degradation in the negative electrode includes MH alloy pulverization due to lattice expansion during hydrogenation [92–95] which results in poor electrical and protonic conduction [49,95,96], alloy surface oxidation hampering electron and proton conduction [36,52–54,93,94,97–101], and surface fluoride formation [36]. The corrosion processes of AB_5 MH alloys have been characterized by Maurel and his coworkers using XRD, SEM, and TEM [102]. In the La-only A_2B_7 superlattice MH alloy, the pulverization due to different sequences of hydrogenation between Mg-containing A_2B_7 and Mg-free AB_5 phases dominates the failure mode [25,103].

Degradation in the positive electrode includes swelling from γ-NiOOH formation [67,104], breaking of the Co-conductive network [105], formation of less electrochemically rechargeable γ-NiOOH [25,93,106], Co dissolution and migration from the conductive network in the positive electrode [107], contamination from leach-out products (Al and Mn) in the negative electrode, deteriorating Co-conductive coating [27,104], and pulverization of positive electrode spherical particles causing detachment of active material [68,108]. The increased surface area in the positive electrode as a result of pulverization also deprives electrolyte from the separator, which increases cell resistance [109].

The mechanical disintegration of the negative electrode may include breakage of the NPPS substrate due to increased stress from electrode expansion/distortion and MH alloy powder detachment from the substrate. The mechanical disintegration of the positive electrode may include breakage of the Ni-foam substrate due to large amounts of stress from electrode expansion/distortion [110], especially in a small wounded cylindrical cell [111], and separation of spherical particles from the substrate [112]. Venting occurs when high pressure (mostly H_2) is built up inside the cell primarily from inadequate gas recombination capabilities of the MH alloy surface and/or unbalanced capacity distribution [113], which results in reduced electrolyte content [114].

2.2. Capacity Loss During Long-Term Room Temperature Storage

The irreversible capacity loss during long-term room temperature storage can be attributed to the dissolution of the surface CoOOH conducting network [115,116], corrosion/passivation of the negative electrode [23,117–119], decomposition of the positive electrode [115], decomposition of the separator [116], and poisoning of the positive electrode from cations that originate from the negative electrode [68,80,115].

2.3. Capacity Loss During High-Temperature Storage

Temperature is one of the key factors affecting cycle stability [120]. In addition to the regular capacity losses described in Section 2.1, high-temperature environments ($\geqslant$45 °C) will accelerate the cell degradation through the following pathways: oxidation rate increases at the surface of the MH alloy particles [121,122], dissolution of Co-compounds in the Co-conductive network [113,123], higher self-discharge rates that lower the cell voltage and result in further alloy oxidation, and separator degradation [124]. The charging method used in the high-temperature range has to be specially designed. First, the cell voltage tends to be lower at higher temperature, which demands that a lower cut-off voltage be adopted during charging to prevent over-charge [122] as it can be directly correlated to capacity degradation [125]. Next, the oxygen gas evolution potential in the positive electrode tends to decrease with increased temperature, which forces the positive electrode to finish charging prematurely and for which the $-\Delta V$ cut-off method is less effective [122,126,127]. Ni/MH batteries are also more sensitive to over-charge at elevated temperatures. Ni/MH batteries overcharged at rates of 0.2C, 0.5C, and 1.0C for one month show irreversible capacity losses of 12%, 30%, and 40%, respectively [126]. Different from the irreversible capacity losses during high-temperature cycling, losses in capacity observed during low-temperature cycling are recoverable when returned to room temperature [74].

2.4. Capacity Loss Due to Low-Temperature Cycling

As stated above, low-temperature storage of Ni/MH batteries causes no apparent damage to performance. However, Chen *et al.* [128] reported capacity degradation during a −20 °C cycling experiment with MH alloy pulverization, but the alloy corrosion was less serious compared to results from room temperature and high temperature. At low temperatures, a special "surface icing" appears to form on the MH alloy, further hindering electrochemical reactions and then disappearing at higher temperature [129].

2.5. Capacity Loss Due to High-Rate Cycling

Fast charge acceptance is controlled by solid-state hydrogen diffusion [130], and the diffusion coefficient of hydrogen decreases with increasing current density [131]. The increase in the degradation rate with fast charging typically originates from an improper termination method for detecting the end of charge, which leads to a large degree of over-charge especially within an aged cell [132]. The heat generated from the internal resistance of the cell and the hydrogen-oxygen recombination reaction cannot be dissipated quickly enough, and this results in an increase in the cell temperature. Both the high rate and the high temperature conditions reduce charging efficiency [133] and therefore both conditions facilitate similar failure mechanisms, except that a high-rate cycled cell also shows electrode disintegration from extraordinarily fast gas release [134] (mostly H_2 [135]) as well as gas venting due to the insufficient time for hydrogen-oxygen recombination [136,137]. As such, fast charging of a large-sized Ni/MH battery is not recommended unless special temperature monitoring devices are installed [138,139].

2.6. Capacity Loss in a Multi-Cell Module

Thus far, the discussion in this section has focused on the cell-level where most of the capacity degradation occurs. In a single Ni/MH cell, both the over-charge (with a state-of-charge (SOC) greater than 100%) and over-discharge (depth of discharge (DOD) greater than 100%) conditions can be avoided by the proper monitoring of the cell voltage. Because of the low risk of operating Ni/MH cells under disadvantageous conditions, a multi-cell module or pack does not require voltage monitoring at the cell-level whereas the Li-ion battery does. With the proper design of the negative-to-positive capacity (n/p) ratio, the size of the over-discharge reservoir [113] and the anticipated rates of capacity degradation in both electrodes, the over-charge or the over-discharge of the cells only results in small amounts of oxygen gas or hydrogen gas evolution, respectively, in the positive electrode [17]. The small amounts of generated oxygen gas can be recombined with the hydrogen stored in the negative electrode in case of over-charge, and the small amounts of generated hydrogen gas can be stored in the negative electrode in case of over-discharge [113]. Repetitive gas evolutions from the positive electrode can result in both mechanical disintegration of the electrode and cell venting to relieve the pressure, causing a loss in capacity and an increase in cell impedance. The DOD in a multi-cell pack also plays an important role in the cycle life performance. For example, an increase of

DOD from 10% to 90% in a HEV Ni/MH pack can reduce cycle life from 5000 cycles to 500 cycles [61]. For high-rate operation, as in a HEV, large swings in the SOC can result in premature MH alloy pulverization.

3. Methods to Improve Cycle Stability

There are many academic publications and issued patents offering, at least, partial solutions to the capacity loss problem during cycling. While patents addressing cycle stability are reviewed in two other papers [1,2], the strategies issued from the academic research community reviewed here fall under six general categories: (1) cell designs guided mainly by n/p ratio, electrolyte loading, and electrode thickness parameters; (2) active binder and additive material designs in the negative electrode; (3) composition, coating, and paste additives in the positive electrode; (4) choice of separator; (5) electrolyte; and (6) other components. Other systematic maintenance protocols for battery packs using Ni/MH cells were reported by Zhu and his coworkers [140].

3.1. Cell Design

In good Ni/MH cell design, an appropriate n/p ratio is critical to the balance of the various performance requirements in a specific application. For instance, a high energy consumer cell, a general purpose cell, and a high-rate cell may have n/p ranges of 1.05–1.2, 1.4–1.6, and 1.8–2.2, respectively. Adequate distribution of the extra negative electrode capacity into the over-charge-reservoir (OCR) and the over-discharge-reservoir (ODR) to avoid cell-venting is especially critical, particularly near the end of service life [116]. Nearly all cases of venting are due to short-circuits in the OCR that arises from material oxidation and γ-NiOOH formation that overwhelm the ODR. Other important design parameters that impact cycle life performance are electrolyte loading and electrode thickness. The amount of electrolyte added to the cell is proportional to cycle life, but too much electrolyte will eliminate the gas recombination centers and cause venting during formation. Optimal electrolyte loading is about 1.7–1.9 $g \cdot A^{-1} \cdot h^{-1}$ [141]. Thicker electrodes can improve the gravimetric and volumetric energy densities of the battery at the expense of high-rate discharge capability and mechanical integrity of the electrode. Methods for improving cycle performance through cell design are summarized in Table 4.

Table 4. Summary of cycle stability improvement methods related to cell design. ODR: over- discharge-reservoir; and n/p: negative-to-positive capacity.

Method	Direct impact	Environmental impact	Cost impact	Effectiveness	References
Pre-charge of the positive electrode	Reduction of ODR	None	Negligible	★★★★★	[142,143]
Increase in the n/p ratio	Trade-off of capacity for longer life	None	None	★★★★★	[134,144]
Optimization of electrolyte loading	Balance between cycle life and production yield	None	None	★★★★	[141]
Optimization of positive electrode thickness	Reduction in electrode breakage	None	None	★★★★	[145]
Pre-charge during the formation process	Protection of MH alloy	None	Negligible	★★★	[146]

3.2. Negative Electrode

While studies of degradation in MH alloys such as AB_2 [147–153], Mg-Ni [154–157], and V-based body-center-cubic [158] are available, we will singularly focus on the discussion of misch-metal based AB_5 and A_2B_7 superlattice MH alloys and their related electrode properties. In this section, improvements in cycle stability related to the negative electrodes are summarized in Table 5 and are categorized by alloy formula, alloy preparation, alloy post-treatment, electrode additives, and different electrode types.

Table 5. Summary of cycle stability improvement methods related to negative electrode. PVA: polyvinel alcohol; HEC: hydroxyethyl cellulose; and RE: rare earth metal.

Method		Direct impact	Environmental impact	Cost impact	Effectiveness	References
A. Alloy formula	Increase in Al-content	Increase in unit cell volume and reduction in lattice expansion during hydrogenation. Formation of Al_2O_3 protection layer on MH alloy.	None	None	*****	[159–162]
	Increase in Co-content	Reduction in hardness and prevention of La-migration onto surface	None	Modest	*****	[163]
	Use of misch-metal instead of pure La	Increase in degree of disorder	None	Reduction	*****	[164,165]
	Increase in Ce and Nd content	Increase in oxidation resistance	None	Modest	*****	[166]
	Zr addition	Decrease in pulverization rate	None	Negligible	*****	[167,168]
	Ti addition	Decrease in pulverization rate	None	Negligible	****	[168,169]
	Use of hyper-stoichiometry	Reduction in pressure-concentration-temperature hysteresis and pulverization	None	None	****	[92,163]
B. Alloy preparation	Fast quenching-gas atomization	Distribution of stress from lattice expansion	None	Modest	*****	[40,170–173]
	Fast quenching-melt spin	Improvement in alloy homogeneity	None	Modest	*****	[174,175]
C. Surface treatment	Ni surface plating	Protection of alloy surface from oxidation and reduction in inner pressure	None	Modest	*****	[176,177]
	Cu coating	Protection of alloy surface from oxidation	None	Modest	****	[178–181]
	Co coating	Protection of alloy surface from oxidation	None	Modest	****	[182]
	Pd coating	Protection of alloy surface from oxidation	None	High	****	[183]
	Ni-B alloy coating	Protection of alloy surface from oxidation	None	Modest	****	[89]
	Ni-P alloy coating	Protection of alloy surface from oxidation	None	Modest	****	[184]
	Ni-S alloy coating	Protection of alloy surface from oxidation	None	Modest	****	[185]
	Ni-Cu alloy coating	Protection of alloy surface from oxidation	None	Modest	****	[186]
	Alkaline pre-activation	Formation of a Ni-rich surface	None	Modest	*****	[187]
	KBH_4 treatment	Formation of a Ni-rich surface	Toxic in contact with skin	Modest	*****	[187,188]
	Surface fluorination	Protection of alloy surface from oxidation	None	Modest	*****	[189–192]
	Cu and HF surface treatment	Formation of CuF_2 protective layer on the surface	None	Modest	***	[193]
D. Other treatments	AB_5 annealing	Improvement in Mn homogeneity and reduction in inner pressure	None	Modest	*****	[166,194–196]
	La-A_2B_7 annealing	Improvement in phase homogeneity	None	Modest	*****	[197]
	Magnetization	Improvement in mechanical integrity	None	Modest	***	[198]
	Ultrasound treatment	Reduction in pulverization	None	Modest	***	[128]

Table 5. *Cont.*

Method	Direct impact	Environmental impact	Cost impact	Effectiveness	References
E. Additives					
Ni fine powder	Increase in mechanical integrity	None	Negligible	****	[199]
Cu fine powder	Increase in mechanical integrity	None	Negligible	***	[200]
Co-compounds	Increase in oxidation resistance	None	Modest	****	[60,201–203]
CMC:PVA (3:2)	Increase in mechanical integrity	None	Negligible	****	[204]
Ratio of binder to conductive additives	Increase in mechanical integrity	None	None	****	[205]
PTFE	Improvement in hydrogen gas absorption capability to reduce pressure	None	Negligible	****	[206]
Teflonized carbon	Creation of 3D conductive network	None	Negligible	****	[207]
HEC	Improvement in hydrogen gas absorption capability to reduce pressure	Very low toxicity if swallowed	Negligible	****	[127,195]
BC-1 (irigenin)	Improvement in gas recombination rate	None	Negligible	****	[208]
Carbon nanotube	Increase in mechanical integrity	None	Modest	****	[209,210]
Y_2O_3	Improvement in corrosion resistance	None	Modest	****	[211]
Oxides of light RE	Improvement in corrosion resistance	None	Modest	****	[212]
Oxides of heavy RE	Improvement in corrosion resistance	None	Modest	****	[213,214]
F. Electrode type					
Use of a pellet electrode	Increase in mechanical integrity	None	Reduction	***	[215]
Use of a sintered type electrode	Increase in mechanical integrity	None	Reduction	****	[216]

128

3.3. Positive Electrode

Currently, the most commonly used positive active material is a spherical hydroxide co-precipitated from sulphates [45] of Ni, Co, and Zn [217]. Ni has been in active use for more than one hundred years due to the chemical reversibility between Ni^{2+} and Ni^{3+} and a voltage slightly above the oxygen gas evolution potential that maximizes energy density for aqueous chemistries. Both Zn and Cd [218] are good suppressors of γ-NiOOH formation, which causes swelling of the positive electrode and consequently premature failure, but Cd is highly toxic to the environment. The element Co is interesting in that it has oxides with different oxidation states (CoO, Co_2O_3, Co_3O_4, β-CoOOH, β-$H_{0.5}CoO_2$ [219], and Co^{4+} [220]). The mechanism of reaction for Co in alkaline solution is rather complicated [38,221], but a simplified version for electrochemical engineers can be used as a guideline. Co in a +2 state is not a good conductor for electrons or protons, and it is only slightly soluble in 30% KOH. Co can be oxidized into the +3 state through solid-state reaction [222], and it is a good conductor for both electrons and protons due to the half-filled proton plane between two Co-O layers in the CoOOH crystal structure; however, the reaction is not easily reversible. The presence of Co^{4+} through a solid-state reaction can be detected at charge rates greater than C/5, and it can be reduced back to Co^{3+} at a potential of 1.05 V $versus$ Cd-electrode [220]. There are three general methods for incorporating Co into the positive electrode of Ni/MH batteries, which leverage the irreversible oxidation of Co^{2+} in the normal operation voltage range (>0.63 V $versus$ Cd-electrode [220]). First, Co co-precipitated with the spherical hydroxide particles form Co^{3+} to enhance the electron and proton conductivities for $Ni(OH)_2$. Second, the addition of Co, CoO, or other Co-compound into the electrode paste allows the formation of a CoOOH conductive network that surround the spherical $Ni(OH)_2$ particles. This Co-conductive network is crucial for the operation of Ni/MH batteries, especially at high rate conditions, but they can have issues with distribution, thickness uniformity, and severe degradation at high-temperatures [223]. A third method involves adding a pre-coating of CoOOH onto the spherical particles prior to making the slurry for the electrode paste, which can involve a wet-precipitation [224–226], a mud-slurry [227], or a dry mixing method. The use of Co in the pre-coating form is the most effective and economical method, and thus is indispensable in high-end Ni/MH consumer products. Suggestions to improve cycle stability related to the positive electrode are summarized in Table 6 and are categorized by spherical particle composition and size, coatings, additives, fabrication process, and substrates.

Table 6. Summary of cycle stability improvement methods related to the positive electrode. NPPS: nickel plated perforated stainless steel.

Method		Direct impact	Environmental impact	Cost impact	Effectiveness	References
A. Composition and particle size	Co-precipitation of Co	Increase in intrinsic conductivity	None	Modest	*****	[228]
	Co-precipitation of Zn	Prevention of γ-NiOOH formation	None	Negligible	*****	[93,229]
	Co-precipitation of Mg and/or Ca	Improvement in high-temperature performance	None	Negligible	***	[230]
	New type of Ni-Al double layered hydroxide	High capacity α-Ni(OH)$_2$/γ-NiOOH	None	Negligible	****	[58]
	Increase in Ni(OH)$_2$ crystallite size	Trade-off in activation	None	None	****	[231]
B. Surface coating	CoOOH coating	Enhancement in survival rate after long-term storage	None	Modest	*****	[121,139,223]
	Yb(OH)$_3$ coating	Improvement in high-temperature performance	None	Modest	****	[232]
	Electrode-less plating of Co	Improvement in Co-conductive network	None	Modest	****	[233]
	Co/Yb hydroxide coating	Improvement in high-temperature performance	None	Modest	****	[234]
C. Additives	Nano-sized Ni(OH)$_2$	Increase in electrochemical reaction reversibility	None	None	****	[235]
	Nano-sized ZnO	Increase in the flexibility of the electrode	None	None	****	[236]
	Co in paste	Formation of conductive Co-network	None	Modest	****	[237–239]
	CoO in paste	Formation of conductive Co-network	None	Modest	****	[110,240]
	Co(OH)$_2$ in paste	Formation of conductive Co-network	None	Modest	*****	[195,241,242]
	CoOOH in paste	Formation of conductive Co-network	None	Modest	*****	[243,244]
	CoSO$_4$ in paste	Formation of conductive Co-network	None	Modest	****	[245]
	Co$_3$O$_4$ in paste	Formation of conductive Co-network	None	Modest	****	[246]
	Co and CaCo$_3$	Prevention of oxygen evolution	None	Modest	****	[247,248]
	CuO in paste	Uniform dispersion of Co-conductive network	None	Negligible	***	[249]
	ZnO in paste	Prevention of oxygen evolution	None	Negligible	***	[250,251]
	Zn(OH)$_2$ in paste	Prevention of electrode swelling	None	Negligible	***	[252]
	Na$_{0.6}$CoO$_2$	Formation of better conductive Co-network	None	Modest	****	[253–255]
	RE	Decrease in oxidation rate of MH alloy	None	Modest	*****	[256–259]

Table 6. *Cont.*

Method	Direct impact	Environmental impact	Cost impact	Effectiveness	References
	Y_2O_3	Decrease in oxidation rate of MH alloy	None	Modest	*****	[23,250,260–262]
	$Y(OH)_3$	Decrease in oxidation rate of MH alloy	None	Modest	*****	[263,264]
	Oxides of heavy RE	Improvement in corrosion resistance	None	Modest	****	[213,265,266]
	Calcium metal borate	Prevention of oxygen evolution	None	Negligible	****	[267]
	CaF_2	Improvement in high-temperature performance	None	Negligible	***	[116]
	$Ca(OH)_2$	Improvement in high-temperature performance	None	Negligible	***	[268,269]
	CaS	Improvement in high-temperature performance	Reacts with acid and releases toxic H_2S gas	Negligible	***	[270]
	$Ca_3(PO_4)_2$	Improvement in high-temperature performance	None	Negligible	***	[271]
D. Electrode process	Use of sintered electrode	Enhancement in survival rate after long-term storage	None	Reduction	******	[272]
	Use of pasted electrode on NPPS	Increase in mechanical integrity	None	Reduction	***	[273]
	Use of granulated particles	Suppression of electrode swelling	None	None	*****	[274]
E. Substrate	Use of 3D Ni-plated steel sheet	Increase in power and cycle stability	None	Modest	****	[275]
	Use of Ni fiber felt	Increase in surface area and flexibility	None	Modest	****	[276]
	Pre-coating of Co-Ce alloy	Increase in contact area between substrate and $Ni(OH)_2$	None	Modest	***	[277]

3.4. Separator

The selection of the separator has a strong impact on the discharge capacity, voltage, and cycle stability [278]. Degradation related to the separator under storage and cycling conditions includes: (1) lower rates of electrolyte permeation in the separator; (2) lower electrolyte holding capability; (3) reduction in the separator volume due to electrode expansion; and (4) reduced gas recombination abilities [96]. Degradations (1) and (2) can be attributed to the debris formed in the separator as precipitation products ($ZnMn_2O_4$) of ions leached from the negative and positive electrodes [23,67]. These deposits not only offer a path for self-discharge, but also reduce the ionic conductivity and electrolyte holding capacity by filling the fine pores in the separator [26,29]. Degradation (3) can be traced to swelling of the positive electrode active material that accompanies over-charging, converting β-NiOOH to γ-NiOOH with Al-contamination leached from the negative electrode [29]. Methods to address separator degradation are listed in Table 7.

Table 7. Summary of cycle stability improvement methods related to the separator. EVOH: ethylene- vinyl alcohol copolymer; and AMPE: alkaline microporous polymer electrolyte.

Method	Direct impact	Environmental impact	Cost impact	Effectiveness	References
Sulfonated separator	Reduction in N-compound shuttling effects	None	Modest	★★★★★	[23,279–282]
Grafted acrylic acid/PP	Improvement in electrolyte holding capability	None	Negligible	★★★	[283]
Polymer gel-type	Improvement in durability	None	Negligible	★★★	[284,285]
Hydroentangled CMC composite	Improvement in integrity	None	Negligible	★★★	[286]
EVOH	Improvement in integrity	Cytotoxic	Modest	★★★	[287,288]
AMPE	Improvement in voltage window	None	Modest	★★	[289]
Addition of a K-conducting solid oxide film	Elimination of cross-contamination from the negative electrode	None	High	★★	New idea

3.5. Electrolyte

The earliest indication of performance degradation is a decrease in cell voltage, which can be traced to a reduction in the amount of electrolyte stored in the separator [50]. Electrolyte losses can be traced to: (1) electrode active material expansion and pulverization, causing an increase in surface area and the wicking of electrolyte away from the separator [54,290]; (2) venting of the cell; and (3) oxidation of metal [98,290]. The contamination in/through the electrolyte is also crucial for the life of both electrodes. Strategies involving the modification of electrolyte that can enhance cycle life performance in Ni/MH batteries are listed in Table 8.

Table 8. Summary of cycle stability improvement methods related to the electrolyte.

Method	Direct impact	Environmental impact	Cost impact	Effectiveness	References
Reduction in KOH concentration	Slow-down in alloy oxidation	None	None	★★★★	[291]
Replacement with NaOH	Slow-down in alloy oxidation	None	Negligible	★★★★★	[292]
ZnO additives	Slow-down in alloy oxidation	None	Negligible	★★★	[293]
LiOH additives	Prevention of K^+ migrating into $Ni(OH)_2$ and suppression of Fe-poisoning	None	Negligible	★★★	[93]
$Al_2(SO_4)_3$ additives	Slow-down in alloy oxidation	None	Negligible	★★★	[91]
NaH_2PO_4 additives	Formation of a Ni-rich surface on MH alloy	None	Negligible	★★★	[294]
$NaBO_2$ additives	Improvement of high-temperature cycle stability	None	Negligible	★★★	[295]
Na_2WO_4 additives	Increase in oxygen evolutionary potential	None	Negligible	★★★	[296]
$K_4Fe(CN)_6$ additives	Prevention of electrolyte decomposition	Highly toxic	Modest	★★★	[297]
Use of gel-type electrolyte	Reduction in corrosion and pulverization in the positive electrode	None	Modest	★★★	[298,299]
Use of polymer electrolyte	Wide voltage window and better mechanical integrity	None	Modest	★★★★	[300–310]

3.6. Other Components

Strategies to improve cycle stability not covered in Sections 3.1–5 are summarized in Table 9, which include charging processes, formation processes, storage conditions, and hardware modifications.

Table 9. Summary of cycle stability improvement methods related to other components. OCV: open-circuit voltage.

Method	Direct impact	Environmental Impact	Cost impact	Effectiveness	References
Install super water absorbing material at cell bottom	Reservoir for additional electrolyte	None	Negligible	★★★★★	[104]
Maintain cell OCV above 1.0 V	Prevention of Co dissolution and migration from the conductive network in the positive electrode	None	None	★★★★	[107,311,312]
Maintain cell OCV above 1.1 V	Prevention of Co dissolution and migration from the conductive network in the positive electrode	None	None	★★★★★	[313]
Reduction of depth of discharge	Prevention of swelling in the positive electrode	None	None	★★★★★	[110]
Reduction of number of shallow depth discharge	Prevention of memory effect	None	None	★★★★★	[314]
Implementation of an improved battery management system	Prevention of abuse	None	Modest	★★★★★	[315]
Pulse charging	Reduction in heat generated	None	Negligible	★★★	[316]
Optimization of formation parameters	Reduction in cell performance variation	None	None	★★★★	[317,318]
Battery sealing under vacuum	Reduction in inner pressure	None	Modest	★★	[319]
Improvement in sealing technology	Prevention of electrolyte leak	None	Negligible	★★	[320,321]

4. Revival of Degraded/Failed Battery

After long-term storage, a few small-current charge/discharge cycles can bring back some of the lost capacity in Ni/MH batteries [322,323]. A more complicated method proposed by Li and Meng [324] involves 33% SOC small-current charge, high-temperature storage (45–60 °C for 20–24 h), and a small current charge/discharge cycle to restore at least part of the lost capacity. Its strategy is to redistribute the Co-conductive network that was destroyed during storage. An alternative method uses ultrasound to disperse active materials from both electrodes in order to create a fresh surface and increase the capacity and power of the used cells [325].

Since the most common failure mode for Ni/MH batteries is electrolyte dry-out, opening cycled cells and refilling with fresh electrolyte can restore the capacity almost to the level before cycling [204]. For re-activation of MH alloy, a patent describes a method of recycling a deteriorated nickel-hydrogen battery by cleaning the cells with a concentrated sulfuric acid containing at least one type of Ni ion, Co ions, and La ions [326]. The concentrated sulfuric acid is poured into the deteriorated nickel-hydrogen battery and maintained at a temperature of 60 ± 10 °C while an electric current is applied to charge the nickel-hydrogen battery. After cleaning, the interior of the nickel-hydrogen battery is filled with an alkaline electrolyte containing a reducing agent. Consequently, γ-NiOOH converts to β-NiOOH, which restores the capacity of the positive electrode, and $RE(OH)_3$, $Al(OH)_3$, $Mn(OH)_2$, and $Co(OH)_2$ dissolve in the concentrated sulfuric acid to activate the negative electrode surface. In addition, the hydrophilic properties of the separator are restored following this method. Recycling used negative electrodes is also possible through the removal of oxide by acetic acid [327]. At the end of usable cycle life, procedures of dismantling, recovery, and reuse of spent Ni/MH batteries have been reported by Nan *et al.* [328, 329], Tenorio and Espinosa [330], Bertuol *et al.* [331], Zhang *et al.* [332], Rodrigues and Mansur [333], Muller and Friedrich [334], Rabah *et al.* [335], Santos *et al.* [336], and Larsson *et al.* [337]. U.S. Patents regarding recycling Ni/MH batteries are reviewed in a separate article [2].

5. Conclusions

Various failure modes and capacity degradation mechanisms are reviewed here. Solutions to enhance the cycle stability have been summarized in seven tables covering cell design, negative and positive electrodes, separator, electrolyte, and other hardware. After investigating the capacity-fade issue in a single cell, the next step is to study the consistency in the capacity degradation in a battery module composed of multiple cells.

Acknowledgments: The authors would like to thank the following people for their help: from Wayne State University (Simon Ng, Shuli Yan, and Ms. Shiuan Chang), BASF-Ovonic (Michael A. Fetcenko, Taihei Ouchi, Benjamin Reichman, Cristian A. Fierro, John Koch, Mr. Michael Zelinsky, Benjamin Chao, Jean Nei, Baoquan Huang, Tiejun Meng, Lixin Wang, Jun Im, Haoting Shen, Diana F. Wong, David Pawlik, Allen Chan, Ryan J. Blankenship, Nathan English, and Su Cronogue), FDK (Masazumi Tsukada, Hiroaki Yanagawa, Hirohito Teraoka,

Kazuta Takeno, and Jun Ishida), Japan Metals & Chemicals Company (Jin Nakamura), and Shenzhen High Power Battery Company (Wenliang Li, Lingkun Kong, and Xingqun Liao).

Author Contributions: All the authors made significant contribution to the writing of this manuscript.

Conflicts of Interest: The authors declare no conflict of interest.

References

1. Ouchi, T.; Young, K. Reviews on the Japanese Patents about nickel/metal hydride batteries. *Batteries* **2016**, submitted for publication.
2. Chang, S.; Nei, J.; Young, K. Reviews on the U.S. Patents about nickel/metal hydride batteries. *Batteries* **2016**, submitted for publication.
3. Young, K. Metal Hydrides. In *Reference Module in Chemistry, Molecular Sciences and Chemical Engineering*; Reedijk, J., Ed.; Elsevier: Waltham, MA, USA, 2012.
4. Yasuoka, S.; Magari, Y.; Murata, T.; Tanaka, T.; Ishida, J.; Nakamura, H.; Nohma, T.; Kihara, M.; Baba, Y.; Teraoka, H. Development of high-capacity nickel-metal hydride batteries using superlattice hydrogen-absorbing alloys. *J. Power Sources* **2006**, *156*, 662–666.
5. Watada, M.; Kuzuhara, M.; Oshitani, M. Development trend of rechargeable nickel-metal hydride battery for replacement of dry cell. *GS Yuasa Tech. Rep.* **2006**, *3*, 46–53.
6. Teraoka, H. Development of Low Self-Discharge Nickel-Metal Hydride Battery. Available online: http://www.scribd.com/doc/9704685/Teraoka-Article-En (accessed on 30 September 2015).
7. Noréus, D. Substitution of Rechargeable NiCd Batteries. Available online: http://ec.europa.eu/ environment/waste/studies/batteries/nicd.pdf (accessed on 30 September 2015).
8. Yonezu, I.; Yasuoka, S.; Kitaoka, K.; Nagae, T.; Takee, M. Advanced Ni-MH Batteries for Hybrid Electric Vehicles Using Super-Lattice Alloys as A Negative Electrode Materials. In Proceedings of the 8th International Advanced Automotive Battery and Ultracapacitor Conference (AABC-08), Tampa, FL, USA, 14–16 May 2008.
9. Young, K.; Wang, C.; Wang, L.Y.; Strunz, K. Electric Vehicle Battery Technologies. In *Electric Vehicle Integration into Modern Power Networks*; Garcia-Valle, R., Peças Lopes, J.A., Eds.; Springer: New York, NY, USA, 2013; pp. 15–56.
10. Yano Economic Research Institute Ltd. The Investigation Results from World Market of Batteries for Vehicle Use. Available online: https://www.yano.co.jp/press/pdf/1346.pdf (accessed on 30 September 2015).
11. Young, K.; Ng, K.Y.S.; Bendersky, L. A Technical Report of the Robust Affordable Next Generation Energy Storage System-BASF Program. *Batteries* **2016**, *2*.
12. Meng, T.; Young, K.; Beglau, D.; Yan, S.; Zeng, P.; Cheng, M.M. Hydrogenated Amorphous Silicon Thin Film Anode for Proton Conducting Batteries. *J. Power Sources* **2016**, *302*, 31–38.

13. Takasaki, T.; Nishimura, K.; Saito, M.; Fukunaga, H.; Iwaki, T.; Sakai, T. Cobalt-free nickel-metal hydride battery for industrial applications. *J. Alloys Compd.* **2013**, *580*, S378–S381.

14. Nishimura, K.; Takasaki, T.; Sakai, T. Introduction of large-sized nickel-metal hydride battery GIGACELL® for industrial applications. *J. Alloys Compd.* **2013**, *580*, S353–S358.

15. Zelinsky, M.; Fetcenko, M.A.; Kusay, J.; Koch, J. Storage-Integrated PV Systems Using Advanced NiMH Battery Technology. In Proceedings of the 5th International Renewable Energy Storage Conference, Berlin, Germany, 22–24 November 2010.

16. Wendler, M.; Hübner, G.; Krebs, M. Development of Printed Thin and Flexible Batteries. Available online: https://www.hdmstuttgart.de/international_circle/circular/issues/11_01/ICJ_04_32_wendler_huebner_ krebs.pdf (accessed on 4 October 2015).

17. Fetcenko, M.A.; Koch, J. Nickel-Metal Hydride Batteries. In *Linden's Handbook of Batteries*, 4th ed.; Reddy, T.B., Ed.; McGraw-Hill: New York, NY, USA, 2011; pp. 22.1–22.50.

18. Hall, D.S.; Lockwood, D.J.; Bock, C.; MacDougall, B.R. Nickel Hydroxides and Related Materials: A Review of Their Structures, Synthesis and Properties. *Proc. Math. Phys. Eng. Sci.* **2015**, *471*.

19. Young, K.; Nei, J. The current status of hydrogen storage alloy development for electrochemical applications. *Materials* **2013**, *6*, 4574–4608.

20. Ruiz, F.C.; Martinez, P.S.; Castro, E.B.; Humana, R.; Peretti, H.A.; Visintin, A. Effect of electrolyte concentration on the electrochemical properties of an AB_5-type alloy for Ni/MH batteries. *Int. J. Hydrog. Energy* **2013**, *38*, 240–245.

21. Karwowska, M.; Jarson, T.; Fijalkowski, K.J.; Leszczynski, P.J.; Rogilski, Z.; Czerwinski, A. Influence of electrolyte composition and temperature on behaviour of AB_5 hydrogen storage alloy used as negative electrode in Ni/MH Batteries. *J. Power Sources* **2014**, *263*, 304–309.

22. Li, X.; Chen, B.; Li, X.; Liu, J. Study on the storage performance improvement of Ni-MH cell. *Chin. Battery Ind.* **2001**, *6*, 18–20.

23. Shinyama, K.; Harada, Y.; Maeda, R.; Nakamura, H.; Matsuta, S.; Nohma, T.; Yonezu, I. Effect of separators on the self-discharge reaction in nickel-metal hydride batteries. *Res. Chem. Intermed.* **2006**, *32*, 447–452.

24. Kong, L.; Chen, B.; Young, K.; Koch, J.; Chan, A.; Li, W. Effects of Al- and Mn-contents in the negative MH alloy on the self-discharge and long-term storage properties of Ni/MH battery. *J. Power Sources* **2012**, *213*, 128–139.

25. Zhou, X.; Young, K.; West, J.; Regalado, J.; Cherisol, K. Degradation mechanisms of high-energy bipolar nickel metal hydride battery with AB_5 and A_2B_7 alloys. *J. Alloys Compd.* **2013**, *580*, S373–S377.

26. Young, K.; Koch, J.; Yasuoka, S.; Shen, H.; Bendersky, L.A. Mn in misch-metal based superlattice metal hydride alloy—Part 2 Ni/MH battery performance and failure mechanism. *J. Power Sources* **2015**, *277*, 433–442.

27. Bernard, P. Effects on the positive electrode of the corrosion of AB_5 alloys in nickel-metal-hydride batteries. *J. Electrochem. Soc.* **1998**, *145*, 456–458.

28. Maurel, F.; Knosp, B.; Backhaus-Ricoult, M. Characterization of corrosion products of AB_5-type hydrogen storage alloys for nickel-metal hydride batteries. *J. Electrochem. Soc.* **2000**, *147*, 78–86.

29. Wang, L.; Young, K.; Meng, T.; English, N.; Yasuoka, S. Partial substitution of cobalt for nickel in mixed rare earth metal based superlattice hydrogen absorbing alloy—Part 2 battery performance and failure mechanism. *J. Alloys Compd.* **2016**, *664*, 417–427.

30. Rajamathi, A.; Kamath, V.P.; Seshadri, R. Polymorphism in nickel hydroxide: Role of interstratification. *J. Mater. Chem.* **2000**, *10*, 503–506.

31. Young, K.; Chao, B.; Liu, Y.; Nei, J. Microstructures of the oxides on the activated AB_2 and AB_5 metal hydride alloys surface. *J. Alloys Compd.* **2014**, *606*, 97–104.

32. Young, K.; Chao, B.; Pawlik, D.; Shen, H. Transmission electron microscope studies in the surface oxide on the La-containing AB_2 metal hydride alloy. *J. Alloys Compd.* **2016**, submitted for publication.

33. Young, K.; Chang, S.; Chao, B.; Nei, J. Microstructures of the activated Si-containing AB_2 metal hydride alloy surface by transmission electron microscope. *Batteries* **2016**, *2*.

34. Young, K.; Huang, B.; Regmi, R.K.; Lawes, G.; Liu, Y. Comparisons of metallic clusters imbedded in the surface of AB_2, AB_5, and A_2B_7 alloys. *J. Alloys Compd.* **2010**, *506*, 831–840.

35. Ayari, M.; Paul-Boncour, V.; Lamloumi, J.; Percheron-Guégan, A.; Guillot, M. Study of the aging of $LaNi_{3.55}Mn_{0.4}Al_{0.3}(Co_{1-x}Fe_x)_{0.75}$ ($0 \leqslant x \leqslant 1$) compounds in Ni-MH batteries by SEM and magnetic measurements. *J. Magn. Magn. Mater.* **2005**, *288*, 374–383.

36. Qi, D.; Yang, Y.; Lin, Z. Investigation on degradation of storage characteristics of Ni-MH batteries. *Chin. J. Power Sources* **1998**, *22*, 236–239.

37. Watanabe, N.; Arakawa, T.; Sasaki, Y.; Yamashita, T.; Koiwa, I. Influence of the memory effect on X-ray photoelectron spectroscopy and Raman scattering in positive electrode of Ni-MH batteries. *J. Electrochem. Soc.* **2012**, *159*, A1949–A1953.

38. Higuchi, E.; Otsuka, H.; Chiku, M.; Inoue, H. Effect of pretreatment on the surface structure of a $Co(OH)_2$ electrode. *J. Power Sources* **2014**, *248*, 762–768.

39. Yan, D.; Suda, S. Electrochemical characteristics of $LaNi_{4.7}Al_{0.3}$ alloy activated by alkaline solution containing hydrazine. *J. Alloys Compd.* **1995**, *223*, 28–31.

40. Imoto, T.; Kato, K.; Higashiyama, N.; Kimoto, M.; Itoh, Y.; Nishio, K. Influence of surface treatment by HCL aqueous solution on electrochemical characteristics of a $Mm(Ni-Co-Al-Mn)_{4.76}$ alloy for nickel-metal hydride battery. *J. Alloys Compd.* **1999**, *282*, 274–278.

41. Young, K.; Ouchi, T.; Banik, A.; Koch, J.; Fetcenko, M.A. Improvement in the electrochemical properties of gas atomized AB_2 metal hydride alloys by hydrogen annealing. *Int. J. Hydrog. Energy* **2011**, *36*, 3547–3555.

42. Etiemble, A.; Idrissi, H.; Meille, S.; Roué, L. *In situ* investigation of the volume change and pulverization of hydride materials for Ni-MH batteries by concomitant generated force and acoustic emission measurements. *J. Power Sources* **2012**, *205*, 500–505.

43. Etiemble, A.; Roue, L.; Idrissi, H. Study of Metal Electrodes for Ni-MH Batteries by Acoustic Emission. In Proceedings of the European Working Group on Acoustic Emission (EWAGE 2010), Vienna, Austria, 8–10 September 2010.

44. Oliva, P.; Leonardi, J.; Laurent, J.F.; Delmas, C.; Braconnier, J.J.; Figlarz, M.; Fievet, F.; Guibert, A. Review of the structure and the electrochemistry of nickel hydroxides and oxy-hydroxides. *J. Power Sources* **1982**, *8*, 229–255.

45. Yang, S.; Yin, Y.; Chen, H.; Jia, J.; Zhang, M.; Ding, L. Charge/discharge characteristics and structural changing of spherical β-Ni(OH)$_2$ prepared from different nickel salt. *J. Electrochem.* **2001**, *7*, 310–315.

46. Bantignies, J.L.; Deabate, S.; Righi, A.; Rols, S.; Hermet, P.; Sauvajol, J.L.; Henn, F. New insight into the vibration behavior of nickel hydroxide and oxyhydroxide using inelastic neutron scattering, far/mid-infrared and Raman spectroscopies. *J. Phys. Chem. C* **2008**, *112*, 2193–2201.

47. Hall, D.S.; Lockwood, D.J.; Poirier, S.; Bock, C.; McDougall, B.R. Raman and infrared spectroscopy of α and β phases of thin nickel hydroxide films electrochemically formed on nickel. *J. Phys. Chem. A* **2012**, *116*, 6771–6784.

48. Hall, D.S.; Lockwood, D.J.; Poirier, S.; Book, C.; MacDougall, B.R. Applications of *in situ* Raman spectroscopy for identifying nickel hydroxide materials and surface layers during chemical aging. *ACS Appl. Mater. Interface* **2014**, *6*, 3141–3149.

49. Kuriyama, N.; Sakai, T.; Miyamura, H.; Uehara, I.; Ishikawa, H. Electrochemical impedance spectra and deterioration mechanism of metal hydride electrodes. *J. Electrochem. Soc.* **1992**, *139*, L72–L73.

50. Cheng, S.; Zhang, J.; Liu, H.; Leng, Y.; Yuan, A.; Cao, C. Study on cycling deterioration of Ni-MH battery by electrochemical impedance spectroscopy (EIS). *Chin. J. Power Sources* **1999**, *23*, 62–63.

51. Mo, Z.; Xu, P.; Han, X.; Zhao, L. Electrochemical impedance behavior of Ni-MH battery under different depth of discharge. *Chin. J. Power Sources* **2006**, *30*, 388–390.

52. Guenne, L.L.; Benard, P. Life duration of Ni-MH cells for high power applications. *J. Power Sources* **2002**, *105*, 134–138.

53. Geng, M.; Han, J.; Feng, F.; Northwood, D.O. Charging/discharging stability of a metal hydride battery electrode. *J. Electrochem. Soc.* **1999**, *146*, 2371–2375.

54. Yuan, J.; Qi, J.; Tu, J.; Feng, H.; Chen, C.; Chen, W. Electrochemical analysis of failed Ni-MH batteries. *Chin. J. Power Sources* **2001**, *25*, 279–282.

55. Yuan, X.; Xu, N. Electrochemical and hydrogen transport kinetic performance of MlNi$_{3.75}$Co$_{0.63}$Mn$_{0.4}$Al$_{0.2}$ metal hydride electrodes at various temperatures. *J. Electrochem. Soc.* **2002**, *149*, A407–A413.

56. Sequeira, C.A.C.; Chen, Y.; Santos, D.M.F. Effects of temperature on the performance of the MmNi$_{3.6}$Co$_{0.7}$Mn$_{0.4}$Al$_{0.3}$ metal hydride electrode in alkaline solution. *J. Electrochem. Soc.* **2006**, *153*, A1863–A1867.

57. Gamboa, S.A.; Sebastian, P.J.; Feng, F.; Geng, M.; Northwood, D.O. Cyclic voltammetry investigation of a metal hydride electrode for nickel metal hydride batteries. *J. Electrochem. Soc.* **2002**, *149*, A137–A139.

58. Béléké, A.B.; Higuchi, E.; Inoue, H.; Mizuhata, M. Durability of nickel-metal hydride (Ni-MH) battery cathode using nickel-aluminum layered double hydroxide/carbon (Ni-AL LDH/C) composite. *J. Power Sources* **2014**, *247*, 572–578.

59. Lyons, M.E.G.; Doyle, R.L.; Godwin, I.; O'Brien, M.; Russell, L. Hydrous nickel oxide: Redox switching and the oxygen evolution reaction in aqueous alkaline solution. *J. Electrochem. Soc.* **2012**, *159*, H932–H944.

60. Zhang, Q.; Su, G.; Li, A.; Liu, K. Electrochemical performance of AB_5-type hydrogen storage alloy modified with Co_3O_4. *Trans. Nonferrous Met. Soc. Chin.* **2011**, *21*, 1428–1434.

61. Serrao, L.; Chehab, Z.; Guezennec, Y.; Rizzoni, G. An Aging Model of Ni-MH Batteries for Hybrid Electric Vehicles. In Proceedings of the IEEE Vehicle Power and Propulsion Conference, Chicago, IL, USA, 7–9 September 2005; pp. 78–85.

62. Somogye, R.H. An Aging Model of Ni-MH Batteries for Use in Hybrid-Electric Vehicles. Master's Thesis, Ohio State University, Columbus, OH, USA, 2004.

63. Picciano, N. Battery Aging and Characterization of Nickel Metal Hydride and Lead Acid Batteries. Bachelor's Thesis, Ohio State University, Columbus, OH, USA, 2007.

64. Ikoma, M.; Hoshina, Y.; Matsumoto, I. Self-discharge mechanism of sealed-type nickel/metal-hydride battery. *J. Electrochem. Soc.* **1996**, *143*, 1904–1907.

65. Young, K.; Ouchi, T.; Koch, J.; Fetcenko, M.A. Compositional optimization of vanadium-free hypo-stoichiometric AB_2 metal hydride alloy for Ni/MH battery application. *J. Alloys Compd.* **2012**, *510*, 97–106.

66. Guo, H.; Qiao, Y.; Zhang, H. Fast determination of micro short circuit in sintered MH-Ni battery. *Chin. J. Power Sources* **2010**, *34*, 608–609.

67. Shinyama, K.; Magari, Y.; Kumagae, K.; Nakamura, H.; Nohma, T.; Takee, M.; Ishiwa, K. Deterioration mechanism of nickel metal-hydride batteries for hybrid electric vehicles. *J. Power Sources* **2005**, *141*, 193–197.

68. Zhu, W.H.; Zhu, Y.; Tatarchuk, B.J. Self-discharge characteristics and performance degradation of Ni-MH batteries for storage applications. *Int. J. Hydrog. Energy* **2014**, *39*, 19789–19798.

69. Lee, J.-H.; Lee, K.-Y.; Lee, J.-Y. Self-discharge behaviour of sealed Ni-MH batteries using $MmNi_{3.3+x}Co_{0.7}Al_{1.0-x}$ anodes. *J. Alloys Compd.* **1996**, *232*, 197–203.

70. Wang, C.; Marrero-Rivera, M.; Serafini, D.A.; Baricuatro, J.H.; Soriaga, M.P.; Srinivasan, S. The self-discharge mechanism of AB_5-type hydride electrodes in Ni/MH batteries. *Int. J. Hydrog. Energy* **2006**, *31*, 603–611.

71. Lee, J.; Lee, K.; Lee, S.; Lee, J. Self-discharge characteristics of sealed Ni-MH batteries using $Zr_{1-x}Ti_xV_{0.8}Ni_{1.6}$ anodes. *J. Alloys Compd.* **1995**, *221*, 174–179.

72. Jang, K.; Jung, J.; Kim, D.; Yu, J.; Lee, J. Self-discharge mechanism of vanadium-titanium metal hydride electrodes for Ni-MH rechargeable battery. *J. Alloys Compd.* **1998**, *268*, 290–294.

73. Yang, X.; Liaw, B.Y. Self-discharge and charge retention in AB_2-based nickel metal hydride batteries. *J. Electrochem. Soc.* **2004**, *151*, A137–A143.

74. Fan, M.; Liao, W.; Wu, B.; Chen, H.; Jian, X. Temperature characteristic of Ni-MH battery used in EVs. *Chin. Battery Ind.* **2004**, *9*, 287–289.

75. Huang, J.; Li, X. Factors affecting the self-discharge of Ni-MH batteries. *Chin. Battery Ind.* **2005**, *10*, 290–294.

76. Zhu, W.H.; Zhu, Y.; Davis, Z.; Tatarchuk, B.J. Energy efficiency and capacity retention of Ni-MH batteries for storage applications. *Appl. Energy* **2013**, *106*, 307–313.

77. NiMH Battery Charging Algorithm. Available online: http://www.schaffler.com/whitepapers/NiMH Battery Algorithm.pdf (accessed on 4 October 2015).

78. Kritzer, P. Separators for nickel metal hydride and nickel cadmium batteries designed to reduce self-discharge rates. *J. Power Sources* **2004**, *137*, 317–321.

79. Liu, Y.; Tang, Z.; Xu, Q.; Zhang, X.; Liu, Y. Assessment of separators in high rate discharge performance of Ni−MH battery. *Chin. J. Process Eng.* **2006**, *6*, 114–119.

80. Li, X.; Song, Y.; Wang, L.; Xia, T.; Li, S. Self-discharge mechanism of Ni-MH battery by using acrylic acid grafted polypropylene separator. *Int. J. Hydrog. Energy* **2010**, *35*, 3798–3801.

81. Takasaki, T.; Nishimura, K.; Saito, M.; Iwaki, T.; Sakai, T. Cobalt-free materials for nickel-metal hydride battery: Self-discharge suppression and overdischarge resistance improvement. *Electrochemistry* **2013**, *81*, 553–558.

82. Li, X.; Li, J.; Tan, P.; Li, Y. Effects of electrolyte weight on self-discharge of Ni-MH battery. *Chin. Battery Ind.* **2008**, *13*, 299–302.

83. Zai, Y.; Han, J.; Li, Y.; Lai, X.; Duan, Q. Influence of copper on the micro-short in Ni-MH battery. *Chin. Battery Ind.* **2012**, *17*, 78–80.

84. Zha, Y.; Chen, X.; Li, L.; Yang, Z.; Wang, L. Effects of the negative current collector on the performance of the Ni-MH battery. *Chin. Battery Ind.* **2006**, *11*, 91–93.

85. Cheng, S.; Zhang, J.; Yuan, A.; Ding, W.; Cao, C. Effects of separator on discharge capacity and cycle-life of Ni-MH battery. *Chin. J. Power Sources* **1999**, *23*, 10–12.

86. Li, X.; Wang, X.; Dong, H.; Xia, T.; Wang, L.; Song, Y. Self-discharge performance of Ni-MH battery by using electrodes with hydrophilic/hydrophobic surface. *J. Phys. Chem. Solids* **2013**, *74*, 1756–1760.

87. Li, X.; Li, J. Effects of the surface treatment of formed positive electrode on self-discharge performance of Ni-MH batteries. *Chin. Battery Ind.* **2009**, *14*, 298–301.

88. Feng, F.; Northwood, D.O. Self-discharge characteristics of a metal hydride electrode for Ni-MH rechargeable batteries. *Int. J. Hydrog. Energy* **2005**, *30*, 1367–1370.

89. Lv, Y.; Zhang, J. Study on the performance of metal-hydride electrodes coated with Ni-B alloy. *Contemp. Chem. Ind.* **2014**, *43*, 1453–1455, 1463.

90. Deng, Z. Approach on influence factors of discharge oneself MH-Ni battery. *Jiangxi Metal.* **2003**, *23*, 137–139.

91. Wang, Z.M.; Tsai, P.; Chan, S.L.I.; Zhou, H.Y.; Lin, K.S. Effects of electrolytes and temperature on high-rate discharge behavior of MmNi$_5$-based hydrogen storage alloys. *Int. J. Hydrog. Energy* **2010**, *35*, 2033–2039.

92. Notten, P.H.L.; Einerhand, R.E.F.; Daams, J.L.C. How to achieve long-term electrochemical cycling stability with hydride-forming electrode materials. *J. Alloys Compd.* **1995**, *231*, 604–610.

93. Dong, Q.; Yan, G.; Yu, C. Research on the cycle life of Ni-MH battery. *Chin. J. Power Sources* **2000**, *24*, 306–310.

94. Yan, J.; Zhou, Z.; Li, Y.; Song, D.; Zhang, Y. Structure and property changes of positive and negative electrodes in Ni/MH batteries during charge/discharge cycles. *J. Inorg. Chem.* **1998**, *14*, 74–78.

95. Tliha, M.; Mathlouthi, H.; Lamloumi, J.; Percheron-Guegan, A. AB_5-type hydrogen storage alloy used as anodic materials in Ni-MH batteries. *J. Alloys Compd.* **2007**, *436*, 221–225.

96. Lin, J.; Cheng, Y.; Liang, F.; Sun, L.; Yin, D.; Wu, Y.; Wang, L. High temperature performance of $La_{0.6}Ce_{0.4}Ni_{3.45}Co_{0.75}Mn_{0.7}Al_{0.1}$ hydrogen storage alloy for nickel/metal hydride batteries. *Int. J. Hydrog. Energy* **2014**, *39*, 13231–13239.

97. Meli, F.; Schlapbach, L. Surface analysis of AB_5-type electrodes. *J. Less Common Met.* **1991**, *172*, 1252–1259.

98. Li, L.; Chen, Y.; Wu, F.; Chen, S. Development of recycling and rusing process for nickel metal hydride batteries for HEV. *J. Funct. Mater.* **2007**, *38*, 1928–1932.

99. Geng, M.; Hah, J.; Feng, F.; Northwood, D.O. Characteristics of the high-rate discharge capability of a nickel/metal hydride battery electrode. *J. Electrochem. Soc.* **1999**, *146*, 3591–3595.

100. Wang, C.; Jin, H.; Li, G.; Wang, R. Study on microstructures of $La_{0.8}Nd_{0.2}Ni_{3.55}Co_{0.75}Mn_{0.4}Al_{0.3}$ alloy during the charge-discharge cycle process. *Chin. J. Power Sources* **1998**, *22*, 65–67.

101. Chen, Y.; Wei, J.; Gao, F.; Zhou, W.; Zhang, Y.; Zhou, Z. Discussing of invalidation factors of Ni-MH battery at high charge and discharge rate. *Chin. J. Power Sources* **2001**, *25*, 142–145.

102. Maurel, F.; Hytch, M.J.; Knosp, B.; Backhaus-Ricoult, M. Formation of rare earth hydroxide nanotubes and whiskers as corrosion product of $LaNi_5$-type alloys in aqueous KOH. *Eur. Phys. J. Appl. Phys.* **2000**, *9*, 205–213.

103. Liu, Y.; Pan, H.; Gao, M.; Lei, Y.; Wang, Q. Degradation mechanism of the La-Mg-Ni-based metal hydride electrode $La_{0.7}Mg_{0.3}Ni_{3.4}Mn_{0.1}$. *J. Electrochem. Soc.* **2005**, *152*, A1089–A1095.

104. Xie, D.; Cheng, H. Study on the fading of MH-Ni batteries. *Chin. Battery Ind.* **2000**, *5*, 7–10.

105. Xie, J.; Wang, S.; Xia, B.; Zhang, Q.; Shi, P. Research on the degradation of Ni-MH batteries (I)—Swelling of nickel electrode. *Chin. J. Power Sources* **1997**, *21*, 22–27.

106. Singh, D. Characteristics and Effects of γ-NiOOH on cell performance and a method to quantify it in nickel electrodes. *J. Electrochem. Soc.* **1998**, *145*, 116–120.

107. Deng, X.; Huang, S.; Wang, Y.; Wang, J. Study on the shelf performance of Ni-MH batteries. *Chin. Battery Ind.* **2006**, *11*, 172–174.

108. Durairajan, A.; Haran, B.S.; White, R.E.; Popov, B.N. Pulverization and corrosion studies of bare and cobalt-encapsulated metal hydride electrodes. *J. Power Sources* **2000**, *87*, 84–91.

109. Li, X.; Li, X.; He, X.; Liu, J.; Chen, B.; Li, C. Investigation on internal resistance of Ni-MH battery during cycle life. *Chin. Battery Ind.* **2001**, *6*, 69–71.

110. Wu, J.; Li, W.; Zai, Y. Relations of specific capacity of Ni-MH battery with self discharge and cycle life. *Chin. Battery Ind.* **2004**, *9*, 197–199.

111. Sastry, A.M.; Choi, S.B.; Cheng, X. Damage in composite NiMH positive electrodes. *J. Eng. Mater. Technol.* **1998**, *120*, 280–283.

112. Zhou, Z.Q.; Lin, G.W.; Zhang, J.L.; Ge, J.S.; Shen, J.R. Degradation behavior of foamed nickel positive electrodes of Ni-MH batteries. *J. Alloys Compd.* **1999**, *293–295*, 795–798.

113. Young, K.; Wu, A.; Qiu, Z.; Tan, J.; Mays, W. Effects of H_2O_2 addition to the cell balance and self-discharge of Ni/MH batteries with AB_5 and A_2B_7 alloys. *Int. J. Hydrog. Energy* **2012**, *37*, 9882–9891.

114. Chen, W.; Xu, Z.; Tu, J. Electrochemical investigations of activation and degradation of hydrogen storage alloy electrodes in sealed Ni/MH battery. *Int. J. Hydrog. Energy* **2007**, *27*, 439–444.

115. Singh, P.; Wu, T.; Wendung, M.; Bendale, P.; Ware, J.; Ritter, D.; Zhang, L. Mechanisms causing capacity loss on long term storage in NiMH system. *Mater. Res. Soc. Symp. Proc.* **1998**, *496*, 25–36.

116. Zhang, X.; Gong, Z.; Zhao, S.; Geng, M.; Wang, Y.; Northwood, D.O. High-temperature characteristics of advanced Ni-MH batteries using nickel electrodes containing CaF_2. *J. Power Sources* **2008**, *175*, 630–634.

117. Iwakura, C.; Kajiya, Y.; Yoneyama, H.; Sakai, T.; Oguro, K.; Ishikawa, H. Self-discharge mechanism of nickel-hydrogen batteries using metal hydride anodes. *J. Electrochem. Soc.* **1989**, *136*, 1351–1355.

118. Leblanc, P.; Blanchard, P.; Senyarich, S. Self-discharge of sealed nickel-metal hydride batteries. *J. Electrochem. Soc.* **1998**, *145*, 844–847.

119. Wang, C.S.; Marrero-Rivera, M.; Baricuatro, J.H.; Soriaga, M.P.; Serafini, D.; Srinivasan, S. Corrosion behavior of AB_5-type hydride electrodes in alkaline electrolyte solution. *J. Appl. Electrochem.* **2003**, *33*, 325–331.

120. Rukpakavong, W.; Phillips, I.; Guan, L. Lifetime Estimation of Sensor Device with AA NiMH Batteries. Available online: http://www.ipcsit.com/vol55/018-ICICM2012-Contents.pdf (accessed on 04 October 2015).

121. Bäuerlein, P.; Antonius, A.; Löffler, J.; Kümpers, J. Progress in high-power nickel–metal hydride batteries. *J. Power Sources* **2008**, *176*, 547–554.

122. Shi, F. The research of voltage's accuracy and temperature effect on the MH-Ni battery group. *Microcomput. Appl.* **2013**, *32*, 60–62.

123. Li, X.; Ma, L.; Lou, Y.; Li, F.; Xia, B. Storage performance of Ni-MH battery at state of discharge. *Chin. J. Power Sources* **2004**, *28*, 364–368.

124. Notten, P.H.L.; Van Beek, J.R.G. Nickel-metal hydride batteries: From concept to characteristics. *Chem. Ind.* **2000**, *54*, 102–115.

125. Li, L.; Wu, F.; Yang, K.; Wang, J.; Chen, S. The effects of overcharge on electrochemical performance of MH-Ni batteries. *Mater. Rev.* **2004**, *18*, 101–102.

126. Hu, W.K.; Geng, M.M.; Gao, X.P.; Burchardt, T.; Gong, Z.X.; Noréus, D.; Nakstad, N.K. Effect of long-term overcharge and operated temperature on performance of rechargeable NiMH cells. *J. Power Sources* **2006**, *159*, 1478–1483.

127. Ye, Z.; Noréus, D. Oxygen and hydrogen gas recombination in NiMH cells. *J. Power Sources* **2012**, *208*, 232–236.

128. Chen, R.; Li, L.; Wu, F.; Qiu, X.; Chen, S. Effects of low temperature on performance of hydrogen-storage alloys and electrolyte. *Min. Metal. Eng.* **2007**, *27*, 44–46.

129. Guo, H.; Chao, M.; Chen, X.; Zhang, B.; Zhang, J. Study on the low temperature performance of MH-Ni battery. *J. Henan Normal Univ. Nat. Sci.* **2003**, *31*, 76–78.

130. Liaw, B.Y.; Yang, X. Limiting mechanism on rapid charging Ni-MH batteries. *Electrochim. Acta* **2001**, *47*, 875–884.

131. Raju, M.; Ananth, M.V.; Vijayaraghavan, L. Rapid charging characterization of $MmNi_{3.03}Si_{0.85}Co_{0.60}Mn_{0.31}Al_{0.08}$ alloy used as anodes in Ni-MH batteries. *Int. J. Hydrog. Energy* **2009**, *34*, 3500–3505.

132. Chen, S.; Zheng, Q.; Wang, F.; Hu, D. Effects of fast charging on efficiency of Ni-MH batteries. *J. Beijing Technol. Bus. Univ. Nat. Sci. Ed.* **2006**, *24*, 5–8.

133. Taheri, P.; Yazdanpour, M.; Bahrami, M. Analytical assessment of the thermal behavior of nickel-metal hydride batteries during fast charging. *J. Power Sources* **2014**, *245*, 712–720.

134. Hou, X.; Xue, J.; Nan, J.; Xia, X.; Yang, M.; Cui, Y. The cycle life performance of MH-Ni batteries for high power applications. *Acta Sci. Nat. Univ. Sunyats.* **2005**, *44*, 77–80.

135. Ayeb, A.; Otten, W.M.; Mank, J.G.; Notten, P.H.L. The hydrogen evolution and oxidation kinetics during overdischarging of sealed nickel-metal hydride batteries. *J. Electrochem. Soc.* **2006**, *153*, A2055–A2065.

136. Cha, C.; Yu, J.; Zhang, J. Comparative experimental study of gas evolution and gas consumption reactions in sealed Ni-Cd and Ni-MH cells. *J. Power Sources* **2004**, *129*, 347–357.

137. Cuscueta, D.J.; Salva, H.R.; Ghilarducci, A.A. Inner pressure characterization of a sealed nickel-metal hydride cell. *J. Power Sources* **2011**, *196*, 4067–4071.

138. Li, N. Methods to prevent explosion in MH-Ni battery. *Chin. Battery Ind.* **2003**, *8*, 277.

139. Li, N.; Xu, Y. Discussion on explosion of MH-Ni batteries. *Chin. Battery Ind.* **2008**, *13*, 372–373.

140. Zhu, X.; Ni, J.; Xu, W.; Huang, L.; Feng, S. Analysis of common malfunctions and maintenance of nickel metal hydride battery used in AGV. *Log. Mater. Handl.* **2013**, *10*, 156–158.

141. Liu, C.; Li, J.; Chen, H.; Liu, J. The development of high performance SubC MH-Ni battery. *J. Hunan Univ. Technol.* **2012**, *26*, 64–67.

142. Fetcenko, M.A.; Young, K.; Fierro, C. Hydrogen Storage Battery; Positive Nickel Electrode; Positive Electrode Active Material and Methods of Making. U.S. Patent 7,396,379, 8 July 2008.

143. Li, X.; Tian, J.; Xia, T. A new way to reduce the consumption of hydrogen-absorbing alloy in Ni-MH battery. *Chin. Battery Ind.* **2011**, *16*, 74–78.

144. Hu, W.; Shan, Z.; Tian, J. Impact of ratio of anode/cathode active materials on internal pressure of MH-Ni batteries. *Chem. Ind. Eng.* **2006**, *23*, 95–97.

145. Sastry, A.M.; Cheng, X.; Wang, C. Mechanics of stochastic fibrous networks. *J. Thermoplast. Compos. Mater.* **1998**, *3*, 288–296.

146. Yu, T.; Zhai, Y.; Zhou, S.; Duan, Q. Effect of precharge on performance of Ni-MH battery. *Chin. Battery Ind.* **2005**, *10*, 74–76.

147. Guiose, B.; Cuevas, F.; Décamps, B.; Leroy, E.; Percheron-Guégan, A. Microstructural analysis of the ageing of pseudo-binary (Ti,Zr)Ni intermetallic compounds as negative electrodes of Ni-MH batteries. *Electrochim. Acta* **2009**, *54*, 2781–2789.

148. Liu, B.H.; Li, Z.P.; Matsuyama, Y.; Kitani, R.; Suda, S. Corrosion and degradation behavior of Zr-based AB_2 alloy electrodes during electrochemical cycling. *J. Alloys Compd.* **2000**, *296*, 201–208.

149. Liu, B.H.; Li, Z.P.; Kitani, R.; Suda, S. Improvement of electrochemical cyclic durability of Zr-based AB_2 alloy electrodes. *J. Alloys Compd.* **2002**, *330–332*, 825–830.

150. Lee, H.; Lee, K.; Lee, J. Degradation mechanism of Ti-Zr-V-Mn-Ni metal hydride electrodes. *J. Alloys Compd.* **1997**, *260*, 201–207.

151. Yu, J.; Lee, S.; Cho, K.; Lee, J. The cycle life of $Ti_{0.8}Zr_{0.2}V_{0.5}Mn_{0.5-x}Cr_xNi_{0.8}$ ($x = 0$ to 0.5) alloys for metal hydride electrodes of Ni-metal hydride rechargeable battery. *J. Electrochem. Soc.* **2000**, *147*, 2013–2017.

152. Liu, B.H.; Li, Z.P.; Suda, S. Electrochemical cycle life of Zr-based Laves phase alloys influenced by alloy stoichiometry and composition. *J. Electrochem. Soc.* **2002**, *149*, A537–A542.

153. Martíez, P.S.; Ruiz, F.C.; Visintin, A. Influence of different electrolyte concentrations on the performance of an AB_2-type alloy. *J. Electrochem. Soc.* **2014**, *161*, A326–A329.

154. Rongeat, C.; Roué, L. On the cycle life improvement of amorphous MgNi-based alloy for Ni-MH batteries. *J. Alloys Compd.* **2005**, *404–406*, 679–681.

155. Ruggeri, S.; Roué, L. Correlation between charge input and cycle life of MgNi electrode for Ni-MH batteries. *J. Power Sources* **2003**, *117*, 260–266.

156. Rongeat, C.; Grosjean, M.-H.; Ruggeri, S.; Dehmas, M.; Bourlot, S.; Marcotte, S.; Roué, L. Evaluation of different approaches for improving the cycle life of MgNi-based electrodes for Ni-MH batteries. *J. Power Sources* **2006**, *158*, 747–753.

157. Goo, N.H.; Woo, J.H.; Lee, K.S. Mechanism of rapid degradation of nanostructured Mg_2Ni hydrogen storage alloy electrode synthesized by mechanical alloying and the effect of mechanically coating with nickel. *J. Alloys Compd.* **1999**, *288*, 286–293.

158. Tsukahara, M.; Takahashi, K.; Isomura, A.; Sakai, T. Improvement of the cycle stability of vanadium-based alloy for nickel-metal hydride (Ni-MH) battery. *J. Alloys Compd.* **1999**, *287*, 215–220.

159. Bliznakov, S.; Lefterova, E.; Dimitrov, N.; Petrov, K.; Popov, A. A study of the Al content impact on the properties of $MmNi_{4.4-x}Co_{0.6}Al_x$ alloys as precursors for negative electrodes in NiMH batteries. *J. Power Sources* **2008**, *176*, 381–386.

160. Zhou, W.; Ma, Z.; Wu, C.; Zhu, D.; Huang, L.; Chen, Y. The mechanism of suppressing capacity degradation of high-Al-AB$_5$-type hydrogen storage alloys at 60 °C. *Int. J. Hydrog. Energy* **2016**, *41*, 1801–1810.

161. Nakamura, Y.; Oguro, K.; Uehara, I.; Akiba, E. X-ray diffraction peak broadening and degradation in LaNi$_5$-based alloys. *Int. J. Hydrog. Energy* **2000**, *25*, 531–537.

162. Wang, X.; Liu, B.; Yuan, H.; Zhang, Y. Effect of Al in Mm(NiCoMnAl)$_5$ on the performance of the sealed MH/Ni cell. *Chin. J. Appl. Chem.* **1999**, *16*, 54–57.

163. Zhang, Y.; Wang, X.; Chen, M.; Li, P.; Lin, Y.; Li, R. Design of chemical compositions on AB$_5$-type hydrogen storage alloy with high quality. *Met. Funct. Mater.* **2002**, *9*, 1–6.

164. Chandra, D.; Chien, W.; Talekar, A. Metal Hydrides for NiMH Battery Applications. Available online: http://www.sigmaaldrich.com/technical-documents/articles/material-matters/metal-hydrides-for.html (accessed on 1 October 2015).

165. Qingxue, Z.; Joubert, J.-M.; Latroche, M.; Jun, D.; Percheron-Guégan, A. Influence of the rare earth composition on the properties of Ni-MH electrodes. *J. Alloys Compd.* **2003**, *360*, 290–293.

166. Guo, J.H.; Jiang, Z.T.; Duan, Q.S.; Liu, G.Z.; Zhang, J.H.; Xia, C.M.; Guo, S.Y. Influence of Ce and Nd contents of Ml(NiCoMnAl)$_5$ alloy on the performances of Ni-MH batteries. *J. Alloys Compd.* **1999**, *293–295*, 821–824.

167. Sakai, T.; Miyamura, H.; Kuriyama, N.; Kato, A.; Oguro, K.; Ishikawa, H. Metal hydride anodes for nickel-hydrogen secondary battery. *J. Electrochem. Soc.* **1990**, *137*, 795–799.

168. Seo, C.; Choi, S.; Choi, J.; Park, C.; Lee, P.S.; Lee, J. Effect of Ti and Zr additions on the characteristics of AB$_5$-type hydride electrode for Ni-MH secondary battery. *Int. J. Hydrog. Energy* **2003**, *28*, 317–327.

169. Srivastava, S.; Upadhyay, R.K. Investigations of AB$_5$-type negative electrode for nickel-metal hydride cell with regard to electrochemical and microstructural characteristics. *J. Power Sources* **2010**, *195*, 2996–3001.

170. Young, K.; Ouchi, T.; Banik, A.; Koch, J.; Fetcenko, M.A.; Bendersky, L.A.; Wang, K.; Vaudin, M. Gas atomization of Cu-modified AB$_5$ metal hydride alloys. *J. Alloys Compd.* **2011**, *509*, 4896–4904.

171. Spodaryk, M.; Shcherbakova, L.; Sameljuk, A.; Wichser, A.; Zakaznova-Herzog, V.; Holzer, M.; Braem, B.; Khyzhun, O.; Mauron, P.; Remhof, A.; *et al.* Description of the capacity degradation mechanism in LaNi$_5$-based alloy electrodes. *J. Alloys Compd.* **2015**, *621*, 225–231.

172. Anderson, M.L.; Anderson, I.E. Gas atomization of metal hydrides for Ni-MH battery applications. *J. Alloys Compd.* **2000**, *313*, 47–52.

173. Bowman, R.C., Jr.; Witham, C.; Fultz, B.; Ratnakumar, B.V.; Ellis, T.W.; Anderson, I.E. Hydriding behavior of gas-atomized AB$_5$ alloys. *J. Alloys Compd.* **1997**, *253–254*, 613–616.

174. Li, C.; Wang, F.; Cheng, W.; Li, W.; Zhao, W. The influence of high-rite quenching on the cycle stability and the structure of the AB$_5$-type hydrogen storage alloys with different Co content. *J. Alloys Compd.* **2001**, *315*, 218–223.

175. Li, C.; Wang, X.; Wang, C. Investigations on ML(NiCoMnTi)$_5$ alloys prepared with different solidification rates. *J. Alloys Compd.* **1998**, *266*, 300–306.

176. Zou, J.; Gao, J.; Zhong, F.; Wang, X. Effect of nickel electroplating on negative electrode surface on cycle life and inner pressure of nickel hydride battery. *Electroplat. Finish.* **2009**, *28*, 9–13.

177. Yang, K.; Wu, F.; Chen, S.; Zhang, C. Influences of Ni plated on electrodes on the cycle life and inner pressure of batteries. *Min. Metal. Eng.* **2006**, *26*, 74–76.

178. Zhang, D.; Yuan, H.; Wang, G.; Zhou, Z.; Zhang, Y. Study on electrochemical characteristic of Cu-coated hydrogen storage alloy. *J. Funct. Mater. Dev.* **1998**, *4*, 167–174.

179. Feng, F.; Northwood, D.O. Effect of surface modification on the performance of negative electrodes in Ni/MH batteries. *Int. J. Hydrog. Energy* **2004**, *29*, 955–960.

180. Huang, J.; Huang, H. Effect of Cu addition on discharge performance of metal hydride electrode for Ni-MH battery. *Rare Met. Lett.* **2000**, *3*, 18–19.

181. Raju, M.; Ananth, M.V.; Vijayaraghavan, L. Influence of electroless coatings of Cu, Ni-P and Co-P on MmNi$_{3.25}$Al$_{0.35}$Mn$_{0.25}$Co$_{0.66}$ alloy used as anodes in Ni-MH batteries. *J. Alloys Compd.* **2009**, *475*, 664–671.

182. Manimaran, K.; Ananth, M.V.; Raju, M.; Renganathan, N.G.; Ganesan, M.; Nithya, G. Electrochemical investigations on cobalt-microencapsulated MmNi$_{2.38}$Al$_{0.82}$Co$_{0.66}$Si$_{0.77}$Fe$_{0.13}$Mn$_{0.24}$ hydrogen storage alloys for Ni-MH batteries. *Int. J. Hydrog. Energy* **2010**, *35*, 4630–4637.

183. Visintin, A.; Tori, C.A.; Garaventta, G.; Triaca, W.E. The effect of palladium coating on hydrogen storage alloy electrodes for nickel/metal hydride batteries. *J. Braz. Chem. Soc.* **1997**, *8*, 125–129.

184. Lv, Y. A study on decorative electrochemical coating of hydrogen storage alloy powder with nickel-phosphorus alloy. *J. Tangshan Coll.* **2014**, *27*, 42–44.

185. Bi, X.; Shan, Z.; Tian, J. Study on the performance of metal-hydride electrodes coated with Ni-S alloy in Ni-MH battery. *Chin. J. Power Sources* **2005**, *29*, 746–749.

186. Casella, I.G.; Gatta, M. Electrodeposition and characterization of nickel-copper alloy films as electrode material in alkaline media. *J. Electrochem. Soc.* **2002**, *149*, B465–B471.

187. Chen, W.; Chen, Y.; Pan, H.; Chen, C. Surface treatments of hydrogen storage alloy improve charge-discharge performances of Ni/MH battery. *Acta Phys. Chim. Sin.* **1998**, *14*, 742–746.

188. Lee, H.; Yang, D.; Park, C.; Park, C.; Jang, H. Effects of surface modifications of the LMNi$_{3.9}$Co$_{0.6}$Mn$_{0.3}$Al$_{0.2}$ alloy in a KOH/NaBH$_4$ solution upon its electrode characteristics within a Ni-MH secondary battery. *Int. J. Hydrog. Energy* **2009**, *34*, 481–486.

189. Li, C. Hydrogen storage alloy and fluorinated hydrating alloy batteries. *Chin. J. Power Sources* **1995**, *19*, 34–37.

190. Sakashita, M.; Li, Z.P.; Suda, S. Fluorination mechanism and its effects on the electrochemical properties of metal hydrides. *J. Alloys Compd.* **1997**, *253–254*, 500–505.

191. Sun, Y.; Iwata, K.; Chiba, S.; Matsuyama, Y.; Suda, S. Studies on the properties and characteristics of the fluorinated AB$_5$ hydrogen-absorbing electrode alloys. *J. Alloys Compd.* **1997**, *253–254*, 520–524.

192. Sun, Y.; Suda, S. Studies on the fluorination method for improving surface properties and characteristics of AB_5-types of hydrides. *J. Alloys Compd.* **2002**, *330–332*, 627–631.

193. Chao, D.; Shi, P.; Zhang, S. Effect of surface modification on the electrochemical performances of $LaNi_5$ hydrogen storage alloy in Ni/MH batteries. *Mater. Chem. Phys.* **2006**, *98*, 514–518.

194. Lou, Y.; Ma, L.; Xia, B.; Yang, C. XRD study on microstructure of rare earth containing hydrogen storage alloy in Ni-MH battery. *Rare Met. Mater. Eng.* **2006**, *35*, 412–417.

195. Pang, L.; Ye, P.; Guo, J. Methods of reducing MH-Ni battery internal pressure and increasing cycle life. *Liaoning Chem. Ind.* **1998**, *27*, 228–230.

196. Sakai, T.; Miyamura, H.; Kuriyama, N.; Ishikawa, H.; Uehara, I. Rare-earth-based hydrogen storage alloys for rechargeable nickel-metal hydride batteries. *J. Alloys Compd.* **1993**, *192*, 155–157.

197. Gao, J.; Yan, X.L.; Zhao, Z.Y.; Chai, Y.J.; Hou, D.L. Effect of annealed treatment on microstructure and cyclic stability for La-Mg-Ni hydrogen storage alloys. *J. Power Sources* **2012**, *209*, 257–261.

198. Ma, J.; Pan, H.; Zhu, Y.; Li, S.; Chen, C. A novel method to improve the electrochemical properties of hydrogen storage electrode alloy. *Acta Metal. Sin.* **2001**, *37*, 57–60.

199. Zhou, C.; Tao, M.; Yan, K.; Chen, Y.; Wu, C. Effects of nickel additive on the performance of the negative electrode for Ni-MH battery. *Chin. J. Power Sources* **2003**, *27*, 301–304.

200. Jung, J.; Lee, S.; Kim, D.; Jang, K.; Lee, J. Effect of Cu powder as compacting material on the discharge characteristics of negative electrodes in Ni–MH batteries. *J. Alloys Compd.* **1998**, *266*, 271–275.

201. Mao, L.; Tong, J.; Shan, Z.; Yin, S.; Wu, F. Effect of Co additives on the cycle life of Ni-MH batteries. *J. Alloys Compd.* **1999**, *293–295*, 829–832.

202. Di, L.; Chen, L. Influence of the CoO in MH electrode on the performance of battery. *Bull. Chin. Ceram. Soc.* **2003**, *3*, 88–90.

203. Chen, L.; Tian, Y.; Zhang, F. Influence of CoO addition on the MH electrode's performance. *J. Tianjin Inst. Technol.* **2003**, *19*, 54–56.

204. Meng, M.; Liu, M.; Deng, X.; Huang, S.; Zhan, J. Research on the cycle life of Ni-MH batteries. *Chin. J. Power Sources* **1998**, *22*, 155–157.

205. Guo, J.; Li, W.; Meng, M. Study on effect of technology factors on the properties of Ni-MH battery. *Chin. J. Power Sources* **2000**, *24*, 319–321.

206. Pang, L.; Zhang, J.; Guo, J.; Wang, B.; Li, K.; Duan, Q. Method of reducing internal pressure of Ni-MH battery and increasing cycle life. *Chin. J. Power Sources* **1999**, *23*, 158–160.

207. Petrov, K.; Rostami, A.A.; Visintin, A.; Srinivasan, S. Optimization of composition and structure of metal-hydride electrodes. *J. Electrochem. Soc.* **1994**, *141*, 1747–1750.

208. Guo, W.; Xue, J.; Xu, Q.; Tang, Z. Effect of additive BC-1 on the performance of Ni-MH battery. *Chin. J. Power Sources* **2007**, *31*, 151–153.

209. Cheng, Y.; Zhang, H.; Zhang, G.; Chen, Y.; Zhu, Q. Influence of addition of carbon nanotubes in the hydrogen storage alloy electrode on the performance of SC high power batteries. *Acta Phys. Chim. Sin.* **2008**, *24*, 527–532.

210. Xie, F.; Dong, C.; Qian, W.; Zhai, Y.; Li, L.; Li, D. Electrochemical properties of nickel-metal hydride battery based on directly grown multiwalled carbon nanotubes. *Int. J. Hydrog. Energy* **2015**, *40*, 8935–8940.

211. Arnaud, O.; Le Guenne, L.; Audry, C.; Bernard, P. Effect of yttria content on corrosion of AB_5-type alloys for nickel-metal hydride batteries. *J. Electrochem. Soc.* **2005**, *152*, A611–A616.

212. Tao, M.; Chen, Y.; Yan, K.; Zhou, C.; Wu, C.; Tu, M. Effect of rare earth oxides on the electrochemical performance of hydrogen storage electrode. *Rare Met. Mater. Eng.* **2005**, *34*, 552–556.

213. Tanaka, T.; Kuzuhara, M.; Watada, M.; Oshitani, M. Effect of rare earth oxide additives on the performance of NiMH batteries. *J. Alloys Compd.* **2006**, *408–412*, 323–326.

214. Shinyama, K.; Nakamura, H.; Nohma, T.; Yonezu, I. Improvement of high- and low-temperature characteristics of nickel-metal hydride secondary batteries using rare-earth compounds. *J. Alloys Compd.* **2006**, *408–412*, 288–293.

215. Gamboa, S.A.; Sebastian, P.J.; Geng, M.; Northwood, D.O. Temperature, cycling, discharge current and self-discharge electrochemical studies to evaluate the performance of a pellet metal-hydride electrode. *Int. J. Hydrog. Energy* **2001**, *26*, 1315–1318.

216. Ma, G.; Guo, J.; Yan, Y.; Zhang, J.; Duan, Q.; Liu, G. Application of sintered metal hydride in Ni-MH battery. *Chin. Battery Ind.* **2003**, *8*, 113–115.

217. Oshitani, M.; Yufu, H.; Takashima, K.; Tsuji, S.; Matsumaru, Y. Development of a pasted nickel electrode with high active material utilization. *J. Electrochem. Soc.* **1989**, *136*, 1590–1593.

218. Oshitani, M.; Sasaki, Y.; Takashima, K. Development of a nickel electrode having stable performance at various charge and discharge rates over a wide temperature range. *J. Power Sources* **1984**, *12*, 219–231.

219. Friebel, D.; Bajdich, M.; Yao, B.S.; Louie, M.W.; Miller, D.J.; Casalongue, H.S.; Mbuga, F.; Weng, T.; Nordlund, D.; Sokaras, D.; *et al.* On the chemical state of Co oxide electrocatalysts during alkaline water splitting. *Phys. Chem. Chem. Phys.* **2013**, *15*, 17460–17467.

220. Pralong, V.; Delahaye-Vidal, A.; Beaudoin, B.; Leriche, J.-B.; Tarascon, J.-M. Electrochemical behavior of cobalt hydroxide used as additive in the nickel hydroxide electrode. *J. Eletrochem. Soc.* **2000**, *147*, 1306–1313.

221. Otsuka, H.; Chiku, M.; Higuchi, E.; Inoue, H. Characterization of pretreated $Co(OH)_2$-coated $Ni(OH)_2$ positive electrode for Ni-MH batteries. *ECS Trans.* **2012**, *41*, 7–12.

222. Pralong, P.; Chabre, Y.; Delahaye-Vidal, A.; Tarascon, J.-M. Study of the contribution of cobalt additive to the behavior of the nickel oxy-hydroxide electrode by potentiodynamic techniques. *Solid State Ion.* **2002**, *147*, 73–84.

223. Li, X.; Xia, T.; Dong, H.; Wie, Y. Study on the reduction behavior of CoOOH during the storage of nickel/metal-hydride battery. *Mater. Chem. Phys.* **2006**, *100*, 486–489.

224. Fu, Z.; Jiang, W.; Yu, L. Preparation of nickel hydroxide coated by cobalt hydroxide. *Chin. J. Nonferr. Met.* **2005**, *15*, 1775–1779.

225. Chang, Z.; Tang, H.; Peng, P. Studies of electrochemical characteristics of nickel hydroxide with surface deposited cobalt. *J. Funct. Mater.* **2001**, *32*, 487–489.

226. Yin, T. Surface modification of nickel hydroxide particles by micro-sized cobalt oxide hydroxide and properties as electrode materials. *Surf. Coat. Technol.* **2005**, *200*, 2376–2379.

227. Ovshinsky, S.R.; Fetcenko, M.A.; Fierro, C.; Gifford, P.R.; Corrigan, D.A.; Benson, P.; Martin, F.J. Enhanced Nickel Hydroxide Positive Electrode Materials for Alkaline Rechargeable Electrochemical Cells. U.S. Patent 5,523,182, 4 June 1996.

228. Yin, X.; Li, J.; Fu, J.; Gong, L.; Xia, X. Preparation, characterization and electrochemical performance of cobalt-substituted nickel hydroxide. *Chem. Res. Appl.* **2002**, *14*, 405–408.

229. Tessier, C.; Faure, C.; Guerlou-Demourgues, L.; Denage, C.; Nabias, G.; Delmas, C. Electrochemical study of zinc-substituted nickel hydroxide. *J. Electrochem. Soc.* **2002**, *149*, A1136–A1145.

230. Fierro, C.; Zallen, A.; Koch, J.; Fetcenko, M.A. The influence of nickel-hydroxide composition and microstructure on the high-temperature performance of nickel metal hydride batteries. *J. Electrochem. Soc.* **2006**, *153*, A492–A496.

231. Zhang, S.; Han, Z. Influence of average grain size on the electrochemical properties of nickel hydroxide. *Battery Ind.* **1999**, *4*, 133–135.

232. Li, W.; Jiang, C.; Wan, C. High-temperature performances of spherical nickel hydroxide coated with Yb(OH)$_3$. *J. Inorg. Mater.* **2006**, *21*, 121–127.

233. Ortiz, M.G.; Real, S.G.; Castro, E.B. Preparation and characterization of positive electrode of Ni-MH batteries with cobalt additives. *Int. J. Hydrog. Energy* **2014**, *39*, 8661–8666.

234. He, X.M.; Jiang, C.Y.; Li, W.; Wan, C.R. Co/Yb hydroxide coating of spherical Ni(OH)$_2$ cathode materials for Ni-MH batteries at elevated temperatures. *J. Electrochem. Soc.* **2006**, *153*, A566–A569.

235. Han, E.; Xu, H.; Kang, H.; Feng, Z. Performance comparison between nano-sized and micro-sized spherical nickel hydroxide. *J. Dalian Marit. Univ.* **2008**, *34*, 87–90.

236. Chen, Q.; Yi, S.; Zhao, M. Influence of nanometric zinc oxide on the electrochemical performance of nickel metal hydride batteries. *Rare Met. Mater. Eng.* **2010**, *39*, 39–42.

237. Yuan, A.; Cheng, S.; Zhang, J.; Cao, C. Effects of metallic cobalt addition on the performance of pasted nickel electrodes. *J. Power Sources* **1999**, *77*, 178–182.

238. Wang, X.; Yan, J.; Yuan, H.; Zhou, Z.; Song, D.; Zhang, Y.; Zhu, L. Surface modification and electrochemical studies of spherical nickel hydroxide. *J. Power Sources* **1998**, *72*, 221–225.

239. Wang, X.; Leo, H.; Yang, H.; Sebastian, P.J.; Gamboa, S.A. Oxygen catalytic evolution reaction on nickel hydroxide electrode modified by electroless cobalt coating. *Int. J. Hydrog. Energy* **2004**, *29*, 967–972.

240. Li, Q. Study on capacity loss of Ni-MH batteries after storage at low potential. *Chin. Battery Ind.* **2006**, *11*, 303–306.

241. Chang, Z.R.; Tang, H.; Chen, J.G. Surface modification of spherical nickel hydroxide for nickel electrodes. *Electrochem. Commun.* **1999**, *1*, 513–516.

242. Cheng, S.; Wenhau, L.; Zhang, J.; Cao, C. Electrochemical properties of the pasted nickel electrode using surface modified $Ni(OH)_2$ powder as active material. *J. Power Sources* **2001**, *101*, 248–252.

243. Hu, W.K.; Gao, X.P.; Gong, Z.X. Synthesis of CoOOH nanorods and application as coating materials of nickel hydroxide for high temperature Ni-MH cells. *J. Phys. Chem. B* **2005**, *109*, 5392–5394.

244. Hu, Q.D.; Gao, X.P.; Li, G.R.; Pan, G.L.; Yan, T.Y. Microstructure and electrochemical properties of Al-substituted nickel hydroxides modified with CoOOH nanoparticles. *J. Phys. Chem. C* **2007**, *111*, 17082–17087.

245. Xia, X.; Zhang, W.; Huang, H.; Gan, Y.; Zhang, L. Effects of $CoSO_4$ as additive on the electrochemical properties of nickel hydroxide electrode. *Zhejiang Chem. Ind.* **2007**, *38*, 1–4.

246. Su, G.; He, Y.; Liu, K. Effects of Co_3O_4 as additive on the performance of metal hydride electrode and Ni–MH battery. *Int. J. Hydrog. Energy* **2012**, *37*, 11994–12002.

247. Nathira Begum, S.; Muralidharan, V.S.; Ahmed Basha, C. The influences of some additives on electrochemical behaviour of nickel electrodes. *Int. J. Hydrog. Energy* **2009**, *34*, 1548–1555.

248. Yuan, A.; Cheng, S.; Zhang, J.; Cao, C. The influence of calcium compounds on the behaviour of the nickel electrode. *J. Power Sources* **1998**, *76*, 36–40.

249. Li, J.; Li, R.; Wu, J.; Su, H. Effect of cupric oxide addition on the performance of nickel electrode. *J. Power Sources* **1999**, *79*, 86–90.

250. Jung, K.; Yang, D.; Park, C.; Park, C.; Choi, J. Effects of the addition of ZnO and Y_2O_3 on the electrochemical characteristics of a $Ni(OH)_2$ electrode in nickel-metal hydride secondary batteries. *Int. J. Hydrog. Energy* **2010**, *35*, 13073–13077.

251. Yuan, X.; Wang, Y.; Zhan, F. Study on the additives for inhibiting swelling of nickel electrode. *Chin. J. Power Sources* **2000**, *24*, 192–196.

252. Chen, J.; Bradhurst, D.H.; Dou, S.X.; Liu, H.K. Nickel hydroxide as an active material for the positive electrode in rechargeable alkaline batteries. *J. Electrochem. Soc.* **1999**, *146*, 3606–3612.

253. Douin, M.; Guerlou-Demourgues, L.; Goubault, L.; Bernard, P.; Delmas, C. Effect of deep discharge on the electrochemical behavior of cobalt oxides and oxyhydroxides used as conductive additives in Ni-MH cells. *J. Power Sources* **2009**, *193*, 864–870.

254. Douin, M.; Guerlou-Demourgues, L.; Goubault, L.; Bernard, P.; Delmas, C. Influence of the electrolyte alkaline ions on the efficiency of the $Na_{0.6}CoO_2$ conductive additive in Ni-MH cells. *J. Electrochem. Soc.* **2008**, *155*, A945–A951.

255. Tronel, F.; Guerlou-Demourgues, L.; Basterreix, M.; Delmas, C. The $Na_{0.60}CoO_2$ phase, a potential conductive additive for the positive electrode of Ni–MH cells. *J. Power Sources* **2006**, *158*, 722–729.

256. Morioka, Y.; Narukawa, S.; Itou, T. State-of-the-art of alkaline rechargeable batteries. *J. Power Sources* **2001**, *100*, 107–116.

257. Ye, X.; Zhu, Y.; Wu, S.; Zhang, Z.; Zhou, Z.; Zheng, H.; Lin, X. Study on the preparation and electrochemical performance of rare earth doped nano-Ni(OH)$_2$. *J. Rare Earths* **2011**, *29*, 787–792.

258. Ren, J.X.; Zhou, Z.; Gao, X.P.; Yan, J. Preparation of porous spherical α-Ni(OH)$_2$ and enhancement of high-temperature electrochemical performances through yttrium addition. *Eletrochim. Acta* **2006**, *52*, 1120–1126.

259. Mi, X.; Gao, X.P.; Jiang, C.Y.; Geng, M.M.; Yan, Y.; Wan, C.R. High temperature performances of yttrium-doped spherical nickel hydroxide. *Eletrochim. Acta* **2004**, *49*, 3361–3366.

260. Li, F.; Yang, Y. Effects of nano-sized Y$_2$O$_3$ additive on performance of Ni-MH battery. *Chin. Battery Ind.* **2008**, *13*, 226–228.

261. Kaiya, H.; Ookawa, T. Improvement in cycle life performance of high capacity nickel-metal hydride battery. *J. Alloys Compd.* **1995**, *231*, 598–603.

262. Mi, X.; Ye, M.; Yan, J.; Wei, J.; Gao, X. High temperature performance of spherical nickel hydroxide with additive Y$_2$O$_3$. *J. Rare Earths* **2004**, *22*, 422–425.

263. Wu, Q.D.; Liu, S.; Li, L.; Yan, T.Y.; Gao, X.P. High-temperature electrochemical performance of Al-α-nickel hydroxides modified by metallic cobalt or Y(OH)$_3$. *J. Power Sources* **2009**, *186*, 521–527.

264. Fan, J.; Yang, Y.; Yu, P.; Chen, W.; Shao, H. Effects of surface coating of Y(OH)$_3$ on the electrochemical performance of spherical Ni(OH)$_2$. *J. Power Sources* **2007**, *171*, 981–989.

265. Oshitani, M.; Watada, M.; Shodai, K.; Kodama, M. Effect of lanthanide oxide additives on the high-temperature charge acceptance characteristics of pasted nickel electrodes. *J. Electrochem. Soc.* **2001**, *148*, A67–A73.

266. Ren, J.; Wang, X.; Li, Y.; Gao, X.; Yan, J. Effect of Lu$_2$O$_3$ on charge/discharge performances of spherical nickel hydroxide at high temperature. *J. Rare Earths* **2005**, *23*, 732–736.

267. Li, J.; Shangguan, E.; Guo, D.; Li, Q.; Chang, Z.; Yuan, X.; Wang, H. Calcium metaborate as a cathode additive to improve the high-temperature properties of nickel hydroxide electrodes for nickel-metal hydride batteries. *J. Power Sources* **2014**, *263*, 110–117.

268. Li, W.; Zhang, S.; Chen, J. Synthesis, characterization, end electrochemical application of Ca(OH)$_2$-, Co(OH)$_2$- and Y(OH)$_3$-coated Ni(OH)$_2$ tubes. *J. Phys. Chem. B* **2005**, *109*, 14025–14032.

269. Hu, M.; Lei, L.; Chen, J.; Sun, Y. Improving the high-temperature performances of a layered double hydroxide, [Ni$_4$Al(OH)$_{10}$]NO$_3$, through calcium hydroxide coatings. *Eletrochim. Acta* **2011**, *56*, 2862–2869.

270. Matsuda, H.; Ikoma, M. Nickel/metal-hydride battery for electric vehicles. *Electrochemistry* **1997**, *65*, 96–100.

271. He, X.; Ren, J.; Li, W.; Jiang, C.; Wan, C. Ca$_3$(PO$_4$)$_2$ coating of spherical Ni(OH)$_2$ cathode materials for Ni–MH batteries at elevated temperature. *Eletrochim. Acta* **2011**, *56*, 4533–4536.

272. Guo, H.; Wang, W.; Zhang, W.; Wang, J. Effects of long-time storage in discharging state on the performance of Ni-MH batteries. *Chin. Battery Ind.* **2009**, *14*, 166–168.

273. Fukunaga, H.; Kishimi, M.; Igarashi, N.; Ozaki, T.; Sakai, T. Non-foam-type nickel electrodes using various binders for Ni-MH batteries. *J. Electrochem. Soc.* **2005**, *152*, A42–A46.

274. Nishimura, K.; Takasaki, T.; Sakai, T. Development of high-capacity Ni(OH)$_2$ electrode using granulation process. *Stud. Sci. Technol.* **2013**, *2*, 107–112.

275. Fukunaga, H.; Kishimi, M.; Matsumoto, N.; Ozaki, T.; Sakai, T.; Tanaka, T.; Kishimoto, T. A nickel electrode with Ni-coated 3D steel sheet for hybrid electric vehicle applications. *J. Electrochem. Soc.* **2005**, *152*, A905–A912.

276. Xi, Z.; Zhang, J.; Wu, L.; Li, J.; Zhu, R.; Yang, Y.; Feng, P.; Zuo, C.; Shi, D. New type of anode material for the MH-Ni battery. *Rare Met. Mater. Eng.* **1999**, *28*, 371–374.

277. Wang, D.; Wang, C.; Dai, C.; Sun, D. Study of Co-Ce coating and surface on pasted nickel electrodes substrate. *Rare Met.* **2006**, *25*, 47–50.

278. Kritzer, P.; Cook, J.A. Nonwoven as separators for alkaline batteries—An overview. *J. Electrochem. Soc.* **2007**, *154*, A481–A494.

279. Yu, T.; Zhai, Y.; Li, Q.; Duan, Q. Study on storage of Ni-MH battery. *Chin. Battery Ind.* **2005**, *10*, 149–152.

280. Ni, J.; Wan, D.; Zhang, L. Influences of using separators made in China on Ni-MH battery performance in electric bicycle. *Electr. Bicycle* **2013**, *6*, 14–20.

281. Cao, S.; Wang, H.; Jin, X. Study on the Sulfonation Treatment of ES Nonwoven Fabrics as Battery Separator. In Proceedings of the 7th National Conference of Functional Materials and Applications, Changsha, China, 15–18 October 2010; Zhao, G.-M., Ed.; Scientific Research Publishing: Merced, CA, USA; pp. 2075–2078.

282. Furukawa, N. Development and commercialization of nickel-metal hydride secondary batteries. *J. Power Sources* **1994**, *51*, 45–59.

283. Du, H.; Liu, D.; Kong, Y.; Yang, J. Hydrophilic modification of polypropylene MH-Ni battery separator and its characterizations. *Polym. Mater. Sci. Eng.* **2013**, *29*, 101–104.

284. Wang, F.; Wu, F.; Mao, L.; Zou, L.; Liu, Y.; Chen, S. Research on polymer-gel electrolyte bipolar nicker/metal hydride batteries. *Mater. Rev.* **2005**, *19*, 124–126.

285. Cai, Z.; Gong, Z.; Geng, M.; Tang, Z. Study on a novel polymer gel improved separator in nickel/metal hydride battery. *Chin. J. Power Sources* **2005**, *29*, 142–145.

286. Tian, Y.; Gao, H.; Wang, J.; Jin, X.; Wang, H. Preparation of hydroentangled CMC composite nonwoven fabrics as high performance separator for nickel metal hydride battery. *Electrochim. Acta* **2015**, *177*, 321–326.

287. Shigematsu, T.; Oku, Y.; Ishikawa, T. Nonwoven Fabric for Alkaline Battery Separator and Manufacture Thereof. Jpn. Patent Application H06–251760, 9 September 1994.

288. Yao, Y.; Hu, J.; Yang, J.; Wang, Y.; Liang, Y. Preliminary study of making MH-Ni battery separator. *Pap. Sci. Technol.* **2009**, *28*, 76–80.

289. Chen, Z.; Ju, Y.; Zhang, W.; Zhu, Y.; Li, L. Research in preparation and performance of a novel alkaline microporous polymer electrolyte. *Mater. Rev.* **2009**, *23*, 316–318, 334.

290. Leblanc, P.; Jordy, C.; Knosp, B.; Blanchard, P.H. Mechanism of alloy corrosion and consequences on sealed nickel-metal hydride battery performance. *J. Electrochem. Soc.* **1998**, *145*, 860–863.

291. Khaldi, C.; Mathlouthi, H.; Lamloumi, J. A comparative study, of 1 M and 8 M KOH electrolyte concentrations, used in Ni-MH batteries. *J. Alloys Compd.* **2009**, *469*, 464–471.

292. Pei, L.; Yi, S.; He, Y.; Chen, Q. Effect of electrolyte formula on the self-discharge properties of nickel-metal hydride batteries. *J. Guangdong Univ. Technol.* **2008**, *25*, 10–12.

293. Wang, C.; Marrero-Rivera, M.; Soriaga, M.P.; Serafini, D.; Srinivasan, S. Improvement in the cycle life of LaB_5 metal hydride electrodes by addition of ZnO to alkaline electrolyte. *Eletrochim. Acta* **2002**, *47*, 1069–1078.

294. Matsuoka, M.; Asai, K.; Fukumoto, Y.; Iwakura, C. Electrochemical characterization of surface-modified negative electrodes consisting of hydrogen storage alloys. *J. Alloys Compd.* **1993**, *192*, 149–151.

295. Shangguan, E.; Wang, J.; Li, J.; Dan, G.; Chang, Z.; Yuan, X.; Wang, H. Enhancement of the high-temperature performance of advanced nickel-metal hydride batteries with NaOH electrolyte containing $NaBO_2$. *Int. J. Hydrog. Energy* **2013**, *38*, 10616–10624.

296. Shangguan, E.; Li, J.; Guo, D.; Chang, Z.; Yuan, X.; Wang, H. Effects of different electrolytes containing Na_2WO_4 on the electrochemical performance of nickel hydroxide electrodes for nickel–metal hydride batteries. *Int. J. Hydrog. Energy* **2014**, *39*, 3412–3422.

297. Zhu, X.; Yang, H.; Ai, X. Possible use of ferrocyanide as a redox additive for prevention of electrolyte decomposition in overcharged nickel batteries. *Electrochim. Acta* **2003**, *48*, 4033–4037.

298. Mao, L.; Wu, F.; Chen, S.; Wang, F.; Liu, Y. Application of gel polymer electrolytes in Ni/MH battery. *J. Funct. Mater.* **2005**, *36*, 1389–1390.

299. Sun, S.; Song, J.; Shan, Z.; Hu, M.; Lei, L.; Chen, J.; Sun, Y. Electrochemical properties of a low molecular weight gel electrolyte for nickel/metal hydride cell. *Eletrochim. Acta* **2014**, *130*, 689–692.

300. Ju, Y.; Chen, Z.; Zhu, Y.; Li, L. Modification mechanism and research progress of polymer electrolytes in Ni/MH batteries. *Mater. Rev.* **2009**, *23*, 54–57.

301. Yang, C. Polymer Ni-MH battery based on PEO-PVA-KOH polymer electrolyte. *J. Power Sources* **2002**, *109*, 22–31.

302. Vassal, N.; Salmon, E.; Fauvarque, J.-F. Nickel/metal hydride secondary batteries using an alkaline solid polymer electrolyte. *J. Electrochem. Soc.* **1999**, *146*, 20–26.

303. Iwakura, C.; Ikoma, K.; Nohara, S.; Furukawa, N. Charge-discharge and capacity retention characteristics of new type Ni/MH batteries using polymer hydrogel electrolyte. *J. Electrochem. Soc.* **2003**, *150*, A1623–A1627.

304. Wang, C.; Wallace, G.G. Flexible electrodes and electrolytes for energy storage. *Electrochim. Acta* **2015**, *175*, 87–95.

305. Zhang, Y.; Pang, H.; Yang, H.; Zhou, Z. Electrochemical stability of PAA alkaline polymer electrolyte and its application in Ni/MH secondary battery. *Acta Sci. Nat. Univ. Nankaiensis* **2008**, *41*, 57–61.

306. Tang, Z.; Li, C.; Liu, Y. Research and development of polymer Ni/MH battery. *Chem. Ind. Eng. Prog.* **2006**, *25*, 1251–1255.

307. Wu, R.; Lu, X.; Zhu, Y.; Li, L. Preparation and properties of PVA/PAAK alkaline polymer electrolytes for nickel metal hydride batteries. *Polym. Mater. Sci. Eng.* **2013**, *29*, 171–174.

308. Lu, X.; Wu, R.; Zhu, Y.; Li, L. Preparation and properties of PVA-PAA-KOH alkaline polymer electrolyte membrane. *J. Funct. Mater.* **2013**, *44*, 590–594.

309. Lu, X.; Wu, R.; Li, B.; Zhu, Y.; Li, L. Novel PVA/SiO$_2$ alkaline micro-porous polymer electrolytes for polymer Ni-MH batteries. *Acta Chim. Sin.* **2013**, *71*, 427–432.

310. Vignarooban, K.; Dissanayake, M.A.K.L.; Albinsson, I.; Mellander, B.-E. Ionic conductivity enhancement in PEO:CuSCN solid polymer electrolyte by the incorporation of nickel-chloride. *Solid State Ion.* **2015**, *278*, 177–180.

311. Li, X.; Xia, B. Corrosion of AB$_5$ alloy during the storage on the performance of nickel-metal-hydride battery. *J. Alloys Compd.* **2005**, *391*, 190–193.

312. Li, X.; Xia, B. Effect of charged-state storage on Ni-MH battery performance. *Chin. J. Power Sources* **2004**, *28*, 737–739.

313. Zhang, Z.; Jia, C.; Xing, Z.; Li, L.; Ma, Y. Factors on storage performance of MH-Ni battery. *J. Rare Earths* **2004**, *22*, 347–348.

314. Cao, M.; Zhang, J.; Du, P.; Wang, J. Investigation on performance of Ni-MH battery. *Chin. J. Power Sources* **2006**, *30*, 852–855.

315. Jung, D.Y.; Lee, B.H.; Kim, S.W. Development of battery management system for nickel-metal hydride batteries in electric vehicle applications. *J. Power Sources* **2002**, *109*, 1–10.

316. Darcy, E.C. Investigation of the Response of NiMH Cells to Burp Charging. Ph.D. Thesis, University of Houston, Houston, TX, USA, 1998.

317. Ma, S.; Yang, K.; Lui, Y.; Lan, Z. The effect of charge/discharge manner on properties of hydrogen storage alloy electrode. *J. Guangxi Univ. Nat. Sci. Ed.* **2006**, *31*, 205–207.

318. Cao, S.; Ma, Y. Influences of the formation on the storage performance of high power MH/Ni battery. *Sci. Technol. Baotou Steel Group Corp.* **2008**, *34*, 50–52.

319. Chen, B.; Li, X.; Li, X.; Zhang, Q. Method of reducing internal pressure of alkaline rechargeable battery. *Chin. Battery Ind.* **2000**, *5*, 211–213.

320. Bai, Z.; Yu, X.; Lu, F. Sealing technology about cylindrical alkaline battery. *Chin. J. Power Sources* **2003**, *27*, 381–384.

321. Yuan, J.; Chen, W. Research analysis on design optimization of cell sealing structure. *Electr. Bicycl.* **2015**, *3*, 10–11.

322. Song, Q. Performance recovery of Ni-MH battery after long storing. *Chin. Battery Ind.* **2005**, *10*, 77–79.

323. Xie, D.; Li, Q.; Wu, T. Discussion on some aspects of MH-Ni battery. *Chin. Battery Ind.* **2001**, *6*, 114–117.

324. Li, N.; Meng, J. Capacity recovery of Ni-MH batteries after long-term storage. *Chin. Battery Ind.* **2008**, *13*, 95–96.

325. Li, L.; Wu, F.; Chen, R.; Chen, S.; Wang, G. Regeneration of electrochemical performance for MH-Ni batteries. *J. Funct. Mater.* **2006**, *37*, 587–590.

326. Minohara, T. Recycling Method of Nickel-Hydrogen Secondary Battery. U.S. Patent 6,077,622A, 20 June 2000.

327. Wang, R.; Yan, J.; Zhou, Z.; Deng, B.; Gao, X.; Song, D. Regeneration of anode power from waste metal hydride/Ni rechargeable batteries. *Chin. J. Appl. Chem.* **2001**, *18*, 979–982.

328. Nan, J.; Han, D.; Yang, M.; Cui, M. Dismantling, recovery, and reuse of spent nickel-metal hydride batteries. *J. Electrochem. Soc.* **2006**, *153*, A101–A105.

329. Nan, J.; Han, D.; Yang, M.; Cui, M.; Hou, X.L. Recovery of metal values from a mixture of spent lithium-ion batteries and nickel-metal hydride batteries. *Hydrometallurgy* **2006**, *84*, 75–80.

330. Tenorio, J.A.S.; Espinosa, D.C.R. Recovery of Ni-based alloys from spent NiMH battery. *J. Power Sources* **2002**, *108*, 70–73.

331. Bertuol, D.A.; Bernardes, A.M.; Tenorio, J.A.S. Spent NiMH batteries: Characterization and metal recovery through mechanical processing. *J. Power Sources* **2006**, *160*, 1465–1470.

332. Zhang, P.; Yokoyama, T.; Itabashi, O.; Wakui, Y.; Suzuki, T.M. Recovery of metal values from spent nickel-metal hydride rechargeable batteries. *J. Power Sources* **1999**, *77*, 116–122.

333. Rodrigues, L.; Mansur, M.B. Hydrometallurgical separation of rare earth elements, cobalt and nickel from spent nickel-metal-hydride batteries. *J. Power Sources* **2010**, *195*, 3735–3741.

334. Muller, T.; Friedrich, B. Development of a recycling process for nickel-metal hydride batteries. *J. Power Sources* **2006**, *158*, 1498–1509.

335. Rabah, M.A.; Farghaly, F.E.; Abd-El Motaleb, M.A. Recovery of nickel, cobalt and some salts from spent Ni-MH batteries. *Waste Manag.* **2008**, *28*, 1159–1167.

336. Santos, V.E.O.; Celante, V.G.; Lelis, M.F.F.; Freitas, M.B.J.G. Chemical and electrochemical recycling of the nickel, cobalt, zinc and manganese from the positives electrodes of spent Ni–MH batteries from mobile phones. *J. Power Sources* **2012**, *218*, 435–444.

337. Larsson, K.; Ekberg, C.; Ødegaard-Jensen, A. Dissolution and characterization of HEV NiMH batteries. *Waste Manag.* **2013**, *33*, 689–698.

Chapter 2:
Metal Hydride Alloys

Gaseous Phase and Electrochemical Hydrogen Storage Properties of $Ti_{50}Zr_1Ni_{44}X_5$ (X = Ni, Cr, Mn, Fe, Co, or Cu) for Nickel Metal Hydride Battery Applications

Jean Nei and Kwo-Hsiung Young

Abstract: Structural, gaseous phase hydrogen storage, and electrochemical properties of a series of the $Ti_{50}Zr_1Ni_{44}X_5$ (X = Ni, Cr, Mn, Fe, Co, or Cu) metal hydride alloys were studied. X-ray diffraction (XRD) and scanning electron microscopy (SEM) revealed the multi-phase nature of all alloys, which were composed of a stoichiometric TiNi matrix, a hyperstoichiometric TiNi minor phase, and a Ti_2Ni secondary phase. Improvement in synergetic effects between the main TiNi and secondary Ti_2Ni phases, determined by the amount of distorted lattice region in TiNi near Ti_2Ni, was accomplished by the substitution of an element with a higher work function, which consequently causes a dramatic increase in gaseous phase hydrogen storage capacity compared to the $Ti_{50}Zr_1Ni_{49}$ base alloy. Capacity performance is further enhanced in the electrochemical environment, especially in the cases of the $Ti_{50}Zr_1Ni_{49}$ base alloy and $Ti_{50}Zr_1Ni_{44}Co_5$ alloy. Although the TiNi-based alloys in the current study show poorer high-rate performances compared to the commonly used AB_5, AB_2, and A_2B_7 alloys, they have adequate capacity performances and also excel in terms of cost and cycle stability. Among the alloys investigated, the $Ti_{50}Zr_1Ni_{44}Fe_5$ alloy demonstrated the best balance among capacity (394 $mAh \cdot g^{-1}$), high-rate performance, activation, and cycle stability and is recommended for follow-up full-cell testing and as the base composition for future formula optimization. A review of previous research works regarding the TiNi metal hydride alloys is also included.

Reprinted from *Batteries*. Cite as: Nei, J.; Young, K.-H. Gaseous Phase and Electrochemical Hydrogen Storage Properties of $Ti_{50}Zr_1Ni_{44}X_5$ (X = Ni, Cr, Mn, Fe, Co, or Cu) for Nickel Metal Hydride Battery Applications. *Batteries* **2016**, *2*, 24.

1. Introduction

Nickel/metal hydride (Ni/MH) batteries have been used commercially for more than 25 years since their debut in 1989 by Ovonic, Matsushita, and Sanyo. Ni/MH batteries were once very popular for consumer applications, which have since largely shifted to Li-ion battery. Moreover, high-power Ni/MH battery technology is still the choice of today's hybrid electric vehicles. Presently, the same battery chemistry is entering into the market of stationary applications, which require a wide temperature range and a durable cycle stability [1,2]. This market is currently dominated by a less

159

environmentally safe chemistry, lead-acid battery. However, comparison between General Motors EV1 electric vehicles powered by the lead-acid and Ni/MH battery packs reveals that using the Ni/MH battery technology not only nearly doubles the driving range with its higher energy density, but it also offers longer service life [3,4]. Those two proven advantages make Ni/MH battery technology a great contender for the stationary market. In fact, the only advantage lead-acid battery has compared to Ni/MH battery is cheaper price. Therefore, there is a large driving force for cost-efficiency in the development of novel Ni/MH battery technology. One of the major components in Ni/MH battery is the negative electrode, which is an intermetallic compound that is capable of hydrogen storage, or in other words, a metal hydride (MH) alloy. Hydrogen storage intermetallic compounds have been extensively reviewed according to alloy type, application, and hydrogen absorption mechanism recently [5–12]. Among the many candidates of MH alloys [10], both the Ti-Ni and Mg-Ni systems have an advantage with regard to cost. The Mg-Ni-based MH alloys, being the cheapest, suffer from a severe capacity degradation with the current 30 wt% KOH electrolyte, leaving the Ti-Ni alloy system as the next choice for cost reduction. Raw material cost for 99.6% pure Ti sponge is approximately $8/kg, which is a very attractive alternative to the more expensive Zr or unstably priced rare earth elements used in today's Ni/MH battery.

The earliest Ni/MH batteries were demonstrated in the early 1970s and based on the Ti-Ni MH alloy system [13,14]. However, the first commercial Ni/MH battery products were fabricated with the misch metal-based AB_5 and transition metal-based disorder AB_2 MH alloys due to their improvements in cycle life. According to the Ti-Ni binary phase diagram [15], two intermetallic compounds can be used as hydrogen storage alloys: TiNi and Ti_2Ni. To achieve an optimal balance between storage capacity and high-rate dischargeability (HRD) for room-temperature battery operation, a stoichiometry of B/A close to two is ideal for the TiZrNi-based AB_2 MH alloys (Figure 1). With regard to a pure Ti-Ni system, however, eliminating Zr means that the B/A ratio must be reduced since the heat of hydride formation (ΔH_h) of Ti is higher than ΔH_h of Zr [16]. Therefore, TiNi is the better selection of the two available Ti-Ni intermetallic compounds. In addition, chemical modification can be adopted to further adjust the metal-hydrogen bond strength. The introduction of other elements can also increase the chance of secondary phase formation and consequently increase the electrochemical capacity through synergetic effects [17].

TiNi is a shape memory alloy. Upon cooling, TiNi is first solidified into a B2 cubic structure (austenite) and later to a B19' monoclinic structure (martensite). The shape at higher temperatures (austenite) is memorized and restored after thermal treatment to the deformed alloy with a martensite structure. Previous research efforts for electrochemical applications with the TiNi alloys are summarized in Table 1. While earlier works focused on the alloys produced by arc melting, induction melting, or other bulk ingot production methods, later works focused more on the powder produced by mechanical alloying. With the severe cycle stability problems observed in the MgNi-based MH alloys prepared by a combination of melt spinning and mechanical alloying [18], we choose induction melting as our first alloy preparation method for the current study. Furthermore, the substitution effects from various transition metals on the structural, gaseous phase, and electrochemical properties are investigated in this study.

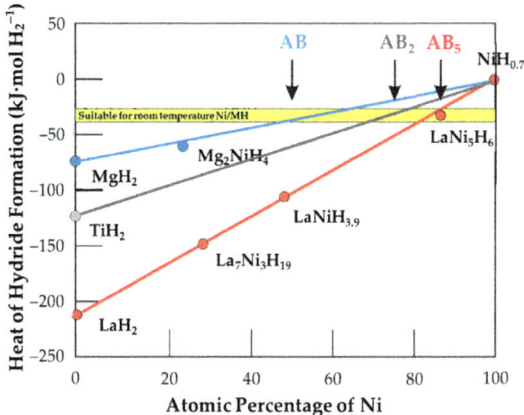

Figure 1. Plot of the heat of hydride formation for metal hydride vs. the corresponding Ni-content. Hydride formers, such as Mg, Ti, and La, have metal-hydrogen bonds that are too strong for room temperature nickel/metal hydride (Ni/MH) battery applications (a ΔH_h between -25 kJ·mol H_2^{-1} and -40 kJ·mol H_2^{-1} is desirable). Therefore, alloying hydride formers with Ni can adjust the bond strength and increase the electrochemical activity.

Table 1. Summaries of prior research on the TiNi-based MH alloys for electrochemical applications in chronological order. Preparation methods SN, AM, PM, IM, CO, MC, MA, ANN, MS, CHR, and MWCNT denote sintering, arc melting, melting in plasma furnace, induction melting, co-precipitation, microencapsulation, mechanical alloying, annealing, melt spinning, calcium hydride reduction, and multiwall carbon nanotube, respectively. TiNi phase structures B2, B19', B19, and R are cubic, monoclinic, orthorhombic, and tetragonal, respectively. HRD: high-rate dischargeability.

Alloy Formula	Preparation Method	TiNi Phase Structure	Main Discoveries	Reference
$Ti_{53.6}Ni_{46.4}$	-	-	• The first TiNi-based metal hydride electrode	[19]
TiNi	SN	B2	• Capacity of 210 to 250 mAh·g^{-1} at C/10 rate • 67% of capacity is obtained at 1C rate at $-30\,°C$	[13]
$TiNi + Ti_2Ni$	SN	B2	• Capacity of 300 to 320 mAh·g^{-1}	[13]
TiNi	SN	B2	• Capacity of 180 mAh·g^{-1} at 50 mA·g^{-1}	[14]
TiNi	AM	-	• Cyclic voltammetry study	[20]
$TiNi_x$ (x = 0.5–1.0)	SN vs. PM	-	• Higher capacity for SN alloy • Higher corrosion resistance for PM alloy • Increased cycle stability with increasing Ni-content	[21]
$Ti_{0.8}Zr_{0.2}Ni_x$ (x = 0.5–1.0)	SN vs. PM	-	• Increased capacity and cycle stability with Zr-substitution	[21]
TiNi	SN	-	• Increased surface reactivity with 1 M HF treatment	[22]
$Ti_{1-y}Zr_yNi_x$ (x = 0.50–1.45, y = 0–1.0)	SN	B2	• As the Zr-content (y) increases, maximized capacity occurs at higher Ni-content (x) • Capacity of 260 mAh·g^{-1} for $Ti_{0.9}Zr_{0.1}Ni_{1.0}$	[23]
$Ti_{0.5}Zr_{0.5}Ni_{0.95-x}Cu_{0.05+x}$ (x = 0–0.05)	SN	B2	• Increased cycle stability with increasing Cu-content	[23]
$TiNi_{0.9}B_{0.1}$	SN	-	• B does not prevent segregation of Ti during cycling	[24]
$Ti_{0.7}Zr_{0.2}V_{0.1}Ni$	IM	B2	• Capacity of 350 mAh·g^{-1} at 30 mA·g^{-1}	[25]
TiNi	CO + SN	-	• Oxygen consumption reaction study	[26]
TiNi	-	B2 vs. B19'	• Capacities of 168 mAh·g^{-1} and 176 mAh·g^{-1} at 50 mA·g^{-1} for B19' and B2 TiNi, respectively	[27]
TiNi	AM	-	• Cycling model determined by both thermal dynamic and kinetic degradations	[28]

162

Table 1. *Cont.*

Alloy Formula	Preparation Method	TiNi Phase Structure	Main Discoveries	Reference
$TiNi + 5\% M$ ($M = Ni, Cu$)	SN + MC	-	• Capacity of 150 mAh·g^{-1} at 30 mA·g^{-1} for TiNi • Increased capacity with Ni- (180 mAh·g^{-1}) or Cu-addition (169 mAh·g^{-1}) • Increased HRD with Cu	[29]
$TiNi_{0.5}Fe_{0.5}$	AM vs. MA	-	• Capacities of 65 and 190 mAh·g^{-1} at 20 mA·g^{-1} for AM and MA alloys, respectively	[30]
$TiNi_{0.6}Fe_{0.4}$	AM vs. AM + MA vs. MA	B2	• Capacities of 108, 138, 161 mAh·g^{-1} at 30 mA·g^{-1} for AM, AM + MA, and MA alloys, respectively • MA alloy demonstrates the best activation and the worst cycle stability	[31]
$Ti_{50}Ni_{41}Nb_9$	AM	B2	• Increased capacity and cycle stability with Nb-substitution	[32]
TiNi	AM	-	• Study on parameters affecting discharge capacity and cycle life	[33]
TiNi	AM	-	• Study on parameters affecting discharge kinetics	[34]
$Ti_{0.7}Zr_{0.2}V_{0.1}Ni$	AM + ANN	B2	• Capacity of 180 mAh·g^{-1} at 50 mA·g^{-1} for TiNi • Increased capacity due to secondary Laves phase formation but harder activation for $Ti_{0.7}Zr_{0.2}V_{0.1}Ni$ (205 mAh·g^{-1})	[35]
$Mg_2Ni + TiNi$	MA	B2	• Increased cycle stability compared to Mg_2Ni	[36]
$Ti_{50-x}Zr_xNi_{50}$ ($x = 0–24$)	IM vs. MS	B19' for IM alloy vs. B2 for MS alloy	• Capacity of 335 mAh·g^{-1} at C/10 rate for B19' $Ti_{32}Zr_{18}Ni_{50}$	[37]
$TiNi_xFe_{1-x}$ ($x = 0–1.0$)	MA + ANN	B2	• Capacity of 67 mAh·g^{-1} at 40 mA·g^{-1} for TiNi • Increased capacity but decreased cycle stability with Fe-substitution (155 mAh·g^{-1} for $TiNi_{0.75}Fe_{0.25}$)	[38]
(Ti + Ni)	MA under H_2	-	• Capacity of 150 mAh·g^{-1} at 50 mA·g^{-1} for 45-h-MA alloy with good cycle stability	[39]
$Ti_{0.5}Ni_{0.25}Al_{0.25}$	IM	B2	• Two discharge plateaus observed due to alloy's multi-phase nature	[40]
$Ti(Ni,Fe,Mo,Cr,Co)$	MA + ANN	B2	• Increased cycle stability for $TiNi_{0.6}Fe_{0.1}Mo_{0.1}Cr_{0.1}Co_{0.1}$ (135 mAh·g^{-1} at 40 mA·g^{-1}) compared to $TiNi_{0.75}Fe_{0.25}$	[41]

Table 1. *Cont.*

Alloy Formula	Preparation Method	TiNi Phase Structure	Main Discoveries	Reference
Ti(Ni,Fe,Mo,Cr,Co)	AM + ANN vs. MA + ANN	B2	• Increased capacity with MA + ANN • Increased cycle stability for $TiNi_{0.6}Fe_{0.1}Mo_{0.1}Cr_{0.1}Co_{0.1}$ (150 mAh·g^{-1} at 40 mA·g^{-1}) compared to $TiNi_{0.75}Fe_{0.25}$	[42]
Mg_2Ni + TiNi	MA vs. MA + ANN	B2	• Increased cycle stability with ANN	[43]
(Ti,Zr,V,Cr,Mn)Ni	AM	B2	• Increased capacity and HRD due to secondary Laves phase formation but harder activation with Mn- and/or Cr-substitutions (232 mAh·g^{-1} at 50 mA·g^{-1} for $Ti_{0.6}Zr_{0.2}V_{0.1}Mn_{0.1}Ni$)	[44]
$TiNi_{0.75}Fe_{0.25}$	MA + ANN	B2	• 1.5 fold increased capacity with Fe-substitution in sealed battery	[45]
Ti(Ni,Fe,Zr)	MA + ANN	B2	• Capacity of 79 mAh·g^{-1} at 40 mA·g^{-1} for TiNi • Increased capacity with Zr-substitution (135 mAh·g^{-1} for $TiNi_{0.875}Zr_{0.125}$) • Increased capacity but decreased cycle stability with Fe-substitution (158 mAh·g^{-1} for $TiNi_{0.75}Zr_{0.125}Fe_{0.125}$)	[46]
TiNi	CHR vs. CHR + ANN	B2 + B19′ for CHR alloy vs. B2 for CHR + ANN alloy	• Capacity of 125 mAh·g^{-1} at 50 mA·g^{-1} for B2 TiNi • Increased capacity due to dual-phase nature for B2 + B19′ TiNi (160 mAh·g^{-1})	[47]
$TiNi_{1-x}M_x$ (M = Mg, Mn, Zr, x = 0–0.25)	MA + ANN	B2	• Increased capacity but slightly decreased cycle stability with Mg- (152 mAh·g^{-1} at 40 mA·g^{-1}) or Mn-substitution (153 mAh·g^{-1}) compared to Zr-substitution (135 mAh·g^{-1}) • Decreased cycle stability with higher Mn-substitution	[48]
Ti (Ni, Fe, Zr, Mo, Cr, Co, Al)	MA + ANN	B2	• Increased capacity with all substitutions in the order of Co (113 mAh·g^{-1} at 40 mA·g^{-1}) < Mo < Al < Cr < Fe (155 mAh·g^{-1}) • Decreased cycle stability with all substitutions in the order of Co < Cr < Fe ≈ Al < Mo (worst) • 1.5 fold increased capacity with Fe-substitution or Fe-, Co-, and Zr-substitutions in sealed battery	[49]

Table 1. *Cont.*

Alloy Formula	Preparation Method	TiNi Phase Structure	Main Discoveries	Reference
TiNi$_{1-x}$M$_x$ (M = Co, Fe, Sn, x = 0–0.2)	MA vs. MA + ANN	-	• Capacities of 52 and 67 mAh·g⁻¹ at 40 mA·g⁻¹ for MA and MA + ANN TiNi, respectively, with good cycle stability • Increased capacity with ANN; Harder activation with ANN • Increased capacity with Co- (83 mAh·g⁻¹ with ANN) or Fe-substitution (79 mAh·g⁻¹ with ANN)	[50]
Ti (Ni, Fe, Zr, Mo, Cr, Co)	MA	B2	• Increased capacity and cycle stability with Fe-substitution in sealed battery • Increased cycle stability with Zr-substitution in sealed battery	[51]
Ti$_{0.8}$M$_{0.2}$Ni (M = Zr, V)	MA vs. MA + ANN	B2	• Decreased capacity with Zr- or V-substitution	[52]
TiNi$_{0.8}$M$_{0.2}$ (M = Cu, Mn)	MA vs. MA + ANN	-	• Decreased capacity with Cu-substitution • Increased capacity with Mn-substitution (75 mAh·g⁻¹ at 40 mA·g⁻¹ with ANN)	[52]
TiNi$_{1-x}$Mn$_x$ (x = 0.2–1.0)	MA vs. MA + ANN	-	• Decreased capacity with higher Mn-substitution	[52]
Ti$_{1.02-x}$Zr$_x$Ni$_{0.98}$ (x = 0–0.48)	IM + ANN	B19'	• Capacity of 150 mAh·g⁻¹ at C/10 rate for TiNi • Increased capacity with Zr-substitution due to the difference in martensitic transformation (350–370 mAh·g⁻¹ for Ti$_{0.78}$Zr$_{0.24}$Ni$_{0.98}$)	[53]
TiNi$_{0.8}$B$_{0.2}$, Ti$_{0.8}$B$_{0.2}$Ni,	MA vs. MA + ANN	B2	• Decreased capacity with B-substitution, and further decrease with ANN	[54]
Ti$_{1.02-x}$Zr$_x$Ni$_{0.98}$ (x = 0–0.48)	IM + ANN	B19'	• Harder activation with higher Zr-substitution • Decreased cycle stability with Zr-substitution	[55]
TiNi	SN (750–950 °C)	B2	• Increased capacity and HRD with higher SN temperature (179–211 mAh·g⁻¹ at 60 mA·g⁻¹)	[56]
TiNi	SN	B2	• Capacity of 205 mAh·g⁻¹ at 60 mA·g⁻¹	[57]
TiNi	MA (20–60 h)	-	• Capacity of 102 mAh·g⁻¹ at 60 mA·g⁻¹ with bad cycle stability	[58]
TiNi	MA vs. MA + ANN	B2	• Higher capacity for crystalline TiNi (150 mAh·g⁻¹ with ANN).	[59]

165

Table 1. *Cont.*

Alloy Formula	Preparation Method	TiNi Phase Structure	Main Discoveries	Reference
$Mg_2Ni + TiNi$	MA	B2	• Decreased capacity compared to Mg_2Ni.	[60]
$Ti_{1.04}Ni_{0.96-x}Pd_x$ (x = 0–0.5)	IM + ANN	As x increases, B2 + B19′ → B2 + B19′ + R → B2 + B19	• Decreased capacity with Pd-substitution (148 and 84 mAh·g^{-1} at C/5 rate for x = 0.1 and 0.4, respectively)	[61]
Sn-doped TiNi	Thin film sputtering	-	• TiNi-based anode for Li-ion battery	[62]
TiNi	MA	-	• TiNi-based anode for Li-ion battery	[63]
TiNi-5 wt% Pd + 5 wt% MWCNT	MA + ANN + MA with MWCNT	B2	• Capacity of 171 mAh·g^{-1} at 40 mA·g^{-1} for TiNi • Increased capacity and cycle stability with Pd- (186 mAh·g^{-1}), MWCNT- (183 mAh·g^{-1}), or Pd + MWCNT-addition (266 mAh·g^{-1})	[64]
$(TiNi)_{1-x}Mg_x$ (x = 0–0.3)	MA (10 to 40 h)	B2	• Increased and then decreased capacity and increased cycle stability with increasing MA time for TiNi (~132 mAh·g^{-1} at 40 mA·g^{-1}) • Decreased capacity with Mg-incorporation	[65]
$Ti_{1.01}Ni_{0.99-x}Cu_x$ (x = 0–0.5)	IM + ANN	As x increases, B19′ → B19	• Increased and then decreased capacity with increasing Cu-content (300 mAh·g^{-1} at C/10 rate for $Ti_{1.01}Ni_{0.79}Cu_{0.2}$) • Increased HRD with Cu-substitution	[66]
$MgTiNi_2$	MA	-	• Capacity of 93 mAh·g^{-1} at 40 mA·g^{-1}	[67]
$Ti_{1-x}Zr_xNi$ (x = 0–0.5)	MA + ANN	B2	• Increased and then decreased capacity and increased cycle stability with increasing Zr-content (192 mAh·g^{-1} at 40 mA·g^{-1} for $Ti_{0.75}Zr_{0.25}Ni$)	[68]
$Ti_{0.75}Zr_{0.25}Ni$-5 wt% Pd vs. $Ti_{0.75}Zr_{0.25}Ni$ + 5 wt% Pd	MA + ANN vs. MA + ANN + MA with Pd	B2	• Increased capacity with Pd-addition • Early addition of Pd (223 mAh·g^{-1}) has higher capacity than later addition (208 mAh·g^{-1})	[68]

2. Experimental Setup

Each alloy sample was prepared in a 2-kg induction furnace with an $MgAl_2O_4$ crucible, an alumina tundish, and a pancake-shaped steel mold under an argon atmosphere. Chemical composition for the ingot was analyzed using a Varian Liberty 100 inductively coupled plasma optical emission spectrometer (ICP-OES, Agilent Technologies, Santa Clara, CA, USA). A Philips X'Pert Pro X-ray diffractometer (XRD, Amsterdam, The Netherlands) was used to study the microstructure, and a JEOL-JSM6320F scanning electron microscope (SEM, Tokyo, Japan) with energy dispersive spectroscopy (EDS) capability was used to study the phase distribution and composition. Gaseous phase hydrogen storage characteristics for each sample were measured using a Suzuki-Shokan multi-channel pressure-concentration-temperature (PCT, Tokyo, Japan) system. For the PCT analysis, each sample (a single piece of ingot with newly cleaved surfaces and a weight of about 4–5 g) was first activated by several thermal cycles between 300 °C and room temperature under 2.5 MPa H_2 pressure. PCT isotherms at 90 °C and 120 °C were then measured. Details of the electrode and cell preparations, as well as the electrochemical measurement methods, have been previously reported [18,69]. Magnetic susceptibility was measured using a Digital Measurement Systems Model 880 vibrating sample magnetometer (MicroSense, Lowell, MA, USA).

3. Results

3.1. Alloy Preparation

Six alloys with the design compositions $Ti_{50}Zr_1Ni_{44}X_5$ (X = Ni, Cr, Mn, Fe, Co, or Cu) were prepared by induction melting. A slightly hypostoichiometric TiNi formulation with a small addition of Zr (has a higher metal-hydrogen bond strength compared to Ti) in the A-site was adopted to increase the degree of disorder (DOD) and consequently improve the electrochemical properties. Zr is also an oxygen scavenger, forming oxide slag in the melt that can be separated by the tundish [70]. Furthermore, adding Zr can possibly reduce capacity degradation by suppressing passivation caused by the thick TiO_2 surface oxide layer [71]. Compositions of the six alloys, verified by ICP and shown in Table 2, are very close to their design values.

3.2. X-Ray Diffraction Analysis

XRD patterns of the six alloys are shown in Figure 2. Two sets of diffraction peaks, the main TiNi phase with a B2 cubic structure (belonging to the space group $Pm\bar{3}m$) and the secondary Ti_2Ni phase with an $E9_3$ face-centered cubic structure (belonging to the space group $Fd\bar{3}m$), are observed in all alloys. Crystal structures of TiNi and Ti_2Ni generated by the XCrySDen software [72] are shown in Figure 3.

Table 2. Design compositions (in **bold**) and inductively coupled plasma (ICP) results (in at %). B/A is the atomic ratio of B-atom (elements other than Ti and Zr) to A-atom (Ti and Zr).

Alloy TN-X	Source	Ti	Zr	Ni	X	B/A
TN-Ni	**Design**	**50.0**	**1.0**	**49.0**	**-**	**0.96**
	ICP	50.0	0.6	49.4	0.0	0.98
TN-Cr	**Design**	**50.0**	**1.0**	**44.0**	**5.0**	**0.96**
	ICP	49.7	1.1	44.4	4.7	0.97
TN-Mn	**Design**	**50.0**	**1.0**	**44.0**	**5.0**	**0.96**
	ICP	49.2	1.0	44.9	4.9	0.99
TN-Fe	**Design**	**50.0**	**1.0**	**44.0**	**5.0**	**0.96**
	ICP	49.9	1.0	44.3	4.8	0.96
TN-Co	**Design**	**50.0**	**1.0**	**44.0**	**5.0**	**0.96**
	ICP	49.8	0.9	44.5	4.8	0.97
TN-Cu	**Design**	**50.0**	**1.0**	**44.0**	**5.0**	**0.96**
	ICP	49.5	1.3	44.5	4.7	0.97

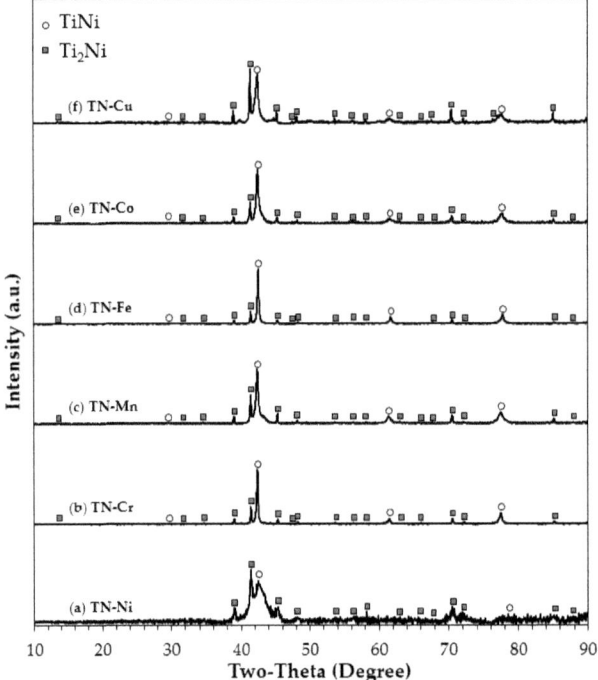

Figure 2. X-ray diffraction (XRD) patterns using Cu-Kα as the radiation source for alloys: (**a**) TN-Ni; (**b**) TN-Cr; (**c**) TN-Mn; (**d**) TN-Fe; (**e**) TN-Co; and (**f**) TN-Cu. Besides the main TiNi phase with a cubic B2 structure, a secondary Ti_2Ni phase with a cubic structure and a larger lattice constant can be also identified.

168

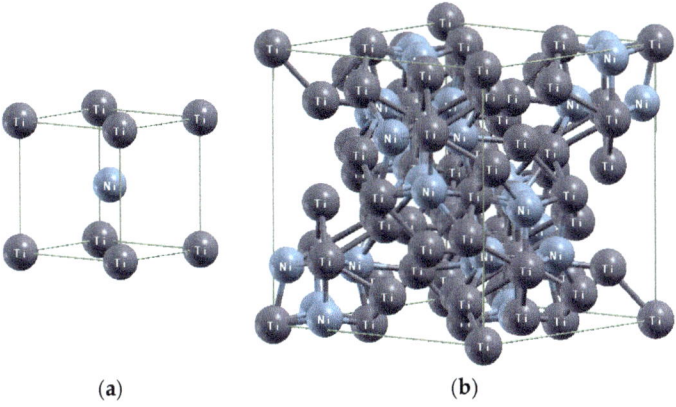

(a)   (b)

Figure 3. Crystal structures of (**a**) TiNi with a B2 cubic structure and (**b**) Ti_2Ni with an $E9_3$ face-centered cubic structure generated using the XCrySDen software [72].

Full XRD pattern fitting was performed using the Rietveld refinement and Jade 9 Software to obtain the lattice parameters, crystallite sizes, and phase abundances, and the results are listed in Table 3. Lattice constant a of the TiNi phase in alloy TN-Ni is smaller than that found in the stoichiometric TiNi alloy [73]. Although the overall alloy formulation is hypostoichiometric, formation of the Ti_2Ni phase increases the B/A ratio in the TiNi phase, which contributes to the reduction in its lattice constant from that of the stoichiometric TiNi phase due to either Ni into Ti-anti-site or Ti-vacancy defects. Lattice constants a's of TiNi in all substituted alloys are larger than that in the base alloy TN-Ni. Since all substituting elements are larger than Ni but much smaller than Ti, they occupy the Ni-site and increase the lattice constant. In Figure 4a, TiNi lattice constant a is correlated with the atomic radius of substituting element in the Laves phase [74] (a value used to simulate the size of substituting element in the TiNi phase), and a linear dependency can be observed. However, such correlation is not seen in the plot of Ti_2Ni lattice constant a versus the atomic radius of substituting element in the Laves phase (Figure 4b), which is due to the change in B/A ratio of the Ti_2Ni phase from 0.51 to 0.70 as revealed by SEM/EDS (see Section 3.3). All alloys show similar $TiNi/Ti_2Ni$ abundance ratios except for alloys TN-Ni and TN-Cu, where the Ti_2Ni phase abundance is higher than those in other alloys. Moreover, crystallite sizes of TiNi and Ti_2Ni were estimated by the Scherrer equation [75] using the full widths at half maximum of the TiNi (110) peak and Ti_2Ni (511) peak in the XRD patterns, and the results are listed in Table 3. Both the crystallite sizes of the TiNi and Ti_2Ni phases in the base alloy TN-Ni are much smaller than those in the substituted alloys. Partial replacement of Ni with other elements may increase the peritectic temperature [76], leaving more time for the crystallites to grow.

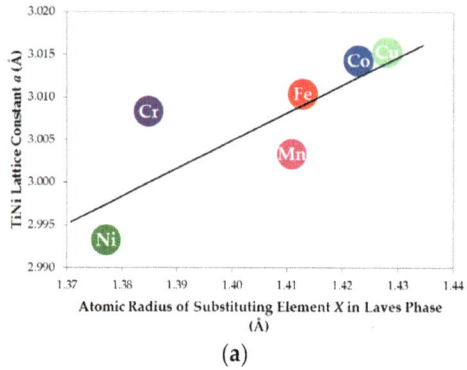
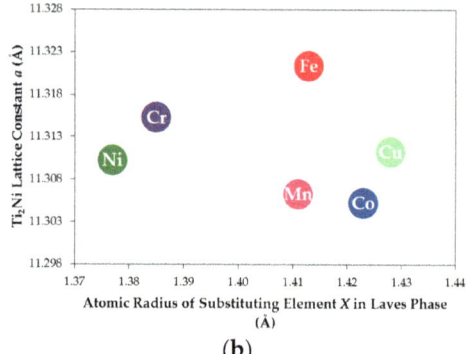

(a)   **(b)**

Figure 4. Plots of the (**a**) TiNi lattice constant a; and (**b**) Ti_2Ni lattice constant a vs. the atomic radius of substituting element X in the Laves phase (data from [74]). A linear correlation is observed for the TiNi lattice constant but not for the Ti_2Ni lattice constant.

Table 3. Summary of XRD analysis (lattice constants, phase abundances, and crystallite sizes).

Alloy TN-X	a of TiNi (Å)	a of Ti_2Ni (Å)	TiNi Abundance (wt%)	Ti_2Ni Abundance (wt%)	TiNi Crystallite Size (Å)	Ti_2Ni Crystallite Size (Å)
TN-Ni	2.993	11.310	68.7	31.3	139	448
TN-Cr	3.014	11.305	75.5	24.5	368	804
TN-Mn	3.015	11.311	78.4	21.6	251	>1000
TN-Fe	3.003	11.306	80.7	19.3	346	812
TN-Co	3.008	11.315	80.6	19.4	212	655
TN-Cu	3.010	11.321	71.6	28.4	206	727

3.3. Scanning Electron Microscopy/Energy Dispersive Spectroscopy Study

Microstructures of the six alloys were studied using SEM, and the resulting $100\times$ and $1000\times$ back-scattering electron images (BEI) are shown in Figures 5 and 6, respectively. BEI images demonstrate the changes in both surface morphology and contrast due to the difference in average atomic weight. EDS was used to study the chemical compositions of several areas with different contrasts identified numerically in the micrographs (Figure 6), and the results are summarized in Table 4. Basically, three different contrasts are observed and assigned to a TiNi-2 (brightest), a TiNi-1 (matrix), and a Ti_2Ni (darkest) phases. Although the grain size of the Ti_2Ni phase varies and increases in the order of TN-Fe < TN-Ni ≈ TN-Cr < TN-Cu < TN-Mn ≈ TN-Co (Figures 5 and 6), the six alloys show similar phase distributions:

- Matrix (TiNi-1): stoichiometric or slightly hyperstoichiometric TiNi with the Zr- and X-contents close to design.

170

- Minor phase (TiNi-2): this phase appears as bright spots in the micrographs and is distributed within the matrix. It is generally hyperstoichiometric, high-Zr, and high-X TiNi except for:

  ○ Hypostoichiometric, high-Zr, and close to the design-X TiNi in alloy TN-Fe and

  ○ Hyperstoichiometric, high-Zr, and low-X TiNi in alloy TN-Co.

- Secondary phase (Ti$_2$Ni): this phase has the darkest contrast in the micrographs and appears next to the main TiNi-1 phase. It is stoichiometric or hyperstoichiometric, low-Zr Ti$_2$Ni.

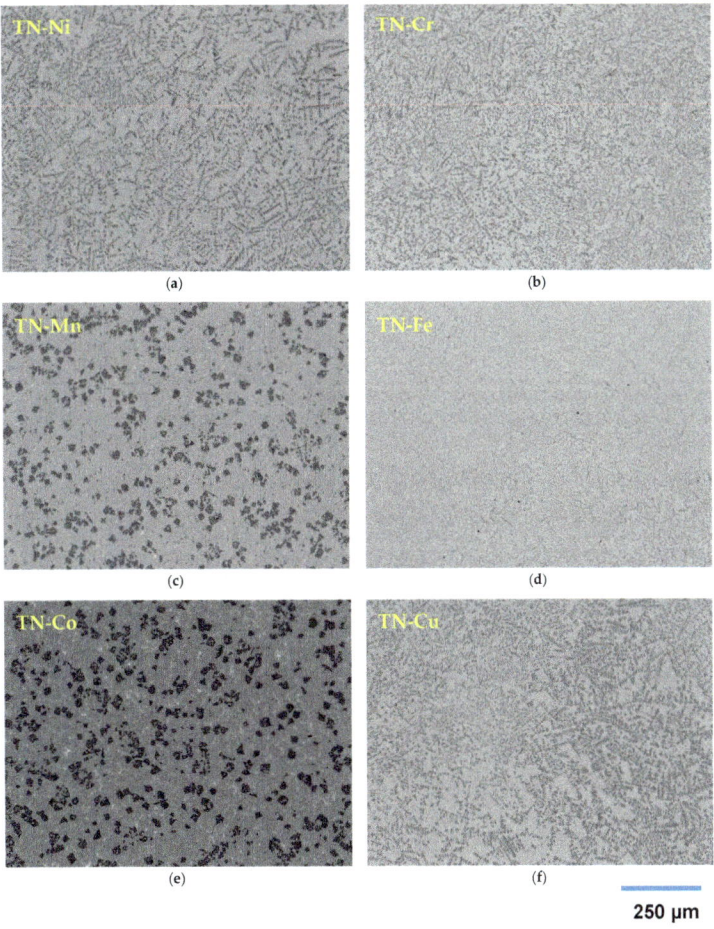

250 μm

Figure 5. Scanning electron microscopy (SEM) × back-scattering electron images (BEI) micrographs from alloys: (**a**) TN-Ni; (**b**) TN-Cr; (**c**) TN-Mn; (**d**) TN-Fe; (**e**) TN-Co; and (**f**) TN-Cu at 100× magnification.

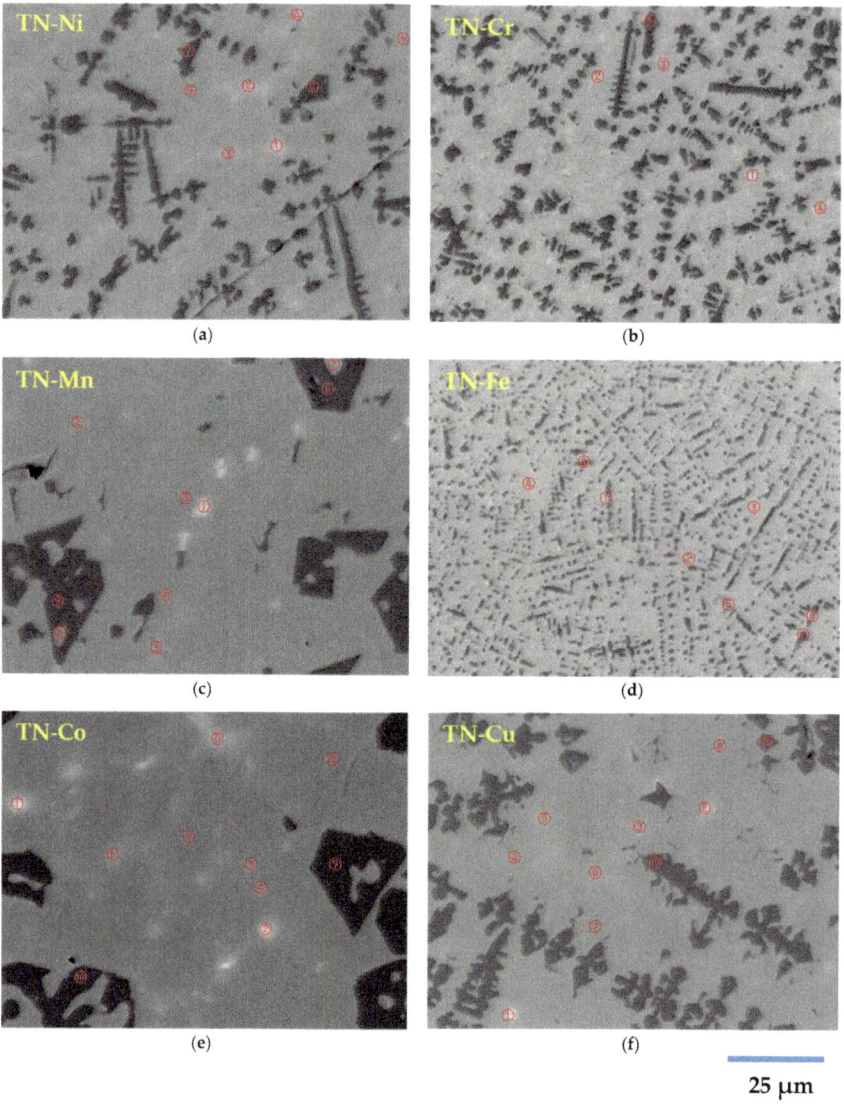

Figure 6. SEM BEI micrographs from alloys: (**a**) TN-Ni; (**b**) TN-Cr; (**c**) TN-Mn; (**d**) TN-Fe; (**e**) TN-Co; and (**f**) TN-Cu at 1000× magnification.

According to the Ti-Ni binary phase diagram [15], as the melt of $Ti_{51}Ni_{49}$ cools to 1310 °C, a hyperstoichiometric TiNi phase, or TiNi-2 in the current study, initially forms due to the TiNi system's preference for hyperstoichiometry (as seen from the wide solubility range of the TiNi phase towards hyperstoichiometric TiNi in the phase diagram). Once the system reaches the temperature of 984 °C, the hyperstoichiometric TiNi-2 phase and rest of the melt goes through a peritectic

172

reaction, which consumes most of the TiNi-2 phase and produces the final stoichiometric TiNi (TiNi-1) and Ti_2Ni phases. Ti_2Ni precipitated within the main TiNi matrix was found to deteriorate the alloy's mechanical properties for engineering applications due to its brittleness [77], but Ti_2Ni can be advantageous for initial pulverization for the use in electrochemical applications. Moreover, the presence of T_2Ni as the secondary phase in the TiNi alloy system was shown to be essential for electrochemical property improvement [14].

Table 4. Summary of EDS results. All compositions are in at%. Compositions of the **main TiNi-1**, minor TiNi-2, and *secondary Ti$_2$Ni* phases are in **bold**, underline, and *italic*, respectively.

Alloy TN-X	Area	Ti	Zr	Ni	X	B/A	Phase(s)
TN-Ni	1	36.3	8.4	55.3	0.0	1.24	TiNi-2
	2	40.9	4.7	54.3	0.0	1.19	TiNi-2
	3	44.2	2.3	53.5	0.0	1.15	TiNi-2
	4	42.4	3.4	54.2	0.0	1.18	TiNi-2
	5	46.9	0.6	52.5	0.0	1.11	**TiNi-1**
	6	47.5	0.5	52.0	0.0	1.08	**TiNi-1**
	7	64.2	0.5	35.3	0.0	0.55	*Ti$_2$Ni*
	8	63.6	0.4	36.0	0.0	0.56	*Ti$_2$Ni*
TN-Cr	1	41.0	4.6	44.4	10.0	1.19	TiNi-2
	2	42.0	2.7	43.9	11.3	1.24	TiNi-2
	3	47.5	0.9	47.6	4.0	1.07	**TiNi-1**
	4	46.3	1.4	46.9	5.4	1.10	**TiNi-1**
	5	62.2	0.8	33.3	3.7	0.59	*Ti$_2$Ni*
TN-Mn	1	33.8	12.0	32.5	21.7	1.18	TiNi-2
	2	44.1	2.7	43.0	10.1	1.13	TiNi-2
	3	43.7	3.1	42.5	10.6	1.13	TiNi-2
	4	46.6	1.0	46.2	6.1	1.10	**TiNi-1**
	5	47.0	0.8	46.7	5.5	1.09	**TiNi-1**
	6	49.0	0.6	45.3	5.0	1.01	**TiNi-1**
	7	46.5	1.0	47.0	5.5	1.11	**TiNi-1**
	8	65.3	0.6	32.4	1.7	0.52	*Ti$_2$Ni*
	9	65.0	0.6	32.5	1.9	0.52	*Ti$_2$Ni*
TN-Fe	1	46.4	7.0	42.1	4.5	0.87	TiNi-2
	2	46.1	6.2	42.9	4.8	0.91	TiNi-2
	3	47.2	1.3	46.3	5.2	1.06	**TiNi-1**
	4	46.8	2.2	45.9	5.2	1.04	**TiNi-1**
	5	52.3	1.1	42.0	4.6	0.87	TiNi + Ti$_2$Ni
	6	52.4	1.2	41.6	4.8	0.87	TiNi + Ti$_2$Ni
	7	62.8	0.9	32.6	3.7	0.57	*Ti$_2$Ni*
	8	57.8	1.0	37.1	4.1	0.70	*Ti$_2$Ni*

Table 4. *Cont.*

Alloy TN-*X*	Area	Ti	Zr	Ni	*X*	B/A	Phase(s)
	1	29.4	13.7	55.6	1.3	1.32	TiNi-2
	2	29.4	13.3	55.8	1.4	1.34	TiNi-2
	3	37.2	26.6	33.1	3.2	0.57	$(TiZr)_2Ni$
	4	41.0	15.1	39.7	4.1	0.78	$(TiZr)_2Ni$
TN-Co	5	40.6	5.0	51.7	2.7	1.19	TiNi-2
	6	43.9	2.8	49.7	3.5	1.14	TiNi-2
	7	47.6	0.5	45.1	6.7	1.08	**TiNi-1**
	8	47.9	0.6	44.9	6.6	1.06	**TiNi-1**
	9	65.5	0.7	29.9	3.9	0.51	Ti_2Ni
	10	65.4	0.6	30.3	3.7	0.51	Ti_2Ni
	1	29.1	10.7	41.0	19.1	1.51	TiNi-2
	2	34.7	9.1	40.1	16.0	1.28	TiNi-2
	3	34.0	30.5	31.0	4.5	0.55	$(TiZr)_2Ni$
	4	42.1	14.2	39.4	4.4	0.78	$(TiZr)_2Ni$
TN-Cu	5	43.7	2.9	46.5	6.8	1.14	TiNi
	6	44.5	2.4	47.1	6.0	1.13	TiNi
	7	47.2	1.0	47.3	4.4	1.07	**TiNi-1**
	8	46.8	1.0	47.6	4.6	1.09	**TiNi-1**
	9	63.5	0.8	34.0	1.7	0.55	Ti_2Ni
	10	65.6	0.8	32.4	1.2	0.51	Ti_2Ni

Unlike the relatively constant B/A ratio in the main TiNi-1 phase among alloys, that of the secondary Ti_2Ni phase ranges significantly, resulting in the inconsistency observed between the size of substituting element and Ti_2Ni lattice constant. Solubility of Zr in the Ti_2Ni phase is less than those in the TiNi-1 and TiNi-2 phases. Such observations can be explained by the pseudo-binary phase diagrams for the TiNi-ZrNi and Ti_2Ni-Zr_2Ni systems [78], where the TiNi phase shows a very high solubility for Zr (up to approximately 30%) while Zr is nearly insoluble in the Ti_2N phase.

3.4. Pressure-Concentration-Temperature Measurement

Gaseous phase hydrogen storage properties of the six alloys were studied by PCT. Due to the slow reaction kinetics, PCT isotherm at 30 °C or 60 °C cannot be measured (absorption weight is approximately 0% at all applied H_2 pressures). Therefore, the absorption and desorption isotherms were measured at 90 °C and 120 °C and are shown in Figure 7. Information obtained from the PCT study is summarized in Table 5. At either temperature, both the maximum and reversible capacities following the trend of TN-Ni ≈ TN-Co < TN-Cu < TN-Mn < TN-Cr < TN-Fe, which demonstrates a very weak correlation to the main TiNi phase lattice constant as seen in Figure 8a. 90 °C gaseous phase maximum capacity is also plotted against ΔH_h of substituting element (Figure 8b), a value that is often taken into consideration during alloy design to achieve balance between the amounts of A-site hydride formers ($\Delta H_h < 0$) and B-site modifiers ($\Delta H_h > 0$). In the current study, the modification is performed on the B-site, and the expected trend is increasing capacity with decreasing ΔH_h of substituting element; however, no clear correlation is observed. ΔH_h of the AB_n alloy can be calculated with the equation [79]:

$$\Delta H_h \left(AB_n H_{2m} \right) = \Delta H_h \left(AH_m \right) + \Delta H_h \left(B_n H_m \right) - \Delta H \left(AB_n \right) \qquad (1)$$

where ΔH is the heat of alloy formation. The difference in trends for ΔH_h of substituting element and gaseous phase capacity is possibly caused by the heat of alloy formation. Finally, 90 °C gaseous phase maximum capacity is plotted against the work function (W, the difference between the electron potentials in vacuum (E_{VAC}) and the Fermi level (E_F) of substituting element in Figure 8c, which illustrates a linear relationship. W's of various phases in a multi-phase MH system have been used to explain the synergetic effects observed in the gaseous phase hydrogen interaction [17]. In the current study, the secondary Ti_2Ni phase with a lower ΔH_h and a higher W is hydrogenated first, which expands the lattice in the TiNi phase near the Ti_2Ni phase (pink region in Figure 9) due to the stress transmitted through the "coherent interface" and eases the hydrogenation of the main TiNi phase [80]. When a substituting element with a smaller W compared to Ni is added into the system, W of the main TiNi phase is reduced while W of the Ti_2Ni phase remains the same (as seen from Table 4, solubility of substituting element in Ti_2Ni is lower than that in TiNi), leading to an increase in difference between W's of the two phases (Figure 9). This larger difference in W's causes an increase in volume of initial hydrogenation of the secondary Ti_2Ni phase and therefore enlarges the volume of the expanded lattice region, resulting in a higher gaseous phase capacity. Moreover, alloy TN-Mn seems to fall off the linear trend slightly in Figure 8. Although Fe and Cr have higher W's than Mn, their higher densities of interface between the main TiNi and secondary Ti_2Ni (Figures 5 and 6) contributes to the amount of distorted lattice zone and facilitates alloy hydrogenation.

Table 5. Summary of gaseous phase properties.

Alloy TN-X	Maximum Capacity at 90 °C (wt%)	Reversible Capacity at 90 °C (wt%)	Maximum Capacity at 120 °C (wt%)	Reversible Capacity at 120 °C (wt%)
TN-Ni	0.13	0.09	0.15	0.13
TN-Cr	1.18	0.57	1.08	0.67
TN-Mn	0.98	0.48	0.92	0.57
TN-Fe	1.21	0.75	1.06	0.85
TN-Co	0.16	0.13	0.19	0.14
TN-Cu	0.87	0.54	0.81	0.60

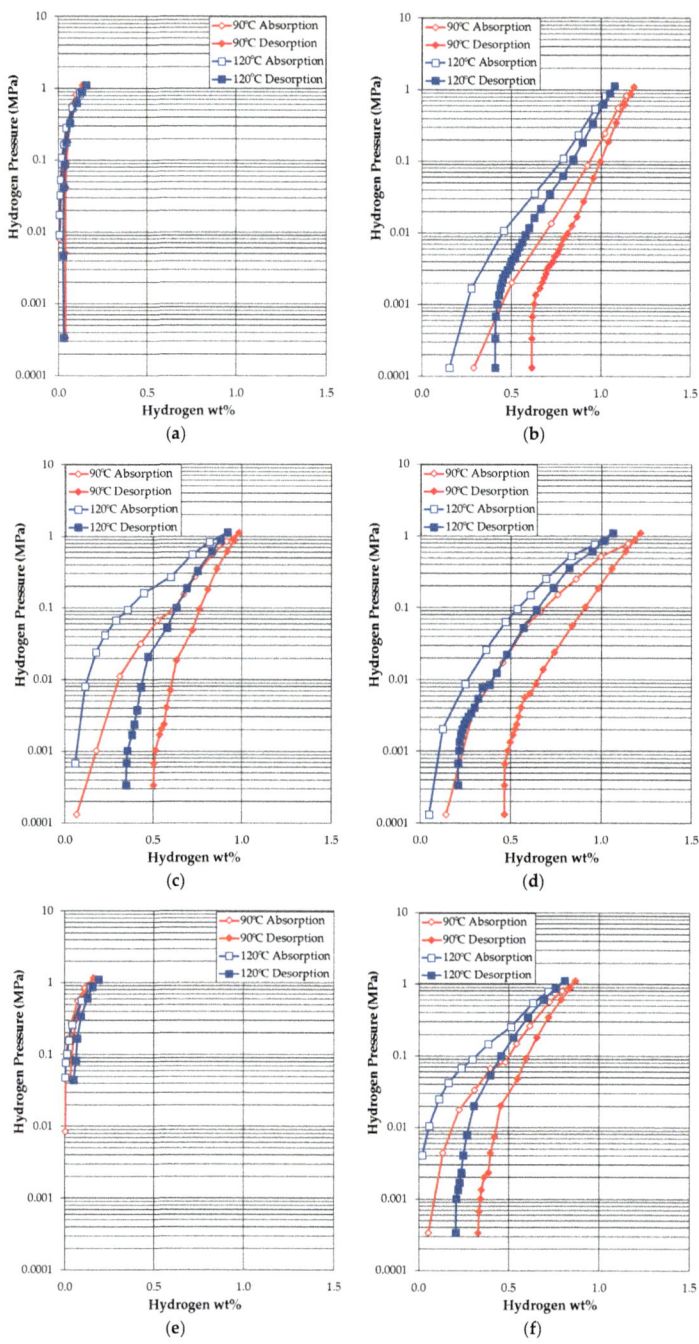

Figure 7. The 90 and 120 °C pressure-concentration-temperature (PCT) isotherms of alloys: (**a**) TN-Ni; (**b**) TN-Cr; (**c**) TN-Mn; (**d**) TN-Fe; (**e**) TN-Co; and (**f**) TN-Cu. Open and solid symbols are for absorption and desorption curves, respectively.

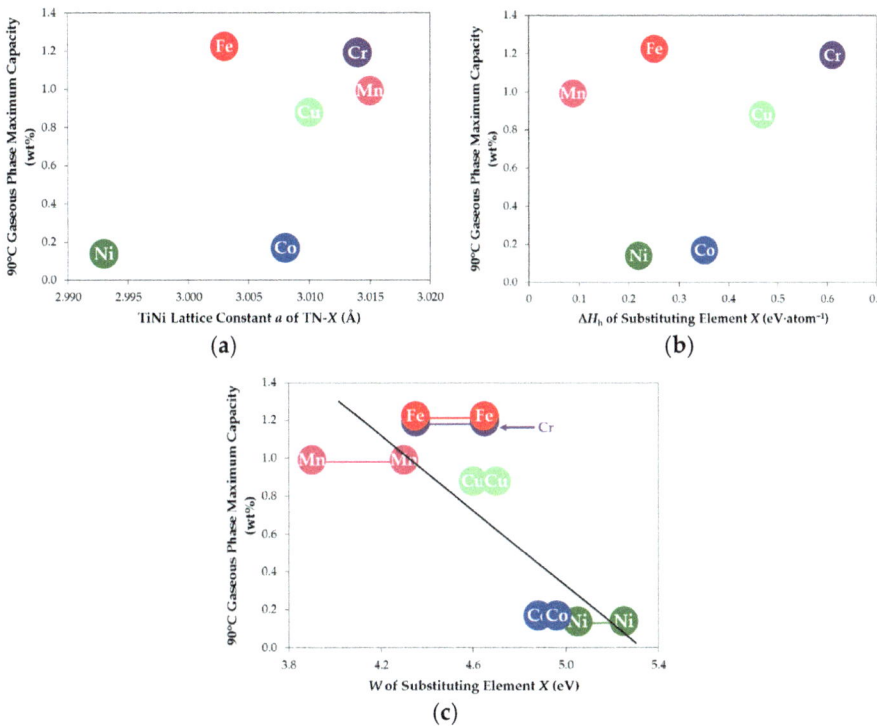

Figure 8. Plots of the 90 °C gaseous phase maximum capacity vs. (**a**) the corresponding TiNi lattice constant a; (**b**) ΔH_h of substituting element X (data from [16]); and (**c**) W of substituting element X (data from [81]). While the correlations with the TiNi lattice constant and ΔH_h of substituting element are weak, the increase in W of substituting element appears to have a negative effect on gaseous phase hydrogen storage.

3.5. Electrochemical Measurement

Discharge capacity performances of the six alloys were measured in a flooded cell configuration against a partially pre-charged $Ni(OH)_2$ positive electrode with 30 wt% KOH electrolyte. Electrodes were made with powder after PCT measurements. No alkaline pretreatment was applied before the half-cell measurement. For the discharge capacity measurement, the half-cell was first charged at a current density of $50\ mA \cdot g^{-1}$ for 10 h and then discharged at a current density of $50\ mA \cdot g^{-1}$ until a cut-off voltage of 0.9 V was reached. Then, the cell was discharged at a current density of $12\ mA \cdot g^{-1}$ until a cut-off voltage of 0.9 V was reached, and finally discharged at a current density of $4\ mA \cdot g^{-1}$ until a cut-off voltage of 0.9 V was reached. Full discharge capacities, specifically the sum of capacities measured at 50, 12, and $4\ mA \cdot g^{-1}$ for each cycle, from the first 10 cycles of the six alloys are plotted in Figure 10a to demonstrate activation and early cycling behaviors. Maximum full capacities, activation performances (the number of cycle needed to reach 95% of the maximum full capacity), and degradation performances (the ratio of the difference in maximum full capacity and full capacity at the tenth cycle to the maximum full capacity) are listed in Table 6. Capacity from the base alloy TN-Ni with a nominal composition of $Ti_{50}Zr_1Ni_{49}$ is higher than those from the stoichiometric TiNi alloys previously reported [13,14,27,29,38,39,46,47,50,53,57,58,64] due to the increase in DOD and formation of a considerable amount of the Ti_2Ni secondary phases. Most alloys are activated during the first cycle while alloys TN-Ni and TN-Cr take longer to fully activate, and alloys TN-Mn, TN-Co, and TN-Cu show more severe degradation. The lowest cycle stability is observed in alloy TN-Mn, and Mn's detrimental effects were also previously shown in the TiNi alloy [48,52]. Among all alloys, alloy TN-Fe does not only demonstrate the best balance between easy activation and cycle stability, it also has the highest full capacity of $397\ mAh\cdot g^{-1}$, corresponding well with its highest gaseous phase capacity. Furthermore, capacities obtained from the gaseous phase and electrochemical measurements are compared in Figure 11, where the gaseous phase capacities are converted to their equivalent electrochemical capacities using the conversion factor 1 wt% $H_2 = 268\ mAh\cdot g^{-1}$. Gaseous phase capacities obtained at 30 °C are usually used for such comparison, but those cannot be measured due to the slow reaction kinetics of the alloys used in the current study. Therefore, 90 °C gaseous phase capacities are used. With the increasing atomic number of substituting elements, the electrochemical capacity first increases and then decreases. Increase in capacity due to the substitutions of Cr [49], Mn [48,52], Co [49,50], or Fe [38,50] and decrease in capacity from the substitution of Cu [52,66] in the TiNi-based alloys have been reported previously, and a comparative study of various substitutions has also shown a similar trend in electrochemical capacity compared to the results in this study. Moreover, electrochemical capacities of the six alloys are above the corresponding gaseous phase maximum capacities. Gaseous phase maximum capacity is composed of reversible and irreversible capacities and considered to be the upper bound for the electrochemical capacity, which usually falls between the boundaries set by the gaseous phase maximum and reversible capacities in most alloy systems [17,69,82–88]. Although increasing the temperature for the PCT measurements causes a reduction in gaseous phase maximum capacity [89] and consequently explains the out-of-bounds electrochemical capacities observed in most alloys, the tremendously large gaps between the electrochemical and gaseous phase maximum capacities for alloys TN-Ni and TN-Co indicate that the electrochemical environment is able to reduce the alloy system's equilibrium pressure and increases its capacity. Such a phenomenon has also been seen in the $ZrNi_5$-based alloys [90,91].

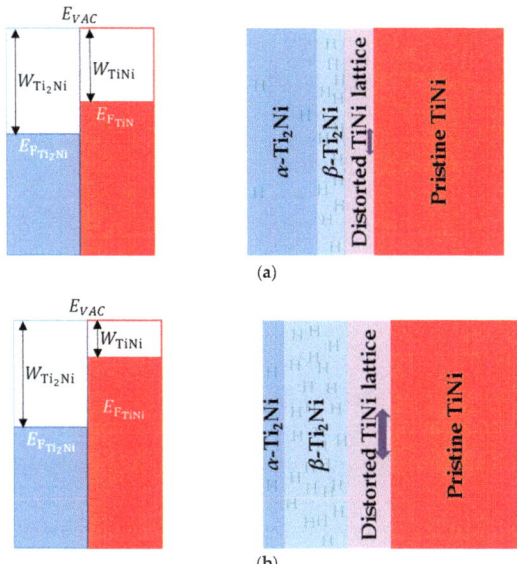

Figure 9. MH system composed of Ti_2Ni with higher W and TiNi with lower W before and after hydrogenation for (**a**) incorporation of substituting element with higher W (for example, Ni) in TiNi; and (**b**) incorporation for substituting element with lower W (for example, Cr) in TiNi.

Half-cell HRDs, defined as the ratio of the discharge capacity measured at $50 \text{ mA} \cdot \text{g}^{-1}$ to that measured at $4 \text{ mA} \cdot \text{g}^{-1}$, from the first 10 cycles of the six alloys are plotted in Figure 10b. All alloys achieve a stabilized HRD by the third cycle. HRDs at the second or third cycle (depends on where it is fully activated) of all alloys are listed in Table 6. HRDs of most alloys are better than that of alloy TN-Ni, with HRD of alloy TN-Cu being the highest. The Cu-substitution in the TiNi alloy formula [66] and later Cu-addition to the TiNi alloy [29] were shown to be beneficial to the high-rate performance. No obvious correlations can be found between HRD and structural properties (e.g., lattice constants, phase abundances, and crystallite sizes). Although these HRDs are relatively low compared to those measured from the AB_2 [71,92], AB_5 [71], A_2B_7 [71], and Laves phase-related body-centered-cubic (bcc) solid solution MH alloys [17,87,88,93,94], some of the six alloys in the current study with higher capacities, e.g., alloy TN-Fe, can be used for high energy Ni/MH battery applications without strict power requirements.

179

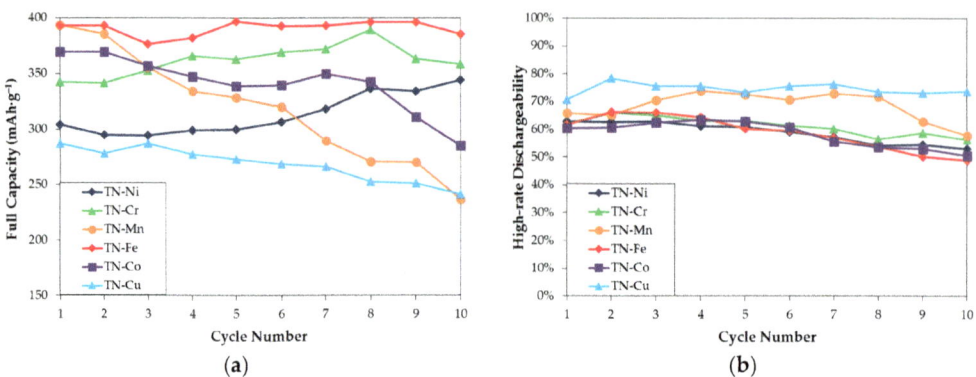

Figure 10. Activation and degradation behaviors observed from: (**a**) half-cell capacity measured at $4 \text{ mA} \cdot \text{g}^{-1}$; and (**b**) half-cell HRD for the first 10 cycles.

Table 6. Summary of room temperature electrochemical and magnetic susceptibility results.

Alloy TN-X	TN-Ni	TN-Cr	TN-Mn	TN-Fe	TN-Co	TN-Cu
Maximum Full Capacity @4 mA·g^{-1} (mAh·g^{-1})	345	389	394	397	370	308
HRD @2nd or 3rd cycle (%)	63	66	71	66	63	79
Number of Cycles Needed to Reach 95% of Maximum Full Capacity	10	6	1	1	1	1
Degradation Performance (%)	0	8	40	3	23	21
D (10^{-10} cm$^2 \cdot$ s^{-1})	3.15	2.71	2.68	1.87	2.37	1.94
I_o (mA·g^{-1})	22.15	24.77	34.08	37.47	26.19	36.43
M_s (emu·g^{-1})	0.187	0.219	0.509	0.511	0.586	0.542
$H_{1/2}$ (kOe)	0.172	0.221	0.170	0.159	0.151	0.484

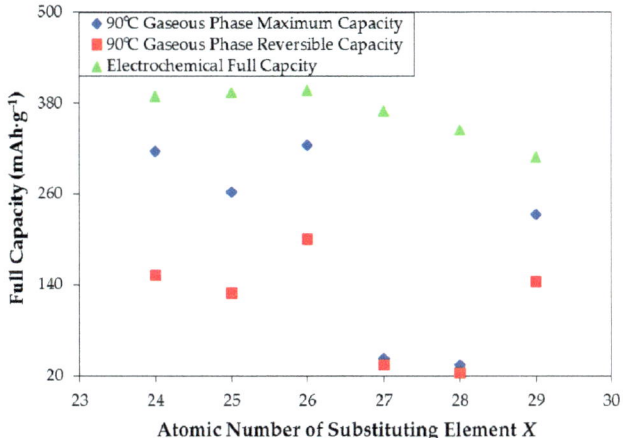

Figure 11. Comparison of capacities obtained from the gaseous phase and electrochemical measurements. Gaseous phase capacity can be converted by $1 \text{ wt}\% = 268 \text{ mAh} \cdot \text{g}^{-1}$.

Surface reaction exchange currents (I_0) and bulk hydrogen diffusion coefficients (D), two dominating factors in determining HRD, were measured electrochemically for the six alloys [69] and are listed in Table 6. All substitutions increase I_0 but decrease D compared to the base alloy TN-Ni. Alloy TN-Fe, which exhibits the lowest D, also has the highest density of interface between the main TiNi and secondary Ti$_2$Ni phases. The channels for hydrogen transport between phases are abundant, however, they are also more tortuous in alloy TN-Fe and may negatively affect the bulk diffusion. By substituting the B-site with other transition metals that are more corrosion susceptible, not only does the activation become easier (Figure 10a), but more of the alloy surface is also dissolved away during activation, leaving more Ni embedded in the surface and consequently improving the surface reaction, although the Ni-content in the alloy formula is reduced. Both I_0 and D are then correlated to HRD, and the resulting correlation factors are 0.24 and 0.46, respectively. Therefore, we conclude that the surface properties are more influential in determining HRD in the TiNi-based MH alloys. In addition, the D and I_0 values of the six alloys in the current study are comparable to those obtained from the AB$_2$ [10,71,92], AB$_5$ [10,71], A$_2$B$_7$ [10,71], and Laves phase-related bcc solid solution MH alloys [17,87,88,94] and cannot explain their relatively low HRDs. Further investigation into other contributing factors affecting HRD of the TiNi-based MH alloys is needed.

3.6. Magnetic Properties

Magnetic susceptibility was used to characterize the nature of metallic nickel particles present in the surface layer of the alloy following an alkaline activation treatment [71]. Details of the background and experimental method have been reported earlier [95]. Metallic Ni is an active catalyst for the water splitting and recombination reactions that contributes to the I_o in the electrochemical system. This technique allows us to obtain the saturated magnetic susceptibility (M_s), a quantification of the amount of metallic Ni (the product of preferential oxidation) in the surface oxide, and the magnetic field strength at one-half of the M_s value ($H_{1/2}$), a measure of the averaged reciprocal number of Ni atoms in a metallic cluster. Magnetic susceptibility plots for the six alloys are shown in Figure 12, and the calculated M_s and $H_{1/2}$ values are listed in Table 6. Compared to alloy TN-Ni, M_ss of all substituted alloys are larger and correspond well with the overall observation in I_o. However, the largest M_s of 0.586 emu·g^{-1} is obtained from alloy TN-Co, which has a relatively low I_o among the substituted alloy. The $H_{1/2}$ values listed in Table 6 indicate that the sizes of metallic nickel are similar among most alloys except for alloy TN-Cu, which has much smaller metallic nickel particles in the surface and very high M_s. Both of these contribute to its high I_o, leading to alloy TN-Cu's impressive HRD performance among all alloys.

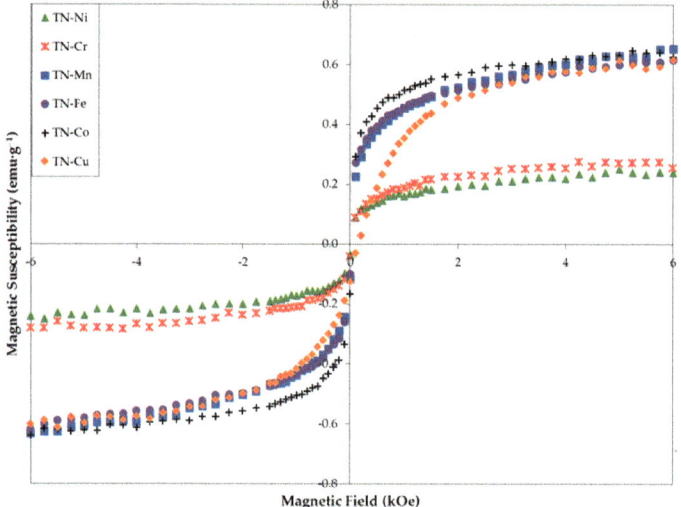

Figure 12. Magnetic susceptibility plots of the six alloys in the current study.

3.7. Comparison among Various Sbustitutions

Properties characterized from the gaseous phase and electrochemical measurements of the five substituted alloys compared to those of the base alloy TN-Ni are summarized in Table 7. Most of the properties are improved by substitution except for the bulk diffusion performance. Some general trends can also be observed: electrochemical capacity is inversely proportional to HRD, and HRD shows a similar trend compared to I_o and M_s. Among all substituted alloys, alloy TN-Fe has the best balance among capacity, HRD, activation and degradation performances and is recommended as the base alloy for further optimization.

Table 7. Performance comparison for the six alloys in the current study. +, −, or ≈ denote improved, deteriorated, or equal performance compared to the base alloy TN-Ni, respectively. Number of symbols indicates the degree of change compared to the base alloy TN-Ni.

Alloy TN-X	Gaseous Phase Capacity	Electrochemical Full Capacity	HRD	Activation	Degradation	D	I_o	M_s
TN-Cr	+ + + +	+ + + +	+	+ +	−	− −	+	+
TN-Mn	+ + +	+ + + + +	+ +	+ + + + +	− − − − −	− −	+ + + +	+ + + +
TN-Fe	+ + + + +	+ + + + +	+	+ + + + +	≈	− − − − −	+ + + + +	+ + + +
TN-Co	≈	+ +	≈	+ + + + +	− − −	− − −	+	+ + + + +
TN-Cu	+ + +	− − −	+ + + + +	+ + + + +	− −	− − − −	+ + + + +	+ + + + +

3.8. Property Comparison among Various Metal Hydride Alloy Systems

The general battery performances using various MH alloys are summarized in Table 8. In the table, AB$_5$, with a representative composition of La$_{10.5}$Ce$_{4.3}$Pr$_{0.5}$Nd$_{1.4}$Ni$_{60.0}$Co$_{12.7}$Mn$_{5.9}$Al$_{4.7}$, is the most widely used in the current consumer and transportation markets, and it has the best overall performance except for its relatively low capacity (320–330 mAh·g^{-1}). Significant research efforts on substitution have been performed with the AB$_5$ alloy to lower the cost and fulfill other specific requirements [10,96,97]. The AB$_2$ multi-phase alloy, with a very high capacity (420–436 mAh·g^{-1} [98]) and flexibility in composition [99,100], has great potential in ultra-low-temperature [101] and high-temperature applications due to the non-passive nature of its surface oxide [71]. The A$_2$B$_7$ superlattice-based multi-phase alloy family is the MH alloy currently used by FDK (Tokyo, Japan) for their Eneloop Ni/MH products. The A$_2$B$_7$ alloy, with a representative composition of La$_{3.3}$Ce$_8$Pr$_8$Mg$_{3.9}$Ni$_{72.8}$Al$_{4.0}$, has a marginally improved capacity (355 mAh·g^{-1}) over the conventional AB$_5$ but also a tradeoff with regard to cycle stability due to the incorporation of Mg, which can be solved by adding the proper binder to the negative electrode [102]. Works related to the improvement in capacity and low-temperature performance have been previously reported by authors [103–107]. The bcc alloy has a very high theoretical capacity (1072 mAh·g^{-1}) but a much lower reported capacity (247 mAh·g^{-1}) for the composition Ti$_{40}$V$_{30}$Cr$_{15}$Mn$_{13}$Mo$_2$, and its stability is very poor, requiring an electrolyte with much lower corrosion capability, for example, ionic liquid [108]. The high cost of V is also a concern for utilizing the bcc alloy in any practical applications. The Laves phase-related bcc alloy was developed through a U.S. Department of Energy sponsored program, and the resulted alloy has a composition of $Ti_{14.5}Zr_{2.7}V_{46.6}Cr_{11.9}Mn_{6.5}Co_{1.5}Ni_{16.9}Al_{0.4}$ and a capacity of 414 mAh· g^{-1} [17,94,109], which was recommended for electric vehicle applications running at a C/3 discharge rate. The MgNi-based amorphous alloy made by mechanical alloying has a very attractive cost model and an ultra-high capacity (780 mAh· g^{-1} for the composition $Mg_{52}Ni_{39}Co_3Mn_6$), but it suffers in cycle stability in 30 wt% KOH electrolyte [18]. Studies that address changes in anions of the hydroxides used [18] and additions of various salts [110] have been developed to extend the usable cycles of the MgNi-based alloy. The last row in Table 8 is the main subject of this paper—the TiNi alloy family. TiNi at the current stage is not suitable for either consumer electronic or propulsion applications that require a decent HRD. However, its low cost and long cycle life make the TiNi alloy perfect for stationary applications. Therefore, testing of the TiNi-based alloys at different temperatures will be our next task.

Table 8. Performance comparison among various MH alloy families. The number of stars indicates the superiority in the performance category. TBD and Temp. denote to-be-determined and temperature, respectively.

Alloy	Cost	Capacity	HRD	Activation	Low Temp.	High Temp.	Charge Retention	Cycle Life
AB_5	★★★★	★	★★★★★	★★★★★	★★★★	★★	★★★★	★★★★★
AB_2	★★★	★★★★	★★★★	★★★	★★★★★	★★★★★	★★★	★★★★★
A_2B_7	★★★	★★	★★★★★	★★★★★	★★★★★	★★★	★★★★★	★★★★
bcc	★★	★★★★★	★★	★★★★★	TBD	TBD	★	★
Laves-bcc	★	★★★★	★★★	★★★★	★★★	★★	★	★★★
MgNi	★★★★★	★★★★★	★★	★★★★★	TBD	TBD	TBD	★
TiNi	★★★★★	★★★	★	★★	TBD	TBD	TBD	★★★★★

4. Conclusions

Structural, magnetic, gaseous phase hydrogen storage, and electrochemical properties of a series of $Ti_{50}Zr_1Ni_{44}X_5$ (X = Ni, Cr, Mn, Fe, Co, or Cu) metal hydride alloys were investigated. All alloys show similar multi-phase distributions composed of a stoichiometric TiNi matrix, a hyperstoichiometric TiNi minor phase, and a Ti_2Ni secondary phase. Compared to the $Ti_{50}Zr_1Ni_{49}$ base alloy, substituting element with higher work function enhances the synergetic effects between the main TiNi and secondary Ti_2Ni phases and increases the gaseous phase hydrogen storage capacity substantially. Moreover, the electrochemical environment is able to reduce the alloy system's equilibrium pressure and further improves the capacity. The TiNi-based alloys have a superb cost model and exhibit satisfactory capacity and cycle life performances, however, these alloys score low in high-rate performance despite their similar surface reaction, bulk diffusion, and magnetic properties to the commonly used AB_5, AB_2, and A_2B_7 alloys. Among the alloys investigated, alloy with the composition $Ti_{50}Zr_1Ni_{44}Fe_5$ demonstrates the best balance among capacity, high-rate performance, activation, and cycle stability and is recommended for full-cell testing at various temperatures to validate its practicality for high-energy or stationary applications. $Ti_{50}Zr_1Ni_{44}Fe_5$ will also be used as the base composition for formula optimization in the future.

Acknowledgments: The authors would like to thank the following from BASF-Ovonic for their technical assistance: Alan Chen, Ryan Blankenship, Su Cronogue, Taihei Ouchi, Diana Wong, and Tiejun Meng.

Author Contributions: Jean Nei conceived and performed the experiments, and Kwo-Hsiung Young participated in data interpretation and paper preparation.

Conflicts of Interest: The authors declare no conflicts of interest.

Abbreviations

Ni/MH	Nickel/metal hydride
MH	Metal hydride
HRD	High-rate dischargeability
ΔH_h	Heat of hydride formation
SN	Sintering
AM	Arc melting
PM	Melting in a plasma furnace
IM	Induction melting
CO	Co-precipitation
MC	Microencapsulation
MA	Mechanical alloying
ANN	Annealing
MS	Melt spinning
CHR	Calcium hydride reduction
MWCNT	Multiwall carbon nanotube
ICP	Inductively coupled plasma spectrometer/spectrometry
XRD	X-ray diffractometer/diffraction
SEM	Scanning electron microscope/microscopy
EDS	Energy dispersive spectroscopy
PCT	Pressure-concentration-temperature
W	Work function
E_{VAC}	Electron potential in vacuum (E_{VAC}) and the Fermi level (E_F)
E_F	Electron potential in the Fermi level

bcc	Body-centered-cubic
I_o	Surface reaction exchange current
D	Bulk hydrogen diffusion coefficient
M_s	Saturated magnetic susceptibility
$H_{1/2}$	Magnetic field strength at one-half of the saturated magnetic susceptibility value

References

1. Zelinsky, M.; Koch, J.; Fetcenko, M. Heat Tolerant NiMH Batteries for Stationary Power. Available online: http://www.battcon.com/PapersFinal2010/ZelinskyPaper2010Final_12.pdf (accessed on 5 May 2016).

2. Zelinsky, M.; Koch, J. Batteries and Heat—A Recipe for Success? Available online: http://www.battcon.com/PapersFinal2013/16-Mike%20Zelinsky%20-%20Batteries%20and%20Heat.pdf (accessed on 5 May 2016).

3. Wikipedia, the Free Encyclopedia. General Motors EV1. Available online: https://en.wikipedia.org/wiki/General_Motors_EV1 (accessed on 20 June 2016).

4. The Jaffes. EV1. Available online: http://thejaffes.org/content/ev1 (accessed on 20 June 2016).

5. Sakintuna, B.; Lamari-Darkrim, F.; Hirscher, M. Metal hydride materials for solid hydrogen storage: A review. *Int. J. Hydrog. Energy* **2007**, *32*, 1121–1140.

6. Paul-Boncour, V. Metal hydrides for hydrogen storage. *J. Adv. Sci.* **2007**, *19*, 16–21.

7. Zhao, X.; Ma, L. Recent progress in hydrogen storage alloys for nickel/metal hydride secondary batteries. *Int. J. Hydrog. Energy* **2009**, *34*, 4788–4796.

8. Pukszhselvan, D.; Kumar, V.; Singh, S.K. High capacity hydrogen storage: Basic aspects, new developments and milestones. *Nano Energy* **2012**, *1*, 566–589.

9. Klebanoff, L.E.; Keller, J.O. 5 Years of hydrogen storage research in the U.S. DOE Metal Hydride Center of Excellence (MHCoE). *Int. J. Hydrog. Energy* **2013**, *38*, 4533–4576.

10. Young, K.; Nei, J. The current status of hydrogen storage alloy development for electrochemical applications. *Materials* **2013**, *6*, 4574–4608.

11. Young, K. Metal Hydrides. In *Reference Module in Chemistry, Molecular Sciences and Chemical Engineering*; Elsevier B.V.: Waltham, MA, USA, 2013.

12. Liu, W.; Webb, C.J.; Gray, E.M. Review of hydrogen storage in AB_3 alloys targeting stationary fuel cell applications. *Int. J. Hydrog. Energy* **2016**, *41*, 3485–3507.

13. Gutjahr, M.A.; Buchner, H.; Beccu, K.D.; Säufferer, H. A New Type of Reversible Negative Electrode for Alkaline Storage Batteries Based on Metal Alloy Hydrides. In *Power Sources*; Collins, D.H., Wada, K., Hiraki, A., Eds.; Oriel Press: Newcastle upon Tyne, UK, 1973; Volume 4, pp. 79–91.

14. Beccu, K. Negative Electrode of Titanium-Nickel Alloy Hydride Phases. U.S. Patent 3,824,131, 16 July 1974.

15. Murray, J.L. Ni-Ti (Nickel-Titanium). In *Binary Alloy Phase Diagram*, 2nd ed.; Massalski, T.B., Okamoto, H., Subramanian, P.R., Kacprzak, L., Eds.; ASM International: Materials Park, OH, USA, 1990; Volume 3, pp. 2874–2876.

16. Osumi, Y. *Hydrogen Absorbing Alloy—The Physical Properties and Applications*, 1st ed.; Agune Technology Center: Tokyo, Japan, 1993; p. 73.

17. Young, K.; Ouchi, T.; Meng, T.; Wong, D.F. Studies on the synergetic effects in multi-phase metal hydride alloys. *Batteries* **2016**, *2*.

18. Nei, J.; Young, K.; Rotarov, D. Studies on MgNi-based metal hydride electrode with aqueous electrolytes composed of various hydroxides. *Batteries* **2016**, *2*.

19. Justi, E.W.; Ewe, H.H.; Kalberlah, A.W.; Saridakis, N.M.; Schaefer, M.H. Electrocatalysis in the nickel-titanium system. *Energy Convers.* **1970**, *10*, 183–187.

20. Miles, M.H. Evaluation of electrocatalysts for water electrolysis in alkaline solutions. *J. Electroanal. Chem. Interfacial Electrochem.* **1975**, *60*, 89–96.

21. Wakao, S.; Yonemura, Y.; Nakano, H.; Shimada, H. Electrochemical capacities and corrosion of $TiNi_x$ and its zirconium-substituted alloy hydride electrodes. *J. Less Common Met.* **1984**, *104*, 365–373.

22. Wakao, S.; Nakano, H.; Chubachi, S. Behaviour of hydrogen-absorbing metal alloys in an alkaline solution containing hydrazine. *J. Less Common Met.* **1984**, *104*, 385–393.

23. Wakao, S.; Sawa, H.; Nakano, H.; Chubachi, S.; Abe, M. Capacities and durabilities of Ti-Zr-Ni alloy hydride electrodes and effects of electroless plating on their performances. *J. Less Common Met.* **1987**, *131*, 311–319.

24. Song, D.; Gao, X.; Zhang, Y.; Lin, D.; Zhou, Z.; Wang, G.; Shen, P. Surface analysis of a Ti-Ni-B hydrogen storage electrode. *J. Alloys Compd.* **1993**, *199*, 161–163.

25. Jordy, C.; Latroche, M.; Percheron-Guégan, A.; Achard, J.C. Effect of partial substitution in TiNi on its structural and electrochemical hydrogen storage properties. *Z. Phys. Chem.* **1994**, *185*, 119–130.

26. Yan, D.-Y. Catalytic effects of alloy surface on the oxygen consumption reaction in a sealed Ni/TiNiH battery. *J. Alloys Compd.* **1994**, *209*, 257–261.

27. Wang, C.; Lei, Y.; Yang, X.; Jiang, J.; Wu, J.; Wang, Q. Effects of phase structures of TiNi on the electrochemical properties. *Acta Metall. Sin.* **1995**, *31*, 440–444.

28. Lei, Y.Q.; Wang, C.S.; Yang, X.G.; Pan, H.G.; Wu, J.; Wang, Q.D. A mathematical model for the cycle life of hydride electrodes. *J. Alloys Compd.* **1995**, *231*, 611–615.

29. Liu, J.; Gao, X.; Song, D.; Zhang, Y.; Ye, S. The characteristics of the microencapsulated Ti-Ni alloys and their electrodes. *J. Alloys Compd.* **1995**, *231*, 852–855.

30. Jung, C.B.; Lee, K.S. Electrode characteristics of metal hydride electrodes prepared by mechanical alloying. *J. Alloys Compd.* **1997**, *253–254*, 605–608.

31. Jung, C.B.; Kim, J.H.; Lee, K.S. Electrode characteristics of nanostructured TiFe and $ZrCr_2$ type metal hydride prepared by mechanical alloying. *Nanostructured Mater.* **1997**, *8*, 1093–1104.

32. Wang, C.S.; Lei, Y.Q.; Wang, Q.D. Effects of Nb and Pd on the electrochemical properties of a Ti-Ni hydrogen-storage electrode. *J. Power Sources* **1998**, *70*, 222–227.

33. Wang, C.S.; Lei, Y.Q.; Wang, Q.D. Studies of electrochemical properties of TiNi alloy used as an MH electrode—I. Discharge capacity. *Electrochim. Acta* **1998**, *43*, 3193–3207.

34. Wang, C.S.; Lei, Y.Q.; Wang, Q.D. Studies of electrochemical properties of TiNi alloy used as an MH electrode. II. Discharge kinetics. *Electrochim. Acta* **1998**, *43*, 3209–3216.

35. Zhang, Q.A.; Lei, Y.Q.; Wang, C.S.; Wang, F.S.; Wang, Q.D. Structure of the secondary phase and its effects on hydrogen-storage properties in a $Ti_{0.7}Zr_{0.2}V_{0.1}Ni$ alloy. *J. Power Sources* **1998**, *75*, 288–291.

36. Han, S.S.; Goo, N.H.; Jeong, W.T.; Lee, K.S. Synthesis of composite metal hydride alloy of A_2B and AB type by mechanical alloying. *J. Power Sources* **2001**, *92*, 157–162.

37. Cuevas, F.; Latroche, M.; Ochin, P.; Dezellus, A.; Fernández, J.F.; Sánchez, C.; Percheron-Guégan, A. Influence of the martensitic transformation on the hydrogenation properties of $Ti_{50-x}Zr_xNi_{50}$ alloys. *J. Alloys Compd.* **2002**, *330–332*, 250–255.

38. Jurczyk, M.; Jankowska, E.; Nowak, M.; Jakubowicz, J. Nanocrystalline titanium-type metal hydride electrodes prepared by mechanical alloying. *J. Alloys Compd.* **2002**, *336*, 265–269.

39. Bobet, J.; Chevalier, B. Reactive mechanical grinding applied to a (Ti + Ni) mixture and to a TiNi compound. *Intermetallics* **2002**, *10*, 597–601.

40. Xu, Y.H.; Chen, C.P.; Wang, X.L.; Wang, Q.D. The analysis of the two discharge plateaus for Ti-Ni-based metal hydride electrode alloys. *J. Power Sources* **2002**, *112*, 105–108.

41. Szajek, A.; Jurczyk, M.; Jankowska, E. The electronic and electrochemical properties of the TiFe-based alloys. *J. Alloys Compd.* **2003**, *348*, 285–292.

42. Jurczyk, M.; Jankowska, E.; Makowiecka, M.; Wieczorek, I. Electrode characteristics of nanocrystalline TiFe-type alloys. *J. Alloys Compd.* **2003**, *354*, L1–L4.

43. Han, S.S.; Goo, N.H.; Lee, K.S. Effects of sintering on composite metal hydride alloy of Mg_2Ni and TiNi synthesized by mechanical alloying. *J. Alloys Compd.* **2003**, *360*, 243–249.

44. Zhang, Q.A.; Lei, Y.Q. Multi-component TiNi-based hydrogen storage alloys with the secondary Laves phase. *J. Alloys Compd.* **2004**, *368*, 362–366.

45. Jankowska, E.; Jurczyk, M. Electrochemical properties of sealed Ni-MH batteries using nanocrystalline TiFe-type anodes. *J. Alloys Compd.* **2004**, *372*, L9–L12.

46. Makowiecka, M.; Jankowska, E.; Okonska, I.; Jurczyk, M. Effect of Zr additions on the electrode characteristics of nanocrystalline TiNi-type hydrogen storage alloys. *J. Alloys Compd.* **2005**, *388*, 303–307.

47. Shcherbakova, L.G.; Solonin, S.M.; Kolomiets, L.L.; Katashinskii, V.P. Effect of phase composition and activation of a titanium nickelide surface by electrochemical cycling on its hydrogen sorption capacity. *Powder Metall. Met. Ceram.* **2005**, *44*, 389–395.

48. Szajek, A.; Makowiecka, M.; Jankowska, E.; Jurczyk, M. Electrochemical and electronic properties of nanocrystalline $TiNi_{1-x}M_x$ (M = Mg, Mn, Zr, x = 0, 0.125, 0.25) ternary alloys. *J. Alloys Compd.* **2005**, *403*, 323–328.

49. Jankowska, E.; Makowiecka, M.; Jurczyk, M. Nickel-metal hydride battery using nanocrystalline TiFe-type hydrogen storage alloys. *J. Alloys Compd.* **2005**, *404–406*, 691–693.

50. Drenchev, B.; Spassov, T. Electrochemical hydriding of amorphous and nanocrystalline TiNi-based alloys. *J. Alloys Compd.* **2007**, *441*, 197–201.

51. Jankowska, E.; Makowiecka, M.; Jurczyk, M. Electrochemical performance of sealed Ni-MH batteries using nanocrystalline TiNi-type hydride electrodes. *Renew. Energy* **2008**, *33*, 211–215.

52. Drenchev, B.; Spassov, T.; Radev, D. Influence of alloying and microstructure on the electrochemical hydriding of TiNi-based ternary alloys. *J. Appl. Electrochem.* **2008**, *38*, 437–444.

53. Guiose, B.; Cuevas, F.; Décamps, B.; Percheron-Guégan, A. Solid-gas and electrochemical hydrogenation properties of pseudo-binary (Ti,Zr)Ni intermetallic compounds. *Int. J. Hydrog. Energy* **2008**, *33*, 5795–5800.

54. Drenchev, B.; Spassov, T. Influence of B substitution for Ti and Ni on the electrochemical hydriding of TiNi. *J. Alloys Compd.* **2009**, *474*, 527–530.

55. Guiose, B.; Cuevas, F.; Décamps, B.; Leroy, E.; Percheron-Guégan, A. Microstructural analysis of the ageing of pseudo-binary (Ti,Zr)Ni intermetallic compounds as negative electrodes of Ni-MH batteries. *Electrochim. Acta* **2009**, *54*, 2781–2789.

56. Qu, X.; Ma, L.; Yang, M.; Ding, Y. Effect of sintering temperature on electrochemical properties of TiNi hydrogen storage alloy. *Chin. J. Rare Met.* **2010**, *34*, 331–335.

57. Yang, M.; Zhao, X.; Ding, Y.; Ma, L.; Qu, X.; Gao, Y. Electrochemical properties of titanium-based hydrogen storage alloy prepared by solid phase sintering. *Int. J. Hydrog. Energy* **2010**, *35*, 2717–2721.

58. Zhao, X.; Ma, L.; Yang, M.; Ding, Y.; Shen, X. Electrochemical properties of Ti-Ni-H powders prepared by milling titanium hydride and nickel. *Int. J. Hydrog. Energy* **2010**, *35*, 3076–3079.

59. Jiang, X.; Liu, Q.; Zhang, L. Electrochemical hydrogen storage property of NiTi alloys with different Ti content prepared by mechanical alloying. *Rare Met.* **2011**, *30*, 63–67.

60. Zlatanova, Z.; Spassov, T.; Eggeler, G.; Spassova, M. Synthesis and hydriding/dehydriding properties of Mg_2Ni-AB (AB = TiNi or TiFe) nanocomposites. *Int. J. Hydrog. Energy* **2011**, *36*, 7559–7566.

61. Emami, H.; Cuevas, F. Hydrogenation properties of shape memory Ti(Ni,Pd) compounds. *Intermetallics* **2011**, *19*, 876–886.

62. Hu, R.; Liu, H.; Zeng, M.; Liu, J.; Zhu, M. Influence of Sn content on microstructure and electrochemical properties of Sn-NiTi film anodes in lithium ion batteries. *J. Power Sources* **2012**, *244*, 456–462.

63. Bououdina, M.; Oumellal, Y.; Dupont, L.; Aymard, L.; Al-Gharni, H.; Al-Hajry, A.; Maark, T.A.; De Sarkar, A.; Ahuja, R.; Deshpande, M.D.; et al. Lithium storage in amorphous TiNi hydride: Electrode for rechargeable lithium-ion batteries. *Mater. Chem. Phys.* **2013**, *141*, 348–354.

64. Balcerzak, M.; Nowak, M.; Jakubowicz, J.; Jurczyk, M. Electrochemical behavior of nanocrystalline TiNi doped by MWCNTs and Pd. *Renew. Energy* **2014**, *62*, 432–438.

65. Li, X.D.; Elkedim, O.; Nowak, M.; Jurczyk, M. Characterization and first principle study of ball milled Ti-Ni with Mg doping as hydrogen storage alloy. *Int. J. Hydrog. Energy* **2014**, *39*, 9735–9743.

66. Emami, H.; Cuevas, F.; Latroche, M. Ti(Ni,Cu) pseudobinary compounds as efficient negative electrodes for Ni-MH batteries. *J. Power Sources* **2014**, *265*, 182–191.

67. Zhang, Z.; Elkedim, O.; Balcerzak, M.; Jurczyk, M. Structural and electrochemical hydrogen storage properties of $MgTiN_x$ ($x = 0.1, 0.5, 1, 2$) alloys prepared by ball milling. *Int. J. Hydrog. Energy* **2016**. in press.

68. Balcerzak, M. Electrochemical and structural studies on Ti-Zr-Ni and Ti-Zr-Ni-Pd alloys and composites. *J. Alloys Compd.* **2016**, *658*, 576–587.

69. Nei, J.; Young, K.; Salley, S.O.; Ng, K.Y.S. Effects of annealing on $Zr_8Ni_{19}X_2$ (X = Ni, Mg, Al, Sc, V, Mn, Co, Sn, La, and Hf): Hydrogen storage and electrochemical properties. *Int. J. Hydrog. Energy* **2012**, *37*, 8418–8427.

70. Hasson, D.F.; Arsenault, R.J. Substitutional-Interstitial Interactions in bcc Alloys. In *Treatise on Materials Science and Technology: Materials Science Series*; Herman, H., Ed.; Academic Press: New York, NY, USA, 1972; Volume 1, p. 218.

71. Young, K.; Huang, B.; Regmi, R.K.; Lawes, G.; Liu, Y. Comparisons of metallic clusters imbedded in the surface of AB_2, AB_5, and A_2B_7 alloys. *J. Alloys Compd.* **2010**, *506*, 831–840.

72. Kokalj, A. Computer graphics and graphical user interfaces as tools in simulations of matter at the atomic scale. *Comput. Mater. Sci.* **2003**, *28*, 155–168.

73. Dwight, A.E. CsCl-type equiatomic phases in binary alloys of transition elements. *Trans. Am. Inst. Min. Metall. Pet. Eng.* **1959**, *215*, 283–286.

74. The Japan Institute of Metals and Materials. *Non-Stoichiometric Metal Compounds*; Maruzen: Tokyo, Japan, 1975; p. 296.

75. Klug, H.P.; Alexander, L.E. *X-Ray Diffraction Procedures: For Polycrystalline and Amorphous Materials*, 2nd ed.; John Wiley & Sons: New York, NY, USA, 1974; p. 656.

76. Bohnenstiehl, S.D.; Susner, M.A.; Dregia, S.A.; Sumption, M.D.; Donovan, J.; Collings, E.W. Experimental determination of the peritectic transition temperature of MgB_2 in the Mg-B phase diagram. *Thermochim. Acta* **2014**, *576*, 27–35.

77. Yen, F.; Hwang, K. Shape memory characteristics and mechanical properties of high-density powder metal TiNi with post-sintering heat treatment. *Mater. Sci. Eng. A* **2011**, *528*, 5296–5305.

78. Gupta, K.P. The Ni-Ti-Zr system (nickel-titanium-zirconium). *J. Phase Equilib.* **1999**, *20*, 441–448.

79. Osumi, Y. *Hydrogen Absorbing Alloy—The Physical Properties and Applications*, 1st ed.; Agune Technology Center: Tokyo, Japan, 1993; p. 57.

80. Liu, Y.; Young, K. Microstructure investigation on metal hydride alloys by electron backscatter diffraction technique. *Batteries* **2016**, *2*.

81. Drummond, T.J. *Work Functions of the Transition Metals and Metal Silicides*; SAND99-0391J; Sandia National Labs.: Albuquerque, NM, USA; Livermore, CA, USA, 1999.

82. Sun, D.; Jiang, J.; Lei, Y.; Liu, W.; Wu, J.; Wang, Q.; Yang, G. Effects of measurement factor on electrochemical capacity of some hydrogen storage alloys. *Mater. Sci. Eng. B* **1995**, *30*, 19–22.

83. Young, K.; Fetcenko, M.A.; Li, F.; Ouchi, T. Structural, thermodynamic, and electrochemical properties of $Ti_xZr_{1-x}(VNiCrMnCoAl)_2$ C14 Laves phase alloys. *J. Alloys Compd.* **2008**, *464*, 238–247.

84. Young, K.; Nei, J.; Huang, B.; Ouchi, T.; Fetcenko, M.A. Studies of $Ti_{1.5}Zr_{5.5}$ $V_{0.5}(M_xNi_{1-x})9.5$ (M = Cr, Mn, Fe, Co, Cu, Al): Part 2. Hydrogen storage and electrochemical properties. *J. Alloys Compd.* **2010**, *501*, 245–254.

85. Young, K.; Chao, B.; Huang, B.; Nei, J. Studies on the hydrogen storage characteristic of $La_{1-x}Ce_x(NiCoMnAlCuSiZr)_{5.7}$ with a B2 secondary phase. *J. Alloys Compd.* **2014**, *585*, 760–770.

86. Young, K.; Wong, D.F.; Wang, L. Effect of Ti/Cr content on the microstructures and hydrogen storage properties of Laves phase-related body-centered-cubic solid solution alloys. *J. Alloys Compd.* **2015**, *622*, 885–893.

87. Young, K.; Ouchi, T.; Nei, J.; Meng, T. Effects of Cr, Zr, V, Mn, Fe, and Co to the hydride properties of Laves phase-related body-centered-cubic solid solution alloys. *J. Power Sources* **2015**, *281*, 164–172.

88. Young, K.; Ouchi, T.; Nei, J.; Wang, L. Annealing effects on Laves phase-related body-centered-cubic solid solution metal hydride alloys. *J. Alloys Compd.* **2016**, *654*, 216–225.

89. Young, K. Stoichiometry in Inter-Metallic Compounds for Hydrogen Storage Applications. In *Stoichiometry and Materials Science—When Numbers Matter*; Innocenti, A., Kamarulzaman, N., Eds.; Intech: Rijeka, Crotia, 2012; p. 150.

90. Young, K.; Young, M.; Chang, S.; Huang, B. Synergetic effects in electrochemical properties of $ZrV_xNi_{4.5-x}$ (x = 0.0, 0.1, 0.2, 0.3, 0.4, and 0.5) metal hydride alloys. *J. Alloys Compd.* **2013**, *560*, 33–41.

91. Mosavati, N.; Young, K.; Meng, T.; Ng, K.Y.S. Electrochemical open-circuit voltage and pressure-concentration-temperature isotherm comparison for metal hydride alloys. *Batteries* **2016**, *2*.

92. Young, K.; Ouchi, T.; Lin, X.; Reichman, B. Effects of Zn-addition to C14 metal hydride alloys and comparisons to Si, Fe, Cu, Y, and Mo-additives. *J. Alloys Compd.* **2016**, *655*, 50–59.

93. Young, K.; Nei, J.; Wong, D.F.; Wang, L. Structural, hydrogen storage, and electrochemical properties of Laves phase-related body-centered-cubic solid solution metal hydride alloys. *Int. J. Hydrog. Energy* **2014**, *39*, 21489–21499.

94. Young, K.; Wong, D.F.; Nei, J. Effects of vanadium/nickel contents in Laves phase-related body-centered-cubic solid solution metal hydride alloys. *Batteries* **2015**, *1*, 34–53.

95. Nei, J.; Young, K.; Regmi, R.; Lawes, G.; Salley, S.O.; Ng, K.Y.S. Gaseous phase hydrogen storage and electrochemical properties of Zr_8Ni_{21}, Zr_7Ni_{10}, Zr_9Ni_{11}, and ZrNi metal hydride alloys. *Int. J. Hydrog. Energy* **2012**, *37*, 16042–16055.

96. Young, K.; Ouchi, T.; Reichman, B.; Koch, J.; Fetcenko, M.A. Improvement in the low-temperature performance of AB_5 metal hydride alloys by Fe-addition. *J. Alloys Compd.* **2011**, *509*, 7611–7617.

97. Young, K.; Yasuoka, S. Capacity degradation mechanisms in nickel/metal hydride batteries. *Batteries* **2016**, *2*.

98. Young, K.; Ouchi, T.; Koch, J.; Fetcenko, M.A. The role of Mn in C14 Laves phase multi-component alloys for NiMH battery application. *J. Alloys Compd.* **2009**, *477*, 749–758.

99. Chang, S.; Young, K.; Ouchi, T.; Meng, T.; Nei, J.; Wu, X. Studies on incorporation of Mg in Zr-based AB_2 metal hydride alloys. *Batteries* **2016**, *2*.

100. Young, K.; Ouchi, T.; Nei, J.; Moghe, D. Importance of rare-earth additions in Zr-based AB_2 metal hydride alloys. *Batteries* **2016**, *2*.

101. Young, K.; Wong, D.F.; Ouchi, T.; Huang, B.; Reichman, B. Effects of La-addition to the structure, hydrogen storage, and electrochemical properties of C14 metal hydride alloys. *Electrochim. Acta* **2015**, *174*, 815–825.

102. Ouchi, T.; Young, K.; Moghe, D. Reviews on the Japanese patent applications regarding nickel/metal hydride batteries. *Batteries* **2016**, *2*.

103. Young, K.; Wong, D.F.; Wang, L.; Nei, J.; Ouchi, T.; Yasuoka, S. Mn in misch-metal based superlattice metal hydride alloy—Part 1 structural, hydrogen storage and electrochemical properties. *J. Power Sources* **2015**, *277*, 426–432.

104. Young, K.; Koch, J.; Yasuoka, S.; Shen, H.; Bendersky, L.A. Mn in misch-metal based superlattice metal hydride alloy—Part 2 Ni/MH battery performance and failure mechanism. *J. Power Sources* **2015**, *277*, 433–442.

105. Wang, L.; Young, K.; Meng, T.; Ouchi, T.; Yasuoka, S. Partial substitution of cobalt for nickel in mixed rare earth metal based superlattice hydrogen absorbing alloy—Part 1 Structural, hydrogen storage and electrochemical properties. *J. Alloys Compd.* **2016**, *660*, 407–415.

106. Wang, L.; Young, K.; Meng, T.; English, N.; Yasuoka, S. Partial substitution of cobalt for nickel in mixed rare earth metal based superlattice hydrogen absorbing alloy—Part 2 Battery performance and failure mechanism. *J. Alloys Compd.* **2016**, *664*, 417–427.

107. Meng, T.; Young, K.; Koch, J.; Ouchi, T.; Yasuoka, S. Failure mechanisms of nickel/metal hydride batteries with cobalt-substituted superlattice hydrogen-absorbing alloy anodes at 50 °C. *Batteries* **2016**, *2*.

108. Young, K.; Ouchi, T.; Huang, B.; Nei, J. Structure, hydrogen storage, and electrochemical properties of body-centered-cubic $Ti_{40}V_{30}Cr_{15}Mn_{13}X_2$ alloys (X = B, Si, Mn, Ni, Zr, Nb, Mo, and La). *Batteries* **2015**, *1*, 74–90.

109. Young, K.; Ng, K.Y.S.; Bendersky, L.A. A technical report of the robust affordable next generation energy storage system-BASF program. *Batteries* **2016**, *2*.

110. Yan, S.; Young, K.; Ng, K.Y.S. Effects of salt additives to the KOH electrolyte used in Ni/MH batteries. *Batteries* **2015**, *1*, 54–73.

Structure, Hydrogen Storage, and Electrochemical Properties of Body-Centered-Cubic $Ti_{40}V_{30}Cr_{15}Mn_{13}X_2$ Alloys (X = B, Si, Mn, Ni, Zr, Nb, Mo, and La)

Kwo-Hsiung Young, Taihei Ouchi, Baoquan Huang and Jean Nei

Abstract: Structure, gaseous phase hydrogen storage, and electrochemical properties of a series of TiVCrMn-based body-centered-cubic (BCC) alloys with different partial substitutions for Mn with covalent elements (B and Si), transition metals (Ni, Zr, Nb, and Mo), and rare earth element (La) were investigated. Although the influences from substitutions on structure and gaseous phase storage properties were minor, influences on electrochemical discharge capacity were significant. The first cycle capacity ranged from 16 mAh·g^{-1} (Si-substituted) to 247 mAh·g^{-1} (Mo-substituted). Severe alloy passivation in 30% KOH electrolyte was observed, and an original capacity close to 500 mAh·g^{-1} could possibly be achieved by Mo-substituted alloy if a non-corrosive electrolyte was employed. Surface coating of Nafion to the Mo-substituted alloy was able to increase the first cycle capacity to 408 mAh·g^{-1}, but the degradation rate in mAh·g^{-1}·cycle^{-1} was still similar to that of standard testing. Electrochemical capacity was found to be closely related to BCC phase unit cell volume and width of the an extra small pressure plateau at around 0.3 MPa on the 30 °C pressure-concentration-temperature (PCT) desorption isotherm. Judging from its high electrochemical discharge capacity, Mo was the most beneficial substitution in BCC alloys for Ni/metal hydride (MH) battery application.

Reprinted from *Batteries*. Cite as: Young, K.-H.; Ouchi, T.; Huang, B.; Nei, J. Structure, Hydrogen Storage, and Electrochemical Properties of Body-Centered-Cubic $Ti_{40}V_{30}Cr_{15}Mn_{13}X_2$ Alloys (X = B, Si, Mn, Ni, Zr, Nb, Mo, and La). *Batteries* **2015**, *1*, 74–90.

1. Introduction

Among all metal hydride (MH) alloy families, body-centered-cubic (BCC) solid solution alloy has the highest reversible hydrogen storage at ambient temperature. Although its gaseous phase hydrogen storage capacity is very high (up to 4.0 wt%, equivalent to 1072 mAh·g^{-1} [1]), few electrochemical studies have been performed on the pure BCC phase MH alloy due to its strong metal-hydrogen bonding and low surface reaction activity [2–5]. Inoue and his coworker reported a $TiV_{3.4}Ni_{0.6}$ alloy achieving 360 mAh·g^{-1} at room temperature with a discharge rate of 50 mA·g^{-1} [3]. Mori and Iba improved both the capacity and cycle stability by adding Y, lanthanoids,

Pd, or Pt into a TiCrVNi BCC alloy and reached 462 mAh·g^{-1} [4]. Yu and his coworker reported a Ti$_{40}$V$_{30}$Cr$_{15}$Mn$_{15}$ alloy with an initial capacity of 814 mAh·g^{-1} measured with a rate of 10 mA·g^{-1} at 80 °C; however, degradation was high due to surface cracking, preferential leaching of V into the KOH electrolyte, and formation of TiO$_x$ on the surface that further blocks electrochemical reaction [5]. One or more secondary phases, such as C14, C15, and/or B2, with a high grain boundary density was introduced to improve the absorption kinetics [6], facilitate formation due to its brittleness [7–9], and increase the surface catalytic activity [10–15] by enhancing the synergetic effect between the main and secondary phases. High phase boundary density also promotes the formation of coherent and catalytic interfaces between the BCC and secondary phases and, therefore, improves hydrogen absorption and desorption kinetics [16].

In this experiment, we focus on continuing the work on Ti$_{40}$V$_{30}$Cr$_{15}$Mn$_{15}$ alloy with an electrochemical study performed at room temperature and an examination of substitution effects from covalent elements, transition metals, and rare-earth elements on structure, gaseous phase, and electrochemical properties. The alloy formula in the current study can be summarized as Ti$_{40}$V$_{30}$Cr$_{15}$Mn$_{13}$X$_2$, where X = B, Si, Mn, Ni, Zr, Nb, Mo, and La.

2. Experimental Setup

In this experiment, an arc melting technique was chosen for the sample preparation. The ingot size was about 12 g and the melting was performed in an Ar environment. A Varian Liberty 100 inductively-coupled plasma optical emission spectrometer (ICP-OES, Agilent Technologies, Santa Clara, CA, USA) was used to verify the chemical composition of the ingot comparing to the ratios in the raw materials. A Philips X'Pert Pro X-ray diffractometer (XRD, Philips, Amsterdam, The Netherlands) was used to study the microstructure, and a JEOL-JSM6320F scanning electron microscope (SEM, JEOL, Tokyo, Japan) with energy dispersive spectroscopy (EDS) capability was used to study the phase distribution and composition. Gaseous phase hydrogen storage characteristics for each sample were measured using a Suzuki-Shokan multi-channel pressure-concentration-temperature (PCT, Suzuki Shokan, Tokyo, Japan) system. A piece of ingot was freshly cleaved before putting in the PCT ample holder. PCT sample was first hydrided and dehydrided at 30 °C, followed by a 2 h, 400 °C degassing with a vacuum pump. PCT isotherms at 90 °C, 30 °C, and 60 °C were then measured with a 2 h, 400 °C degassing between measurements. Details of electrode preparations as well as measurement methods have been reported previously [17,18].

3. Results and Discussion

3.1. X-Ray Diffraction Structure Analysis

Eight alloys were prepared by arc melting, and their compositions were verified by ICP. XRD patterns of the alloys are shown in Figure 1. Three major peaks are detected in all alloys and belong to a BCC structure. Most of the peak intensity ratios are similar except for $I(200)/I(110)$ in Alloy-Nb (alloy with partial replacement of Nb). Among all substitutions, elemental Nb and Mo are similar in size and both have a BCC structure; however, only Alloy-Nb has the unusually larger (200) peak. The reason for such phenomenon is not clear and requires further structural refinement analysis. In addition to the main phase, one or more secondary phases can be found in the XRD pattern of each alloy apart from Alloy-Nb.

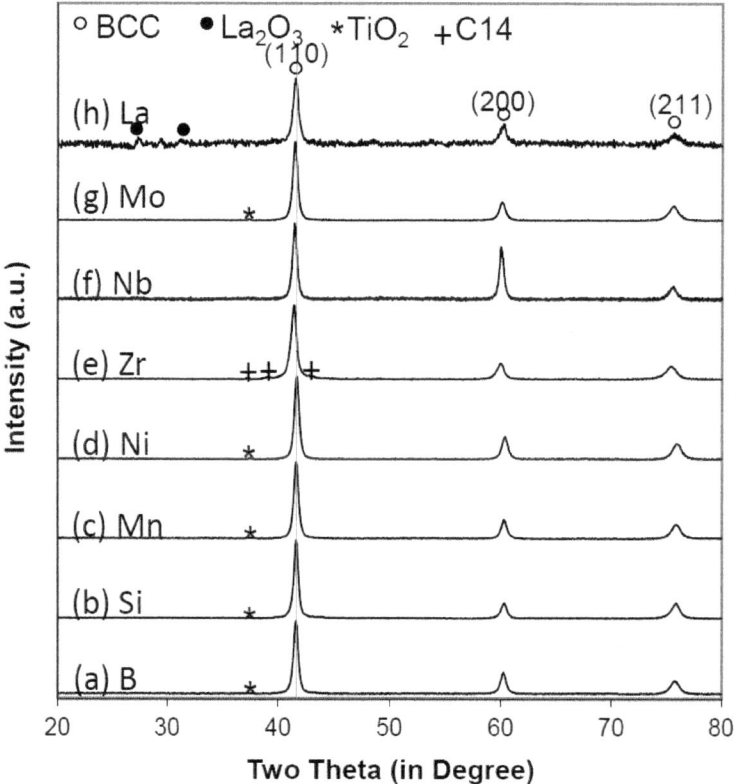

Figure 1. XRD patterns using Cu-K$_\alpha$ as the radiation source for Ti$_{40}$V$_{30}$Cr$_{15}$Mn$_{13}$X$_2$ alloys, where X = (**a**) B; (**b**) Si; (**c**) Mn; (**d**) Ni; (**e**) Zr; (**f**) Nb; (**g**) Mo; and (**h**) La. The vertical line is used to indicate shifts in the body-centered-cubic (BCC) peak (110) with respect to that in Ti$_{40}$V$_{30}$Cr$_{15}$Mn$_{15}$ alloy.

Rietveld refinement results from the XRD analysis are summarized in Table 1. Lattice parameter a of the BCC phase ranges from 3.0679 Å to 3.0839 Å, which is larger than the optimized value of 3.042 Å corresponding to the maximized hydrogen storage capacity [19], leaving room for potential improvement in storage capacity for future studies.

Table 1. Summary of X-ray diffraction (XRD) results from alloys $Ti_{40}V_{30}Cr_{15}Mn_{13}X_2$.

X	a of BCC phase	BCC phase abundance	Secondary phase	a of secondary phase	c of secondary phase	Secondary phase abundance
	Å	wt%		Å	Å	wt%
B	3.0703	98.2	TiO_2	4.1761	-	1.8
Si	3.0679	96.9	TiO_2	4.1472	-	3.1
Mn	3.0687	98.9	TiO_2	4.1687	-	1.1
Ni	3.0649	99.7	TiO_2	4.1567	-	0.3
Zr	3.0839	98.2	C14	4.9895	8.1790	1.8
Nb	3.0790	99.8	TiO_2	4.1743	-	0.2
Mo	3.0774	99.6	TiO_2	4.1706	-	0.4
La	3.0693	98.3	La_2O_3	11.302	-	1.7

The BCC lattice constant a is plotted against the atomic radius of the substituting element in Figure 2. Alloys substituted with transition metals show a linear relationship between the lattice constant and atomic radius (represented by the straight line in Figure 2). B and Si, with smaller atomic radii and larger electronegativity, do not shrink the BCC unit cell volume, which is possibly due to the electrons transferred from neighboring atoms and increases in the radius. Similar behavior has been found in the increase in the radius of B and Si in the Laves phase intermetallic compound [20]. La, with the largest atomic radius, does not change the BCC lattice constant, indicating La does not enter the BCC phase. According to the results of Rietveld refinement, the BCC phase abundances in all alloys are greater than 96.9%. TiO_2 is the dominating secondary phase, with some exceptions, seen in Alloy-Zr (C14), Alloy-Ni (TiNi observed in the SEM/EDS analysis as discussed in the next section), and Alloy-La (La_2O_3). Zr is known to promote the Laves phase in BCC-predominant alloys [21–26]. TiNi is a common phase seen in C14-predominant alloys with high concentration of Ni [27–30]. The main diffraction peak of TiNi overlaps with BCC (110) and, therefore, is indistinguishable in the XRD pattern (Figure 1). La_2O_3 was formed since La is too large to be included in the BCC phase, agreeing with the immiscibility shown in the La-V binary phase diagram [31].

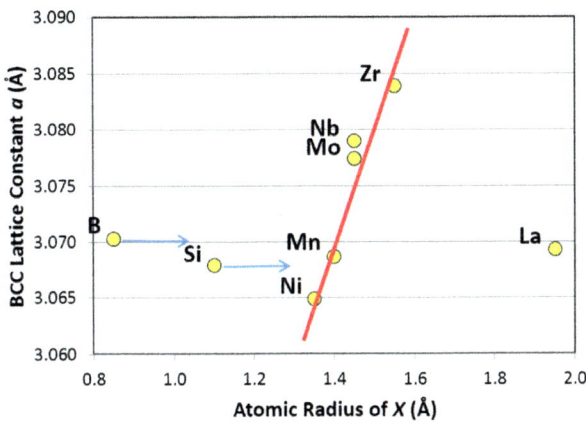

Figure 2. BCC lattice constant a *vs.* atomic radius of the partial substitution element X in $Ti_{40}V_{30}Cr_{15}Mn_{13}X_2$ alloys. There is a linear dependence when X is a transition metal. Addition of the largest La does not change the BCC lattice constant because La does not dissolve in the BCC phase and, instead, forms La_2O_3 secondary phase. Adding relatively small B and Si with higher electronegativity do not shrink the BCC unit cell because these atoms attract electrons from neighboring metallic atoms.

3.2. Scanning Electron Microscope/Energy Dispersive Spectroscopy Microstructure Analysis

Microstructures of the alloys were studied using SEM. The back-scattering electron images (BEI) are presented in Figure 3.

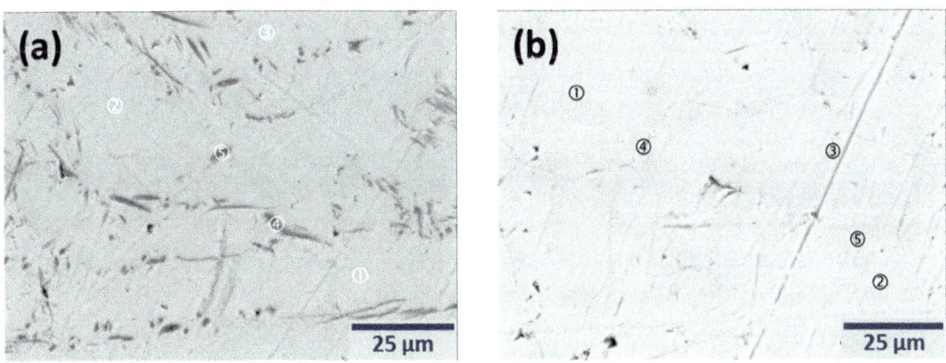

Figure 3. *Cont.*

197

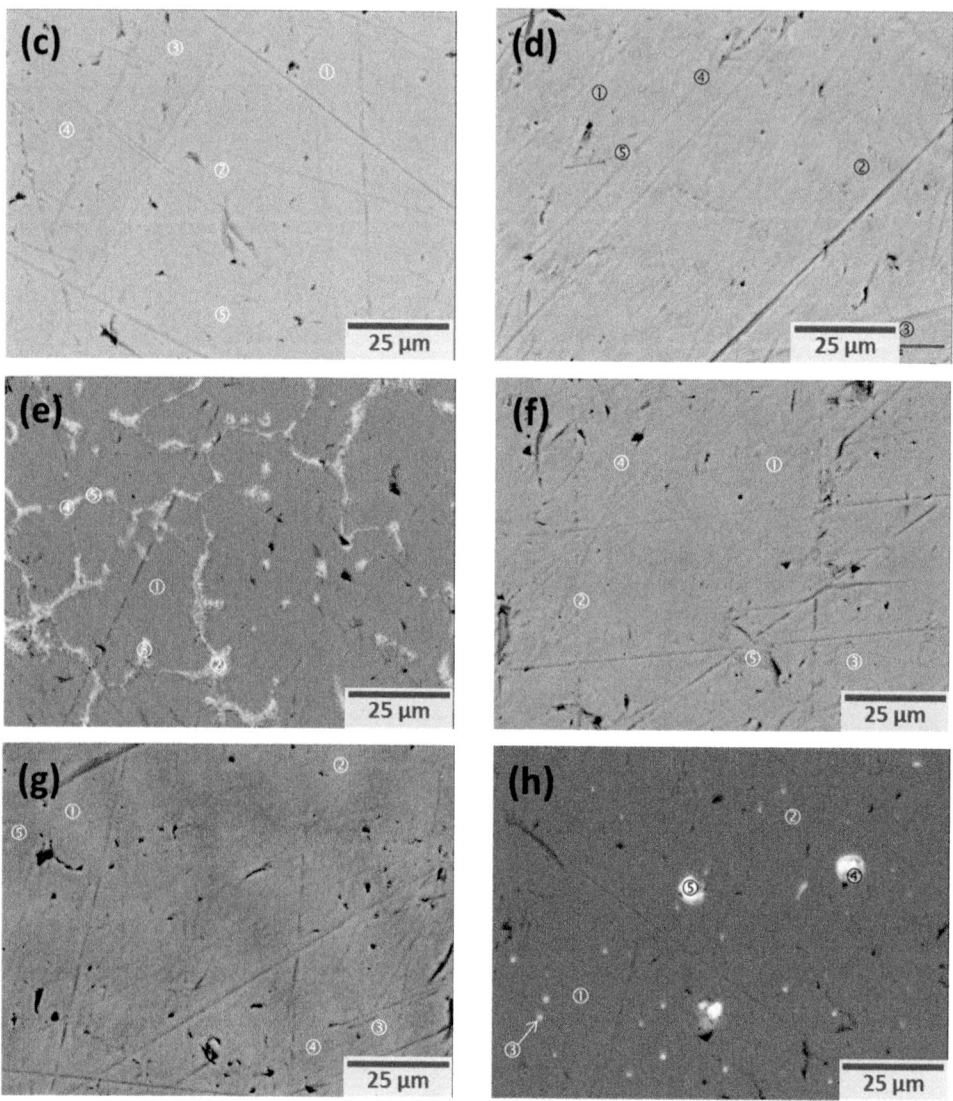

Figure 3. Scanning electron microscope (SEM) back-scattering electron images (BEI) for $Ti_{40}V_{30}Cr_{15}Mn_{13}X_2$ alloys, where X = (**a**) B; (**b**) Si; (**c**) Mn; (**d**) Ni; (**e**) Zr; (**f**) Nb; (**g**) Mo; and (**h**) La. Chemical compositions in the numbered areas measured by energy dispersive spectroscopy (EDS) are listed in Table 2.

EDS, although a semi-quantitative analysis, was used to study the chemical compositions of several spots with different contrasts identified numerically in the micrographs (Figure 3), and the results are summarized in Table 2 due to convenience and availability.

Table 2. Summary of energy dispersive spectroscopy (EDS) results. All compositions are in at%. Compositions of BCC phase are in **bold**.

Location	Ti	V	Cr	Mn	X	Phase
Figure 3a-1	**41.1**	**29.3**	**16.7**	**12.9**	**0.0**	**BCC**
Figure 3a-2	**42.2**	**28.0**	**16.2**	**13.7**	**0.0**	**BCC**
Figure 3a-3	**42.2**	**28.2**	**16.3**	**13.3**	**0.0**	**BCC**
Figure 3a-4	59.5	23.9	9.3	7.3	0.0	Oxide
Figure 3a-5	64.4	22.5	7.6	5.5	0.0	Oxide
Figure 3b-1	**38.7**	**31.9**	**15.8**	**12.1**	**1.6**	**BCC**
Figure 3b-2	**38.4**	**32.2**	**15.8**	**12.1**	**1.5**	**BCC**
Figure 3b-3	49.7	15.4	12.0	15.4	7.5	Oxide
Figure 3b-4	55.8	15.3	10.5	12.6	5.8	Oxide
Figure 3b-5	55.6	17.4	10.2	11.5	5.2	Oxide
Figure 3c-1	**38.8**	**30.1**	**15.8**	**15.3**	**0.0**	**BCC**
Figure 3c-2	**38.7**	**29.9**	**15.9**	**15.5**	**0.0**	**BCC**
Figure 3c-3	41.6	26.4	14.9	17.0	0.0	BCC
Figure 3c-4	43.3	26.2	14.6	15.9	0.0	BCC
Figure 3c-5	42.1	25.9	14.9	17.1	0.0	BCC
Figure 3d-1	**36.9**	**33.8**	**16.6**	**11.4**	**1.3**	**BCC**
Figure 3d-2	**38.6**	**31.9**	**16.1**	**12.0**	**1.3**	**BCC**
Figure 3d-3	42.6	28.2	14.7	12.4	2.1	BCC
Figure 3d-4	51.1	16.8	10.0	12.6	9.4	TiNi
Figure 3d-5	57.9	9.1	5.9	11.2	16.0	TiNi
Figure 3e-1	**39.7**	**31.3**	**15.7**	**12.1**	**1.1**	**BCC**
Figure 3e-2	43.1	19.9	13.3	15.9	7.7	$C14/Zr_xNi_y$
Figure 3e-3	32.9	19.3	16.7	20.4	10.6	$C14/Zr_xNi_y$
Figure 3e-4	31.5	17.9	16.4	21.5	12.7	$C14/Zr_xNi_y$
Figure 3e-5	39.0	17.3	14.8	18.4	10.5	$C14/Zr_xNi_y$
Figure 3f-1	**38.7**	**32.8**	**15.1**	**11.3**	**2.1**	**BCC**
Figure 3f-2	**39.6**	**31.6**	**14.9**	**11.8**	**2.1**	**BCC**
Figure 3f-3	**39.6**	**31.8**	**14.9**	**11.6**	**2.1**	**BCC**
Figure 3f-4	43.3	27.8	14.1	13.0	1.9	BCC
Figure 3f-5	45.1	25.9	13.9	13.3	1.9	BCC
Figure 3g-1	**40.5**	**29.9**	**16.1**	**12.0**	**1.5**	**BCC**
Figure 3g-2	41.4	29.0	16.1	12.3	1.3	BCC
Figure 3g-3	42.1	27.9	15.9	12.9	1.2	BCC
Figure 3g-4	44.9	25.5	15.3	13.6	0.7	BCC
Figure 3g-5	46.7	23.6	14.9	14.3	0.6	BCC
Figure 3h-1	**45.5**	**26.1**	**13.9**	**14.5**	**0.0**	**BCC**
Figure 3h-2	**43.6**	**28.2**	**14.5**	**13.6**	**0.0**	**BCC**
Figure 3h-3	34.0	24.2	10.5	9.1	22.2	La_2O_3
Figure 3h-4	16.1	10.4	1.0	5.6	66.9	La_2O_3
Figure 3h-5	5.0	2.9	1.0	0.0	91.2	La

Except for B and La, the substituting element is present in the BCC phase, ranging in content from 1.1 at% to 2.1 at%. EDS, although a semi-quantitative analysis, was used to study the chemical compositions of several spots with different contrasts identified numerically in the micrographs (Figure 3), and the results are summarized in Table 2 due to convenience and availability. Except for B and La, the substituting element is present in the BCC phase, ranging in content from 1.1 at% to 2.1 at%. The EDS system used for the current study cannot quantify the amount of lighter elements, such as B. According to the XRD and SEM-BEI analyses, the B-predominating phase does not exist; therefore, it is assumed that B is distributed in the BCC phase. Area with darker contrast in Alloy-B and Alloy-Si (Figure 3a-4,3a-5,3b-3,3b-4,3b-5) are small TiO_2 particles embedded in the BCC matrix. Alloy-Mn, Alloy-Nb, and Alloy-Mo are uniform in composition. In Alloy-Ni, the TiNi secondary phase was found (Figure 3d-4,3d-5). The C14/Zr_xNi_y phase in Alloy-Zr distributes inter-granularly since the BCC phase solidifies first and pushes Zr into the C14 phase. Average electron density (e/a) of the secondary phase in Alloy-Zr is 5.06, which is below the C14/C15 threshold [32,33] and is, therefore, another piece of evidence that the secondary phase is C14 rather than C15 in addition to the findings in XRD analysis. B/A in this C14 phase is in the range of 0.97 to 1.3, which is way too low for an AB_2 with a perfect B/A of 2.0. Since there is no major shift in XRD peaks of C14, these areas are not hypo-stoichiometric AB_2. Therefore, other Zr_xNi_y secondary phase must also co-exist in this C14 phase, as in the case of AB_2-predominated alloys [34,35]. Since the B/A ratios of the components of Zr_xNi_y (Zr_7Ni_{10}, Zr_9Ni_{11}, TiNi, and ZrNi) are all below 2.0, their existence will lower the B/A ratio in this region. In Alloy-La, La does not precipitate in the main BCC phase. La either forms a large metallic inclusion (Figure 3h-5) or an oxide suspended uniformly in the BCC matrix (Figure 3h-3)/near the edge of La metallic clusters (Figure 3h-4). The zero-solubility of La in BCC explains why the addition of La does not change the BCC lattice constant (Figure 2).

3.3. Gaseous Phase Study

PCT analysis was used to characterize the gaseous phase hydrogen storage properties of alloys in this study. The chamber containing the sample was filled with 7 MPa of hydrogen at 30 °C, and then the absorption amount was calculated, followed by a PCT desorption measurement at the same temperature. The sample was degassed at 400 °C for 2 h with a mechanical vacuum pump, and then a full 60 °C absorption-desorption PCT was measured. The sample was degassed at 400 °C for 2 h again, followed by a 90 °C PCT measurement. Finally, it was degassed at 400 °C for 2 h, and a last 30 °C PCT measurement was conducted. Absorption and desorption isotherms measured at 30 °C, 60 °C, and 90 °C together with the initial 30 °C desorption isotherm are shown in Figure 4. Information obtained from the

PCT study is summarized in Table 3. Most of the alloys show similar gaseous phase properties. Except for Alloy-Zr (3.12 wt%), the pristine alloys have similar maximum storage capacities in the range of 3.30 wt% to 3.55 wt%. A storage capacity of 3.50 wt% can be translated into an electrochemical discharge capacity 938 mAh·g^{-1} based on 1 wt% of hydrogen storage is equivalent to 268 mAh·g^{-1}. Maximum storage capacities measured at 30 °C and 60 °C after 400 °C degassing show the following trend: substitution of B > Mo ~Nb ~La > Ni ~Mn ~Si > Zr, which demonstrate very weak correlations to the BCC unit cell volume (correlation factors $R^2 = 0.18$ and 0.22 for storage capacities at 30 °C and 60 °C, respectively, indicating larger BCC unit cell corresponds to lower capacity) that were opposite to was expected. In general, reversibility of these alloys (ratio of revisable capacity down to 0.001 MPa and maximum capacity) is much worse than that of AB$_2$ or AB$_5$ MH alloy because of the fact the first pressure plateau between BCC and body-center-tetragonal (BCT) phases is too low to be observed with our PCT apparatus. While Alloy-B shows the best reversibility at 30 °C, Alloy-Mn and Alloy-Ni have better reversibility at 60 °C than others. Average reversible 30 °C storage capacity is about 0.7 wt%, which is equivalent to an electrochemical discharge capacity of 188 mAh·g^{-1}. The 90 °C desorption plateau pressure of Alloy-Ni is much higher than those of other alloys. Hysteresis of the PCT isotherm is defined as $\ln(P_a/P_d)$, where P_a and P_d are the absorption and desorption equilibrium pressures at 2.0 wt% hydrogen storage, respectively. In this series of alloys, only PCT hysteresis at 90 °C can be measured. All substitutions show similar or slightly lower hysteresis, except for Si. PCT hysteresis is mainly from the energy required to elastically deform the lattice near the metal/MH interface during hydrogenation. Most substitutions increase the chemical disorder and reduce the PCT hysteresis. Nb has the same BCC crystal structure as Mn, therefore its effects on the degree of disorder and PCT hysteresis are limited. Adding Si with covalent bonding may stiffen the lattice, requiring higher energy to expand the MH phase in the host metal.

Due to the low desorption plateau pressure in these alloys, the regular thermodynamic calculation cannot be performed. Instead, the absorption equilibrium pressures at 2.0 wt% hydrogen storage at 60 °C and 90 °C were used to estimate the changes in enthalpy (ΔH) and entropy (ΔS) by the equation:

$$\Delta G = \Delta H - T\Delta S = RT\ln P \tag{1}$$

where R is the ideal gas constant and T is the absolute temperature. Results of these calculations are listed in Table 3. Compared to the base Alloy-Mn, all substitutions decrease ΔH except for Zr, which indicates that Zr decreases the hydride stability. In the case of Alloy-Zr, addition of the C14 phase in the alloy facilitates hydrogen absorption through the synergetic effect between the storage and catalytic phases [36] and destabilizes the hydride. ΔS, usually calculated with the desorption isotherm, is an indication of how far the MH system is from a perfect and ordered situation. The theoretical value of ΔS is the entropy of hydrogen gas, which is close to -130 J·mol^{-1}·K^{-1} [37]. In our calculation, all substitutions decrease ΔS to below -135 J·mol^{-1}·K^{-1} except for Zr, which is an indication that a more ordered MH system was formed. Alloy-Zr shows a relatively high value of ΔS, suggesting a more disordered MH system was formed, possibly due to the interaction between the main BCC and C14 secondary phases.

One interesting feature in the PCT isotherms caught one of the authors' (Ouchi) attention: several alloys—Alloy-B, Alloy-Zr, Alloy-Nb, and Alloy-Mo—show a small plateau near 0.3 MPa on the 30 °C desorption curve while others do not. This plateau, although very small (about 0.10 wt% to 0.16 wt%), is at a pressure just above one atmosphere (0.1 MPa) and can be from a catalytic phase that has not been reported previously. The importance of this phase with respect to electrochemical performance will be discussed in the discussion section of this paper.

Table 3. Summary of gaseous phase hydrogen storage properties. PCT: pressure-concentration-temperature.

X	Initial maximum capacity	30 °C maximum capacity	30 °C reversible capacity	60 °C maximum capacity	60 °C reversible capacity	90 °C desorption pressure @2.0 wt%	90 °C hysteresis @2.0 wt%	$-\Delta H$	$-\Delta S$	PCT plateau @0.3 MPa
	wt%	wt%	wt%	wt%	wt%	MPa		kJ·mol^{-1}	J·mol^{-1}·K^{-1}	
B	3.48	3.38	1.73	3.43	1.00	0.011	1.7	67	181	Yes
Si	3.30	3.08	0.52	3.08	1.12	0.011	2.5	56	156	No
Mn	3.47	3.11	0.49	3.02	1.40	0.013	2.0	37	107	No
Ni	3.39	3.16	0.63	3.16	1.56	0.028	1.8	58	165	No
Zr	3.12	2.76	0.53	2.59	1.09	0.016	1.9	22	63	Yes
Nb	3.48	3.25	0.59	3.19	0.86	0.011	2.0	64	176	Yes
Mo	3.42	3.24	0.56	3.29	1.00	0.012	1.8	66	178	Yes
La	3.55	3.19	0.49	3.19	0.74	0.009	1.6	67	178	No

3.4. Electrochemical Measurement in 30% KOH

In a flooded half-cell, the electrochemical properties of MH alloys in this study were studied. Electrodes were made with powder after the PCT measurement and degassed four times at 400 °C for 2 h. No alkaline pretreatment was applied before the half-cell measurement. The charge condition was 10 h with a current density of 50 mA·g^{-1} and discharged at the same rate initially and followed by two pulls at 12 mA·g^{-1} and 4 mA·g^{-1} with a cut-off voltage at 0.9 V *versus* the counter electrode. The 500 mAh total charge input was based on the maximum reversible gaseous phase capacity being 1.73% (429 mAh·g^{-1}). The charge and discharge voltage curves for Alloy-Mo are shown in Figure 5. The high resistance through the poor-conducting TiO$_2$ surface resulted from the highly-corrosive 30% KOH electrolyte may cause the large charge and discharge overpotentials [5].

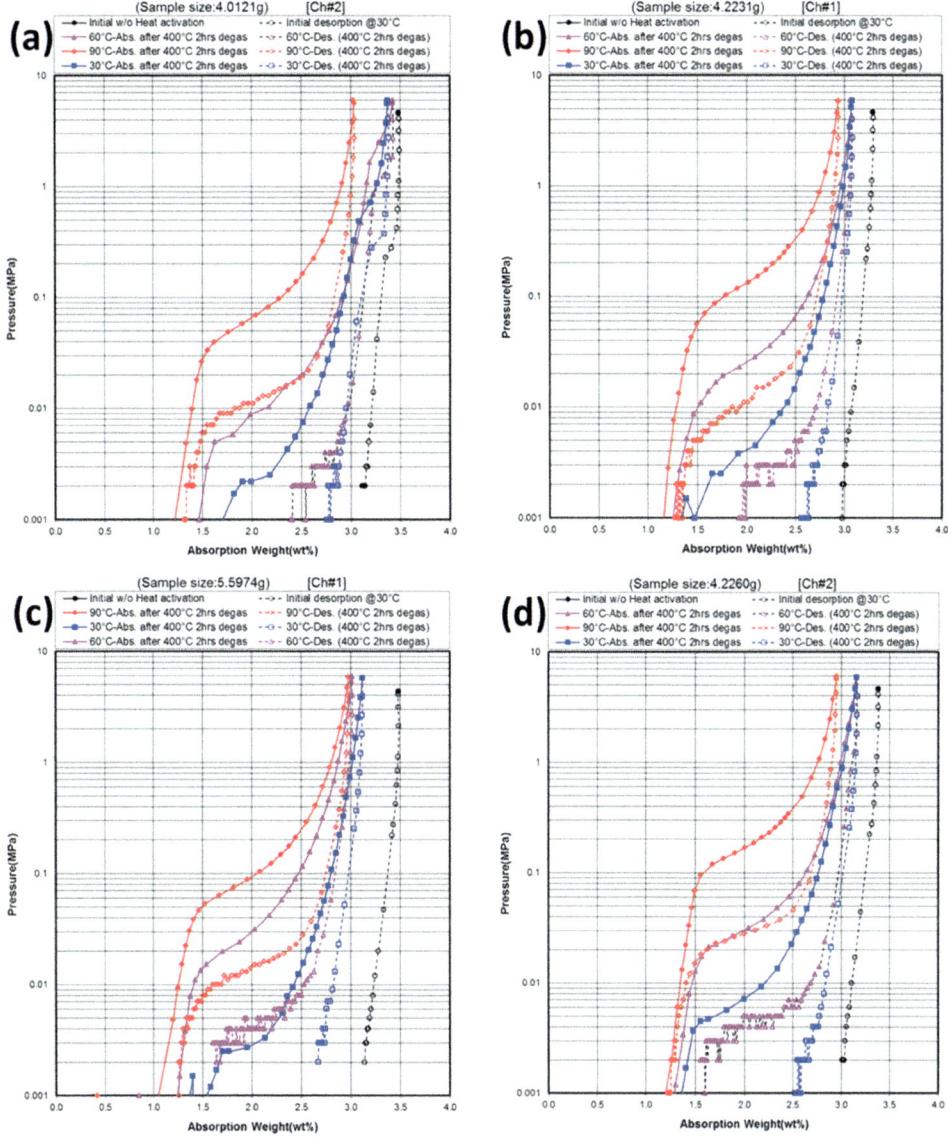

Figure 4. *Cont.*

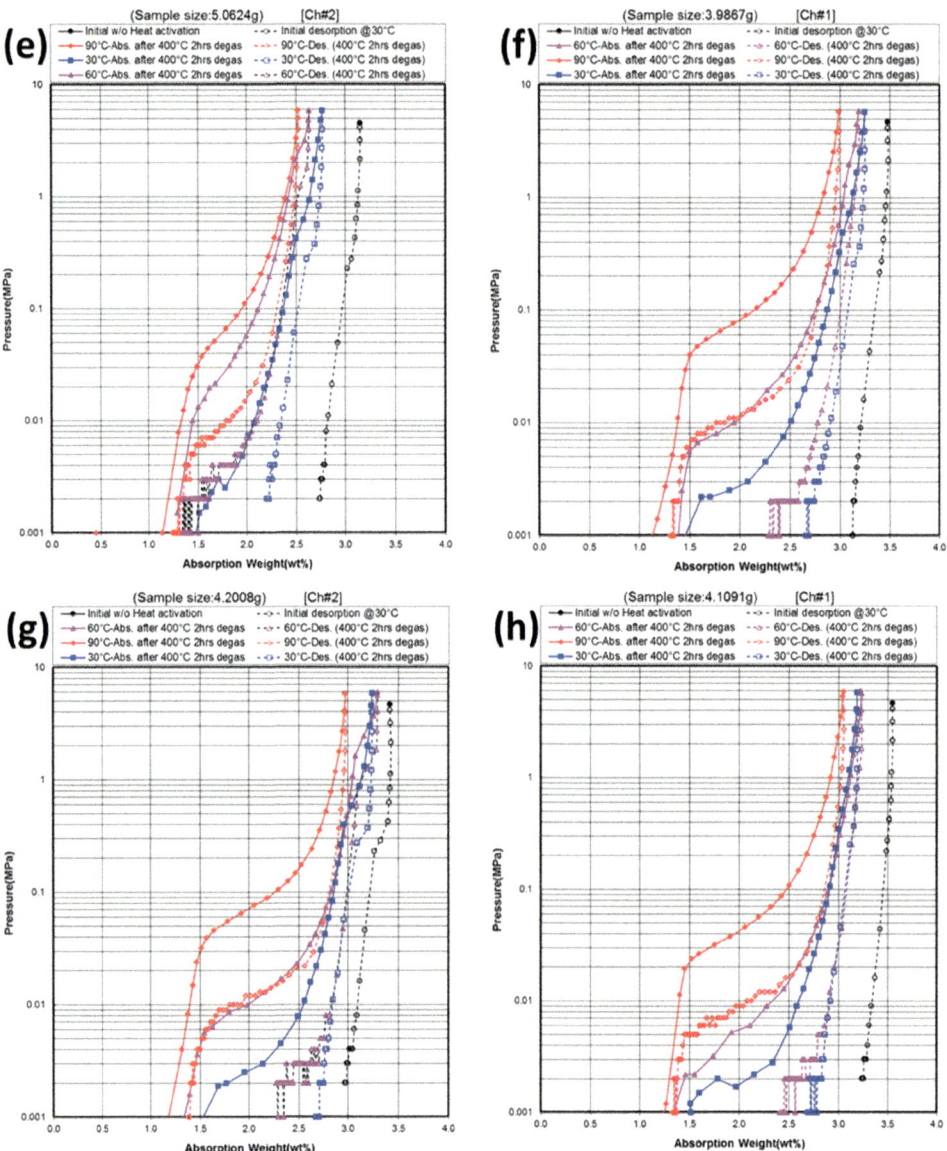

Figure 4. PCT isotherms measured at 30 °C (both before and after 400 °C degassing), 60 °C, and 90 °C for $Ti_{40}V_{30}Cr_{15}Mn_{13}X_2$ alloys, where X = (**a**) B; (**b**) Si; (**c**) Mn; (**d**) Ni; (**e**) Zr; (**f**) Nb; (**g**) Mo; and (**h**) La. Open and solid symbols are for absorption and desorption curves, respectively.

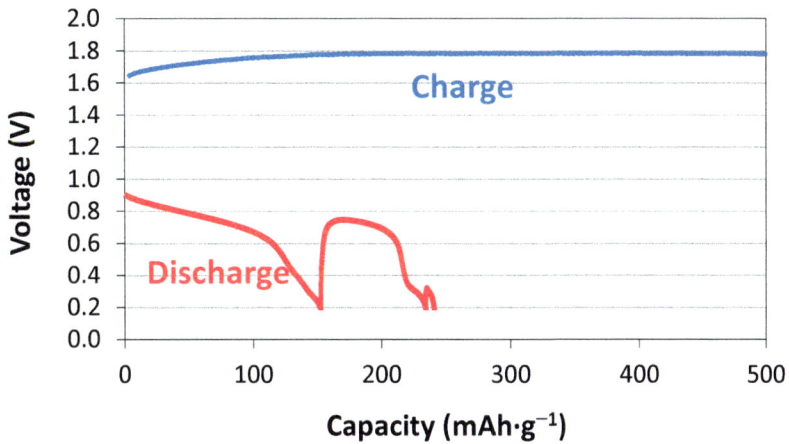

Figure 5. The first cycle charge and discharge voltage profiles for $Ti_{40}V_{30}Cr_{15}Mn_{13}Mo_2$.

Capacities totaled at 50, 12, and 4 mA·g^{-1} are listed in Table 4.

Table 4. Summary of electrochemical hydrogen storage properties.

X	1st cycle capacity @ 50 mA·g^{-1}	1st cycle capacity @ 12 mA·g^{-1}	1st cycle capacity @ 4 mA·g^{-1}
	mAh·g^{-1}	mAh·g^{-1}	mAh·g^{-1}
B	81	163	179
Si	8	14	16
Mn	12	20	24
Ni	33	47	61
Zr	64	130	144
Nb	41	68	79
Mo	152	234	247
La	23	39	41

About 50% of the capacity was obtained at the highest rate. The total capacity (totaled at 4 mA·g^{-1}) for the first six cycles of each alloy is plotted in Figure 6a. All substitutions for Mn, except for Si, show improvement in the first cycle capacity. The first cycle capacity demonstrates the trend of substitution of Mo > B > Zr > Nb > Ni > La > Mn > Si. Alloy-Si has the highest hysteresis, indicating its proneness to pulverization [38] and, thus, poor electrochemical performance. Alloy-Mo shows the highest discharge capacity at 247 mAh·g^{-1}. Partial Mo substitution in Ti-Cr MH alloy was reported previously, and it stabilized the BCC structure and improved the gaseous phase properties [39]. As seen in Figure 6a, capacity drops to almost nothing at the second cycle due mainly to the highly corrosive nature of 30% KOH electrolyte. The large amount of over-charge may also contribute to the severe capacity degradation. The original capacity of Alloy-Mo without corrosion from KOH can be extrapolated and is about double that obtained from the first cycle; in other words, electrochemical capacity close to 500 mAh·g^{-1} is possible if corrosion and passivation can be prevented from the use of non-corrosive electrolyte. In cycles two to six, Alloy-Ni with the TiNi phase shows the highest discharge capacity since the TiNi phase protects some portions of the bulk from being completely corroded. TiNi was found to increase the cycle stability of Laves phase MH alloys in a previous study [32].

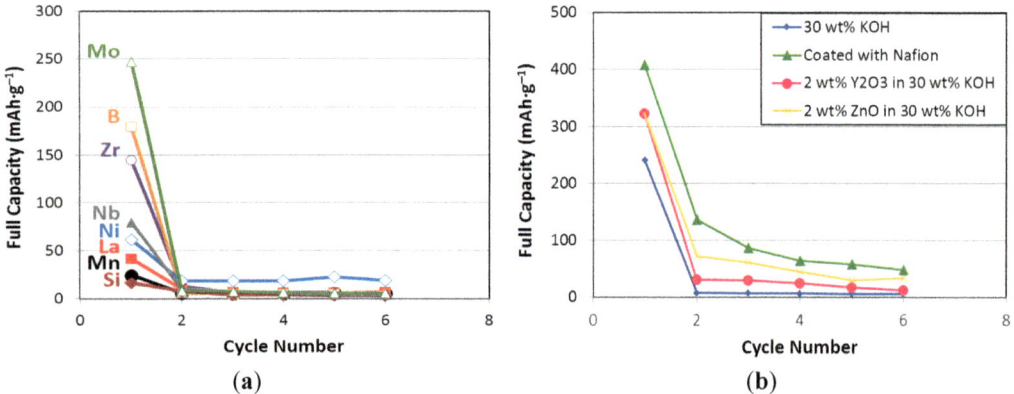

Figure 6. (a) Discharge capacities measured at 4 mA·g^{-1} in 30% KOH for Ti$_{40}$V$_{30}$Cr$_{15}$Mn$_{13}$X$_2$ alloys and (b) discharge capacities measured at 4 mA·g^{-1} for Ti$_{40}$V$_{30}$Cr$_{15}$Mn$_{13}$Mo$_2$ in modified electrolytes and with Nafion treatment.

A SEM micrograph taken from the surface of Alloy-Mo after six electrochemical cycles reveals severe pulverization (Figure 7a), and the EDS spectrum taken from the surface shows no trace of oxide (Figure 7b). It may be too thin and compact to be detected in the current study, but surface TiO$_2$ formed after cycling was reported previously in a BCC alloy [5] and can hinder the surface electrochemical reaction completely.

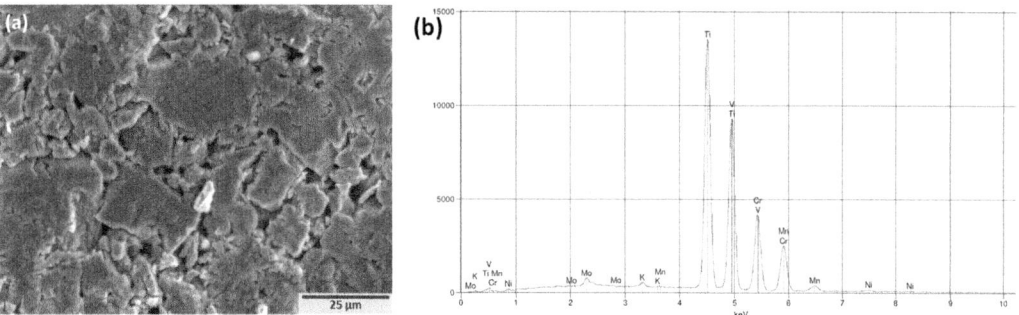

Figure 7. (**a**) SEM surface micrograph exhibiting the severe pulverization and (**b**) EDS spectrum showing negligible amount of oxygen on the surface after six electrochemical cycles with $Ti_{40}V_{30}Cr_{15}Mn_{13}Mo_2$.

In order to improve the cycle stability, preferential leaching, and surface passivation issues need to be addressed. Therefore, the effects of Y_2O_3 and ZnO additions in electrolyte and Nafion surface coating on electrode are investigated. It was found that by mixing Y_2O_3 powder to AB_5 alloy powder, dissolution of the alloy's constituent elements and formation of rare earth hydroxides were suppressed by an yttrium protective film on the alloy surface [40,41]. A small amount of Y_2O_3 dissolved in the alkaline electrolyte, and the yttrium complex ions were then adsorbed on or chemically bound to the surface of the alloy powder. ZnO, an amphoteric species, is nearly insoluble in water but soluble in acid and base. ZnO's effect as an electrolyte additive on electrochemical degradation performance will be interesting to observe. One drawback of Y_2O_3 addition is that the yttrium protective layer lowers the catalytic activity of the alloy surface and, consequently, reduces the surface charge transfer reaction [40]. In order to possibly solve such an issue, Nafion, which is hydrogen-permeable with great chemical stability, was also adopted in the current study. Nafion applied as a protective coating increased the first cycle capacity by up to 75% in an MgNi-based alloy [42]. For the electrolyte modification study, two electrolytes were made by adding 2 wt% Y_2O_3 and 2 wt% ZnO (both based on the amount of KOH) into 30% KOH and used in the half-cell measurements with Alloy-Mo. For the surface modification study, Alloy-Mo electrode was coated with Nafion (perfluorosulfonic acid-PTFT copolymer, 5 wt% in water and 1-propanol) by dipping it for 2 min in ethanol diluted Nafion solution, where the ethanol to Nafion solution ratio was 5:1 by weight, and then heat treating for 2 h at 120 °C under argon atmosphere. The total capacity of each treatment compared to that of standard 30% KOH is plotted in Figure 6b. The first cycle capacity is improved with the additions of Y_2O_3 and ZnO from 247 mAh·g^{-1} to 320 mAh·g^{-1}, but it still drops dramatically at the second cycle due to pulverization. The additives were incorporated differently than they were the previous study, where pasted electrode was made using the

207

mixture of additive, alloy powder, and water [40]. The wet method can perhaps provide a better distribution of additive and more complete protection for the alloy. Among all treatments, Nafion coating on electrode is the most effective in improving the first cycle capacity and achieves 408 mAh·g^{-1}, but degradation after the first cycle is still severe. After cycling, alloy particle pulverization (as seen from the SEM micrograph in Figure 7a) creates new surfaces that are not protected by the Nafion coating. Combination of surface coating (original robust protection) and electrolyte modification (continuous protection throughout cycling) may be advantageous for further corrosion and passivation inhibitions.

3.5. Discussion

To further study the correlations between electrochemical discharge capacity and various properties, correlation factors R^2 from linear regression are calculated and listed in Table 5. All correlations with gaseous phase properties are insignificant.

Table 5. Correlation factor (R^2) between electrochemical discharge capacity and various gaseous phase hydrogen storage properties. A R^2 value closer to one indicates better linear correlation between the two variables.

Correlation factor	Initial maximum capacity	30 °C maximum capacity	30 °C reversible capacity	60 °C maximum capacity	60 °C reversible capacity	90 °C desorption pressure @2 wt%	90 °C hysteresis @2 wt%	ΔH	BCC lattice constant a	Plateau width
$R^2 =$	0.02	0.03	0.17	0.04	0.06	0.01	0.19	0.02	0.29	0.70

Correlation with BCC lattice constant is only marginally significant ($R^2 = 0.29$) and plotted in Figure 8a, showing the enlargement of unit cell results in an increase in electrochemical capacity but not in a strictly linear relationship.

One significant correlation can be established between electrochemical capacity and occurrence of a small plateau near 0.3 MPa on the 30 °C desorption isotherm. Among all alloys, Alloy-B, Alloy-Zr, Alloy-Nb, and Alloy-Mo have this plateau at around 0.3 MPa and also the highest electrochemical discharge capacity. This plateau is from an intermediate hydride phase that can be catalytic and improve the electrochemical reaction. Electrochemical discharge capacity is plotted against width of the plateau at around 0.3 MPa (0.16 wt% for Alloy-B, 0.09 wt% for Alloy-Zr, 0.08 wt% Alloy-Nb, and 0.10 wt% for Alloy-Mo, which is defined as the width of concentration difference between the extrapolations from two neighboring isotherms) in Figure 8b and shows a strong correlation with $R^2 = 0.70$. Furthermore, correlation with transition metal substitution is even stronger as seen from the straight line connecting points from Alloy-Mo, Alloy-Zr, and Alloy-Nb. Electrochemical discharge capacity is mainly dominated by the catalytic phase formed during hydrogenation.

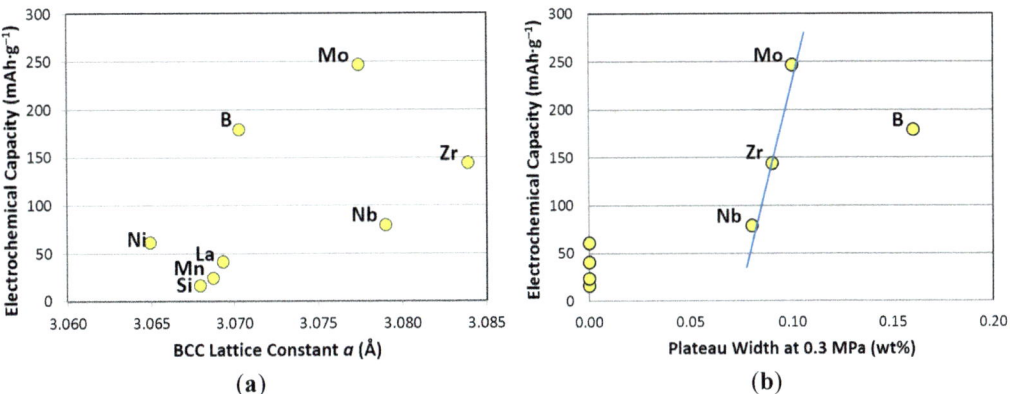

Figure 8. (**a**) Discharge capacities measured at 4 mA·g^{-1} *vs.* BCC lattice constant and (**b**) *vs.* width of 0.3 MPa pressure plateau. The straight line in Figure 8b is to illustrate the linear correlation between capacity and plateau width at 0.3 MPa of the transition metal substitution.

4. Conclusions

Various hydrogen storage properties in gaseous phase and in electrochemistry of a series of TiVCrMn-based BCC alloys with different partial substitutions for Mn with B, Si, Ni, Zr, Nb, Mo, and La were investigated. All substitutions went into the BCC phase except for La. While Ni promoted the formation of TiNi secondary phase that provided better cycle stability, Zr promoted the formation of C14 secondary phase, which did not affect any of the properties significantly due to its small abundance. Correlations between gaseous phase properties and lattice constant were not clear.

A newly discovered catalytic phase formed during hydrogenation was found to be very critical for the electrochemical discharge capacity performance. This phase enabled the electrochemical application of BCC-only alloys without contributions from secondary phases. The highest discharge capacity of 247 mAh·g^{-1} was obtained from Ti$_{40}$V$_{30}$Cr$_{15}$Mn$_{13}$Mo$_2$ alloy with both a catalytic hydride phase at around 0.3 MPa and an enlarged BCC unit cell. Further improvement in electrochemical capacity of this alloy reached as high as 408 mAh·g^{-1} when a protective Nafion coating was applied on the electrode. Substitutions of B, Nb, and Zr also improved the electrochemical capacity but at a lesser degree.

Acknowledgments: This work is financially supported by Advanced Research Projects Agency-Energy (ARPA-E) under the robust affordable next generation EV-storage (RANGE) program (DE-AR0000386).

Conflicts of Interest: The authors declare no conflict of interest.

References

1. Young, K.; Fetcenko, M.A.; Ouchi, T.; Im, J.; Ovshinsky, S.R.; Li, F.; Reinhout, M. Hydrogen Storage Materials Having Excellent Kinetics, Capacity, and Cycle Stability. U.S. Patent 7,344,676, 18 March 2008.

2. Lee, H.; Chourashiya, M.G.; Park, C.; Park, C. Hydrogen storage and electrochemical properties of the $Ti_{0.32}Cr_{0.43-x-y}V_{0.25}Fe_xMn_y$ ($x = 0$–0.055, $y = 0$–0.080) alloys and their composites with $MmNi_{3.99}Al_{0.29}Mn_{0.3}Co_{0.6}$ alloy. *J. Alloys Compd.* **2013**, *566*, 37–42.

3. Inoue, H.; Arai, S.; Iwakura, C. Crystallographic and electrochemical characterization of $TiV_{4-x}Ni_x$ alloys for nickel-metal hydride batteries. *Electrochim. Acta* **1996**, *41*, 937–939.

4. Mori, T.; Iba, H. Hydrogen-absorbing Alloy and Hydrogen-absorbing Alloy Electrode. U.S. Patent 6,338,764, 15 January 2002.

5. Yu, X.B.; Wu, Z.; Xia, B.J.; Xu, N.X. A Ti-V-based bcc phase alloy for use as metal hydride electrode with high discharge capacity. *J. Chem. Phys.* **2004**, *121*, 987–990. PubMed]

6. Chen, N.; Li, R.; Zhu, Y.; Liu, Y.; Pan, H. Electrochemical hydrogenation and dehydrogenation mechanisms of the Ti-V base multiphase hydrogen storage electrode alloy. *Acta Metal. Sin.* **2004**, *40*, 1200–1204.

7. Iba, H.; Akiba, E. The relation between microstructure and hydrogen absorbing property in Laves phase-solid solution multiphase alloys. *J. Alloys Compd.* **1995**, *231*, 508–512.

8. Rönnebro, E.; Noréus, D.; Sakai, T.; Tsukahara, M. Structural studies of a new Laves phase alloy $(Hf,Ti)(Ni,V)_2$ and its very stable hydride. *J. Alloys Compd.* **1995**, *231*, 90–94.

9. Tsukahara, M.; Takahashi, K.; Mishima, T.; Isomura, A.; Sakai, T. V-based solid solution alloys with Laves phase network: Hydrogen absorption properties and microstructure. *J. Alloys Compd.* **1996**, *236*, 151–155.

10. Qiu, S.; Chu, H.; Zhang, Y.; Sun, D.; Song, X.; Sun, L.; Xu, F. Electrochemical kinetics and its temperature dependence behaviors of $Ti_{0.17}Zr_{0.08}V_{0.35}Cr_{0.10}Ni_{0.30}$ alloy electrode. *J. Alloys Compd.* **2009**, *471*, 453–456.

11. Young, K.; Nei, J.; Wong, D.F.; Wang, L. Structural, hydrogen storage, and electrochemical properties of Laves phase-related body-centered-cubic solid solution metal hydride alloys. *Int. J. Hydrog. Energy* **2014**, *39*, 21489–21499.

12. Young, K.; Wong, D.F.; Wang, L. Effect of Ti/Cr content on the microstructures and hydrogen storage properties of Laves phase-related body-centered-cubic solid solution alloys. *J. Alloys Compd.* **2015**, *622*, 885–893.

13. Young, K.; Ouchi, T.; Nei, J.; Meng, T. Effects of Cr, Zr, V, Mn, Fe, and Co to the hydride properties of Laves phase-related body-centered-cubic solid solution alloys. *J. Power Sources* **2015**, *281*, 164–172.

14. Yan, Y.; Chen, Y.; Liang, H.; Zhou, X.; Wu, C.; Tao, M.; Pang, L. Hydrogen storage properties of V–Ti–Cr–Fe alloys. *J. Alloys Compd.* **2008**, *454*, 427–431.

15. Huot, J.; Akiba, E.; Ogura, T.; Ishido, Y. Crystal structure, phase abundance and electrode performance of Laves phase compounds (Zr, A) $V_{0.5}Ni_{1.1}Mn_{0.2}Fe_{0.2}$ (A = Ti, Nb or Hf). *J. Alloys Compd.* **1995**, *218*, 101–109.

16. Iba, H.; Akiba, E. Hydrogen absorption and modulated structure in Ti–V–Mn alloys. *J. Alloys Compd.* **1997**, *253–254*, 21–24.

17. Young, K.; Fetcenko, M.A.; Li, F.; Ouchi, T. Structural, thermodynamic, and electrochemical properties of $Ti_xZr_{1-x}(VNiCrMnCoAl)_2$ C14 Laves phase alloys. *J. Alloys Compd.* **2008**, *464*, 238–247.

18. Young, K.; Fetcenko, M.A.; Koch, J.; Morii, K.; Shimizu, T. Studies of Sn, Co, Al, and Fe additives in C14/C15 Laves alloys for NiMH battery application by orthogonal arrays. *J. Alloys Compd.* **2009**, *486*, 559–569.

19. Yoshida, M.; Akiba, E. Hydrogen absorbing-desorbing properties and crystal structure of the Zr–Ti–Ni–Mn–V AB_2 Laves phase alloys. *J. Alloys Compd.* **1995**, *224*, 121–126.

20. Gakkai, N.K. *Hi Kagaku Ryouronteki Kinzoku Kagobutu*; Maruzen: Tokyo, Japan, 1975; p. 296. (In Japanese)

21. Huot, H.; Akiba, E.; Ishido, Y. Crystal structure of multiphase alloys $(Zr,Ti)(Mn,V)_2$. *J. Alloys Compd.* **1995**, *231*, 85–89.

22. Kuriiwa, T.; Tamura, T.; Amemiya, T.; Fuda, T.; Kamegawa, A.; Takamura, H.; Okada, M. New V-based alloys with high protium absorption and desorption capacity. *J. Alloys Compd.* **1999**, *293–295*, 433–436.

23. Young, K.; Ouchi, T.; Fetcenko, M.A. Roles of Ni, Cr, Mn, Sn, Co, and Al in C14 Laves phase alloys for NiMH battery application. *J. Alloys Compd.* **2009**, *476*, 774–781.

24. Bendersky, L.A.; Wang, K.; Levin, I.; Newbury, D.; Young, K.; Chao, B.; Creuziger, A. $Ti_{12.5}Zr_{21}V_{10}Cr_{8.5}Mn_xCo_{1.5}Ni_{46.5-x}$ AB_2-type metal hydride alloys for electrochemical storage application: Part 1. Structural characteristics. *J. Power Sources* **2012**, *218*, 474–486.

25. Young, K.; Ouchi, T.; Huang, B.; Reichman, B.; Blankenship, R. Improvement in $-40\,^\circ$C electrochemical properties of AB_2 metal hydride alloy by silicon incorporation. *J. Alloys Compd.* **2013**, *575*, 65–72.

26. Young, K.; Reichman, B.; Fetcenko, M.A. Electrochemical performance of AB_2 metal hydride alloys measured at $-40\,^\circ$C. *J. Alloys Compd.* **2013**, *580*, S349–S352.

27. Smith, J.F.; Lee, K.J. La-V (Lanthanum-Vanadium). In *Binary Alloy Phase Diagram*, 2nd ed.; Massalski, T.B., Okamoto, H., Subramanian, P.R., Kacprzak, L., Eds.; ASM International: Geauga County, OH, USA, 1990; Volume 3, pp. 2437–2439.

28. Nei, J.; Young, K.; Salley, S.O.; Ng, K.Y.S. Determination of C14/C15 phase abundance in Laves phase alloys. *Mater. Chem. Phys.* **2012**, *136*, 520–527.

29. Nei, J.; Young, K.; Regmi, R.; Lawes, G.; Salley, S.O.; Ng, K.Y.S. Gaseous phase hydrogen storage and electrochemical properties of Zr_8Ni_{21}, Zr_7Ni_{10}, Zr_9Ni_{11}, and ZrNi metal hydride alloys. *Int. J. Hydrog. Energy* **2012**, *37*, 16042–16055.

30. Züttle, A. Materials for hydrogen storage. *Mater. Today* **2003**, *6*, 24–33.

31. Huang, T.; Li, J.; Yu, J.; Liu, Z.; Mao, S.; Zhang, Y.; Sun, G.; Han, J.; Ren, H.; Chen, J. Influence of partial substitution of Mo for Cr on structure and hydrogen storage characteristics of non-stoichiometric Laves phase $TiCrB_{0.9}$ alloy. *Int. J. Hydrog. Energy* **2013**, *38*, 11955–11963.

32. Young, K.; Nei, J.; Ouchi, T.; Fetcenko, M.A. Phase abundances in AB_2 metal hydride alloys and their correlations to various properties. *J. Alloys Compd.* **2011**, *509*, 2277–2284.

33. Johnston, R.L.; Hoffmann, R. Structure-bonding relationships in the Laves phases. *Z. Anorg. Allg. Chem.* **1992**, *616*, 105–120.

34. Boettinger, W.J.; Newbury, D.E.; Wang, K.; Bendersky, L.A.; Chiu, C.; Kattner, U.R.; Young, K.; Chao, B. Examination of multiphase (Zr,Ti)(V,Cr,Mn,Ni)$_2$ Ni-MH electrode alloys: Part I. Dendritic solidification structure. *Metall. Mater. Trans. A* **2010**, *41*, 2033–2047.

35. Bendersky, L.A.; Wang, K.; Boettinger, W.J.; Newbury, D.E.; Young, K.; Chao, B. Examination of multiphase (Zr,Ti)(V,Cr,Mn,Ni)$_2$ Ni-MH electrode alloys: Part II. Solid-state transformation of the interdendritic B$_2$ phase. *Metall. Mater. Trans. A* **2010**, *41*, 1891–1906.

36. Wong, D.F.; Young, K.; Nei, J.; Wang, L.; Ng, K.Y.S. Effects of Nd-addition on the structural, hydrogen storage, and electrochemical properties of C14 metal hydride alloys. *J. Alloys Compd.* **2015**, *647*, 507–518.

37. Lide, D.R. *CRC Handbook of Chemistry and Physics*, 74th ed.; CRC Press: Boca Raton, FL, USA, 1993; pp. 6–22.

38. Young, K.; Ouchi, T.; Fetcenko, M.A. Pressure-composition-temperature hysteresis in C14 Laves phase alloys: Part 1. Simple ternary alloys. *J. Alloys Compd.* **2009**, *480*, 428–433.

39. Kamegawa, A.; Shirasaki, K.; Tamura, T.; Kuriiwa, T.; Takamura, H.; Okada, M. Crystal structure and protium absorption properties of Ti-Cr-X alloys. *Mater. Trans.* **2002**, *43*, 470–473.

40. Kaiya, H.; Ookawa, T. Improvement in cycle life performance of high capacity nickel-metal hydride battery. *J. Alloys Compd.* **1995**, *231*, 598–603.

41. Kong, L.; Chen, B.; Young, K.; Koch, J.; Chan, A.; Li, W. Effects of Al- and Mn-contents in the negative MH alloy on the self-discharge and long-term storage properties of Ni/MH battery. *J. Power Sources* **2012**, *213*, 128–139.

42. Kim, S.; Chourashiya, M.G.; Park, C.; Park, C. Electrochemical performance of NAFION coated electrodes of hydriding combustion synthesized MgNi based composite hydride. *Mater. Lett.* **2013**, *93*, 81–84.

Studies on Incorporation of Mg in Zr-Based AB$_2$ Metal Hydride Alloys

Shiuan Chang, Kwo-hsiung Young, Taiehi Ouchi, Tiejun Meng, Jean Nei and Xin Wu

Abstract: Mg, the A-site atom in C14 (MgZn$_2$), C15 (MgCu$_2$), and C36 (MgNi$_2$) Laves phase alloys, was added to the Zr-based AB$_2$ metal hydride (MH) alloy during induction melting. Due to the high melting temperature of the host alloy (>1500 °C) and high volatility of Mg in the melt, the Mg content of the final ingot is limited to 0.8 at%. A new Mg-rich cubic phase was found in the Mg-containing alloys with a small phase abundance, which contributes to a significant increase in hydrogen storage capacities, the degree of disorder (DOD) in the hydride, the high-rate dischargeability (HRD), and the charge-transfer resistances at both room temperature (RT) and −40 °C. This phase also facilitates the activation process in measurement of electrochemical discharge capacity. Moreover, through a correlation study, the Ni content was found to be detrimental to the storage capacities, while Ti content was found to be more influential in HRD and charge-transfer resistance in this group of AB$_2$ metal hydride (MH) alloys.

Reprinted from *Batteries*. Cite as: Chang, S.; Young, K.-h.; Ouchi, T.; Meng, T.; Nei, J.; Wu, X. Studies on Incorporation of Mg in Zr-Based AB$_2$ Metal Hydride Alloys. *Batteries* **2016**, *2*, 11.

1. Introduction

Laves phase-based AB$_2$ metal hydride (MH) alloy is one of the high-capacity negative electrode materials used in nickel/metal hydride (Ni/MH) batteries. Its reversible hydrogen storage capacity can be as high as 3 wt% [1], which is equivalent to an electrochemical capacity of 804 mAh·g^{-1}. The measured electrochemical discharge capacity can reach up to 436 mAh·g^{-1} [2], which is about 25% higher than the conventional AB$_5$ MH alloys based on rare earth metals (330 mAh·g^{-1}) [3,4]. Early in their development, AB$_2$ MH alloys suffered from a harder activation and a shorter cycle life when compared to AB$_5$ MH alloys [5–8]. With composition and process refinement, the activation and cycle stability of AB$_2$ MH alloys as negative electrode active material improved substantially [9]. However, the high-rate dischargeability (HRD) of the AB$_2$ MH alloys, especially at low temperature, is still significantly inferior to the AB$_5$ MH alloys because of the relatively low nickel content in the AB$_2$ alloy [10]. Various additions, including transition metals—Al [11], Cr [12], Co [13], Cu [14,15], Fe [16], Mo [17,18], Zn [19], Pt [20], Pd [21,22], rare earth metals—Y [23], Ce [24], La [25,26], and Nd [27], and others such as Si [28] and B [29],

213

have been used to reduce the surface charge transfer resistance and increase the HRD of AB_2 MH alloys. In this paper, we summarize our findings regarding the use of one of the alkaline earth elements, Mg, as an additive in AB_2 MH alloys.

Mg can form various Laves phase alloys with different transition metals, such as C14 ($MgZn_2$), C15 ($MgCu_2$), and C36 ($MgNi_2$) [30]. Mg-containing Mg_2Ni, with an hP18 hexagonal structure (derivative of AlB_2 type [31]), is an important MH alloy that typically works in a temperature range of 200–250 °C. When the crystalline size decreases to the nanoscale or an amorphous state, Mg_2Ni can be used as the negative electrode material in Ni/MH batteries [32–41]. A more suitable stoichiometry for the Mg-Ni system is MgNi (1:1) but, unfortunately, it is not possible to obtain this material through conventional melt-and-cast, according to the phase diagram [42]. Amorphous MgNi prepared by a combination of melt-spin and mechanical alloying can achieve an electrochemical capacity of 720 mAh·g^{-1} for the first cycle. However, it has very poor cycle stability [43] and is, therefore, the subject of a DoE-funded project [44,45]. Reports of Mg use as a modifier in adjusting the hydrogen storage properties of AB_2 MH alloys are very scarce [46], which is very different from Mg-containing superlattice-based MH alloys (reviewed in [47]). Although Mg alone can form Laves phases, Mg is only slightly solubility in Zr(Ti)-based AB_2 phases (0.3 at%) and segregates into a Mg_2Ni phase [46]. The Mg_2Ni secondary phase reduces the surface reaction current, but increases the charge retention [46]. In addition to adding Mg to AB_2 alloy, we also investigated the role of Mg-addition (9.5 at%) in Zr_8Ni_{21} and found that the Mg-added alloy segregates into Zr_7Ni_{10} matrix, consisting of Zr_2Ni_7 grains with occasional Mg_2Ni inclusions, and Mg has a solubility of about 1.5 at% in the Zr_2Ni_7 phase [48]. Additionally, Mg added in Zr_8NNi_{21} alloy hindered the formation process, but increased the surface reaction exchange current [49].

2. Experimental Setup

An induction melting process involving an $MgAl_2O_4$ crucible, an alumina tundish, a 2-kg furnace under argon atmosphere, and a pancake-shaped steel mold was used to prepare the ingot samples. $MgNi_2$ alloy was used as the Mg source, which was added in the final melting step. A 50% excess of Mg was added to compensate for evaporation loss. The ingots were first hydrided/dehydrided to increase their brittleness and then crushed and ground into −200 mesh powder. The chemical compositions of the ingots were analyzed using a Varian Liberty 100 inductively coupled plasma-optical emission spectrometer (ICP-OES, Agilent Technologies, Santa Clara, CA, USA). A Philips X'Pert Pro X-ray diffractometer (XRD, Amsterdam, The Netherlands) was used to study the phase component. A JEOL-JSM6320F scanning electron microscope (SEM, Tokyo, Japan) with energy dispersive spectroscopy (EDS) was applied in investigating the phase distribution and composition. The hydrogen storage was measured using a Suzuki-Shokan multi-channel pressure-concentration-temperature system (PCT, Tokyo, Japan). In PCT analysis, each sample was first activated by a 2-h thermal cycle between room temperature (RT) and 300 °C under 2.5 MPa H_2 pressure, and then measured at 30, 60, and 90 °C. Details of the electrode and cell preparations, as well as the measurement methods, were previously reported [50,51]. AC impedance measurements were conducted using a Solartron 1250 Frequency Response Analyzer (Solartron Analytical, Leicester, UK) with a sine wave amplitude of 10 mV and frequency range of 0.5 mHz to 10 kHz. Prior to experiments, electrodes were subjected to one full charge/discharge cycle at a rate of 0.1C, using a Solartron 1470 Cell Test galvanostat, discharged to 80% state-of-charge, and then cooled to −40 °C. Magnetic susceptibility was measured using a Digital Measurement Systems Model 880 vibrating sample magnetometer (MicroSense, Lowell, MA, USA).

3. Results and Discussion

Six alloys were prepared by the induction melting technique. The design composition, with the ICP results, is summarized in Table 1. The Mg-free base alloy, Mg0, has been used numerous times in the previous comparison works [15–19]. In the design, Mg-content was varied from 0 to 5 at% in alloys Mg0–Mg5, respectively. However, due to the strong rejection from the major phase, the Mg-content in the final alloys is in the range of only 0.6–0.8 at%. To compensate for the increase in Mg-content in the design, both the Ti- and Ni contents were reduced. While the reduction in Ti content is clearly observed in ICP results, the reduction in Ni content is not obvious in alloys Mg1–Mg3, and the average Ni content actually increases in alloys Mg4 and Mg5 because of Mg loss. The average electron density (e/a), a strong factor in determining the C14/C15 phase abundance ratio [30,52], decreases monotonically in the design, but stabilized in the beginning and then increases in the ICP results, mirroring the evolution of Ni content. The B/A ratio, defined by the ratio of atomic percentage of B-sites (elements other than Zr, Ti, and Mg) and A-site atoms (Ti, Zr, and Mg), decreases in the design (hypo-stoichiometry), but stabilizes and then increases (hyper-stoichiometry) in the ICP results due to the increase in Ni content. The impact of stoichiometry on the performance of AB_2 MH alloys has been previously studied [53,54]. In general, hypo-stoichiometry promotes the C14 phase, lowers the PCT plateau pressure, and decreases HRD.

Table 1. Design compositions and inductively coupled plasma (ICP) results in at%. e/a is the average electron density. B/A is the atomic ratio of B-atom (elements other than Ti, Zr, and Mg) to A-atom (Ti, Zr, and Mg).

Alloy	Source	Zr	Ti	V	Cr	Mn	Co	Ni	Mg	Sn	Al	e/a	B/A
Mg0	Design	21.5	12.0	10.0	7.5	8.1	8.0	32.2	0.0	0.3	0.4	6.82	1.99
	ICP	21.5	12.0	10.0	7.5	8.1	8.0	32.2	0.0	0.4	0.3	6.82	1.99
Mg1	Design	21.5	11.4	10.0	7.5	8.1	8.0	31.8	1.0	0.3	0.4	6.78	1.95
	ICP	21.0	11.3	10.0	7.5	8.3	8.1	32.2	0.6	0.3	0.7	6.82	2.04
Mg2	Design	21.5	10.8	10.0	7.5	8.1	8.0	31.4	2.0	0.3	0.4	6.73	1.92
	ICP	22.5	10.9	9.9	7.2	8.2	8.0	31.8	0.8	0.3	0.4	6.78	1.92
Mg3	Design	21.5	10.2	10.0	7.5	8.1	8.0	31.0	3.0	0.3	0.4	6.69	1.88
	ICP	21.5	10.5	10.3	7.3	8.4	8.4	32.0	0.7	0.3	0.6	6.82	2.06
Mg4	Design	21.5	9.6	10.0	7.5	8.1	8.0	30.6	4.0	0.3	0.4	6.65	1.85
	ICP	21.0	9.6	9.8	6.9	8.3	8.2	34.7	0.6	0.3	0.6	6.96	2.21
Mg5	Design	21.5	9.0	10.0	7.5	8.1	8.0	30.2	5.0	0.3	0.4	6.60	1.82
	ICP	20.5	8.6	9.6	7.0	8.0	7.9	36.7	0.8	0.3	0.6	7.05	2.34

3.1. X-Ray Diffractometer Analysis

XRD analysis is an important tool to study the multi-phase nature of the Laves phase MH alloys [55–58]. The XRD patterns for alloys Mg0–Mg5 are shown in Figure 1. Peaks from the C15 cubic phase overlap with some of those from the C14 hexagonal phase. The TiNi phase with a B2 cubic structure can be seen in most of the XRD patterns. In addition to the C14, C15, and TiNi phases, one more cubic phase was observed in the Mg-containing alloys and it is believed to relate to Mg addition in the alloy. As the alloy number increases (Mg0 → Mg5), the intensities of C14-only peaks (for example, the one near 39.5°) decrease and the main C14/C15 peak (around 42.8°) first shifts to the left (larger unit cell) and then shifts to the right (smaller unit cell), as indicated by the blue vertical line in Figure 1.

The lattice constants of the four phases obtained from the XRD analysis are listed in Table 2 with the crystallite size of the main C14 phase. With the increase in alloy number, the lattice constants of the C14 phase first increase and then decrease. The changes are very isotropic, as seen from the nearly unchanged a/c ratio. The crystallite size of the C14 phase decreases, and the lattice constants of C15 and TiNi follow the same trend as observed in the C14 main phase.

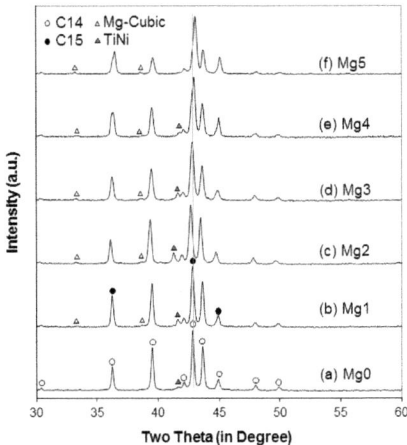

Figure 1. X-ray diffractometer (XRD) patterns using Cu-Kα as the radiation source for alloys: (**a**) Mg0; (**b**) Mg1; (**c**) Mg2; (**d**) Mg3; (**e**) Mg4; and (**f**) Mg5. In addition to two Laves phases, two cubic phases can be also identified. The vertical line indicates the main C14/C15 peak shifting to lower and then higher angles with an increasing alloy number.

Table 2. Lattice constants a and c, a/c ratio, unit cell volume, and crystallite size of the main C14 phase of alloys Mg0–Mg5 from XRD analysis. ND denotes non-detectable.

Alloy	Mg0	Mg1	Mg2	Mg3	Mg4	Mg5
C14 a (Å)	4.9545	4.9566	4.9727	4.9593	4.9555	4.9473
C14 c (Å)	8.0733	8.0781	8.1082	8.0846	8.0738	8.0586
C14 a/c	0.6137	0.6136	0.6133	0.6134	0.6138	0.6139
C14 unit cell volume (Å³)	171.63	171.87	173.64	172.20	171.71	170.82
C14 crystallite size (Å)	685	518	400	385	398	373
C15 a (Å)	6.9932	6.9894	7.0098	6.9926	6.9770	6.9628
TiNi a (Å)	3.0667	3.0674	3.0914	3.0687	3.0590	ND
Mg-related cubic a (Å)	ND	4.6553	4.6655	4.6670	4.6567	4.6711

The phase abundances of four constituent phases, obtained from a Rietveld refinement of the XRD patterns, are listed in Table 3. In general, as the alloy number increases, the C14 phase was replaced by the C15 phase and the TiNi phase abundance first increases and then decreases while the phase abundance of the Mg-related cubic phase remains unchanged. The evolution of the C14/C15 phase agrees with the changes in e/a and B/A (Table 1), because the C14/C15 phase determination threshold of e/a is approximately 6.9 in this case [52].

Table 3. Phase abundances of alloys Mg0–Mg5 from XRD analysis. ND denotes non-detectable.

Alloy	C14 Abundance (%)	C15 Abundance (%)	TiNi Abundance (%)	Cubic Abundance (%)
Mg0	93.7	5.2	1.2	ND
Mg1	81.8	15.9	2.0	0.3
Mg2	88.3	7.9	3.5	0.3
Mg3	72.0	25.1	2.4	0.5
Mg4	56.5	42.2	1.0	0.4
Mg5	36.9	62.6	ND	0.5

3.2. Scanning Electron Microscope/Energy Dispersive Spectroscopy Analysis

SEM back-scattering electron images (BEI) for alloys Mg1–Mg5 are shown in Figure 2. EDS was used to study composition information of representative spots with different contrasts on the BEI micrographs and the results are summarized in Table 4. As we know, EDS is a semi-quantitative analysis and results are only for comparison purpose. The microstructure of the Mg-free alloy (Mg0) was published before (as alloy Mo0 in [17]) and is composed of C14 and TiNi phases. In the microstructures of Mg-containing alloys, a C15 phase (judging from its relatively higher e/a value) with a slightly brighter contrast and an Mg-predominated phase with a darker contrast start to appear. The TiNi phase is usually surrounded by the C15 phase, since the cooling sequence is C14-C15-TiNi [59,60]. According to the EDS results shown in Table 4, the Mg-content in the C14 phase is very small (0.2–0.3 at%), while that of the C15 phase is slightly higher (0.3–0.7 at%). In addition, the B/A ratios in the C14 phase are higher than those in the C15 phases. The initial increase followed by a decrease in the C14 lattice parameters found through XRD analysis can be explained by the balance between the decrease in the content of relatively small Ti (larger C14 unit cell) and the increase in B/A ratio (smaller C14 unit cell [61]). The B/A ratio in the TiNi phase is calculated based on the assumption of V occupying the A-site [27] and the results are still higher than 1, which indicates the possibility of other Zr_xNi_y secondary phases with higher B/A ratios. The nature of the Zr_xNi_y phase was studied before by transmission electron microscopy [62]. The Mg-rich phase (the fourth phase in each BEI micrograph) has an Mg-content from 45.4 at% to 82.1 at%. It is difficult to link this phase to the cubic phase found by XRD since all alloys in Mg-Ni phase diagram are hexagonal (Mg, Mg_2Ni, and $MgNi_2$). An $MgIn_2$ intermetallic alloy with a cubic structure and a lattice constant of 4.60 Å [63] was the prototype used in our XRD analysis. The phase with the bright contrast (the fifth phase in each BEI micrograph) has a relatively higher Zr-content and a B/A ratio close to 1.0, and therefore is identified as the ZrNi phase. It cannot be the mixture of Zr metal and neighboring Laves phase because of a relatively low V-content (similar to the case of TiNi phase). The corresponding XRD peak of this

218

ZrNi phase cannot be identified due to low abundance. The Mg-contents in the TiNi and ZrNi phases are slightly higher than those in the C14 and C15 phases.

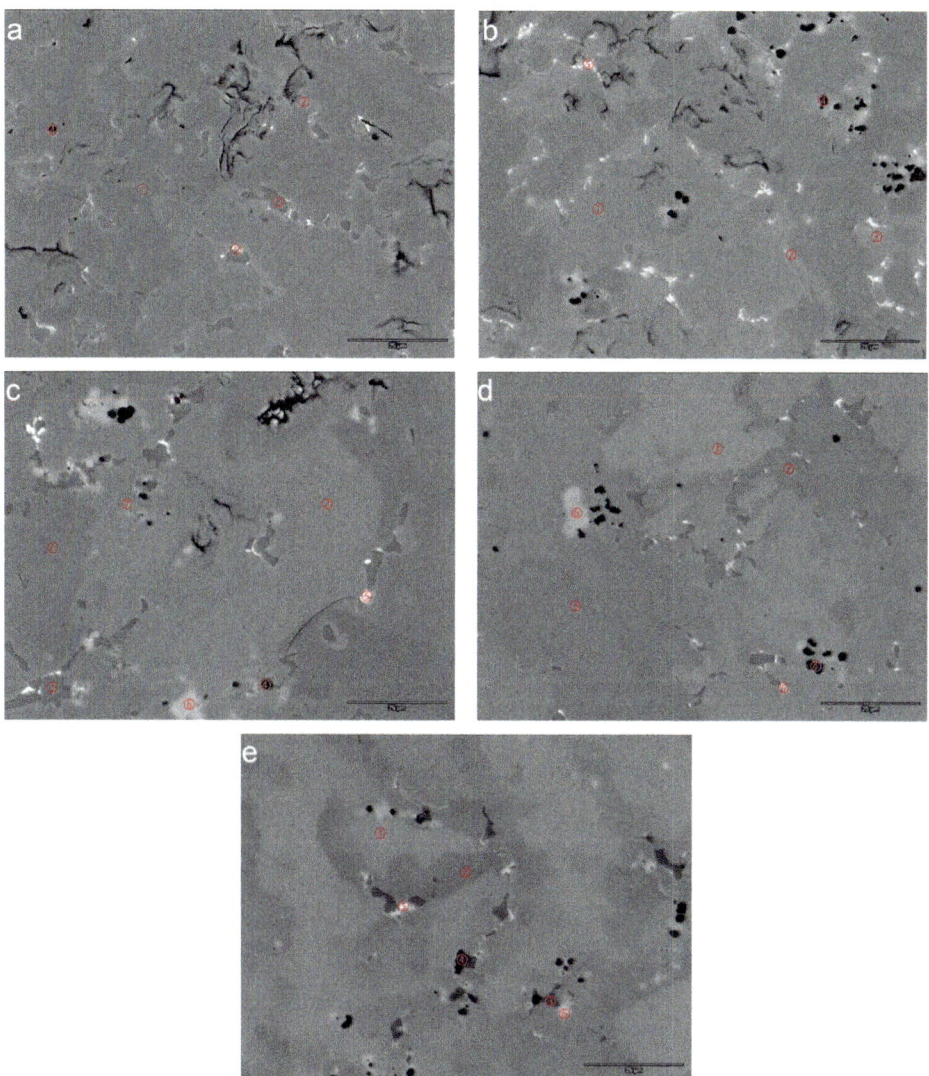

Figure 2. Scanning electron microscope (SEM) back-scattering electron image (BEI) micrographs from alloys: (**a**) Mg1; (**b**) Mg2; (**c**) Mg3; (**d**) Mg4; and (**e**) Mg5. The scale bar at the lower right corner represent 25 μm. The compositions of the numbered areas were analyzed by energy dispersive spectroscopy (EDS), and results are shown in Table 4. Areas 1, 2, 3, 4, 5, and 6 are identified as C14, C15, TiNi, Mg-cubic, ZrNi, and Zr phases, respectively.

Table 4. Summary of EDS results. All compositions are in at%. Compositions of the main C14 and C15 phase are in **bold** and *italic*, respectively.

Area	Zr	Ti	V	Cr	Mn	Co	Ni	Mg	Sn	Al	*e/a*	B/A	Phase
Mg1-1	**20.4**	**10.5**	**11.1**	**8.6**	**8.8**	**8.7**	**31.0**	**0.2**	**0.2**	**0.5**	**6.83**	**2.22**	C14
Mg1-2	*21.0*	*12.2*	*6.9*	*4.2*	*6.2*	*6.1*	*41.9*	*0.1*	*0.8*	*0.6*	*7.15*	*2.00*	C15
Mg1-3	16.9	23.4	2.4	1.3	3.1	5.9	45.2	0.3	1.1	0.4	7.14	1.33	TiNi
Mg1-4	9.9	8.4	4.6	2.7	3.6	3.6	20.0	46.6	0.4	0.2	4.65	0.54	Mg-cubic
Mg1-5	27.5	12.1	2.2	1.0	2.5	3.2	42.0	0.3	8.9	0.3	6.79	1.05	ZrNi
Mg2-1	**21.9**	**9.9**	**11.9**	**8.0**	**9.1**	**8.7**	**29.4**	**0.3**	**0.2**	**0.6**	**6.74**	**2.12**	C14
Mg2-2	*22.8*	*11.6*	*6.7*	*3.7*	*6.4*	*6.2*	*40.9*	*0.4*	*0.8*	*0.5*	*7.08*	*1.87*	C15
Mg2-3	20.7	21.4	1.4	0.5	2.3	6.0	45.5	1.2	0.5	0.5	7.09	1.24	TiNi
Mg2-4	12.9	5.7	2.0	1.0	2.0	2.3	16.1	57.2	0.7	0.0	4.03	0.32	Mg-cubic
Mg2-5	29.0	11.7	1.8	0.9	2.5	3.4	41.5	0.9	7.7	0.5	6.74	1.03	ZrNi
Mg3-1	**20.6**	**9.8**	**11.9**	**7.4**	**9.3**	**9.1**	**31.0**	**0.3**	**0.2**	**0.4**	**6.85**	**2.26**	C14
Mg3-2	*20.4*	*12.2*	*7.2*	*3.2*	*6.5*	*6.3*	*42.5*	*0.5*	*0.7*	*0.5*	*7.18*	*2.02*	C15
Mg3-3	15.6	25.7	1.7	0.4	2.6	5.8	45.6	1.3	0.8	0.6	7.10	1.26	TiNi
Mg3-4	9.2	2.7	0.4	0.2	0.5	0.6	4.2	82.1	0.1	0.0	2.66	0.06	Mg-cubic
Mg3-5	30.7	9.4	1.0	0.6	1.9	2.2	41.1	1.1	11.7	0.4	6.63	0.89	ZrNi
Mg3-6	95.2	1.8	0.3	0.2	0.4	0.3	1.7	0.1	0.0	0.0	4.13	0.03	Zr
Mg3-7	78.1	5.4	2.4	1.3	1.9	2.0	8.4	0.2	0.2	0.0	4.70	0.19	ZrO$_2$
Mg4-1	**22.4**	**7.1**	**12.1**	**12.3**	**10.1**	**9.9**	**25.4**	**0.3**	**0.1**	**0.3**	**6.68**	**2.36**	C14
Mg4-2	*19.3*	*12.6*	*7.9*	*3.9*	*6.7*	*6.6*	*41.7*	*0.3*	*0.5*	*0.5*	*7.18*	*2.11*	C15
Mg4-3	13.8	26.8	2.0	0.6	3.0	5.8	45.8	0.7	1	0.5	7.14	1.31	TiNi
Mg4-4	7.1	8.9	2.0	0.9	2.1	2.6	19.6	56.0	0.6	0.1	4.28	0.39	Mg-cubic
Mg4-5	26.6	11.5	2.7	1.1	3.1	2.8	42.8	0.6	8.4	0.4	6.83	1.12	ZrNi
Mg4-6	95.1	1.6	0.3	0.2	0.3	0.3	1.9	0.2	0.0	0.0	4.14	0.03	Zr
Mg5-1	**21.8**	**6.3**	**11.1**	**11.9**	**9.4**	**9.8**	**28.8**	**0.3**	**0.1**	**0.4**	**6.84**	**2.52**	C14
Mg5-2	*17.6*	*11.7*	*8.6*	*4.0*	*7.2*	*6.4*	*42.8*	*0.7*	*0.4*	*0.6*	*7.25*	*2.33*	C15
Mg5-3	8.4	28.8	3.3	0.8	3.9	6.1	45.4	1.0	1.2	1.0	7.16	1.41	TiNi
Mg5-4	12.9	8.0	3.0	1.0	2.4	2.2	24.4	45.4	0.6	0.2	4.79	0.51	Mg-cubic
Mg5-5	25.4	13.0	1.8	0.7	2.1	2.1	44.5	0.3	9.9	0.3	6.87	1.06	ZrNi
Mg5-6	83.6	4.8	1.0	0.7	0.8	0.7	7.8	0.2	0.5	0.0	4.55	0.13	Zr

3.3. Pressure-Concentration-Temperature Analysis

PCT measurement has been used extensively in the study of Laves phase MH alloys reacting with hydrogen gas [64–69]. PCT isotherms measured at 30 °C and 60 °C for alloys Mg0–Mg5 are compared in Figure 3. These isotherms lacking noticeable plateaus are commonly observed in highly-disordered AB$_2$ MH alloys. The multi-phase nature in this group of alloys lowers the critical temperature (T_c) when the pressure plateau starts to disappear [70–72]. Some hydrogen storage properties detected from the PCT isotherms are listed in Table 5. Both the maximum and reversible hydrogen storage capacities first increase and then decrease as the alloy number increases. Due to the lack of obvious plateau pressure, the desorption pressure at 0.75 wt% of hydrogen storage capacity was used for comparison between equilibrium pressure and calculation of hysteresis, and heat of hydride formation (ΔH_h) and change in entropy (ΔS_h). In the Mg-containing alloys, the equilibrium pressure first decreases and then increases as the alloy number increases, which complies with the general rule that a higher metal-hydrogen bond strength yields a lower plateau pressure and a higher hydrogen storage capability [50]. The slope factor (SF) indicates the degree of disorder (DOD) in an alloy. SF is defined as the ratio of the storage capacity between 0.01 MPa and 0.5 MPa to the total capacity in the desorption isotherm [2,19,51]. An alloy with a large SF has a flatter plateau and less DOD (less components or less variations among the components). As the alloy number increases, the SF decreases, indicating an increase in alloy homogeneity with addition of Mg. The hysteresis of the PCT isotherm is defined as $\ln(P_a/P_d)$, where P_a and P_d are the absorption and desorption equilibrium pressures, respectively, at 0.75 wt% H-storage. The irreversible energy loss during plastic deformation of the hydride phase in the alloy matrix is a common explanation for PCT hysteresis [73–75], and was linked to the a/c ratio and pulverization rate of the alloy [76]. In this study, the addition of Mg does not significantly change PCT hysteresis and should have no impact on the pulverization rate of alloy during cycling.

Table 5. Summary of the solid state (gaseous phase) hydrogen storage properties of the Mg-containing AB$_2$ alloys. SF: slope factor.

Alloy	Mg0	Mg1	Mg2	Mg3	Mg4	Mg5
Maximum capacity @ 30 °C (wt%)	1.45	1.42	1.54	1.41	1.25	0.89
Reversible capacity @ 30 °C (wt%)	1.32	1.32	1.33	1.30	1.16	0.87
Desorption pressure @ 0.75 wt%, 30 °C (MPa)	0.078	0.096	0.026	0.102	0.246	1.203
SF @ 30 °C	0.60	0.54	0.55	0.51	0.38	0.18
PCT hysteresis @ 0.75 wt%, 30 °C	0.04	0.05	0.10	0.01	0.06	0.02
$-\Delta H_h$ (kJ·mol^{-1})	32.0	32.2	36.2	35.2	31.2	-
$-\Delta S_h$ (J·mol^{-1}·K^{-1})	104	106	108	116	110	-

The desorption equilibrium pressures at the midpoint capacity measured at 30, 60, and 90 °C (P) were used to estimate the changes in enthalpy (ΔH_h) and entropy (ΔS_h) using the equation:

$$\Delta G = \Delta H_h - T\Delta S_h = RT\ln P \tag{1}$$

where R is the ideal gas constant and T is the absolute temperature. Since the hydrogenation reaction is exothermic, the heat of hydride formation (ΔH_h) is negative. Both ΔH_h and ΔS_h decrease (become more negative) and then increase with the increase in alloy number. The evolution in $-\Delta H_h$ value correlates to hydrogen storage capacity and is agrees with the strength of hydrogen-metal bond assumption described earlier. ΔS_h is an indicator for showing the DOD in hydride from a completely ordered solid (e.g., solid hydrogen). The difference between ΔS_h and -130.7 J·mol^{-1}·K^{-1} (the S for H$_2$ (g) at 300 K and 0.1 MPa [77]) can be interpreted as the DOD for hydrogen in the hydride form (β-phase). In this study, the trend of $|\Delta S_h|$ increases, and then decreases with the increase in the alloy number, which is similar to that of SF (the indicator for the DOD in the host metal alloy). The same correlation between the DOD of the hydride and the DOD of the occupied hydrogen was previously reported [19].

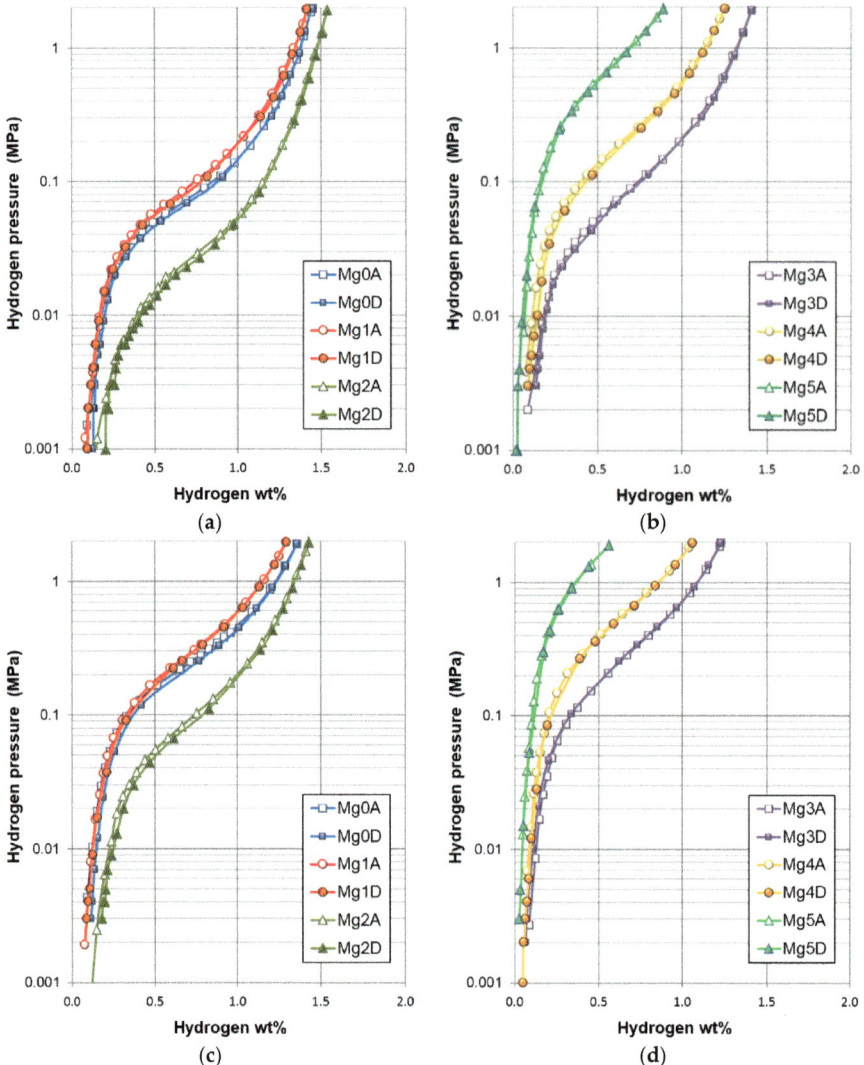

Figure 3. 30 °C pressure-concentration-temperature (PCT) isotherms of alloys: (**a**) Mg0–Mg2 and (**b**) Mg3–Mg5. 60 °C PCT isotherms of alloys: (**c**) Mg0–Mg2 and (**d**) Mg3–Mg5. Open and solid symbols are for absorption and desorption curves, respectively.

3.4. Electrochemical Analysis

Discharge capacities measured at a discharge current of 4 mA·g^{-1} for the first 13 cycles for each alloy in this study are plotted in Figure 4a to show the activation behavior at full capacity. The activation of the Mg-containing alloys appeared to be slightly easier than the Mg-free Mg0 alloy. The results of discharge capacities with other electrochemical tests are listed in Table 6. For Mg-containing alloys, both discharge capacities measured at 4 mA·g^{-1} and 50 mA·g^{-1} rates increase and then decrease with the corresponding increase in the alloy number. The maximum capacities obtained are from alloy Mg2, which demonstrated the lowest e/a value and B/A ratio. Mg2 also has the highest hydrogen storage capacity among the alloys in this study. The HRD values, defined as the ratio of the tenth cycle capacities measured at 50 mA·g^{-1} and 4 mA·g^{-1} rates, are listed in Table 6 and demonstrated an increasing trend correlating to alloy number. The HRD obtained from the first 13 cycles are plotted in Figure 4b, which exhibited easier activation in HRD with higher alloy numbers. From Table 6, the addition of Mg is shown to be effective in improving both activation and HRD. The improvement in HRD with Mg was further investigated by electrochemically measuring the bulk diffusion constant (D) and surface exchange current (I_0). Method details of these two parameters were previously reported [25] and the results are listed in Table 6. In general, as the alloy number increases, D first decreases and then increases while I_0 shows the opposite trend. The increase in HRD with alloy number is related to both the bulk and surface properties of the alloys. The addition of Mg decreases the D value (bulk), except for alloys with very high Ni content (Mg4 and Mg5), and increases the I_0 value (surface) of the alloys. The contribution of Mg to faster activation is similar to that with La in AB$_2$ MH alloys [78–81], where the new Mg-containing phase may absorb a larger amount of hydrogen, causing surface cracking and an increase in the surface area.

Table 6. Summary of the room temperature (RT) electrochemical and magnetic results (capacity, rate, D, I_0, and M_s, $H_{1/2}$,) of the Mg-containing AB$_2$ alloys.

Alloy	Mg0	Mg1	Mg2	Mg3	Mg4	Mg5
10th cycle capacity @ 50 mA·g^{-1} (mAh·g^{-1})	346	327	360	339	311	221
10th cycle capacity @ 4 mA·g^{-1} (mAh·g^{-1})	366	362	377	351	320	225
HRD @ 10th cycle	0.94	0.96	0.96	0.97	0.97	0.98
Number of activation cycles to reach 90% HRD	8	3	4	4	3	2
Diffusion coefficient D (10^{-10} cm^2·s^{-1})	2.5	1.3	1.2	1.8	2.8	2.9
Exchange current I_0 (mA·g^{-1})	22.5	31.6	35.8	34.7	27.4	25.9
M_s (emu·g^{-1})	0.0372	0.0388	0.0148	0.0598	0.0730	0.0563
$H_{1/2}$ (kOe)	0.111	0.261	0.260	0.290	0.140	0.118

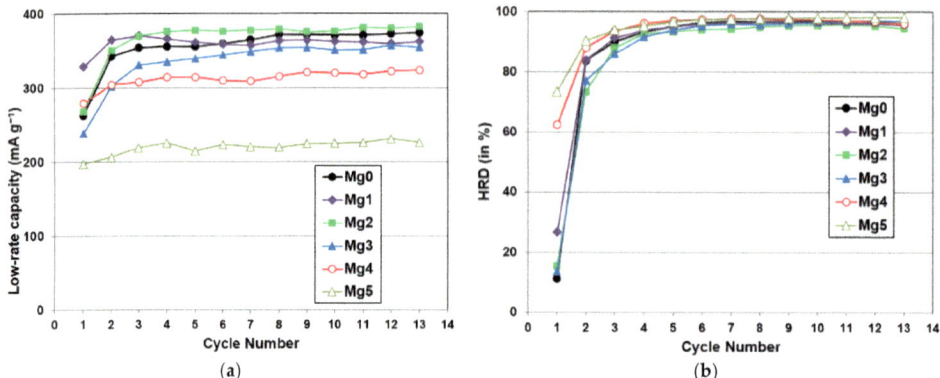

Figure 4. Activation behavior observed from (**a**) half-cell capacity measured at 4 mA·g^{-1} and (**b**) half-cell high-rate dischargeability (HRD) for the first 13 electrochemical cycles.

Low-temperature performance is a very important parameter in propulsion applications, especially in start-stop type micro-hybrid vehicles. The conventional AB$_5$ MH alloy performed poorly below $-25\,^{\circ}$C, but the Co-doped A$_2$B$_7$ superlattice MH alloy can result in significant improvements [82]. Additions of La, Nd, and Si in AB$_2$ MH alloys can lower the surface charge-transfer resistance (R) at $-40\,^{\circ}$C to a comparable level with AB$_5$ MH alloys [23,25,27]. In order to study the effect of Mg addition to the low-temperature performance of AB$_2$ MH alloys, AC impedance at $-40\,^{\circ}$C was measured, and R and the double-layer capacitance (C, closely related to the surface reaction area) were calculated from the obtained Cole-Cole plot. R and C values for alloys in this study are listed in Table 7. The addition of Mg into AB$_2$ MH alloys increases the R value and decreases C. Therefore, different from rare earth elements and Si, the addition of alkaline earth elements in AB$_2$ MH alloys should not be considered when improvement in the low-temperature performance is needed.

Table 7. Summary of the electrochemical results from AC impedance measurement (R: charge transfer resistance, C: double-layer capacitance at $-40\,^{\circ}$C and RT) of the Mg-containing AB$_2$ alloys.

Alloy	R @ $-40\,^{\circ}$C ($\Omega \cdot$g)	C @ $-40\,^{\circ}$C (Farad·g^{-1})	R @ RT ($\Omega \cdot$g)	C @ RT (Farad·g^{-1})
Mg0	29	0.24	0.32	0.34
Mg1	38	0.23	0.52	0.45
Mg2	130	0.22	0.56	0.39
Mg3	118	0.21	0.70	0.38
Mg4	90	0.21	0.92	0.30
Mg5	148	0.15	1.44	0.23

3.5. Magnetic Properties

Magnetic susceptibility was used to characterize the nature of the metallic nickel particles present in the surface layer of the alloy following an alkaline activation treatment [10]. Details on the background and experimental methods were reported earlier [83]. Metallic Ni is an active catalyst for the water splitting and recombination reactions that contributes to the I_o in this electrochemical system. This technique allows us to obtain the saturated magnetic susceptibility (M_s), a quantification of the amount of surface metallic Ni (the product of preferential oxidation), and the magnetic field strength at one-half of the M_s value ($H_{1/2}$), a measurement of the averaged reciprocal number of Ni atoms in a metallic cluster (Figure 5a). The magnetic susceptibility graphs for alloys in this study are shown in Figure 5b and the calculated M_s and $H_{1/2}$ values are listed in Table 6. The M_s decreases and then increases with the increase in the alloy number, which is the opposite trend to the data observed for I_o. The increase in I_o for Mg2 is not from the metallic nickel particles embedded in the surface oxide and may be due to the high content of TiNi phase, which was reported as a catalytic phase for electrochemical reaction [84,85]. The $H_{1/2}$ values listed in Table 6 indicate that the size of metallic nickel decreases and then increases as the alloy number increases. In general, the Mg-containing alloys have smaller metallic nickel clusters in the surface than the Mg-free Mg0.

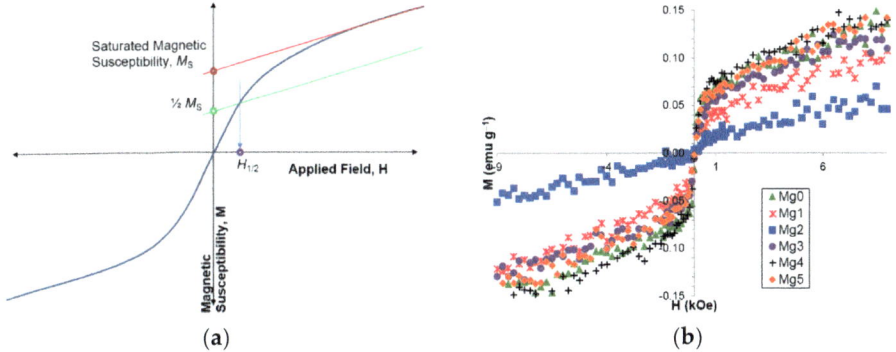

(a) (b)

Figure 5. (a) Representative magnetic susceptibility measurement results showing the saturated magnetic susceptibility (M_s) with paramagnetic component removed and $H_{1/2}$ defined as the magnitude of applied field corresponding to half of maximum magnetic susceptibility (1/2 M_s); (b) the magnetic susceptibility of the alloys in this study.

3.6. Correlations

Due to the limited solubility of Mg in AB_2 MH alloys, the ICP results have some deviations from the original design of $Zr_{21.5}Ti_{21-0.6x}V_{10}Cr_{7.5}Mn_{8.1}$

$Co_8Ni_{32.2-0.4x}Mg_xSn_{0.3}Al_{0.4}$, where $x = 0, 1, 2, 3, 4$, and 5, especially in Mg and Ni contents. As the alloy number increases, Mg content remains at approximately the same level, Ni content remain unchanged and then increases for the last two alloys, Mg4 and Mg5, while the Ti content decreases monotonically. In order to study the influences of different compositions with regard to various alloy properties, the correlation factor (R^2) was calculated between composition (Ni content, Ti content, e/a, and B/A) and properties. The comparison results are listed in Table 8. The findings of the correlation can thus be summarized as demonstrating that Ni content has more influences on the B/A ratio, C14 unit cell volume, hydrogen storage properties, low-rate electrochemical capacity, bulk diffusion, and size of metallic Ni embedded in the surface oxide layer when compared to Ti content.

Table 8. Correlation factors (R^2) between the composition and hydrogen storage properties.

Property	Ni Content	Ti Content	C14 Abundance	TiNi Abundance	B/A	e/a
B/A	0.92	0.79	0.94	0.72	1.00	0.96
C14 unit cell volume	0.51	0.19	0.40	0.91	0.63	0.57
C14 crystallite size	0.17	0.61	0.42	0.00	0.20	0.16
C14 abundance	0.88	0.95	1.00	0.49	0.94	0.90
Maximum capacity	0.94	0.76	0.88	0.71	0.93	0.94
Reversible capacity	0.95	0.80	0.87	0.62	0.87	0.92
Plateau pressure	0.84	0.68	0.75	0.57	0.76	0.80
Low rate capacity	0.93	0.79	0.88	0.64	0.90	0.91
HRD	0.45	0.85	0.76	0.08	0.55	0.46
D	0.61	0.31	0.43	0.80	0.59	0.66
I_0	0.25	0.01	0.05	0.70	0.19	0.26
$R @ -40\,°C$	0.21	0.57	0.37	0.00	0.16	0.18
$C @ -40\,°C$	0.76	0.86	0.84	0.33	0.72	0.72
M_s	0.34	0.36	0.49	0.41	0.56	0.43
$H_{1/2}$	0.44	0.09	0.15	0.69	0.30	0.43

The hydrogen capacities obtained from PCT (converted into electrochemical capacity by 1 wt% = 268 mAh·g^{-1}) and half-cell tests are plotted against Ni content in Figure 6a, showing a decrease in capacity with an increase in the Ni content. The Ti content is more closely related to the C14 phase crystallite size, C14 phase abundance, HRD, and R and C measured at $-40\,°C$. The last three characteristics are plotted against Ti content in Figure 6b, showing that as Ti content increases, the R at $-40\,°C$ decreases and C at $-40\,°C$ increases. However, the RT HRD decreases. The decrease in R is consistent with the increase in C (reaction surface area), but should decrease the RT HRD. The RT R and C follow the trend of $-40\,°C$ R and C (Table 7) and, therefore, the discrepancy is not due to different temperatures and requires further investigation. Both C14 and TiNi phase abundances are correlated to various properties. The results show that the C14 main phase abundance is more influential in hydrogen storage capacities both in solid state and electrochemistry, while the TiNi minor phase abundance has high impacts on D, I_o, $H_{1/2}$, and C14 cell volume. The correlation of TiNi phase abundance to D and I_o are plotted in Figure 6c. High amount in TiNi phase increases the surface reactivity (I_o) but hinders the bulk diffusion of hydrogen (D). While the connections between TiNi phase abundance and $H_{1/2}$ may be true and require future validation, the connection with the C14 cell volume is most likely a coincidence. Both B/A ratio and e/a value are not as influential as other factors in this comparison.

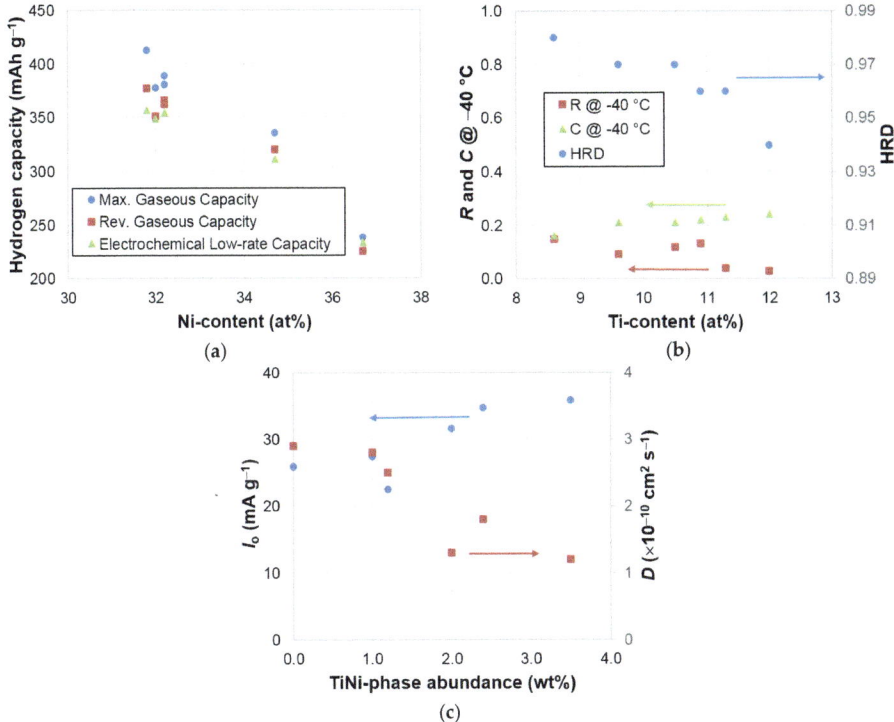

Figure 6. Examples of correlations found in this study showing: (**a**) a high Ni content decreases capacities; (**b**) a higher Ti content decreases the HRD and R, while C increases; and (**c**) a higher TiNi phase abundance increases I_o (surface), but decreases D (bulk).

4. Conclusions

The influence of Mg addition (approximately 0.7 at%) to the structural, solid state, and electrochemical properties of a series of Laves phase based AB$_2$ MH alloys with different Ti and Ni contents were investigated. In general, addition of Mg does not lower the surface charge-transfer resistance, unlike other non-transitional metals, such as Si, Y, La, and Nd. Interestingly, one of the alloys, Mg2 ($Zr_{22.5}Ti_{10.9}V_{9.9}Cr_{7.5}Mn_{8.2}Co_{8.0}Ni_{31.8}Mg_{0.8}Sn_{0.3}Al_{0.4}$) with a phase distribution of 88.3% C14, 7.9% C15, 3.5% TiNi, and 0.3% of a Mg-rich cubic phase, shows improved hydrogen capacities in both solid state and electrochemistry and a higher surface exchange current, but a lower bulk diffusion coefficient.

Acknowledgments: The authors would like to thank the following individuals from BASF-Ovonic for their help: Benjamin Reichman, Benjamin Chao, Baoquan Huang, Diana F. Wong, David Pawlik, Allen Chan, Ryan J. Blankenship, and Su Cronogue.

Author Contributions: Shiuan Chang designed and conducted XRD, PCT, electrochemical, and MS experiments. Taiehi Ouchi prepared the sample. Kwo-hsiung Young, Tiejun Meng, and Xin Wu analyzed the data and participated in the paper preparation.

Conflicts of Interest: The authors declare no conflict of interest.

Abbreviations

MH	Metal hydride
ICP-OES	Inductively coupled plasma-optical emission spectrometer
XRD	X-ray diffractometer
SEM	Scanning electron microscope
EDS	Energy dispersive spectroscopy
PCT	Pressure-concentration-temperature
e/a	Average electron density
ND	Non-detectable
BEI	Back-scattering electron image
DOD	Degree of disorder
D	Diffusion constant
I_o	Surface exchange current
M_s	Saturated magnetic susceptibility
$H_{1/2}$	Magnetic field strength at one-half of the M_s value
RT	Room temperature

References

1. Jacob, I.; Stern, A.; Moran, A.; Shaltiel, D.; Davidov, D. Hydrogen absorption in $(Zr_xTi_{1-x})B_2$ (B = Cr, Mn) and the phenomenological model for the absorption capacity in pseudo-binary Laves-phase compounds. *J. Less Comm. Met.* **1980**, *73*, 1369–1376.

2. Young, K.; Ouchi, T.; Koch, J.; Fetcenko, M.A. The role of Mn in C14 Laves phase multi-component alloys for NiMH battery application. *J. Alloys Compd.* **2009**, *477*, 749–758.

3. Hu, W.; Noréus, D. Rare-earth-based AB$_5$-type hydrogen storage alloys as hydrogen electrode catalysts in alkaline fuel cells. *J. Alloys Compd.* **2003**, *356*, 734–737.

4. Zhang, Y.; Wang, G.; Dong, X.; Guo, S.; Ren, J.; Wang, X. Effect of substituting Co with Fe on the cycle stabilities of the as-cast and quenched AB_5-type hydrogen storage alloys. *J. Power Sources* **2005**, *148*, 105–111.

5. Züttle, A.; Meli, F.; Schlapbach, L. Electrochemical and surface properties of $Zr(V_xNi_{1-x})_2$ alloys as hydrogen absorbing electrodes in alkaline electrolyte. *J. Alloys Compd.* **1994**, *203*, 235–241.

6. Liu, B.; Jung, J.; Lee, H.; Lee, K.; Lee, J. Improved electrochemical performance of AB_2-type metal hydride electrodes activated by the hot-charging process. *J. Alloys Compd.* **1996**, *245*, 132–141.

7. Yang, X.G.; Lei, Y.Q.; Shu, K.Y.; Lin, G.F.; Zhang, Q.A.; Zhang, W.K.; Zhang, X.B.; Lu, G.L.; Wang, Q.D. Contribution of rare-earths to activation property of Zr-based hydride electrodes. *J. Alloys Compd.* **1999**, *293–295*, 632–636.

8. Liu, B.; Li, Z.; Higuchi, E.; Suda, S. Improvement of the electrochemical properties of Zr-based AB_2 alloys by an advanced fluorination technique. *J. Alloys Compd.* **1999**, *293–295*, 702–706.

9. Young, K.H.; Fetcenko, M.A.; Ovshinsky, S.R.; Ouchi, T.; Reichman, B.; Mays, W.C. Improved surface catalysis of Zr-based Laves phase alloys for NiMH Battery. In *Hydrogen at Surface and Interface*; Jerkiewicz, G., Feliu, J.M., Popov, B.N., Eds.; Electrochemical Society: Pennington, NJ, USA, 2000; pp. 60–71.

10. Young, K.; Huang, B.; Regmi, R.K.; Lawes, G.; Liu, Y. Comparisons of metallic clusters imbedded in the surface of AB_2, AB_5, and A_2B_7 alloys. *J. Alloys Compd.* **2010**, *506*, 831–840.

11. Xu, Y.-H.; Chen, C.-P.; Wang, X.-L.; Wang, Q.-D. The electrode properties of the $Ti_{1.0}Zr_{0.2}Cr_{0.4}Ni_{0.8}V_{0.8-x}Al_x$ (x = 0.0, 0.02, 0.4 and 0.6) alloys. *Mater. Chem. Phys.* **2003**, *77*, 1–5.

12. Li, L.; Wang, W.; Fan, X.; Jin, X.; Wang, H.; Lei, Y.; Wang, Q.; Chen, L. Microstructure and electrochemical behavior of Cr-added $V_{2.1}TiNi_{0.4}Zr_{0.06}Cr_{0.152}$ hydrogen storage electrode alloy. *Int. J. Hydrog. Energy* **2007**, *32*, 2434–2438.

13. Zhu, Y.F.; Pan, H.G.; Wang, G.Y.; Gao, M.X.; Ma, J.X.; Chen, C.P.; Wang, Q.D. Phase structure, crystallography and electrochemical properties of Laves phase compounds $Ti_{0.8}Zr_{0.2}V_{1.6}Mn_{0.8-x}M_xNi_{0.6}$ (M = Fe, Al, Cr, Co). *Int. J. Hydrog. Energy* **2001**, *26*, 807–816.

14. Züttle, A.; Meli, F.; Schlapbach, L. Effects of electrode compacting additives on the cycle life and high-rate dischargeability of $Zr(V_{0.25}Ni_{0.75})_2$ metal hydride electrodes in alkaline solution. *J. Alloys Compd.* **1994**, *206*, 31–38.

15. Young, K.; Ouchi, T.; Huang, B.; Reichman, B.; Fetcenko, M.A. Studies of copper as a modifier in C14-predominant AB_2 metal hydride alloys. *J. Power Sources* **2012**, *204*, 205–212.

16. Young, K.; Ouchi, T.; Huang, B.; Reichman, B.; Fetcenko, M.A. The structure, hydrogen storage, and electrochemical properties of Fe-doped C14-predominating AB_2 metal hydride alloys. *Int. J. Hydrog. Energy* **2011**, *36*, 12296–12304.

229

17. Young, K.; Ouchi, T.; Huang, B.; Reichman, B. Effect of molybdenum content on structural, gaseous storage, and electrochemical properties of C14-predominant AB_2 metal hydride alloys. *J. Power Sources* **2011**, *196*, 8815–8821.

18. Erika, T.; Ricardo, F.; Fabricio, R.; Fernando, Z.; Verónica, D. Electrochemical and metallurgical characterization of $ZrCr_{1-x}NiMo_x$ AB_2 metal hydride alloys. *J. Alloys Compd.* **2015**, *649*, 267–274.

19. Young, K.; Ouchi, T.; Lin, X.; Reichman, B. Effects of Zn-addition to C14 metal hydride alloys and comparisons to Si, Fe, Cu, Y, and Mo-additives. *J. Alloys Compd.* **2016**, *655*, 50–59.

20. Ruiz, F.C.; Peretti, H.A.; Visintin, A.; Triaca, W.E. A study on $ZrCrNiPt_x$ alloys as negative electrode components for NiMH batteries. *Int. J. Hydrog. Energy* **2011**, *36*, 901–906.

21. Ruiz, F.C.; Peretti, H.A.; Visintin, A. Electrochemical hydrogen storage in $ZrCrNiPd_x$ alloys. *Int. J. Hydrog. Energy* **2010**, *35*, 5963–5967.

22. Liu, Y.; Zhang, S.; Li, R.; Gao, M.; Zhong, K.; Miao, H.; Pan, H. Electrochemical performance of the Pd-added Ti-V-based hydrogen storage alloy. *Int. J. Hydrog. Energy* **2008**, *33*, 728–734.

23. Young, K.; Wong, D.F.; Nei, J.; Reichman, B. Electrochemical properties of hypo-stoichiometric Y-doped AB_2 metal hydride alloys at ultra-low temperature. *J. Alloys Compd.* **2015**, *643*, 17–27.

24. Qiao, Y.; Zhao, M.; Zhu, X.; Cao, G. Microstructure and some dynamic performances of $Ti_{0.17}Zr_{0.08}V_{0.24}RE_{0.01}Cr_{0.1}Ni_{0.3}$ (RE = Ce, Dy) hydrogen storage electrode alloys. *Int. J. Hydrog. Energy* **2007**, *32*, 3427–3434.

25. Young, K.; Wong, D.F.; Ouchi, T.; Huang, B.; Reichman, B. Effects of La-addition to the structure, hydrogen storage, and electrochemical properties of C14 metal hydride alloys. *Electrochim. Acta* **2015**, *174*, 815–825.

26. Sun, J.C.; Li, S.; Ji, S.J. The effects of the substitution of Ti and La for Zr in $ZrMn_{0.7}V_{0.2}Co_{0.1}Ni_{1.2}$ hydrogen storage alloys on the phase structure and electrochemical properties. *J. Alloys Compd.* **2007**, *446–447*, 630–634.

27. Wong, D.F.; Young, K.; Nei, J.; Wang, L.; Ng, K.Y.S. Effects of Nd-addition on the structural, hydrogen storage, and electrochemical properties of C14 metal hydride alloys. *J. Alloys Compd.* **2015**, *647*, 507–518.

28. Young, K.; Ouchi, T.; Huang, B.; Reichman, B.; Blankensip, R. Improvement in $-40\,^\circ C$ electrochemical properties of AB_2 metal hydride alloy by silicon incorporation. *J. Alloys Compd.* **2013**, *575*, 65–72.

29. Li, S.; Wen, B.; Li, X.; Zhai, J. Structure and electrochemical property of ball-milled $Ti_{0.26}Zr_{0.07}Mn_{0.1}Ni_{0.33}V_{0.24}$ alloy with 3 mass% B. *J. Alloys Compd.* **2016**, *654*, 580–585.

30. Johnson, R.L.; Hoffmann, R.Z. Structure-bonding relationships in the Laves Phases. *Z. Anorg. Allg. Chem.* **1992**, *616*, 105–120.

31. Pearson, W.B. *The Crystal Chemistry and Physics of Metals and Alloys*; Wiley & Sons: New York, NY, USA, 1971; pp. 514–515.

32. Chen, C.; Wang, C.; Lee, P.; Lin, C. Hydrogen absorption performance of mechanically alloys $(Mg_2Ni)_{100-x}Ti_x$ powder. *Mater. Trans.* **2007**, *48*, 3170–3175.

33. Gupta, A.; Shervani, S.; Faisal, M.; Balani, K.; Subramaniam, A. Hydrogen storage in Mg-Mg2Ni-carbon hybrids. *J. Alloys Compd.* **2015**, *601*, 280–288.

34. Zadorozhnyy, V.Y.; Menjo, M.; Zadogozhnyy, M.Y.; Kaloshkin, S.D.; Louzguine-Luzgin, D.D. Hydrogen sorption properties of nanostructured bulk Mg_2Ni intermetallic compound. *J. Alloys Compd.* **2014**, *586*, S400–S404.

35. Zhang, Y.; Li, C.; Cai, Y.; Hu, F.; Liu, Z.; Guo, S. Highly improved electrochemical hydrogen performances of the Nd-Cu-added Mg_2Ni-type alloys by melt spinning. *J. Alloys Compd.* **2014**, *584*, 81–86.

36. Zhu, Y.; Yang, C.; Zhu, J.; Li, L. Structural and electrochemical hydrogen storage properties of Mg_2Ni-based alloys. *J. Alloys Compd.* **2011**, *509*, 5309–5314.

37. Cui, N.; Luo, J.L.; Chuang, K.T. Nickel-metal hydride (Ni-MH) battery using Mg_2Ni-type hydrogen storage alloy. *J. Alloys Compd.* **2000**, *302*, 218–226.

38. Zhang, Y.; Zhang, G.; Yang, T.; Hou, Z.; Guo, S.; Qi, Y.; Zhao, D. Electrochemical and hydrogen absorption/desorption properties of nanocrystalline Mg_2Ni-type alloys prepared by melt spinning. *Rare Metal Mater. Eng.* **2012**, *41*, 2069–2074.

39. Zhang, Y.; Ren, H.; Hou, Z.; Zhang, G.; Li, X.; Wang, X. Investigation on gaseous and electrochemical hydrogen storage kinetics of as-quenched nanocrystalline Mg_2Ni-type alloys. *Rare Metal Mater. Eng.* **2012**, *41*, 1516–1521.

40. Zhang, Y.; Li, B.; Ren, H.; Hu, F.; Zhang, G.; Guo, S. Gaseous and electrochemical hydrogen storage kinetics of nanocrystalline Mg_2Ni-type alloy prepared by rapid quenching. *J. Alloys Compd.* **2011**, *509*, 5604–5610.

41. Shao, H.; Li, X. Effect of nanostructure and partial substitution on gas absorption and electrochemical properties in Mg_2Ni-based alloys. *J. Alloys Compd.* **2016**, *667*, 191–197.

42. Massalski, T.B. *Binary Alloy Phase Diagram*, 2nd ed.; ASM International: Materials Park, OH, USA, 1990; pp. 2529–2530.

43. Ovshinsky, S.R.; Fetcenko, M.A. Electrochemical Hydrogen Storage Alloys and Batteries Fabricated from Mg Containing Base Alloys. U.S. Patent 5,506,069, 9 April 1996.

44. Yan, S.; Young, K.; Ng, S. Effects of salt additives to the KOH electrolyte used in Ni/MH batteries. *Batteries* **2015**, *1*, 54–73.

45. Young, K.; Ng, S.K.Y.; Bendersky, L.A. A technical report of robust affordable next generation energy storage system-BASF program. *Batteries* **2016**, *2*.

46. Young, K.; Ouchi, T.; Huang, B.; Fetcenko, M.A. Effects of B, Fe, Gd, Mg, and C on the structure, hydrogen storage, and electrochemical properties of vanadium-free AB_2 metal hydroxide alloy. *J. Alloys Compd.* **2012**, *511*, 242–250.

47. Liu, Y.; Cao, Y.; Huang, L.; Gao, M.; Pan, H. Rare earth-Mg-Ni-based hydrogen storage alloys as negative electrode materials for Ni/MH batteries. *J. Alloys Compd.* **2011**, *509*, 675–686.

48. Nei, J.; Young, K.; Salley, S.O.; Ng, K.Y.S. Effects of annealing on $Zr_8Ni_{19}X_2$ (X = Ni, Mg, Al, Sc, V, Mn, Co, Sn, La, and Hf): Structural characteristics. *J. Alloys Compd.* **2012**, *516*, 144–152.

49. Nei, J.; Young, K.; Salley, S.O.; Ng, K.Y.S. Effects of annealing on $Zr_8Ni_{19}X_2$ (X = Ni, Mg, Al, Sc, V, Mn, Co, Sn, La, and Hf): Hydrogen storage and electrochemical properties. *Int. J. Hydrog. Energy* **2012**, *37*, 8418–8427.

50. Young, K.; Fetcenko, M.A.; Li, F.; Ouchi, T. Structural, thermodynamics, and electrochemical properties of $Ti_xZr_{1-x}(VNiCrMnCoAl)_2$ C14 Laves phase alloys. *J. Alloys Compd.* **2008**, *464*, 238–247.

51. Young, K.; Fetcenko, M.A.; Koch, J.; Morii, K.; Shimizu, T. Studies of Sn, Co, Al, and Fe additives in C14/C15 Laves alloys for NiMH battery application by orthogonal arrays. *J. Alloys Compd.* **2009**, *486*, 559–569.

52. Nei, J.; Young, K.; Salley, S.O.; Ng, K.Y.S. Determination of C14/C15 phase abundance in Laves phase alloys. *Mat. Chem. Phys.* **2012**, *136*, 520–527.

53. Zhu, Y.; Pan, H.; Gao, M.; Ma, J.; Lei, Y.; Wang, Q. Electrochemical studies on the Ti-Zr-V-Mn-Cr-Ni hydrogen storage electrode alloys. *Int. J. Hydrog. Energy* **2003**, *28*, 311–316.

54. Pan, H.; Zhu, Y.; Gao, M.; Liu, Y.; Li, R.; Lei, Y.; Wang, Q. A study on the cycling stability of the Ti-V-based hydrogen storage electrode alloys. *J. Alloys Compd.* **2004**, *364*, 271–279.

55. Chen, L.; Wu, F.; Tong, M.; Chen, D.M.; Long, R.B.; Shang, Z.Q.; Liu, H.; Sun, W.S.; Yang, K.; Wang, L.B.; *et al.* Advanced nanocrystalline Zr-based AB₂ hydrogen storage electrode materials for NiMH EV batteries. *J. Alloys Compd.* **1999**, *293–295*, 508–520.

56. Li, Z.P.; Higuchi, E.; Liu, B.H.; Suda, S. Electrochemical properties and characteristics of a fluorinated AB₂-alloy. *J. Alloys Compd.* **1999**, *293–295*, 593–600.

57. Young, K.; Ouchi, T.; Yang, J.; Fetcenko, M.A. Studies of off-stoichiometric AB₂ metal hydride alloy: Part 1. Structural characteristics. *Int. J. Hydrog. Energy* **2011**, *36*, 11137–11145.

58. Lee, S.M.; Lee, H.; Yu, J.S.; Fateev, G.A.; Lee, J.Y. The activation characteristics of a Zr-based hydrogen storage alloy electrode surface-modified by ball-milling process. *J. Alloys Compd.* **1999**, *292*, 258–265.

59. Liu, Y.; Young, K. Microstructure investigation on metal hydride alloys by electron backscatter diffraction technique. *Batteries* **2016**. to be submitted for publication.

60. Boettinger, W.J.; Newbury, D.E.; Wang, K.; Bendersky, L.A.; Chiu, C.; Kattner, U.R.; Young, K.; Chao, B. Examination of multiphase (Zr,Ti)(V,Cr,Mn,Ni)₂ Ni-MH electrode alloys: Part I. Dendritic solidification structure. *Metal. Mater Trans. A* **2010**, *41*, 2033–2047.

61. Young, K. Stoichiometry in Inter-Metallic Compounds for Hydrogen Storage Applications. In *Stoichiometry and Materials Science—When Numbers Matter*; Innocenti, A., Kamarulzaman, N., Eds.; Intech: Rijeka, Croatia, 2012; pp. 162–164.

62. Bendersky, L.A.; Wang, K.; Boettinger, W.J.; Newbury, D.E.; Young, K.; Chao, B. Examination of multiphase (Zr,Ti)(V,Cr,Mn,Ni)₂ Ni-MH electrode alloys: Part II. Solid-state transformation of the interdendric B₂ phase. *Metal. Mater Trans. A* **2010**, *41*, 1891–1906.

63. Pearson, W.B. *A Handbook of Lattice Spacings and Structures of Metals and Alloys: International Series of Monographs on Metal Physics and Physical Metallurgy*; Elsevier: Philadelphia, PA, USA, 2013; Volume 4, p. 693.

64. Shaltiel, D.; Jacob, I.; Davidov, D. Hydrogen absorption and desorption properties of AB_2 Laves-phase pseudobinary compounds. *J. Less Common Met.* **1977**, *53*, 117–131.

65. Jacob, I.; Shaltiel, D. Hydrogen sorption properties of some AB_2 Laves phase compounds. *J. Less Common Met.* **1979**, *65*, 117–128.

66. Qian, S.; Northwood, D.O. Thermodynamic characterization of $Zr(Fe_xCr_{1-x})$ systems. *J. Less Common Met.* **1989**, *147*, 149–159.

67. Perevesenzew, A.; Lanzel, E.; Eder, O.J.; Tuscher, E.; Weinzierl, P. Thermodynamics and kinetics of hydrogen absorption in intermetallic compounds $Zr(Cr_{1-x}V_x)_2$. *J. Less Common Met.* **1988**, *143*, 39–47.

68. Park, J.; Lee, J. Thermodynamic properties of the $Zr_{0.8}Ti_{0.2}(Mn_xCr_{1-x})Fe$ system. *J. Less Common Met.* **1991**, *167*, 245–253.

69. Drašner, A.; Blažina, Ž. On the structural and hydrogen desorption properties of the $Zr(Cr_{1-x}Cu_x)_2$ alloys. *J. Less Common Met.* **1991**, *175*, 103–108.

70. Schlapbach, L.; Züttel, A. Hydrogen-storage materials for mobile applications. *Nature* **2001**, *414*, 353–358. PubMed]

71. Züttel, A. Materials for hydrogen storage. *Mater. Today* **2003**, *6*, 24–33.

72. Sastri, M.V.C. Introduction to Metal Hydrides: Basic Chemistry and Thermodynamics of Their Formation. In *Metal Hydride*; Sastri, M.V.C., Viswanathan, B., Murthy, S.S., Eds.; Springer-Verlag: Berlin, Germany, 1998; p. 5.

73. Scholtus, N.A.; Hall, W.K. Hysteresis in the palladium-hydrogen system. *J. Chem. Phys.* **1963**, *39*, 868–870.

74. Makenas, B.J.; Birnbaum, H.K. Phase changes in the niobium-hydrogen system 1: Accommodation effcts during hydrde precipitation. *Acta Metal. Mater.* **1980**, *28*, 979–988.

75. Balasubramaniam, R. Accommodation effects during room temperature hydrogen transformations in the niobium-hydrogen system. *Acta Metal. Mater.* **1993**, *41*, 3341–3349.

76. Young, K.; Ouchi, T.; Fetcenko, M.A. Pressure-composition-temperature hysteresis in C14 Laves phase alloys: Part 1. Simple ternary alloys. *J. Alloys Compd.* **2009**, *480*, 428–433.

77. Lide, D.R. *CRC Handbook of Chemistry and Physics*, 74 ed.; CRC Press: Boca Raton, FL, USA, 1993; pp. 6–22.

78. Kim, S.; Lee, J.; Park, H. A study of the activation behaviour of ZrCrNiLa metal hydride electrodes in alkaline solution. *J. Alloys Compd.* **1994**, *205*, 225–229.

79. Jung, J.; Lee, K.; Lee, J. The activation mechanism of Zr-based alloy electrodes. *J. Alloys Compd.* **1995**, *226*, 166–169.

80. Park, H.Y.; Cho, W.I.; Cho, B.W.; Jang, H.; Lee, S.R.; Yun, K.S. Effect of fluorination on the lanthanum-doped AB_2-type metal hydride electrodes. *J. Power Sources* **2001**, *92*, 149–156.

81. Park, H.Y.; Chang, I.; Cho, W.I.; Cho, B.W.; Jang, H.; Lee, S.R.; Yun, K.S. Electrode characteristics of the Cr and La doped AB_2-type hydrogen storage alloys. *Int. J. Hydrog. Energy* **2001**, *26*, 949–955.

82. Wang, L.; Young, K.; Meng, T.; English, N.; Yasuoka, S. Partial substitution of cobalt for nickel in mixed area earth metal based superlàttice hydrogen absorbing alloy—Part 2 battery performance and failure analysis. *J. Alloys Compd.* **2016**, *664*, 417–427.

83. Nei, J.; Young, K.; Regmi, R.; Lawes, G.; Salley, S.O.; Ng, K.Y.S. Gaseous phase hydrogen storage and electrochemical properties of Zr_8Ni_{21}, Zr_7Ni_{10}, Zr_9Ni_{11}, and ZrNi metal hydride alloys. *Int. J. Hydrog. Energy* **2012**, *37*, 16042–16055.

84. Zlatanova, Z.; Spassov, T.; Eggeler, G.; Spassova, M. Synthesis and hydriding/dehydriding properties of Mg_2Ni-AB (AB = TiNi or TiFe) nanocomposites. *Int. J. Hydrog. Energy* **2011**, *36*, 7559–7566.

85. Balcerzak, M. Electrochemical and structural studies on Ti-Zr-Ni and Ti-Zr-Ni-Pd alloys and composites. *J. Alloys Compd.* **2016**, *658*, 576–587.

The Importance of Rare-Earth Additions in Zr-Based AB$_2$ Metal Hydride Alloys

Kwo-Hsiung Young, Taihei Ouchi, Jean Nei and Dhanashree Moghe

Abstract: Effects of substitutions of rare earth (RE) elements (Y, La, Ce, and Nd) to the Zr-based AB$_2$ multi-phase metal hydride (MH) alloys on the structure, gaseous phase hydrogen storage (H-storage), and electrochemical properties were studied and compared. Solubilities of the RE atoms in the main Laves phases (C14 and C15) are very low, and therefore the main contributions of the RE additives are through the formation of the RENi phase and change in TiNi phase abundance. Both the RENi and TiNi phases are found to facilitate the bulk diffusion of hydrogen but impede the surface reaction. The former is very effective in improving the activation behaviors. $-40\,^{\circ}$C performances of the Ce-doped alloys are slightly better than the Nd-doped alloys but not as good as those of the La-doped alloys, which gained the improvement through a different mechanism. While the improvement in ultra-low-temperature performance of the Ce-containing alloys can be associated with a larger amount of metallic Ni-clusters embedded in the surface oxide, the improvement in the La-containing alloys originates from the clean alloy/oxide interface as shown in an earlier transmission electron microscopy study. Overall, the substitution of 1 at% Ce to partially replace Zr gives the best electrochemical performances (capacity, rate, and activation) and is recommended for all the AB$_2$ MH alloys for electrochemical applications.

Reprinted from *Batteries*. Cite as: Young, K.-H.; Ouchi, T.; Nei, J.; Moghe, D. The Importance of Rare-Earth Additions in Zr-Based AB$_2$ Metal Hydride Alloys. *Batteries* **2016**, *2*, 25.

1. Introduction

Nickel/metal hydride (Ni/MH) rechargeable batteries have been extensively used to replace Ni/Cd rechargeable and alkaline primary batteries in the consumer electronics market. In addition, Ni/MH batteries also dominate the hybrid electric vehicle market [1] and are found in stationary applications [2,3]. Although the volumetric energy density of Ni/MH battery is good, its gravimetric energy density is only one third of that of its Li-ion rival. Therefore, Laves phase AB$_2$ metal hydride (MH) alloys have been proposed to replace the currently used rare earth (RE) element-based AB$_5$ MH alloys to achieve an increase in capacity of the negative electrode [4,5]. Besides its relatively high capacity (420 mAh·g^{-1} [6] versus 330 mAh·g^{-1} of AB$_5$), alloy design for AB$_2$ also has a higher flexibility in constituent elements and phases, making it more adaptable to various requirements, including low cost, ultra-low-temperature operation [7,8], and high-temperature storage [9]. RE elements, such as Y, La, Ce, Pr, and Nd, were added in the AB_2 MH alloys in the early days to improve activation behavior of the negative electrode [10–15]. Recently, we reported our works of adding Y [7,8,16], La [17], and Nd [18] in the AB_2 MH alloys and observed a substantial improvement in $-40\,^\circ$C electrochemical performance. In this paper, results from the partial substitution of Zr by Ce in the AB_2 MH alloys are presented in detail and are compared to those obtained from the AB_2 MH alloys substituted with Y, La, and Nd.

2. Experimental Setup

Ingot samples were prepared by arc melting under a 0.08 MPa Ar protection atmosphere. Samples were flipped five times during the melting–cooling procedure to ensure homogeneity, and they then underwent a hydriding/dehydriding process, which introduced volume expansion/contraction and consequently damages to the crystal structure, leading to an increase in brittleness of the samples before they were crushed and grounded into -200 mesh powder. We used a Varian Liberty 100 inductively coupled plasma-optical emission spectrometer (ICP-OES, Agilent Technologies, Santa Clara, CA, USA) to study the chemical composition, a Philips X'Pert Pro X-ray diffractometer (XRD, Amsterdam, The Netherlands) to perform phase analysis, a JEOL-JSM6320F scanning electron microscope (SEM, Tokyo, Japan) with energy dispersive spectroscopy (EDS) to investigate the phase distribution and composition, and a Suzuki-Shokan multi-channel pressure–concentration–temperature (PCT, Tokyo, Japan) system to measure the gaseous phase hydrogen storage (H-storage) characteristics. PCT measurements at 30, 60, and 90 $^\circ$C were performed after activation, which was a 2 h thermal cycle between room temperature and 300 $^\circ$C under 2.5 MPa H_2 pressure. Details of the electrode and cell preparations, as well as the electrochemical measurement methods, are as previously reported [19,20]. We used a Solartron 1250 Frequency Response Analyzer (Solartron Analytical, Leicester, UK) with a sine wave amplitude of 10 mV and a frequency range of 0.5 mHz–10 kHz to conduct the AC impedance measurements and a Digital Measurement Systems Model 880 vibrating sample magnetometer (MicroSense, Lowell, MA, USA) to measure the magnetic susceptibility (M.S.) of the activated alloy surfaces (4 h in 100 $^\circ$C 30 wt% KOH solution).

3. Property Comparison of Elements and Intermetallic Compounds of Y, La, Ce, Pr, and Nd

Several important properties of Y, La, Ce, Pr, and Nd elements and their intermetallic compounds (RENi and $RENi_5$) with Ni are listed in Table 1. The name "rare earth" originated from the difficult nature in separating one element from another in the ore deposits but not the scarcity. Indeed, Ce is as abundant as Cu in the earth's crust [21]. Compared to the other RE elements, Ce has an additional common 4+ oxidation state [22], which is a perfect example of Hund's rule that states the empty, half-filled, and completely filled electronic levels tend to be at more stable states [23]. Therefore, Ce is different from others with regards to several properties, such as the RE melting point (M.P.), hydroxide and $RENi_5$ heats of formation, $RENi_5$ unit cell volume, and $RENi_5$ plateau pressure. Furthermore, in MH alloy formula design of the misch metal-based AB_5/A_2B_4 stacking superlattice MH alloys, Ce is not incorporated [24] due to its small radius, which cannot maintain the lattice constant in the AB_5 slab and promotes the formation of the AB_2 phase [25].

Comparing the stabilities of $RE(OH)_3$ compounds, represented by their heats of formation (Table 1), reveals that $Ce(OH)_3$ and $Y(OH)_3$ are less stable than the other hydroxides. Also, the ease of oxidation can be judged based on the oxidation potentials of the RE elements (Table 1) and is in the order of $Y \approx La > Pr > Ce > Nd$. Solubility of $RE(OH)_3$ can be represented by the constant A in Equation (1):

$$2RE^{3+} + 3H_2O \leftrightarrows RE_2O_3 + 6H^+, \log(RE^{3+}) = A - 3pH \qquad (1)$$

and is in the order of $La > Pr > Ce > Nd > Y$. Moreover, SEM pictures of the A_2B_7 and AB_5 alloys with La or Nd show the difference in surface morphology: small needles of $La(OH)_3$ covering the entire surface versus large rods of $Nd(OH)_3$ covering any surface with open space [26].

Judging from the comparison of heats of formation for La_3Ni (-13), LaNi (-24.8), $LaNi_2$ (-20), $LaNi_3$ (-21), La_2Ni_7 (-24), and $LaNi_5$ (-21 kJ$\cdot$mol^{-1}) [41,42], the RE-Ni intermetallic compounds are relatively easy to form. In addition, theoretical calculation results of the heats of formation for both the Ce-Ni [39] and Nd-Ni [46] intermetallic compounds show a minimum at 1:1 stoichiometry (RENi). RENi is commonly found in the RE-doped Laves phase AB_2 MH alloys as a segregated phase [7,8,12,16–18] since the larger radius of the RE atom does not fit into the atomic radius ratio range for the Laves phases (atomic radius ratio of the A to B atoms between 1.05 and 1.68) [47]. Crystal structure of YNi (FeB-type) is slightly different from the other RENi intermetallic compounds (CrB-type) [48]. While both types are composed of trigonal prisms with the RE atoms occupying the eight corners and Ni forming a zigzag chain, the placement of these prisms is in either the CrB or FeB structure (Figure 1). Previously, some RENi alloys have been studied as H-storage alloys ($YNiH_3$ and $YNiH_4$ [49], $LaNiH_3$ [50,51], $CeNiH_{2.9}$ [52], $CeNiH_4$ [53], and $PrNiH_{4.3}$ [53]).

Table 1. Properties of the rare earth (RE = Y, La, Ce, Pr, and Nd) elements, RENi and RENi₅ intermetallic compounds. Data are from [27] unless otherwise cited. HCP and DHCP denote hexagonal close packed, and double-*c* hexagonal close packed, respectively.

Properties of RE, RENi, and RENi₅	Y	La	Ce	Pr	Nd
Atomic number	39	57	58	59	60
Content in earth crust (ppm) [28]	33	39	66.5	9.2	41.5
Outer shell electron configuration	$5s^24d^1$	$6s^25d^1$	$6s^25d^14f^1$	$6s^25d^14f^2$	$6s^25d^14f^3$
Electronegativity	1.22	1.10	1.12	1.13	1.14
Ionic radius (Å)	1.04 (Y^{3+})	1.17 (La^{3+})	1.15 (Ce^{3+}); 1.01 (Ce^{4+})	1.13 (Pr^{3+})	1.12 (Nd^{3+})
Atomic radius in Laves phase alloy (Å) [29]	1.990	3.335	2.017	2.013	2.013
Crystal structure at 25°C	HCP	DHCP	DHCP	DHCP	DHCP
Melting point (°C)	1522	918	798	931	1021
Temperature when vapor pressure reaches 0.001 Pa (°C)	1220	1301	1290	1083	995
Oxidation potential (V)	−2.37	−2.37	−2.335	−2.353	−2.246
Heat of hydride formation (kJ·mol⁻¹) [30]	−114	−97	−103	−104	−106
Heat of formation of RE(OH)₃ (kJ·mol⁻¹)	−937.6 [31]	−1415.5 [32]	−1014.5 [33]	−1419 [34]	−1403.6 [35]
Solubility represented by the value of A in Equation (1) [36]	19.86	23.02	22.15	22.50	21.25
Crystal structure of RENi	FeB-(Pnma)	CrB-(Cmcm)	CrB-(Cmcm)	CrB-(Cmcm)	CrB-(Cmcm)
Unit cell volume of RENi (Å³) [37]	162.6	183.9	174.6	174.5	172.6
Melting point of RENi (°C)	1070 [38]	715 [38]	680 [39]	730 [38]	780 [38]
Heat of formation of RENi (kJ·mol⁻¹)	−37 [40]	−24.8 [41]	−30.3 [41]	−28.1 [41]	−25.0 [41]
Crystal structure of RENi₅ [34]	CaCu₅	CaCu₅	CaCu₅	CaCu₅	CaCu₅
Unit cell volume of RENi₅ (Å³) [34]	81.7	86.8	82.8	84.8	84.3
Heat of formation of RENi₅ (kJ·mol⁻¹)	−204.6 [42]	−158.9 [42]	−199 [36]	−160.6 [36]	−151.2 [43]
Plateau pressure at 20 °C of RENi₅ (MPa)	30 [44]	0.15 [44]	4.8 [45]	1.2 [44]	0.62 [44]
Heat of formation of RENi₅H₆ (kJ·mol⁻¹)	-	−30.1 [44]	−14.2 [45]	−30.5 [45]	−29.4 [45]

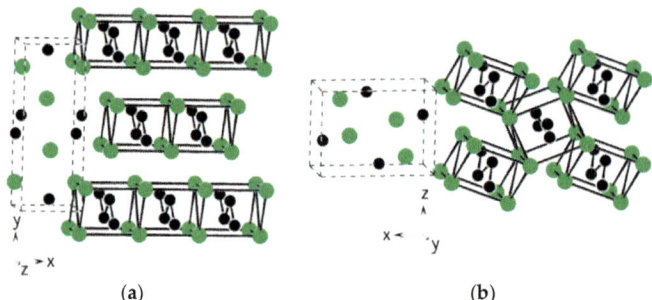

(a) (b)

Figure 1. (a) CrB- and (b) FeB-type structures. Green and black circles represent the RE and Ni atoms, respectively. Both structures are composed of trigonal prisms containing four RE and four Ni atoms. While the RE atoms form a HCP stacking, the Ni atoms form a zigzag chain in both cases. Reprinted from [48].

4. Results and Discussion

Results from the Y- [7], La- [17], and Nd- [18] substitutions have been previously published. In this paper, new data from the Ce-substitution are organized and presented. Five alloys with different Ce-contents were made by arc melting. Their design compositions are listed in Table 2. The base Ce-free alloy, Ce0 ($Ti_{12}Zr_{22.8}V_{10}Cr_{7.5}Mn_{8.1}Co_{7.0}Ni_{32.2}Co_{0.4}$), is the same base alloy used in the La- [17] and Nd- [18] substitution studies and a high-rate derivative of the base alloy in the Y-substitution study, Y0 ($Ti_{12}Zr_{21.5}V_{10}Cr_{7.5}Mn_{8.1}Co_{8.0}Ni_{32.2}Sn_{0.3}Co_{0.4}$) [7]. Compared to alloy Y0, alloy Ce0 contains 1.3 at% higher Zn-content, 1.0 at% lower Co-content, and no Sn. ICP results of the current batch of alloys (alloys Ce0 to Ce5) are also listed in Table 2 for comparison. Most samples show very close compositions to the design values with the exception of a slight deficiency in Cr observed in alloy Ce2. Inconsistent uniformity of Cr in the AB_2 MH alloy has been seen previously [17]. Comparing among Zr, V, and Cr, which are the higher-content elements with the highest melting points, Cr has the highest eutectic temperature with Ni (1345 °C, 960 °C for Zr-Ni, and 1202 °C for V-Ni) [38]. Therefore, Cr's distribution may not be as consistently uniform compared to the other constituent elements among alloys. In combination with the small sample size for ICP, the sampling from alloy Ce2 may exclude a Cr-rich region that is not dissolved in the main phases and results in the observed lower Cr-content compared to the design composition. Average electron densities (e/a) of these alloys are just slightly under the C14/C15 threshold (6.83) [54], so a C14-predominating microstructure is predicted. The B/A ratios are slightly over the design value (1.87) due to the loss of Zr during slag formation. The slight hypo-stoichiometry design is used to balance between the degree of disorder (DOD, from the secondary phase abundances) and electrochemical properties [55].

Table 2. Design compositions (in **bold**) and inductively coupled plasma (ICP) results in at%. e/a is the average electron density, and B/A is the ratio of the B-atom-content (V, Cr, Mn, Co, Ni, and Al) to the A-atom-content (Ti, Zr, and Ce).

Alloy	Source	Ti	Zr	V	Cr	Mn	Co	Ni	Ce	Al	e/a	B/A
Ce0	**Design**	**12.0**	**22.8**	**10.0**	**7.5**	**8.1**	**7.0**	**32.2**	**0.0**	**0.4**	**6.771**	**1.87**
	ICP	11.9	22.9	10.0	7.5	8.0	7.1	32.2	0.0	0.4	6.773	1.87
Ce1	**Design**	**12.0**	**21.8**	**10.0**	**7.5**	**8.1**	**7.0**	**32.2**	**1.0**	**0.4**	**6.771**	**1.87**
	ICP	11.9	21.3	10.3	7.7	8.0	7.0	32.3	1.0	0.5	6.780	1.92
Ce2	**Design**	**12.0**	**20.8**	**10.0**	**7.5**	**8.1**	**7.0**	**32.2**	**2.0**	**0.4**	**6.771**	**1.87**
	ICP	12.0	20.6	10.3	6.8	8.2	7.1	32.6	2.0	0.4	6.792	1.89
Ce3	**Design**	**12.0**	**19.8**	**10.0**	**7.5**	**8.1**	**7.0**	**32.2**	**3.0**	**0.4**	**6.771**	**1.87**
	ICP	12.0	19.4	10.2	7.5	7.8	7.1	32.6	3.0	0.4	6.793	1.91
Ce4	**Design**	**12.0**	**18.8**	**10.0**	**7.5**	**8.1**	**7.0**	**32.2**	**4.0**	**0.4**	**6.771**	**1.87**
	ICP	11.9	18.4	10.2	7.6	8.0	7.0	32.5	3.9	0.5	6.789	1.92
Ce5	**Design**	**12.0**	**17.8**	**10.0**	**7.5**	**8.1**	**7.0**	**32.2**	**5.0**	**0.4**	**6.771**	**1.87**
	ICP	12.1	17.6	10.3	7.6	7.2	7.1	32.8	4.9	0.4	6.790	1.89

4.1. X-Ray Diffraction Analysis

XRD analysis was used to study the microstructures of the alloys. The obtained XRD patterns are shown in Figure 2 with four phases identified: C14, C15, CeNi, and TiNi. Full XRD pattern fitting was performed using the Rietveld refinement and Jade 9 Software to obtain the lattice parameters and crystallite size of the main C14 phase and the abundances of all four phases, and the results are summarized in Table 3. With the increase in Ce-content (and the corresponding decrease in Zr-content) in the alloy, both lattice constants a and c in the C14 phase decrease, and the a/c ratio increases slightly (Figure 3). Lattice constant a of the C15 phase follows the same trend as the one in the C14 phase, and both decreasing trends are caused by the reduction in Zr-content (second largest among all constituent elements) and zero or close to zero solubility of Ce (largest among all constituent elements) in the C14 and C15 phases as the Ce-content in the alloy design increases. Un-shifted positions of the major peaks for the TiNi and CeNi phases, shown in Figure 2, indicate that the unit cell sizes of these two phases stay the same as the Ce-content increases in the alloy design, which can be explained by the relatively unchanged TiNi and CeNi phase compositions as revealed by the EDS analysis (Section 4.2). Crystallite size of the main C14 phase first increases with the initial introduction of Ce in the alloy but then decreases with further increases in Ce-content, and this trend is very similar to that observed in the La-substitution study [17]. One possible explanation for the C14 crystallize size evolution is by the phase abundance of the non-Laves secondary phase. The higher abundance of non-Laves phase reduces the crystallization time of the matrix phase and consequently reduces the crystallite size of the matrix phase. In the beginning, when the CeNi phase abundance is small, the C14 crystallite size increases due to the decrease in the TiNi abundance. After more Ce is added, the CeNi phase abundance started to increase and caused a decrease in the C14 crystallite size.

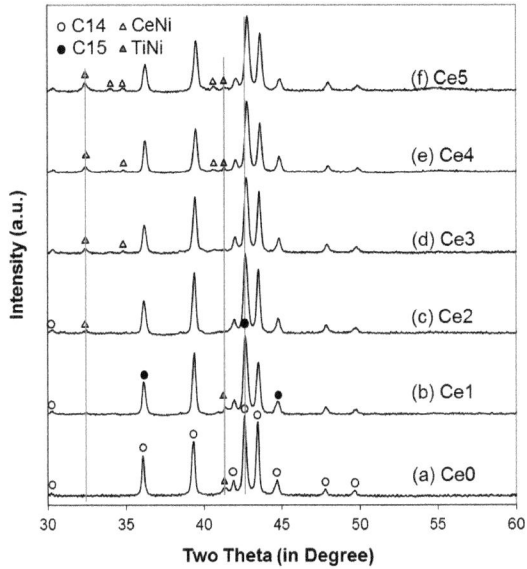

Figure 2. X-ray diffraction (XRD) patterns from alloys: (**a**) Ce0; (**b**) Ce1; (**c**) Ce2; (**d**) Ce3; (**e**) Ce4; and (**f**) Ce5. Vertical lines show the shift exists for the C14 peak but not for the CeNi or TiNi phases.

Table 3. Lattice constants a and c, a/c ratio, unit cell volume (V_{C14}), full-width at half-maximum (FWHM) for the (103) peak, and crystallite size for the C14 phase, lattice constant a for the C15 phase, and phase abundances in wt% calculated from the XRD analysis.

Structural Properties	Ce0	Ce1	Ce2	Ce3	Ce4	Ce5
a, C14 (Å)	4.9739	4.9703	4.9678	4.9616	4.9566	4.9542
c, C14 (Å)	8.1134	8.1067	8.1018	8.0905	8.0824	8.0763
a/c, C14 (Å)	0.61305	0.61311	0.61317	0.61326	0.61326	0.61342
V_{C14} (Å³)	173.83	173.44	173.16	172.48	171.96	171.67
FWHM C14 (103)	0.237	0.216	0.217	0.23	0.245	0.255
C14 crystallite size (Å)	482	554	551	503	458	434
a, C15 (Å)	7.0121	7.003	6.9973	6.9886	6.989	6.9831
C14 abundance (%)	85.4	79.9	78.8	79.9	73.2	78.1
C15 abundance (%)	11.2	16.9	17.5	16	20.9	13.2
TiNi abundance (%)	3.4	2.9	0.0	0.0	0.8	1.2
CeNi abundance (%)	0.0	0.3	3.7	4.1	5.1	7.5

Evolutions of the TiNi and RENi (where RE = Y, La, Ce, and Nd) phase abundances with increasing RE-content in the alloy design are plotted in Figure 4

241

and demonstrate that larger amounts of Ce (>1 at%) is very effective in suppressing the TiNi phase (beneficial to low-temperature performance [7,8]) and promoting the RENi phase (detrimental to low-temperature performance in the case of YNi [7,8]). Furthermore, the RENi phase abundances in various RE-substituted alloys do not correlate very well with the heats of formation or melting temperatures of the corresponding RENi phases.

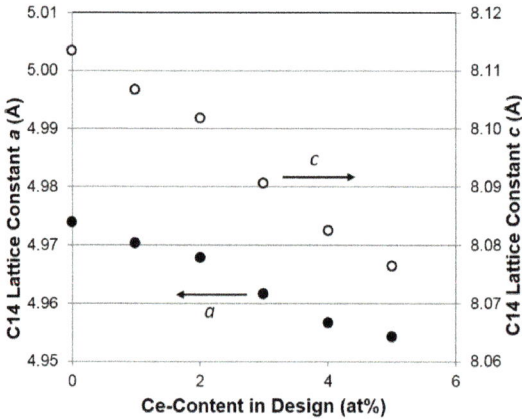

Figure 3. Evolutions of the C14 lattice constants *a* and *c* with increasing Ce-content in the design.

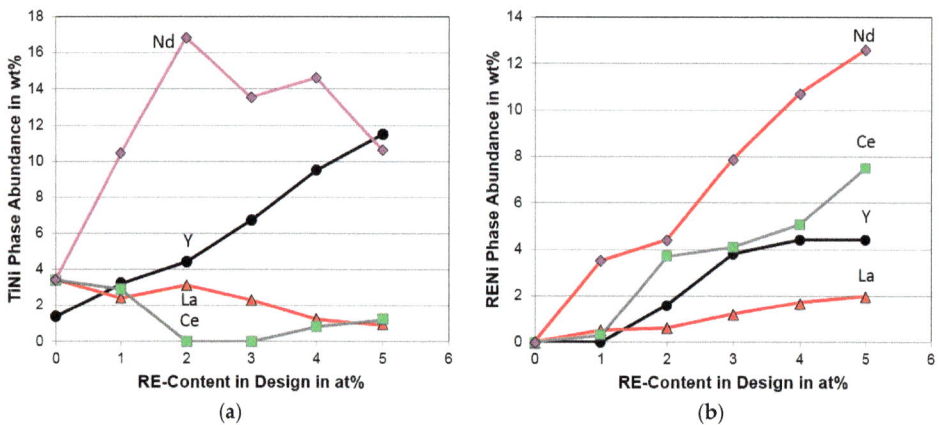

Figure 4. Evolutions of (**a**) TiNi and (**b**) RENi phase abundances with increasing RE-content in the design.

4.2. Scanning Electron Microscope/Energy Dispersive Spectroscopy Analysis

Both SEM secondary electron images (SEI) and back-scattering electron images (BEI) were taken from each alloy in this study. While the former carry surface topological information, the latter demonstrate both topological and compositional information and are shown in Figure 5 for comparison. Compositions of several representative areas (identified by Roman numerals) in Figure 5 were studied by EDS, and the results are summarized in Table 4. The brightest region in each micrograph belongs to Ce metal (Figure 5a-1 in alloy Ce1) or CeNi (Figure 5b-1,c-1,d-1,e-1 in Alloys Ce2–Ce5, respectively). These observations are in agreement with the close-to-zero CeNi-content in alloy Ce1 revealed by the XRD analysis. Areas with the second brightest contrast (Figure 5a-2,b-2,c-2,d-2,e-2) are identified as a Zr-rich phase with a B/A ratio very close to the theoretical stoichiometry of Zr_7Ni_{10} (1.43) by taking into consideration that V resides in the A-site of the crystal [18]. Solubilities of V, Co, Mn, and Cr in the Zr_7Ni_{10} phase are much lower than those in the Laves phases. Moreover, the B/A ratio of the TiNi phase (Figure 5a-5,b-5,c-5,d-5,e-5) is always greater than 1 even with the assignment of V in the A-site [16–18]. The Ti-Ni binary phase diagram shows a solubility range of TiNi on the Ni-rich side at higher temperatures [38]. Therefore, a higher B/A ratio for the TiNi phase is expected due to the quick cooling nature of arc melting technique. The main phase is composed of two regions with very close contrasts. The slightly brighter region (Figure 5a-3,b-3,c-3,d-3,e-3) has a smaller B/A ratio and an e/a value higher than the C14/C15 threshold of 6.83 [54], and it was consequently assigned to the C15 phase. The other phase (Figure 5a-4,b-4,c-4,d-4,e-4) with an e/a value lower than the C14/C15 threshold was assigned to the C14 phase. Since the solubility of Ce, same as those of the other RE elements, in the AB_2 phase is almost zero, the lattice constants of the C14 and C15 phases do not increase with increasing overall Ce-content; on the contrary, they decrease due to the decrease in Zr/Ti ratio (Zr is larger than Ti) in both the C14 and C15 phases as seen in Table 4. Upon a closer examination of the C14 and C15 phase compositions, we found that the main differences occur in the V- and Ni-contents. More specifically, the Ni/V ratio in the C14 phase is lower compared to that in the C15 phase (a lower Ni-content and a higher V-content in the C14 phase compared to those in the C15 phase), which results in a lower e/a in the C14 phase. In addition, the C14 phase is predicted to have a larger H-storage capacity due to its higher V-content but also exhibit a lower electrochemical high-rate dischargeability (HRD) due to its lower Ni-content according to our previous findings from a phase contribution study on the AB_2 MH alloys [56]. Finally, phase distribution analysis reveals the following solidification sequence: AB_2 phase first (M.P. $\approx$ 1450 °C), followed by TiNi (1310 °C), Zr_7Ni_{10} (1158 °C), and finally CeNi (680 °C) as the last solid formed from the liquid.

243

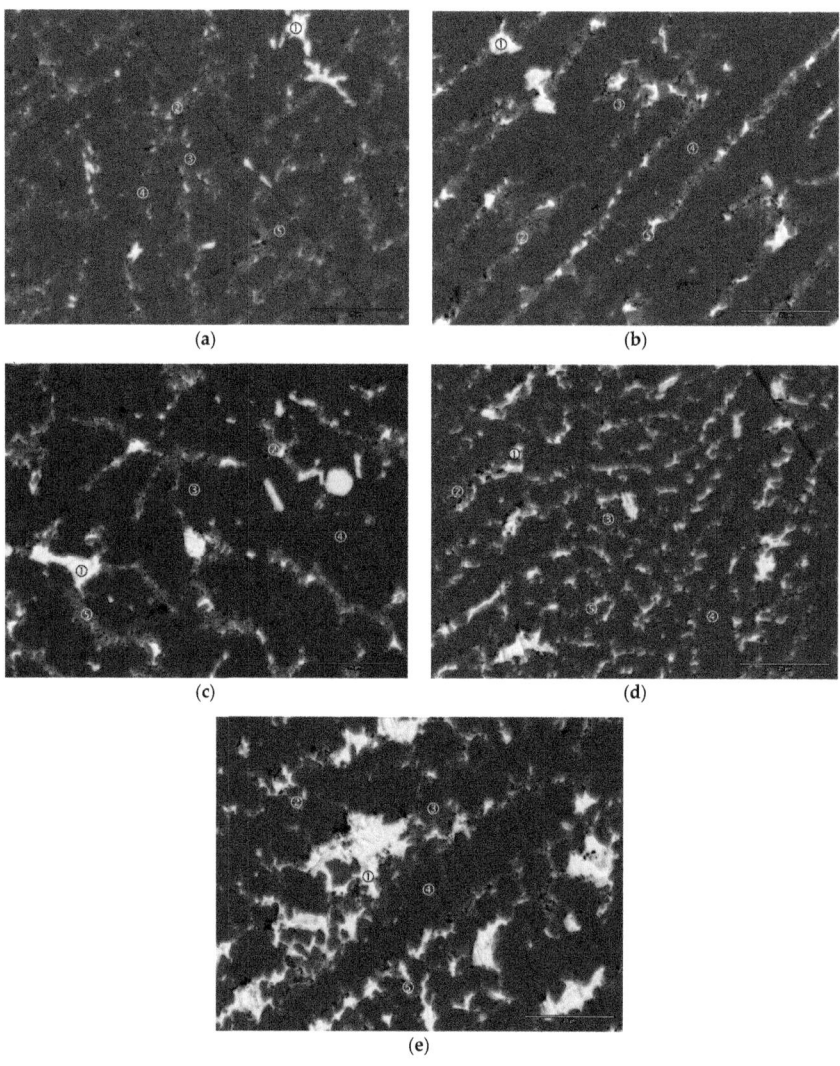

Figure 5. Scanning electron microscope (SEM)-back-scattering electron images (BEI) micrographs from alloys: (**a**) Ce1; (**b**) Ce2; (**c**) Ce3; (**d**) Ce4; and (**e**) Ce5. The bar at the lower right corner represents a scale of 25 μm.

Table 4. Summary of the energy dispersive spectroscopy (EDS) results from several selective spots in the SEM-BEI micrographs shown in Figure 5. All compositions are in at%. The main C14 phase is identified in **bold**.

Alloy	Location	Ti	Zr	V	Ni	Co	Mn	Cr	Al	Ce	B/A	e/a	Phase
	Figure 5a-1	0.9	3.1	1.7	3.8	1.0	1.0	0.0	0.1	88.4	0.08	3.44	Ce
	Figure 5a-2	13.5	26.8	0.7	54.8	1.6	1.2	0.3	0.3	0.8	1.39	7.41	Zr_7Ni_{10}
Ce1	Figure 5a-3	11.9	22.8	6.9	41.7	5.5	6.3	3.9	0.6	0.4	1.85	7.10	C15
	Figure 5a-4	**10.6**	**22.3**	**12.4**	**28.4**	**7.7**	**8.9**	**9.3**	**0.5**	**0.0**	**2.04**	**6.67**	**C14**
	Figure 5a-5	24.7	17.2	1.9	44.2	6.0	2.9	1.1	0.5	1.6	1.20	7.06	TiNi
	Figure 5b-1	2.7	1.8	1.2	41.8	0.8	1.0	0.0	0.2	50.5	0.82	6.08	CeNi
	Figure 5b-2	13.4	27.0	0.6	55.1	1.6	1.1	0.3	0.3	0.6	1.40	7.42	Zr_7Ni_{10}
Ce2	Figure 5b-3	11.8	22.8	6.9	41.0	5.7	7.2	3.6	0.6	0.3	1.86	7.09	C15
	Figure 5b-4	**10.4**	**22.2**	**13.1**	**26.6**	**7.8**	**9.6**	**9.9**	**0.4**	**0.0**	**2.07**	**6.60**	**C14**
	Figure 5b-5	28.8	14.0	1.5	41.9	7.4	3.1	0.7	0.5	2.1	1.16	6.98	TiNi
	Figure 5c-1	0.4	0.4	0.7	49.5	0.3	0.5	0.0	0.3	47.9	1.05	6.53	CeNi
	Figure 5c-2	13.6	26.0	0.5	55.8	1.4	1.0	0.2	0.3	1.1	1.42	7.44	Zr_7Ni_{10}
Ce3	Figure 5c-3	12.1	22.6	6.8	41.3	5.7	6.9	3.7	0.6	0.3	1.86	7.10	C15
	Figure 5c-4	**11.9**	**21.3**	**13.1**	**26.0**	**8.0**	**9.2**	**9.8**	**0.5**	**0.2**	**1.99**	**6.56**	**C14**
	Figure 5c-5	25.7	13.3	1.2	44.3	6.5	3.4	0.5	1.0	4.1	1.26	7.06	TiNi
	Figure 5d-1	1.1	1.3	1.0	49.0	0.7	0.9	0.0	0.0	46.0	1.07	6.55	CeNi
	Figure 5d-2	13.7	24.7	0.5	55.6	1.5	1.2	0.2	0.3	2.3	1.43	7.43	Zr_7Ni_{10}
Ce4	Figure 5d-3	12.3	22.4	6.6	41.2	5.6	7.1	3.8	0.6	0.4	1.85	7.10	C15
	Figure 5d-4	**11.7**	**20.5**	**13.7**	**26.5**	**8.2**	**9.2**	**9.8**	**0.4**	**0.1**	**2.10**	**6.61**	**C14**
	Figure 5d-5	28.3	12.4	2.4	41.8	7.3	3.7	1.5	0.7	2.0	1.22	7.02	TiNi
	Figure 5e-1	0.5	0.7	1.1	48.6	0.5	0.6	0.0	0.3	47.7	1.04	6.49	CeNi
	Figure 5e-2	13.6	24.0	0.6	55.4	1.6	1.2	0.3	0.3	3.0	1.43	7.42	Zr_7Ni_{10}
Ce5	Figure 5e-3	12.5	21.6	6.5	41.7	5.8	6.9	3.9	0.6	0.5	1.89	7.13	C15
	Figure 5e-4	**12.4**	**19.5**	**13.3**	**27.9**	**8.2**	**8.2**	**10.0**	**0.3**	**0.2**	**2.12**	**6.66**	**C14**
	Figure 5e-5	28.8	11.2	2.3	41.8	7.8	3.8	1.2	0.9	2.1	1.25	7.03	TiNi

4.3. Pressure–Concentration–Temperature Analysis

PCT isotherms were measured at 30, 60, and 90 °C, and results from the first two temperatures are shown in Figure 6. These isotherms are typical examples of multi-phase alloys with coherent synergetic effects among constituent phases [18]. Gaseous phase H-storage characteristics obtained from the PCT isotherms are summarized in Table 5. Both the maximum and reversible H-storage capacities first increase (maximized at alloy Ce1) and then decrease as the Ce-content in the alloy increases. Compared to the other RE additives, as shown in Figure 7a, alloy Ce1 has the second highest maximum H-storage capacity (alloy Y4 with 4 at% Y has the highest maximum H-storage capacity capacity). It is interesting to observe that alloys with 5 at% La, Ce, and Nd show very similar maximum H-storage capacities. Since no obvious pressure plateau can be identified, desorption pressures at 0.75 wt% H-storage capacity were compared and used in the calculations for hysteresis, heat of hydride formation (ΔH_h), and change in entropy (ΔS_h). For the Ce-containing alloys, both the 30 °C and 60 °C desorption pressure at 0.75 wt% H-storage first decrease (minimized at alloy Ce1) and then increase as the Ce-content in the alloy increases, and this trend is opposite to the trend observed in H-storage capacity. Desorption pressure is plotted against maximum H-storage capacity for all four RE-substituted series of alloys in Figure 8a, and a trend emerges where alloy with a higher metal–hydrogen (M–H) bond strength demonstrates both a lower plateau pressure and a higher H-storage capability, which was previously proposed [19]. This trend is also related to the evolution of the main phase unit cell volume as seen in Figure 8b, where a larger unit cell makes for a more stable hydride with a stronger M–H bond and increases the maximum H-storage capacity.

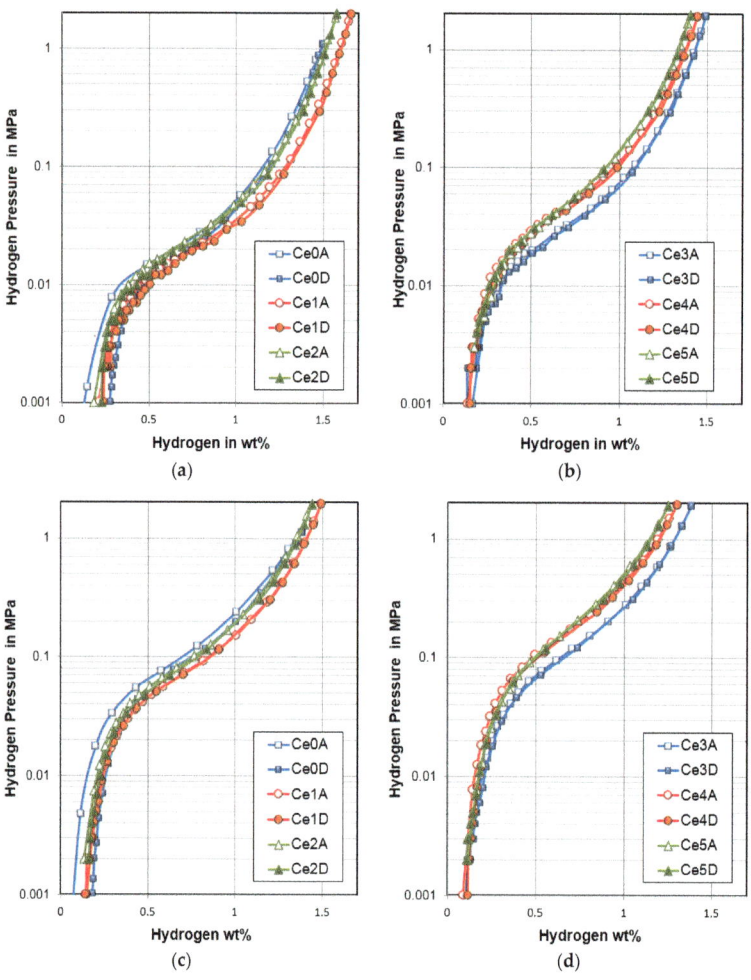

Figure 6. Pressure–concentration–temperature (PCT) isotherms measured at 30 °C for (**a**) alloys Ce0, Ce1, and Ce2; and (**b**) alloys Ce3, Ce4, and Ce5; and at 60 °C for (**c**) alloys Ce0, Ce1, and Ce2; and (**d**) alloys Ce3, Ce4, and Ce5.

Table 5. Summary of gaseous phase properties, including the maximum and reversible capacities, desorption pressure at 0.75 wt% H-storage capacity, slope factor (*SF*), hysteresis, and changes in enthalpy and entropy.

Gaseous Phase Properties	Ce0	Ce1	Ce2	Ce3	Ce4	Ce5
Maximum capacity @30 °C (wt%)	1.49	1.66	1.57	1.49	1.44	1.4
Reversible capacity @30 °C (wt%)	1.22	1.42	1.34	1.35	1.29	1.22
Desorption pressure @30 °C (MPa)	0.021	0.019	0.023	0.035	0.05	0.058
SF @30 °C (%)	78	72	76	76	78	73
Hysteresis @30 °C	0.21	0.05	0.13	0.08	0.08	0.02
Maximum capacity @60 °C (wt%)	1.38	1.49	1.44	1.38	1.3	1.25
Reversible capacity @60 °C (wt%)	1.2	1.35	1.28	1.27	1.19	1.14
Desorption pressure @60 °C (MPa)	0.1	0.08	0.096	0.13	0.2	0.23
SF @60 °C (%)	80	78	79	74	73	72
Hysteresis @60 °C	0.13	0.01	0.02	0.02	0.02	0.01
$-\Delta H_h$ (kJ·mol^{-1})	41.6	40.2	40	39.1	38.3	38.5
$-\Delta S_h$ (J·mol^{-1}·K)	125	119	120	120	121	122

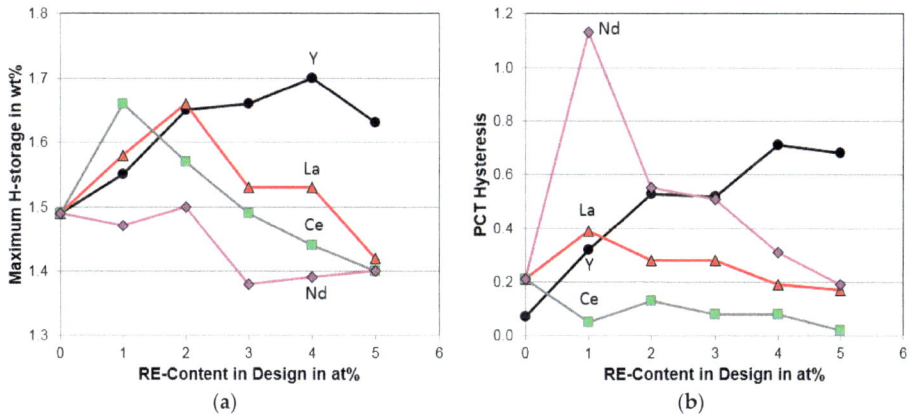

Figure 7. Evolutions of (**a**) maximum H-storage capacity and (**b**) PCT absorption–desorption hysteresis with increasing RE-content in the design.

Slope factor (*SF*) is defined as the ratio of the storage capacity between 0.01 MPa and 0.5 MPa to the total capacity in the desorption isotherm. This ratio can be used as an indicator for DOD in a multi-phase MH alloy system [6,20,57]. A large *SF* corresponds to a flatter plateau and lower DOD (fewer components or variations) in the system. The 30 °C *SF*s of this series of alloys are similar with the exception of alloy Ce1, which has the lowest *SF* and thus the highest DOD. Hysteresis of the PCT isotherm is defined as $\ln(P_a/P_d)$, where P_a and P_d are the absorption and desorption equilibrium pressures, respectively, at 0.75 wt% H-storage. Hysteresis is believed to be associated with the irreversible energy loss during plastic deformation of the hydride phase (β) in the alloy matrix (α) [58–60] and was found to be related to both the a/c ratio and pulverization rate of the alloy [61]. PCT hystereses at 30 °C of the Ce-containing alloys are smaller than that of the Ce-free alloy Ce0, indicating that the addition of the CeNi phase can reduce the stress built from the α to β transformation; however, such effect was not observed with the other RE-substitutions (Figure 7b). 60 °C H-storage characteristics are similar to those measured at 30 °C with the only exception of *SF*. *SF* measured at 60 °C decreases with increasing Ce-content in the alloy. One possible explanation is that the CeNi phase contributes more to DOD at a higher temperature with a higher participation rate (CeNi has a stronger M–H bond than AB$_2$ and will not release hydrogen at room temperature [52]).

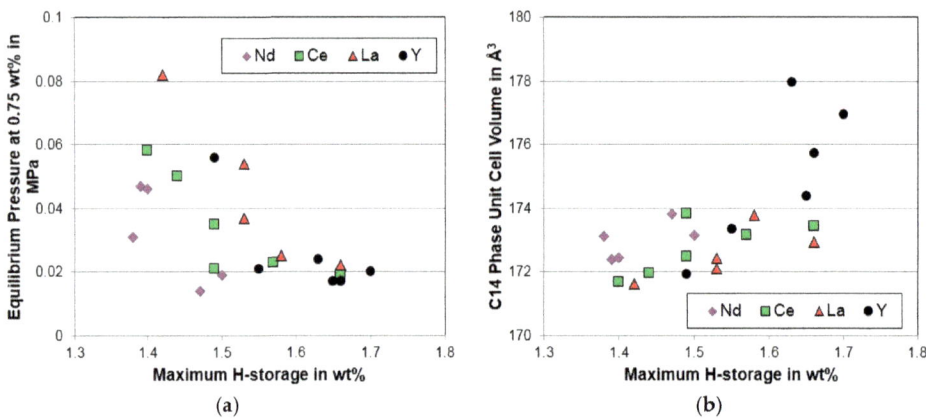

Figure 8. Correlations between the maximum H-storage and (**a**) equilibrium pressure at 0.75 wt% H-storage; and (**b**) C14 phase unit cell volume.

Desorption equilibrium pressures at 0.75 wt% H-storage at 30, 60, and 90 °C were used to estimate ΔH_h and ΔS_h with the equation:

$$\Delta G = \Delta H_h - T\Delta S_h = RT\ln P \qquad (2)$$

where R is the ideal gas constant, and T is the absolute temperature at which the measurement was performed. ΔH_h is negative due to the exothermic nature of the hydrogenation reaction. ΔH_h increases (becomes less negative) with increasing Ce-content in the alloy and can be correlated to the shrinking C14 unit cell volume, resulting in a less stable hydride (lower capacity and higher equilibrium pressure). ΔS_h is related to DOD in a hydride from a completely ordered solid (e.g., solid hydrogen) (-130.7 J·mol^{-1}·K^{-1} for H$_2$ (g) at 300 K and 0.1 MPa [27]). The Ce-containing alloys have slightly higher ΔS_h (less negative), suggesting that the

248

addition of the CeNi phase increases DOD of the hydride, which is another indication of the CeNi phase's participation in H-storage.

4.4. Electrochemical Analysis

Activation characteristics were studied by half-cell capacity measurements in a flooded configuration. Electrode was made from ground and sieved powder without any annealing or surface acid/alkaline treatments. Low-rate capacities (sum of discharge capacity of 50 mA·g^{-1} and discharge capacities from two additional pulls at 12 mA·g^{-1} and 4 mA·g^{-1}) and HRDs (ratio of discharge capacity at 50 mA·g^{-1} to that at 4 mA·g^{-1}) for the first 13 cycles for each alloy in this study are plotted in Figure 9a,b, respectively. Activations in both capacity and HRD were facilitated by the addition of Ce, and therefore alloys with higher Ce-content activated faster. The improvement in the activation by the addition of RE element is commonly seen and was attributed to the formation of RENi secondary phase with a higher solubility in KOH solution [8,16–18]. Both the low-rate discharge capacities and HRDs of the two base alloys (alloys Y0 and Ce0) and alloys with 1 at% RE are compared in Figure 10a,b, respectively. For the low-rate capacity performance, alloy Y1 exhibits the easiest activation due to the ease of oxidation (Y has the lowest oxidation potential in Table 1), and alloy La1 shows a degradation in capacity as a result of the thick and passive surface oxide formation [62]. For the HRD activation, which is more related to the new surface formation from cracking, alloys La1 and Ce1 show the fastest formations due to the relative high solubilities of lanthanum and cerium hydroxides in 30 wt% KOH electrolyte (pH $\approx$ 14.7 in Equation (1)). Discharge capacities measured at 50 mA·g^{-1} and 4 mA·g^{-1} and HRD at the third cycle are listed in Table 6. As the Ce-content in the alloy increases, capacities measured at both rates first increase and then decreases while HRD continues to increase. The highest discharge capacity obtained from this series of alloys is from alloy Ce1 (400 mAh·g^{-1}), which also has the highest gaseous phase storage capacity. Low-rate capacities and HRDs measured at the third cycle for alloys with different RE substitutions are compared in Figure 11a,b, respectively. Alloys Y5 and Ce1 show the highest and second highest low-rate capacities, respectively, and the La- and Ce-substitutions demonstrate the best HRD performances among all the RE-substitutions. Except for YNi, the RENi phases are beneficial for HRD with the trade-off occurring in capacity. By examining both the capacity and HRD results, it is concluded that alloy Ce1 has the best electrochemical properties among all the RE-substituted alloys.

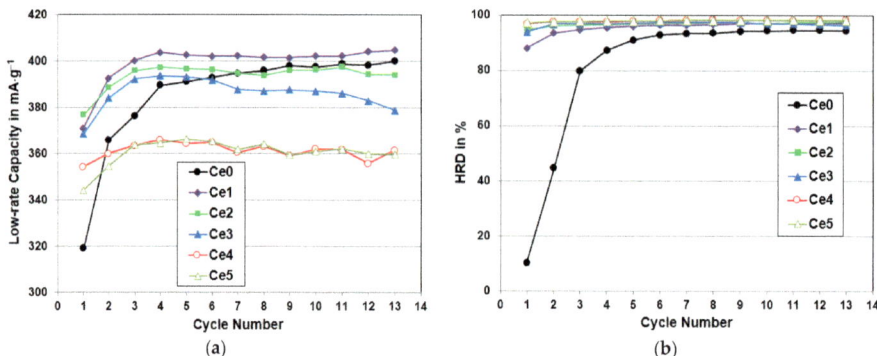

Figure 9. Activation characteristics shown by (**a**) discharge capacities with the lowest discharge current (4 mA· g^{-1}); and (**b**) high-rate dischargeabilities for the first 13 cycles for the alloys in this study.

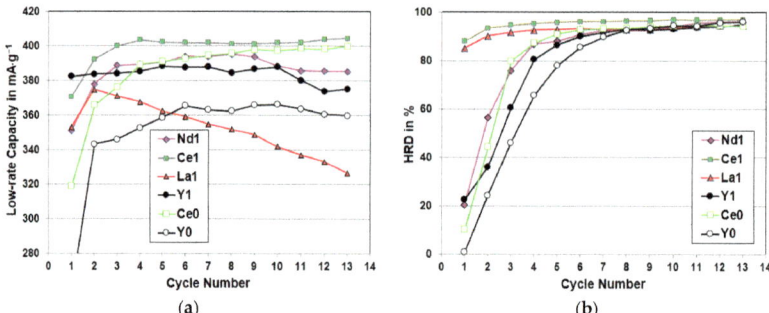

Figure 10. Comparisons of the activation behaviors in (**a**) capacity and (**b**) high-rate dischargeability (HRD) among the RE-free alloys (alloys Y0 and Ce) and 1 at% RE-containing alloys (alloys Y1, La1, Ce1, and Nd1).

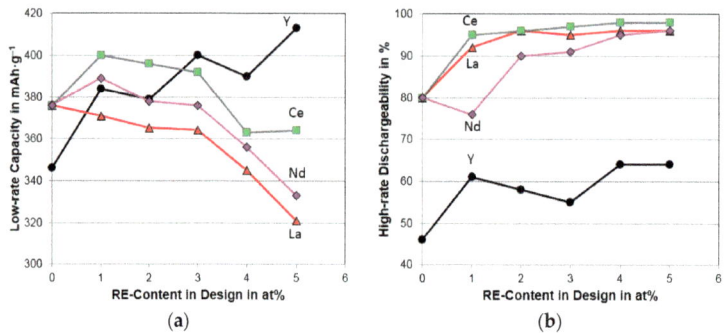

Figure 11. Evolutions of (**a**) low-rate capacity and (**b**) HRD at the third cycle with increasing RE-content in the design.

250

Table 6. Summary of electrochemical half-cell and magnetic measurements, including the capacities at the third cycle, HRD at the third cycle, number of cycles needed to achieve 92% HRD, bulk diffusion coefficient, surface exchange current, results from AC impedance measurements, saturated magnetic susceptibility, and applied field at half of the saturated magnetic susceptibility.

Electrochemical and Magnetics Properties	Ce0	Ce1	Ce2	Ce3	Ce4	Ce5
3rd cycle capacity @50 mA·g^{-1} (mAh·g^{-1})	300	378	381	380	355	355
3rd cycle capacity @4 mA·g^{-1} (mAh·g^{-1})	376	400	396	392	363	364
HRD (%)	80	95	96	97	98	98
Number of activation cycle(s) to reach 92% HRD	6	2	1	1	1	1
Diffusion coefficient, D (10^{-10} cm^2·s^{-1})	2.1	3.1	4.2	4.3	1.5	1.7
Surface reaction current, I_o (mA·g^{-1})	12.8	45.8	32	31.8	40	29
Charge-transfer resistance, R @$-40\,^{\circ}$C (Ω·g)	158.55	14.13	10.89	6.88	5.76	5.84
Double-layer capacitance, C @$-40\,^{\circ}$C (F·g^{-1})	0.18	0.66	1.28	1.73	1.84	2.15
RC product @$-40\,^{\circ}$C (s)	28.4	9.39	13.94	11.93	12.41	12.54
Saturated magnetic susceptibility, M_s (emu·g^{-1})	0.0353	0.197	0.529	0.627	0.534	0.733
Applied field at M.S. = $1/2\,M_s$, $H_{1/2}$ (kOe)	0.500	0.177	0.218	0.196	0.210	0.196

Since the MH alloy can be charged either in the gaseous phase by reacting with hydrogen gas or in the electrochemical environment with protons from water splitting that occurs with voltage, it is always interesting to compare the capacities obtained by these two different paths. After converting the unit of gaseous phase H-storage to that of electrochemical discharge capacity, 1 wt% hydrogen content was found to be equivalent to a discharge capacity of 268 mAh·g^{-1}. Our earlier studies indicated that both the high-rate (50 mA·g^{-1} for AB$_2$) and low-rate (4 mA·g^{-1} for AB$_2$) electrochemical capacities are between the boundaries set by the gaseous phase maximum and reversible H-storage capacities but do not necessarily exhibit strong correlations with the gaseous phase capacities [19]. In order to investigate the connection between the gaseous phase and electrochemical measurements, low-rate discharge capacity vs. maximum H-storage capacity and high-rate discharge capacity vs. reversible H-storage capacity for the RE-substituted AB$_2$ MH alloys are plotted in Figure 12a,b, respectively. From these two subfigures, linear correlations can be established in both cases, and the electrochemical discharge capacities occur between the boundaries set by the maximum and reversible H-storage capacities. It is worthwhile to point out that the Ce-containing alloys have the highest reversible H-storage and high-rate discharge capacities as shown in Figure 12b.

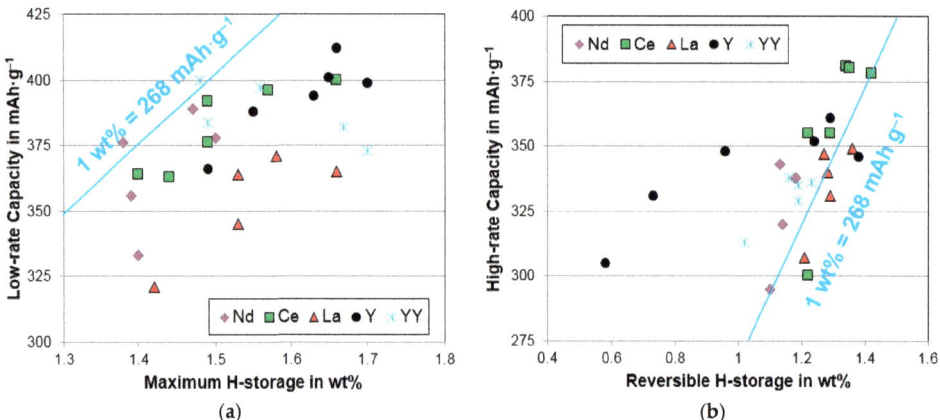

Figure 12. Comparisons of the electrochemical discharge capacities and gaseous phase H-storages of the RE-containing MH alloys by plotting: (**a**) low-rate discharge capacity vs. maximum H-storage capacity; and (**b**) high-rate discharge capacity vs. reversible H-storage capacity. Data set YY is from a previous study on the combination of Y-addition and stoichiometry in the AB_2 MH alloys [8].

Origins of the variations observed in HRD can be studied by the changes in bulk diffusion coefficient (D) and surface reaction current (I_o) measured electrochemically. Details of the measurements and data analysis can be found in our earlier publication [17]. Both D and I_o of the Ce-containing AB_2 MH alloys were measured, and the results are listed in Table 6. In general, both quantities increase and then decrease with increasing Ce-content in the alloy with the exception of I_o from alloy Ce4. Both trends are different from the monotonic increase seen in HRD. Therefore, other factors may also play important roles in affecting HRD for this series of alloys. Comparing among various RE-substitutions, D increases with increasing La-, Y-, and Ce-contents but remains low in the case of Nd-substitution (Figure 13a), and I_o increases with a small amount of Ce, La, and Y but decreases with a small amount of Nd (Figure 13b). Alloy Ce1 shows the highest I_o, which can be attributed to its unique microstructure that contains Ce metal instead of the CeNi phase as in the alloys with higher Ce-contents (Table 4). It appears that Ce metal is more reactive compared to CeNi with KOH electrolyte and thus contributes more to the surface reaction.

In order to study the roles of the TiNi and RENi phases in D and I_o, four sets of correlations are shown in Figures 14 and 15. In general, both the TiNi and RENi phase abundances show positive correlations to D with the exception of the Nd-containing alloys, which have the highest RENi- and TiNi-contents but also the lowest Ds (Figure 4). It is possible that the TiNi and RENi phases are beneficial to D only up to certain percentages (approximately 10 wt% and 5 wt% for TiNi and RENi, respectively). Correlations with the TiNi and RENi phase abundances for I_o

252

are less obvious, but the general decreasing trends with increasing RENi and TiNi phase abundances are observed (Figure 15). In an earlier effort to differentiate the increasing rates of the YNi and TiNi phases, an off-stoichiometric composition design was adopted [8]. In that study, the increase in YNi phase abundance was suppressed, and I_o increased with the increase in TiNi phase abundance (data sets with symbol * in Figures 14 and 15).

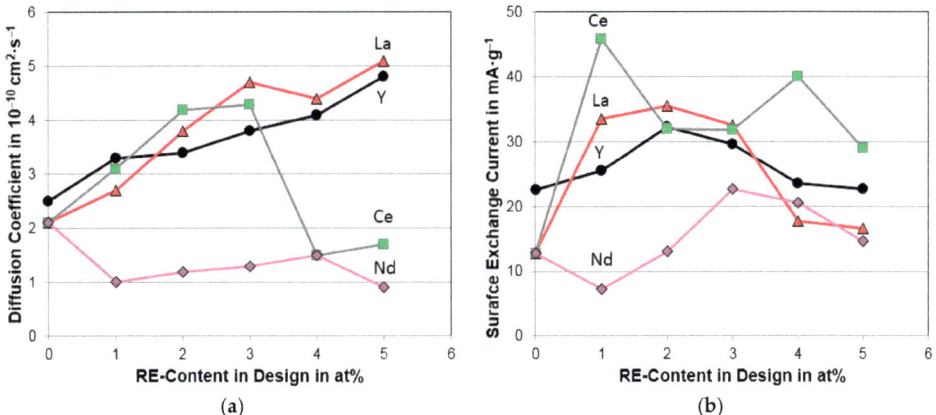

Figure 13. Evolutions of (**a**) bulk diffusion coefficient and (**b**) surface exchange current with increasing RE-content in the design.

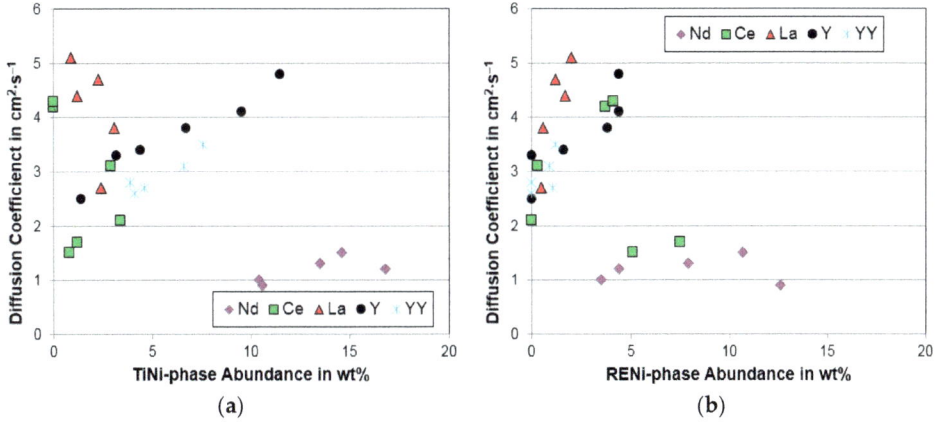

Figure 14. Correlations between diffusion coefficient and (**a**) TiNi phase abundance; and (**b**) RENi phase abundance.

Low-temperature performance of the alloys was studied by AC impedance measured at $-40\,^\circ$C. Surface charge-transfer resistance (R) and surface double-layer capacitance (C) of each alloy were calculated from curve fitting of the Cole-Cole plot and are listed in Table 6. Very similar to the case of Nd-substitution, the Ce-containing alloys have much lower R compared to the RE-free base alloy (Figure 16a). The main reason for the reduction in R in the Ce-containing alloys is the increase in surface area represented by the increase in C as the Ce-content increases in the alloy (Figure 16b). RC product, a quantity inversely proportional to the surface catalytic ability, from the Ce-containing alloys are higher (less catalytic) than those from the La- and Nd-containing alloys.

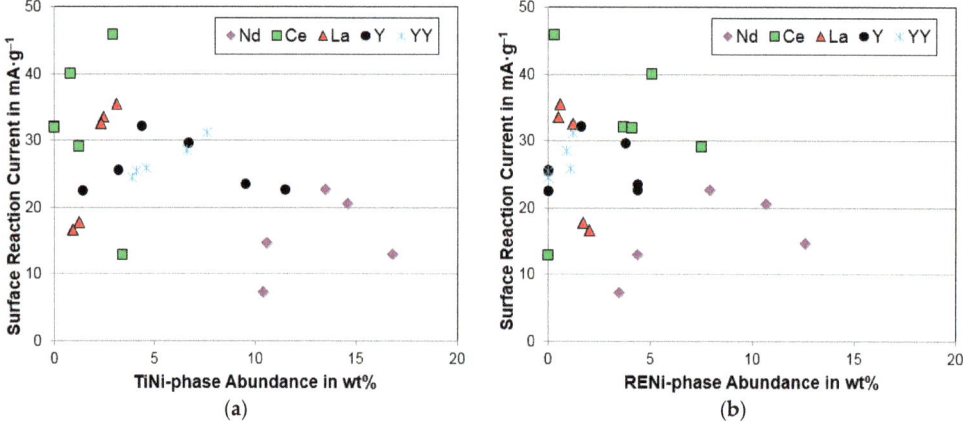

Figure 15. Correlations between surface reaction current and (**a**) TiNi phase abundance; and (**b**) RENi phase abundance.

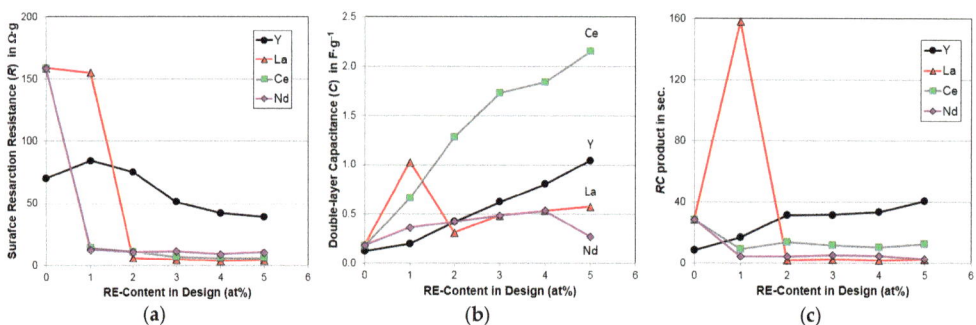

Figure 16. Evolutions of (**a**) surface charge-transfer resistance; (**b**) double-layer capacitance calculated from the AC impedance results measured at $-40\ ^\circ$C; and (**c**) their product with increasing RE-content in the design.

4.5. Magnetic Properties

In most MH alloys, the HRD characteristics are heavily affected by the density of metallic Ni-inclusion embedded in the surface oxide [63–66]. Thus far, the

Si-containing AB$_2$ MH alloys are the only type of alloy where the HRD performance is dominated by the clean oxide/metal interface but not the occurrence of metallic Ni [67]. Both the total saturated magnetic susceptibility (M_s, related to the total volume of metallic Ni-clusters) and applied field corresponding to half of M_s ($H_{1/2}$, inversely proportional to the average number of Ni atoms in a cluster) were calculated from the plot of measured M.S. versus applied magnetic field (as in [65]) and are listed in Table 6. As the Ce-content in the alloy increases, M_s increases, and $H_{1/2}$ drops precipitously and then remains approximately the same. The Ce/CeNi phase in the Ce-containing alloys promotes the formation of metallic Ni-clusters embedded in the surface oxide with a size around two and half times larger than those in the Ce-free base alloy. When these two properties (M_s and $H_{1/2}$) are compared among all the RE-substitutions, as shown in Figure 17, the Ce-doped alloys have the highest M_s values, which is in agreement of their high HRD (Figure 11b) and I_o (Figure 13b) values. The La-substitution shows the lowest M_s but still exibits comparable HRD to that of the Ce-substitution due to the unique clean interface between the alloy and surface catalytic Ni(OH)$_2$ layer [62]. All the Ce- and Nd-doped alloys have approximately the same size of metallic Ni-clusters that are larger than those in the undoped base alloy (Figure 17b), which is very different from the unchanged size of Ni-clusters observed in the La-doped alloys. The relatively high solubility of the La ion promotes the formation of a catalytic Ni(OH)$_2$ layer directly onto the metal surface instead of having the metallic Ni embedded in the surface oxide/hydroixde of Nd or Ce.

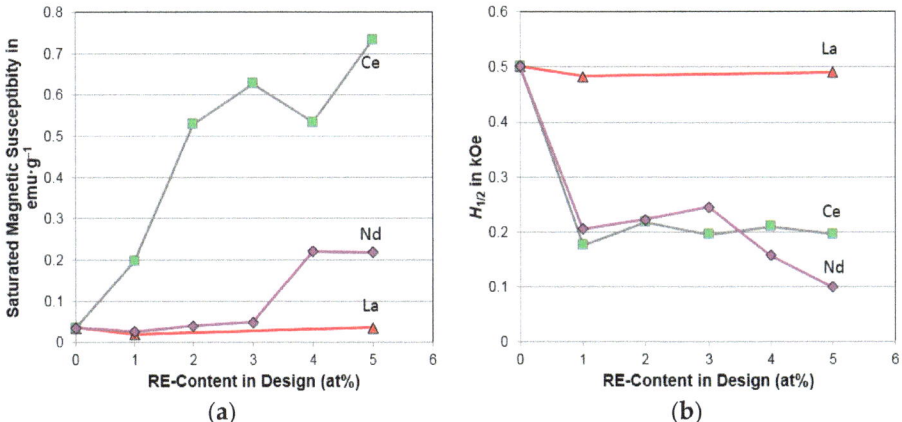

Figure 17. Evolutions of (**a**) saturated magnetic susceptibility and (**b**) strength of the applied field corresponding to half of the saturated magnetic susceptibility with increasing RE-content in the design.

5. Conclusions

We have analyzed and compared the effects of various RE-substitutions to the Zr-based AB$_2$ MH alloys on the structure, gaseous phase H-storage, and electrochemical properties. The additions of RE elements were found to facilitate activation and improve the HRD and low-temperature electrochemical performance. Among all of the substitutions, the Ce-substitution was found to be the most effective in reducing the PCT isotherm hysteresis (improving the cycle stability), increasing the electrochemical capacity, and increasing the total volume of metallic Ni-clusters embedded in the surface oxide, reactive surface area, and consequently surface reaction current. Overall, our observations indicate that a 1 at% Ce-substitution in the AB$_2$ MH alloy can improve the electrochemical properties and is a viable option for Ni/MH rechargeable batteries.

Acknowledgments: The authors would like to thank the following individuals from BASF-Ovonic for their help: Su Cronogue, Baoquan Huang, Diana F. Wong, David Pawlik, Allen Chan, and Ryan J. Blankenship.

Author Contributions: Kwo-Hsiung Young designed the experiments and analyzed the results. Taihei Ouchi prepared the alloy samples and performed the PCT and XRD analyses. Jean Nei prepared the electrode samples and conducted the magnetic measurements. Dhanashree Moghe assisted in data analysis and manuscript preparation.

Conflicts of Interest: The authors declare no conflict of interest.

Abbreviations

Ni/MH	Nickel/metal hydride
MH	Metal hydride
RE	Rare earth
ICP-OES	Inductively coupled plasma-optical emission spectrometer
XRD	X-ray diffractometer
SEM	Scanning electron microscope
EDS	Energy dispersive spectroscopy
PCT	Pressure–concentration–temperature
H-storage	Hydrogen storage
M.S.	Magnetic susceptibility
M.P.	Melting point
HCP	Hexagonal closest-packed
DHCP	Double-c hexagonal close-packed
e/a	Average electron density
DOD	Degree of disorder
V_{C14}	Unit cell volume of the C14 phase
FWHM	Full-width at half-maximum

SEI	Secondary electron image
BEI	Back-scattering electron image
HRD	High-rate dischargeability
ΔH_h	Heat of hydride formation
ΔS_h	Change in entropy
M–H	Metal–hydrogen
SF	Slope factor
P_a	Absorption equilibrium pressure at 0.75 wt% H-storage
P_d	Desorption equilibrium pressure at 0.75 wt% H-storage
β	Hydride phase
α	Metal phase or alloy matrix
R	Ideal gas constant
T	Absolute temperature
D	Bulk diffusion coefficient
I_o	Surface exchange current
R	Surface charge-transfer resistance
C	Surface double-layer capacitance
M_s	Saturated magnetic susceptibility
$H_{1/2}$	Applied magnetic field strength corresponding to half of saturated magnetic susceptibility

References

1. Wikipedia, the Free Encyclopedia. Nickel–Metal Hydride Battery. Available online: https://en.wikipedia.org/wiki/Nickel%E2%80%93metal_hydride_battery (accessed on 14 April 2016).

2. Zelinsky, M.; Koch, J.; Fetcenko, M. Heat Tolerant NiMH Batteries for Stationary Power. Available online: www.battcon.com/PapersFinal2010/ZelinskyPaper2010Final_12.pdf (accessed on 28 March 2016).

3. Zelinsky, M.; Koch, J. Batteries and Heat—A Recipe for Success? Available online: www.battcon.com/PapersFinal2013/16-Mike%20Zelinsky%20-%20Batteries%20and%20Heat.pdf (accessed on 28 March 2016).

4. Young, K.; Nei, J. The current status of hydrogen storage alloy development for electrochemical applications. *Materials* **2013**, *6*, 4574–4608.

5. Chang, S.; Young, K.; Nei, J.; Fierro, C. Reviews on the U.S. Patents regarding nickel/metal hydride batteries. *Batteries* **2016**, *2*.

6. Young, K.; Ouchi, T.; Koch, J.; Fetcenko, M.A. The role of Mn in C14 Laves phase multi-component alloys for NiMH battery application. *J. Alloys Compd.* **2009**, *477*, 749–758.

7. Young, K.; Reichman, B.; Fetcenko, M.A. Electrochemical properties of AB_2 metal hydride alloys measured at -40 °C. *J. Alloys Compd.* **2013**, *580*, S349–S353.

8. Young, K.; Wong, D.F.; Nei, J.; Reichman, B. Electrochemical properties of hypo-stoichiometric Y-doped AB$_2$ metal hydride alloys at ultra-low temperature. *J. Alloys Compd.* **2015**, *643*, 17–27.

9. Huang, B.; Young, K.; BASF-Ovonic. Private Communication, 2016.

10. Kim, S.; Lee, J.; Park, H. A study of the activation behavior of ZrCrNiLa metal hydride electrodes in alkaline solution. *J. Alloys Compd.* **1994**, *205*, 225–229.

11. Jung, J.; Lee, K.; Lee, J. The activation mechanism of Zr-based alloy electrodes. *J. Alloys Compd.* **1995**, *226*, 166–169.

12. Sun, D.; Latroche, M.; Percheron-Guégan, A. Effects of lanthanum or cerium on the equilibrium of ZrNi$_{1.2}$Mn$_{0.6}$V$_{0.2}$Cr$_{0.1}$ and its related hydrogenation properties. *J. Alloys Compd.* **1997**, *248*, 215–219.

13. Park, H.Y.; Cho, W.I.; Cho, B.W.; Lee, S.R.; Yun, K.S. Effect of fluorination on the lanthanum-doped AB$_2$-type metal hydride electrodes. *J. Power Sources* **2001**, *92*, 149–156.

14. Park, H.Y.; Chang, I.; Cho, W.I.; Cho, B.W.; Jang, H.; Lee, S.R.; Yun, K.S. Electrode characteristics of the Cr and La dopes AB$_2$-type hydrogen storage alloys. *Int. J. Hydrog. Energy* **2001**, *26*, 949–955.

15. Gao, M.; Miao, H.; Zhao, Y.; Liu, Y.; Pan, H. Effects of rare earth elements substitution for Ti on the structure and electrochemical properties of Fe-doped Ti-V-based hydrogen storage alloy. *J. Alloys Compd.* **2009**, *484*, 249–255.

16. Young, K.; Young, M.; Ouchi, T.; Reichman, B.; Fetcenko, M.A. Improvement in high-rate dischargeability, activation, and low-temperature performance in multi-phase AB$_2$ alloys by partial substitution of Zr with Y. *J. Power Sources* **2012**, *215*, 279–287.

17. Young, K.; Wong, D.F.; Ouchi, T.; Huang, B.; Reichman, B. Effects of La-addition to the structure, hydrogen storage, and electrochemical properties of C14 metal hydride alloys. *Electrochim. Acta* **2015**, *174*, 815–825.

18. Wong, D.F.; Young, K.; Nei, J.; Wang, L.; Ng, K.Y.S. Effects of Nd-addition on the structural, hydrogen storage, and electrochemical properties of C14 metal hydride alloys. *J. Alloys Compd.* **2015**, *647*, 507–518.

19. Young, K.; Fetcenko, M.A.; Li, F.; Ouchi, T. Structural, thermodynamics, and electrochemical properties of Ti$_x$Zr$_{1-x}$(VNiCrMnCoAl)$_2$ C14 Laves phase alloys. *J. Alloys Compd.* **2008**, *464*, 238–247.

20. Young, K.; Fetcenko, M.A.; Koch, J.; Morii, K.; Shimizu, T. Studies of Sn, Co, Al, and Fe additives in C14/C15 Laves alloys for NiMH battery application by orthogonal arrays. *J. Alloy. Compd.* **2009**, *486*, 559–569.

21. Wikipedia, the Free Encyclopedia. Rare Earth Element. Available online: https://en.wikipedia.org/wiki/Rare_earth_element (accessed on 14 April 2016).

22. Yaroslavtsev, A.; Menushenkov, A.; Chernikov, R.; Clementyev, E.; Lazukov, V.; Zubavichus, Y.; Veligzhanin, A.; Efremova, N.; Gribanov, A.; Kuchin, A. Ce valence in intermetallic compounds by means of XANES spectroscopy. *Z. Kristallogr.* **2010**, *225*, 482–486.

23. Gschneidner, K.A. Rare-Earth Element. Available online: http://www.britannica.com/science/rare-earth-element (accessed on 14 April 2016).

24. Young, K.; Wong, D.F.; Wang, L.; Nei, J.; Ouchi, T.; Yasuoka, S. Mn in misch-metal based superlattice metal hydride alloy—Part 1 Structural, hydrogen storage and electrochemical properties. *J. Power Sources* **2015**, *277*, 426–432.

25. Yasuoka, S.; Ishida, J. Effects of cerium (Ce) on the hydrogen absorption-desorption characteristics of RE-Mg-Ni hydrogen absorbing alloy. *J. Power Sources* **2016**, submitted for publication.

26. Young, K.; Chao, B.; Liu, Y.; Nei, J. Microstructures of the oxides on the activated AB_2 and AB_5 metal hydride alloys surface. *J. Alloys Compd.* **2014**, *606*, 97–104.

27. Lide, D.R. *CRC Handbook of Chemistry and Physics*, 74th ed.; CRC Press Inc.: Boca Raton, FL, USA, 1993; pp. 6–22.

28. Long, K.R.; Van Gosen, B.S.; Foley, N.K.; Cordier, D. *The Principal Rare Earth Elements Deposits of the United States—A Summary of Domestic Deposits and a Global Perspective*; Scientific Investigations Report 2010-5220; U.S. Geological Survey: Reston, VA, USA, 2010.

29. Gakkai, N.K. *Hi Kagaku Ryouronteki Kinzoku Kagobutu*; Maruzen: Tokyo, Japan, 1975; p. 296.

30. Griessen, R.; Riesterer, T. Heat of Formation Models. In *Hydrogen in Intermetallic Compounds I.*; Schlapbach, L., Ed.; Springer-Verlag Berlin Heidelberg: Berlin, Germany, 1988.

31. Courcot, E.; Rebillat, F.; Teyssandier, F. Thermochemical Stability of Rare Earth Sesquioxides under a Most Environment. In *Design, Development, and Applications of Engineering Ceramics and Composites: Ceramic Transaction*; Singh, D., Zhu, D., Zhou, Y., Eds.; John Wiley & Sons: New York, NY, USA, 2010; Volume 215.

32. Cordfunke, E.H.P.; Konings, R.J.M.; Ouweltjes, W. The standard enthalpies of formation of hydroxides IV. La(OH)$_3$ and LaOOH. *J. Chem. Thermodyn.* **1990**, *22*, 449–452.

33. Fang, Z.; Thanthiriwatte, K.S.; Dixon, D.A.; Andrews, L.; Wang, X. Properties of cerium hydroxides from matrix infrared spectra and electronic structure calculations. *Inorg. Chem.* **2015**, *55*, 1702–1714.

34. Ekberg, C.; Brown, P.L. *Hydrolysis of Metal Ions*; Wiley-VCH Verlag GmbH & Co: Weinheim, Germany, 2016; p. 265.

35. Merli, L.; Fuger, J. Thermochemistry of a few neptunium and neodymium oxides and hydroxides. *Radiochim. Acta* **1994**, *66–67*, 109–113.

36. Pourbaix, M. *Atlas of Electrochemical Equilibrium in Aqueous Solutions*; National Association of Corrosion Engineers: Houston, TX, USA, 1974.

37. *Powder Diffraction File (PDF) Database*; MSDS No. 00-014-0481; International Centre for Diffraction Data: Newtown Square, PA, USA, 2011.

38. Massalski, T.B. *Binary Alloy Phase Diagrams*; ASM International: Materials Park, OH, USA, 1990.

39. Du, Z.; Yang, L.; Ling, G. Thermodynamic assessment of the Ce-Ni system. *J. Alloys Compd.* **2004**, *375*, 186–190.

40. Boer, F.R.; Boom, R.; Mattens, W.C.M.; Miedema, A.R.; Niessen, A.K. *Cohesion in Metals Transition Metal Alloys*; North-Holland: Amsterdam, The Netherlands, 1988; p. 302.

41. Chen, N.; Lu, W.; Yang, J.; Li, G. *Support Vector Machine in Chemistry*; World Scientific Pub. Co.: Singapore, 2004; p. 148.
42. Colinet, C.; Pasturel, A. Enthalpies of formation of RNi$_5$ compounds. *Inorg. Chim. Acta* **1984**, *94*, 66–67.
43. Hussain, A.; Ende, A.V.; Kim, J.; Jung, I. Critical thermodynamic evaluation and optimization of the Co-Nd, Cu-Nd and Nd-Ni system. *Calphad* **2013**, *41*, 26–41.
44. Osumi, Y. *Suiso Kyuzou Goukin*; Agune Co. Ltd.: Tokyo, Japan, 1993; p. 54.
45. Lundin, C.E.; Lynch, F.E.; Magg, C.B. A correlation between the interstitial hole sizes in intermetallic compounds and the thermodynamic properties of the hydrides formed from those compounds. *J. Less Common Met.* **1977**, *56*, 19–37.
46. Luo, Q.; Chen, S.; Zhang, J.; Li, L.; Chou, K.; Li, Q. Experimental investigation and thermodynamic assessment of Nd-H and Nd-Ni-H systems. *Calphad* **2015**, *51*, 282–291.
47. Thoma, D.J.; Perepezko, J.H. A geometric analysis of solubility ranges in Laves phases. *J. Alloys Compd.* **1995**, *224*, 330–341.
48. Crystal Structures of CrB and FeB. Available online: http://www.hardmaterials.de/html/crb__feb.html (accessed on 14 April 2016).
49. Matar, S.F.; Nakhl, M.; Alam, A.F.A.; Ouaini, N.; Chevalier, B. YNi and its hydrides: Phase stabilities, electronic structures and chemical bonding properties from first principles. *Chem. Phys.* **2010**, *377*, 109–114.
50. Maeland, A.; Andresen, A.F.; Videm, K. Hydrides of lanthanum-nickel compounds. *J. Less Common. Met.* **1976**, *45*, 347–350.
51. Buschow, K.H.J.; Bouten, P.C.P.; Miedema, A.R. Hydrides formed from inter metallic compounds of two transition metals: A special class of ternary alloy. *Rep. Prog. Phys.* **1982**, *45*, 939–1039.
52. Bobet, J.L.; Grigorova, E.; Chevalier, B.; Khrussanova, M.; Peshev, P. Hydrogenation of CeNi: Hydride formation, structure and magnetic properties. *Intermetallics* **2006**, *14*, 208–212.
53. Kolomiets, A.V.; Miliyanchuk, K.; Galadzhun, Y.; Havela, L.; Vejpravova, J. PrNi and CeNi hydrides with extremely high H-density. *J. Alloys Compd.* **2005**, *402*, 95–97.
54. Nei, J.; Young, K.; Salley, S.O.; Ng, K.Y.S. Determination of C14/C15 phase abundance in Laves phase alloys. *Mater. Chem. Phys.* **2012**, *136*, 520–527.
55. Young, K. Stoichiometry in Inter-Metallic Compounds for Hydrogen Storage Applications. In *Stoichiometry and Materials Science—When Numbers Matter*; Innocenti, A., Kamarulzaman, N., Eds.; Intech: Rijeka, Croatia, 2012.
56. Young, K.; Nei, J.; Ouchi, T.; Fetcenko, M.A. Phase abundances in AB$_2$ metal hydride alloys and their correlations to various properties. *J. Alloys Compd.* **2011**, *509*, 2277–2284.
57. Young, K.; Ouchi, T.; Lin, X.; Reichman, B. Effects of Zn-addition to C14 metal hydride alloys and comparisons to Si, Fe, Cu, Y, and Mo-additives. *J. Alloys Compd.* **2016**, *655*, 50–59.
58. Scholtus, N.A.; Hall, W.K. Hysteresis in the palladium-hydrogen system. *J. Chem. Phys.* **1963**, *39*, 868–870.

59. Makenas, B.J.; Birnbaum, H.K. Phase changes in the niobium-hydrogen system I: Accommodation effects during hydride precipitation. *Acta Metall.* **1980**, *28*, 979–988.

60. Balasubramaniam, R. Accommodation effects during room temperature hydrogen transformations in the niobium-hydrogen system. *Acta Metall. Mater.* **1993**, *41*, 3341–3349.

61. Young, K.; Ouchi, T.; Fetcenko, M.A. Pressure-composition-temperature hysteresis in C14 Laves phase alloys: Part 1. Simple ternary alloys. *J. Alloys Compd.* **2009**, *480*, 428–433.

62. Young, K.; Chao Pawlik, D.; Shen, H.T. Transmission electron microscope studies in the surface oxide on the La-containing AB_2 metal hydride alloy. *J. Alloys Compd.* **2016**, *672*, 356–365.

63. Züttle, A.; Meli, F. Electrochemical and surface properties of $Zr(V_xNi_{1-x})_2$ alloys as hydrogen absorbing electrodes in alkaline electrolyte. *J. Alloys Compd.* **1994**, *203*, 235–241.

64. Young, K.; Huang, B.; Regmi, R.K.; Lawes, G.; Liu, Y. Comparisons of metallic clusters imbedded in the surface of AB_2, AB_5, and A_2B_7 alloys. *J. Alloys Compd.* **2010**, *506*, 831–840.

65. Chang, S.; Young, K.; Ouchi, T.; Meng, T.; Nei, J.; Wu, X. Studies on incorporation of Mg in Zr-based AB_2 metal hydride alloys. *Batteries* **2016**, *2*.

66. Young, K.; Wong, D.F.; Nei, J. Effects of vanadium/nickel contents in Laves phase-related body-centered-cubic solid solution metal hydride alloys. *Batteries* **2015**, *1*, 34–53.

67. Young, K.; Chao, B.; Nei, J. Microstructures of the activated Si-containing AB_2 metal hydride alloy surface by transmission electron microscope. *Batteries* **2016**, *2*.

Effects of Vanadium/Nickel Contents in Laves Phase-Related Body-Centered-Cubic Solid Solution Metal Hydride Alloys

Kwo-hsiung Young, Diana F. Wong and Jean Nei

Abstract: Structural, gaseous phase hydrogen storage, and electrochemical properties of a series of annealed (900 °C for 12 h) Laves phase-related body-centered-cubic (BCC) solid solution metal hydride (MH) alloys with vanadium/nickel (V/Ni) contents ranging from 44/18.5 to 28/34.5 were studied. As the average Ni-content increases, C14 phase evolves into the C15 phase and a new σ-VNi phase emerges; lattice constants in BCC, C14, and TiNi phase all decrease; the main plateau pressure increases; both gaseous phase and electrochemical hydrogen storage capacities decrease; the pressure-concentration-temperature (PCT) absorption/desorption hysteresis decreases; both high-rate dischargeability (HRD) and bulk hydrogen diffusivity increase and then decrease; and the surface reaction current decreases. There is a capacity-rate tradeoff with the change in V/Ni content. Alloys with relatively lower Ni-content show higher capacities but inferior high-rate performance compared to commercially available AB_5 MH alloy. Increasing the Ni-content in this BCC-based multi-phase alloy can improve the high-rate capability over AB_5 alloy but with lower discharge capacities. The inferior surface reaction current in these alloys, compared to AB_5, may be due to the smaller surface area, not the total volume, of the Ni clusters embedded in the surface oxide layer of the activated alloys.

Reprinted from *Batteries*. Cite as: Young, K.-h.; Wong, D.F.; Nei, J. Effects of Vanadium/Nickel Contents in Laves Phase-Related Body-Centered-Cubic Solid Solution Metal Hydride Alloys. *Batteries* **2015**, *1*, 34–53.

1. Introduction

"Laves phase-related body-centered-cubic (BCC) solid solution" is a family of metal hydride (MH) alloys composed mainly of a BCC phase and a Laves phase (mostly C14), which can be used as the negative electrode active material for nickel/metal hydride (Ni/MH) batteries [1,2]. Other minor phases, such as C15, TiNi, Ti_2Ni, and VNi can also be included. Both V and TiCr intermetallic alloys have the same BCC crystal structure, which enable a wide composition solubility range for BCC phase in the Ti-V-Cr ternary phase diagram [3]. Alloys with sole BCC structure can have a hydrogen storage capacity higher than 3.5 wt% [4,5], but the electrochemical performance is poor because of severe cracking and passivation on the surface [5]. Adding Ni into the alloy formula promotes the formation of a C14 phase [6]. The introduction of Laves phases promotes good absorption kinetics [7], easy formation due to its brittleness [8–10], and high surface catalytic activity through a synergetic effect [11]. Many works have been reported in this family of alloys that partially replace Ti, V, and Cr with other transition metals for improvements in both hydrogen storage [9,12–19] and electrochemical applications [20–23]. Our contributions to the field include the introduction of a high hydrogen pressure activation process [24], an optimization of annealing conditions (900 °C for 12 h) [25], and a study examining the contributions of the constituent elements in Laves phase-related BCC solid solutions. The contributions of the elements are summarized as: removal of Fe and decrease in V-content in exchange for higher Ni-content to improve both the electrochemical capacity and high-rate dischargeability (HRD); elimination of the C14 phase by removal of Zr contributing to a reduced discharge capacity, a prolonged activation period, and a less catalytic surface for electrochemical reaction; and the TiNi phase was also found to contribute positively to the bulk diffusion of hydrogen while hindering the surface electrochemical reaction. In the same study, an alloy (P17) achieved a full capacity of 390.8 mAh·g^{-1} at a small discharge rate of 4 mA·g^{-1}; however, the capacity decreases to 365.4 mAh·g^{-1} when the discharge rate was increased to 50 mAh·g^{-1} [26]. The capacity degradation with increased rate in P17 is worse than that in the conventional AB$_5$ MH alloys (from 336 mAh·g^{-1} to 329 mAh·g^{-1}) under the same test condition [27]. The HRD is very important for the high-power applications for Ni/MH battery, such as in hybrid electrical vehicles and power tools. Previously, Ni has been known to increase the HRD of the MH alloys [22,28–32]. In order to further improve the HRD performance, the vanadium/nickel content in alloy P17 was adjusted and the results are presented here.

2. Experimental Section

Ingot sample was prepared by an arc-melting, which includes a non-consumable tungsten electrode, a water-cooled copper tray and a continuous argon flow. A piece of sacrificial titanium underwent a few melting-cooling cycles to reduce the residual oxygen concentration in the system before each run. Each 12 g ingot was re-melted and turned over several times to ensure uniformity in chemical composition. Following previously developed annealing conditions, each ingot was annealed in vacuum at 900 °C for 12-h [25]. Chemical compositions for each ingot was analyzed using a Varian Liberty 100 inductively coupled plasma-optical emission spectrometer (ICP-OES, Agilent Technologies, Santa Clara, CA, USA). A Philips X'Pert Pro X-ray diffractometer (XRD, Philips, Amsterdam, Netherlands) was used to study the microstructure, and a JEOL-JSM6320F scanning electron microscope (SEM, JEOL, Tokyo, Japan) with energy dispersive spectroscopy (EDS) capability was used to study the phase distribution and composition. Gaseous phase hydrogen storage characteristics for each sample were measured using a Suzuki-Shokan multi-channel pressure-concentration-temperature (PCT, Suzuki Shokan, Tokyo, Japan) system. In the PCT analysis, each sample (a single piece of ingot with a newly cleaved surface and a weight of about 2 g) was first activated by a 2-h thermal cycle (between 300 °C and 30 °C) at 5 MPa H_2 pressure. PCT isotherms at 30 and 60 °C were then measured. Magnetic susceptibility was measured using a Digital Measurement Systems Model 880 vibrating sample magnetometer (MicroSense, Lowell, MA, USA).

3. Results and Discussion

3.1. Alloy Composition

Eight alloys with a general formula of $Ti_{15.6}Zr_{2.1}V_{44-x}Cr_{11.2}Mn_{6.9}Co_{1.4}Ni_{18.5+x}Al_{0.3}$, $x = 0, 4, 6, 8, 10, 12, 14$, and 16, were prepared by arc melting for this study. Design compositions of these alloys are listed in Table 1. Alloys P17 ($x = 0$) and P22 ($x = 4$) have been previously studied. P17 shows good balance between capacities and HRD [26], and P22 was the subject of the annealing condition matrix study [25]. The new compositions extend the study of V-content *versus* Ni-content. ICP results from the ingot samples are listed in Table 1. Other than a small but consistent deficiency in Zr (due to formation of ZrO_2 as oxygen scavenger), the ICP results are very close to the target compositions. The measured B/A ratios are slightly higher than the designed values due to the loss in Zr.

Table 1. Design compositions and inductively coupled plasma (ICP) results for alloys in this study. All numbers are in at%.

Alloy		Ti	Zr	V	Cr	Mn	Co	Ni	Al	B/A
P17	Design	15.6	2.1	44.0	11.2	6.9	1.4	18.5	0.3	4.65
	ICP	15.6	2	44.1	11.3	6.4	1.4	18.9	0.3	4.68
P22	Design	15.6	2.1	40.0	11.2	6.9	1.4	22.5	0.3	4.65
	ICP	15.5	1.8	41.0	11.2	6.5	1.4	22.2	0.4	4.78
P23	Design	15.6	2.1	38.0	11.2	6.9	1.4	24.5	0.3	4.65
	ICP	15.2	1.7	38.7	11.4	6.8	1.4	24.3	0.4	4.91
P24	Design	15.6	2.1	36.0	11.2	6.9	1.4	26.5	0.3	4.65
	ICP	15.5	1.8	36.0	11.3	6.4	1.5	27.1	0.4	4.78
P25	Design	15.6	2.1	34.0	11.2	6.9	1.4	28.5	0.3	4.65
	ICP	15.5	1.8	34.4	11.4	6.6	1.4	28.4	0.4	4.77
P26	Design	15.6	2.1	32.0	11.2	6.9	1.4	30.5	0.3	4.65
	ICP	15.5	1.7	33.5	10.8	6.8	1.5	29.8	0.4	4.81
P27	Design	15.6	2.1	30.0	11.2	6.9	1.4	32.5	0.3	4.65
	ICP	15.2	1.8	30.6	11.7	6.6	1.4	32.2	0.4	4.88
P28	Design	15.6	2.1	28.0	11.2	6.9	1.4	34.5	0.3	4.65
	ICP	15.3	1.9	28.1	11.4	6.6	1.4	34.9	0.4	4.81

3.2. X-Ray Diffractometer Structure Analysis

XRD patterns of the eight alloys are shown in Figure 1. Five sets of diffraction peaks are observed: a BCC, two Laves phases (C14 and C15), a TiNi and a σ-VNi. As the Ni-content in the alloy increases, the BCC peaks shift to higher angles, C14 phase is replaced by C15 phase, and σ-VNi phase starts to appear. σ-VNi phase is in a tetragonal structure, and the lattice constants are $a = 8.966$ Å and $c = 4.641$ Å with a composition at $V_{0.6}Ni_{0.4}$ [33]. It has wide solubility in V from 57.5 at% to 75 at% [34]. The V-Ni-Cr ternary phase diagram indicates an extension of σ-VNi phase from 0% to about 40% Cr [35].

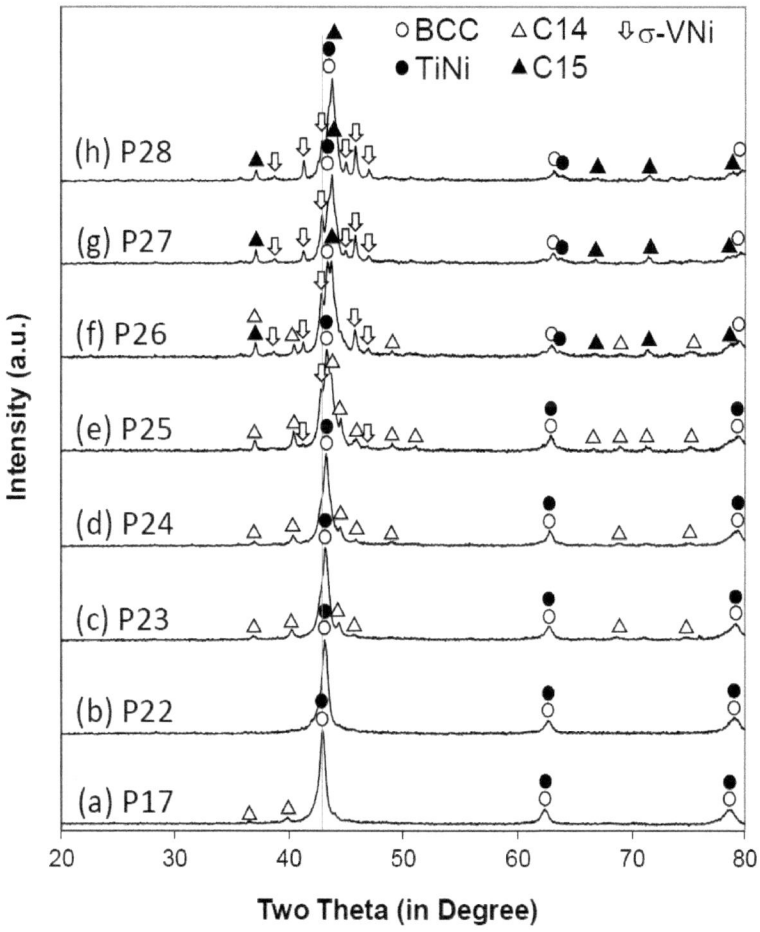

Figure 1. X-ray diffractometer (XRD) patterns using Cu-Kα as the radiation source for alloys: (**a**) P17; (**b**) P22; (**c**) P23; (**d**) P24; (**e**) P25; (**f**) P26; (**g**) P27; and (**h**) P28. A vertical line is used to indicate the shift in body-centered-cubic (BCC) peaks.

Lattice constants, crystallite sizes, and phase abundances of all five constituent phases calculated from the XRD patterns are listed in Table 2. The variations in lattice constants from BCC, C14, and TiNi phases are plotted in Figure 2. As the amount of Ni increases, the lattice constant a in BCC first remains constant and then decreases, both a and c in C14 phase decrease at about the same rate yielding a slight decrease in c/a ratio, while a in TiNi sees a discontinuous two-segment decrease. The changes in lattice constants for C15 and VNi are less prominent. Lattice parameter a in BCC phase decreases from 2.977 Å to 2.946 Å as the Ni-content in the alloy increases, which is further away from the optimized value of 3.042 Å corresponding to a maximized hydrogen storage capacity [36].

Table 2. Lattice parameters, unit cell volumes, phase abundances, and crystallite sizes of phases derived from XRD analysis. XS denotes crystallite size.

Alloy	BCC			C14					
	a (Å)	XS (Å)	Abundance (wt%)	a (Å)	c (Å)	c/a	Unit cell volume (Å³)	XS (Å)	Abundance (wt%)
P17	2.977	171	52.8	4.912	8.010	1.631	167.3	215	13.2
P22	2.974	57	50.7	4.890	7.968	1.629	165.0	275	17.1
P23	2.973	86	57.5	4.878	7.936	1.627	163.5	311	19.4
P24	2.974	72	47.0	4.865	7.915	1.627	162.2	255	20.3
P25	2.954	86	61.2	4.858	7.895	1.625	161.4	287	22.1
P26	2.950	71	55.2	4.844	7.883	1.627	160.2	272	11.1
P27	2.948	68	47.8	-	-	-	-	-	-
P28	2.946	73	34.2	-	-	-	-	-	-

Alloy	C15			TiNi			VNi			
	a (Å)	XS (Å)	Abundance (wt%)	a (Å)	XS (Å)	Abundance (wt%)	a (Å)	c (Å)	XS (Å)	Abundance (wt%)
P17	-	-	-	2.993	170	34.0	-	-	-	-
P22	-	-	-	2.962	175	32.4	-	-	-	-
P23	-	-	-	2.959	273	23.0	-	-	-	-
P24	-	-	-	2.958	204	32.6	-	-	-	-
P25	-	-	-	2.899	105	8.9	9.009	4.686	184	7.7
P26	6.859	637	15.1	2.905	228	4.5	9.010	4.692	219	14.1
P27	6.860	405	26.6	2.912	127	13.0	9.017	4.644	250	12.6
P28	6.858	347	24.7	2.912	115	25.5	9.012	4.642	344	15.7

Phase abundance and crystallite sizes of each phase are listed in Table 2. These values were obtained from full pattern fitting of the XRD data using the Rietveld method and Jade 9 Software (KS Analytical System, Aubrey, TX, USA).

The evolution of phase abundance is plotted in Figure 3. C14 Laves phase abundance first increases and then decreases as it is replaced by another Laves phase—C15 phase. The total Laves phase abundance generally increases, but begins to taper and decrease as it is changed to C15 phase. TiNi phase abundance varies widely as it first decreases, then increases, decreases, and finally increases again. BCC phase constitutes about 50–60 wt% and then drops to below 40 wt%. σ-VNi-phase appears in alloys with higher Ni-content, and the abundance further increases. The crystallite sizes (*XS*) of all five constituent phases have different trends. As the Ni-content in the alloy increases at the expense of V, the *XS* of BCC and C15 phases decrease, the *XS* of C14 and TiNi phases first increase and then decrease, and in VNi phase, the *XS* increases.

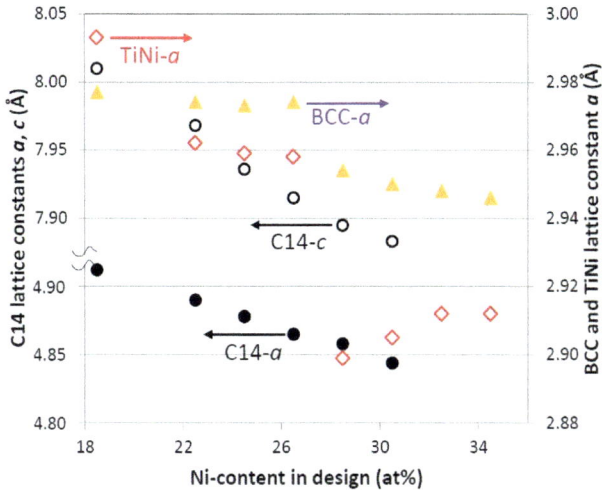

Figure 2. Lattice constants *a* and *c* from C14 phase, and lattice constants *a* from BCC and TiNi phases as functions of Ni-content in design.

3.3. Scanning Electron Microscope/Energy Dispersive Spectroscopy Microstructure Analysis

Microstructures for this series of alloys were studied using SEM, and the resulting back-scattering electron images (BEI) are presented in Figure 4.

EDS analysis was used to study the chemical compositions of several areas showing different contrasts and the results are summarized in Table 3. In the case of BCC-C14 dual-phase alloy, during cooling of the melt, BCC phase with high V-content solidifies first to form a three-dimensional framework while the rest of the liquid solidifies into Laves phase as the alloy cools further. The Laves phase itself is also a three-dimensional framework (Figure 4 in [37]). This interlacing dual-framework can be seen in the alloys in the current study, but the framework has less alignment. The main BCC phase is shown in a darker contrast due to its low Zr- and Ni-contents (lower average atomic weight). The grain size of the main BCC phase increases but the crystallite size decreases (as seen from the XRD analysis) as the Ni-content in the alloy increases. The majority of the secondary phases is in the BCC phase grain boundary, and it evolves from TiNi + C14 + Ti_2Ni (P17, very small amount and not detectable by XRD) to TiNi + C14 + Zr (P22 − P24), TiNi + C14 + C15 + VNi + Zr + TiO_2 (P25, P26), and TiNi + C15 + VNi + Zr + TiO_2 (P27, P28). The average electron density (e/a) calculated from the measured composition was used to distinguish the C14 and C15 Laves phases. The e/a values are listed in Table 3, and in this case, the C14/C15 threshold falls at a value of 7.45 [38,39]. The VNi phase starts at the center of the secondary phase region (Figure 4e-4) and moves to the boundary between BCC main phase and other secondary phases (TiNi and Laves). It is clear from the V-Ni phase diagram that the cooling sequence is first BCC phase, followed by σ-VNi phase and finally ending with a VNi$_2$ ($oI6$ structure) + σ-VNi mixture [34]. In our multi-element case, the final VNi$_2$ is replaced by a combination of TiNi and Laves phases.

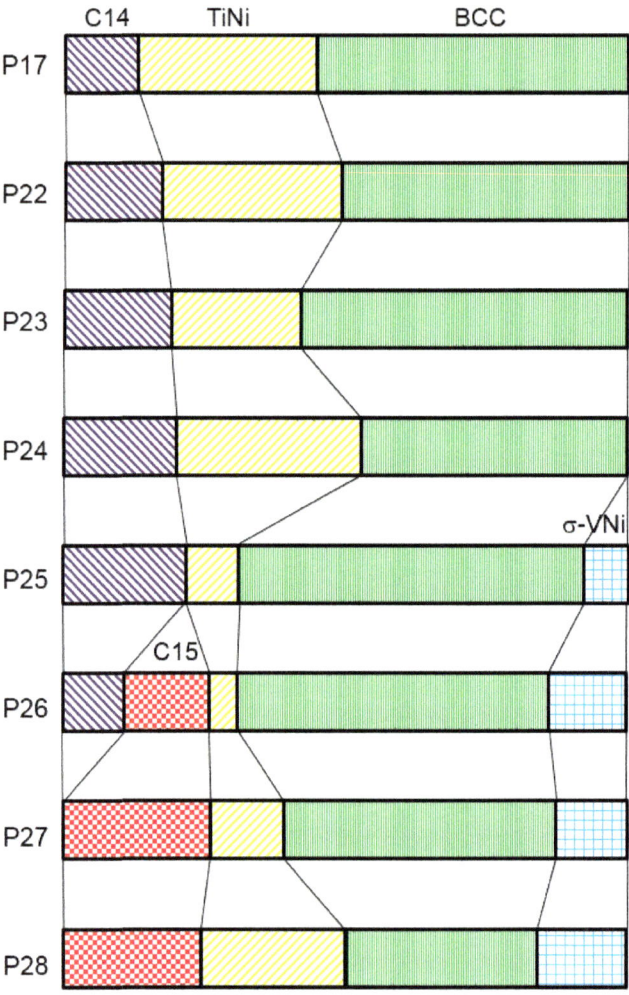

Figure 3. Evolution in the phase abundances of constituent phases from the full-pattern analysis of XRD data.

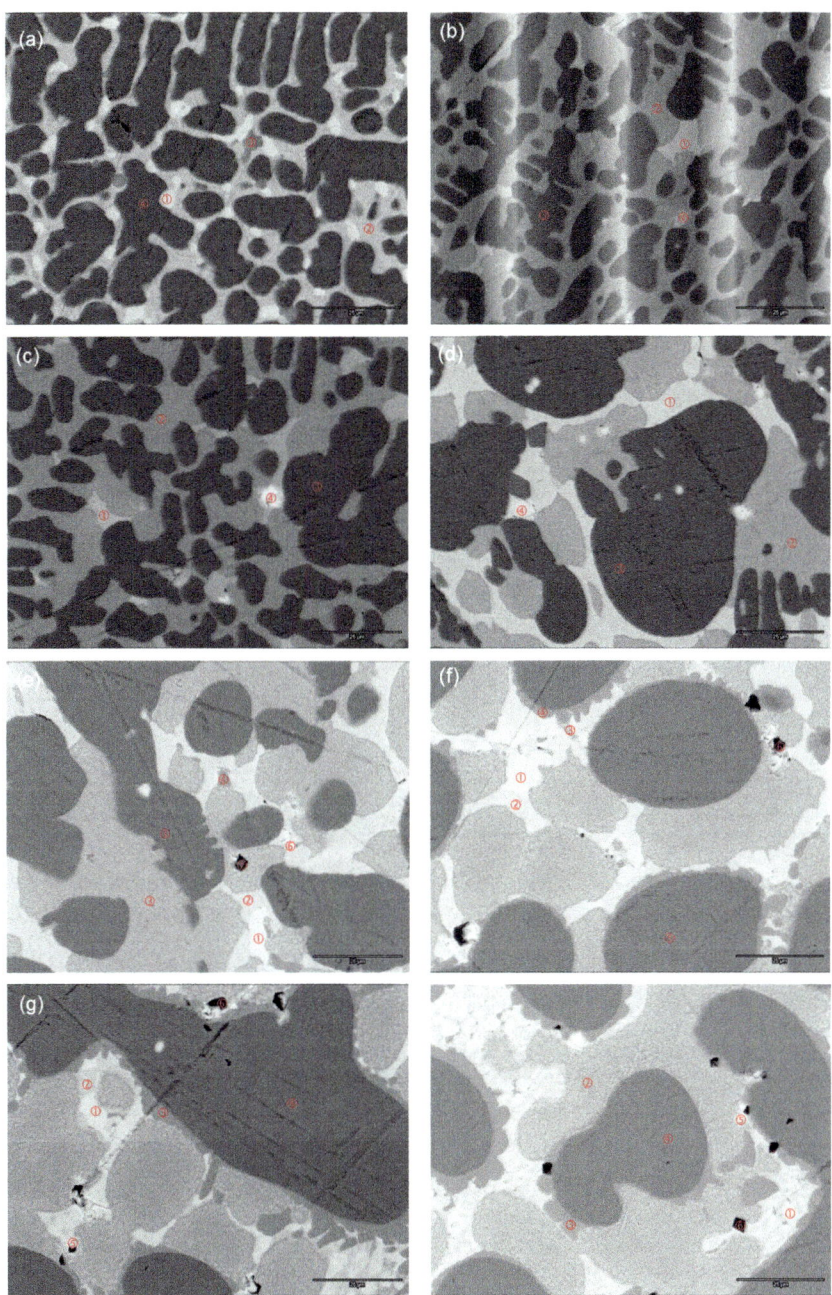

Figure 4. Scanning electron microscope (SEM) back-scattering electron images (BEI) for alloys: (**a**) P17; (**b**) P22; (**c**) P23; (**d**) P24; (**e**) P25; (**f**) P26; (**g**) P27; and (**h**) P28. Chemical compositions in the numbered areas measured by EDS are listed in Table 3.

269

Table 3. Energy dispersive spectroscopy (EDS) composition data from select spots in Figure 3. All numbers are in at%. B/A ratios for C14 and C15 phases were calculated assuming V is in the B-site while those in TiNi and VNi phases were calculated assuming V in the A-site. Compositions in the main BCC phase are highlighted in **bold**.

Location	Zr	Ti	V	Cr	Mn	Fe	Co	Ni	Al	B/A	e/a	Phase
Figure 4a-1	16.2	16.3	7.8	4.4	5.9	0.0	1.8	47.5	0.1	2.08	7.28	C14
Figure 4a-2	4.7	36.4	7.8	0.8	3.6	0.0	2.0	44.4	0.3	1.04	-	TiNi
Figure 4a-3	6.2	48.5	11.7	1.2	2.3	0.0	1.8	28.3	0.1	0.51	-	Ti_2Ni
Figure 4a-4	**0.1**	**5.5**	**63.5**	**16.9**	**7.4**	**0.0**	**0.8**	**5.6**	**0.2**	**-**	**-**	**BCC**
Figure 4b-1	14.3	17.3	9.4	0.6	4.8	0.0	0.8	52.5	0.3	2.16	7.44	C14
Figure 4b-2	2.8	34.3	7.3	0.9	4.7	0.1	2.7	46.4	0.8	1.25	-	TiNi
Figure 4b-3	**0.1**	**3.0**	**60.7**	**17.7**	**11.8**	**0.0**	**0.9**	**5.7**	**0.1**	**-**	**-**	**BCC**
Figure 4b-4	87.2	3.0	3.1	0.5	0.8	0.1	0.4	4.8	0.1	-	-	Zr
Figure 4c-1	12.0	18.6	12.4	0.7	3.5	0.0	0.9	51.6	0.3	2.27	7.38	C14
Figure 4c-2	2.4	34.5	8.4	1.0	3.7	0.0	2.3	47.2	0.5	1.21	-	TiNi
Figure 4c-3	**0.1**	**4.5**	**58.1**	**18.2**	**8.6**	**0.0**	**1.1**	**9.3**	**0.1**	**-**	**-**	**BCC**
Figure 4c-4	93.9	1.1	2.5	0.4	0.3	0.1	0.1	1.5	0.1	-	-	Zr
Figure 4d-1	10.6	17.9	16.5	2.5	4.7	0.0	1.4	46.2	0.2	2.51	7.20	C14
Figure 4d-2	1.9	33.9	8.6	1.1	3.5	0.0	2.4	47.6	1.0	1.25	-	TiNi
Figure 4d-3	**0.0**	**2.3**	**60.0**	**21.1**	**8.3**	**0.0**	**1.0**	**7.1**	**0.2**	**-**	**-**	**BCC**
Figure 4d-4	94.7	0.7	1.8	0.4	0.1	0.0	0.2	1.9	0.2	-	-	Zr
Figure 4e-1	10.8	16.0	14.2	1.8	4.2	0.0	1.0	51.7	0.3	2.73	7.45	C15
Figure 4e-2	8.5	17.7	18.2	2.5	5.0	0.0	1.5	46.2	0.3	2.81	7.22	C14
Figure 4e-3	1.6	33.8	8.6	1.1	4.1	0.1	2.5	47.5	0.8	1.28	-	TiNi
Figure 4e-4	0.2	7.0	48.2	6.6	5.2	0.1	1.3	31.3	0.1	0.81	-	VNi
Figure 4e-5	**0.0**	**2.4**	**58.7**	**21.8**	**8.5**	**0.0**	**1.0**	**7.4**	**0.1**	**-**	**-**	**BCC**
Figure 4e-6	89.0	3.2	2.4	0.4	0.2	0.0	0.2	4.5	0.1	-	-	Zr
Figure 4e-7	4.7	69.5	7.8	1.3	1.6	0.0	0.6	14.1	0.3	-	-	TiO_2
Figure 4f-1	10.5	16.2	13.3	1.9	4.4	0.0	1.0	52.5	0.2	2.75	7.50	C15
Figure 4f-2	8.9	17.3	15.7	2.5	5.1	0.0	1.4	48.8	0.2	2.81	7.35	C14
Figure 4f-3	1.4	30.7	10.3	1.1	5.1	0.0	2.0	48.6	0.8	1.36	-	TiNi
Figure 4f-4	0.2	6.8	48.4	7.1	6.8	0.0	1.5	29.0	0.1	0.80	-	VNi
Figure 4f-5	**0.0**	**2.1**	**58.3**	**21.6**	**8.7**	**0.0**	**1.1**	**8.0**	**0.1**	**-**	**-**	**BCC**
Figure 4f-6	6.0	79.0	8.3	0.1	0.5	0.0	0.2	6.0	0.0	-	-	TiO_2
Figure 4g-1	10.4	16.5	11.9	2.1	5.2	0.1	1.2	52.4	0.3	2.72	7.53	C15
Figure 4g-2	4.5	30.2	9.2	1.0	5.0	0.0	1.0	48.8	0.2	1.28	-	TiNi
Figure 4g-3	0.1	6.6	46.2	9.4	6.6	0.0	1.8	29.3	0.1	0.89	-	VNi
Figure 4g-4	**0.0**	**1.8**	**55.4**	**25.3**	**8.6**	**0.0**	**1.1**	**7.6**	**0.1**	**-**	**-**	**BCC**
Figure 4g-5	85.5	5.1	2.1	0.5	0.6	0.2	0.1	5.8	0.1	-	-	Zr
Figure 4g-6	3.5	82.8	7.6	0.1	0.6	0.0	0.1	5.2	0.1	-	-	TiO_2
Figure 4h-1	10.2	16.2	11.8	2.4	5.5	0.0	1.1	52.5	0.3	2.79	7.53	C15
Figure 4h-2	1.3	30.1	9.1	1.0	6.3	0.0	2.3	49.0	0.9	1.47	-	TiNi
Figure 4h-3	0.1	6.6	43.5	11.6	6.8	0.2	1.8	29.2	0.2	0.99	-	VNi
Figure 4h-4	**0.0**	**1.7**	**53.0**	**27.4**	**8.3**	**0.0**	**1.2**	**8.3**	**0.1**	**-**	**-**	**BCC**
Figure 4h-5	86.1	3.9	2.2	0.4	0.7	0.0	0.2	6.2	0.2	-	-	Zr
Figure 4h-6	2.0	90.6	3.3	0.1	0.5	0.0	0.2	3.3	0.1	-	-	TiO_2

The atomic percentages of a few key elements in BCC, TiNi, and C14 phases examined by SEM/EDS are plotted in Figure 5 as functions of Ni-content in the overall alloy design. Since the sampling volume of one spot in the EDS analysis is very limited (up to a few microns in diameter), the results may not be very representative and should be interpreted with caution. In the main BCC phase, as the Ni-content increases at the expense of V, the V-content decreases, the Cr-content increases, and the Mn- and Ni-contents remain at about the same values (Figure 5a). Because Cr is smaller than V, this decreases the lattice constant in BCC phase. V is a hydride former, and a smaller unit cell with less V-content has direct impacts on several properties: the metal-hydrogen bond strength will be weaker (better HRD), the plateau pressure will be higher, and the storage capacity will be lower. In the TiNi phase, as the Ni-content in design increases, the Ni-content increases, the Ti-content decreases, and the V- and Mn-contents remain relatively unchanged (Figure 5b). Because Ni is smaller than Ti, this results in a smaller unit cell, which is also beneficial in the TiNi catalytic phase for easier hydrogen diffusion. In the C14 phase, as the Ni-content in design increases, the Zr-content decreases, the V-content increases, and Ti and Ni remain unchanged. The decrease in Zr is due to the fact that Laves phase contains most of the Zr and its abundance in Laves phase increases with the increase in Ni. The overall Zr-content is fixed; therefore, the Zr-content in Laves phase is diluted and the heavily over-stoichiometry (B/A > 2.0) invites more V into the Laves phase. Both the lattice constants a and c decrease as more Zr (larger) is replaced by V (smaller). The relatively constant c/a indicates that most of the V sits in the A-site since a relatively large substitution in the B-site will cause an anisotropic enlargement in the unit cell [40,41]. The shrinkage in catalytic C14 phase unit cell also facilitates hydrogen diffusion in the alloy bulk.

3.4. Gaseous Phase Study

Gaseous phase hydrogen storage properties of the alloys were studied by PCT. All samples are activated with one thermal cycle in the presence of hydrogen. Consequent measurements did not change the PCT characteristic significantly. The resulting absorption and desorption isotherms measured at 30 °C and 60 °C are shown in Figure 6. Information obtained from the PCT study is summarized in Table 4. A two-phase characteristic can be observed in alloys P22–P24. The heat of hydride formation of each of the phases can be estimated from the atomic percentages and the heats of hydride formation (ΔH_h) for the three main hydride-former atoms, i.e., Zr ($\Delta H_h = -163$ kJ·mol H_2^{-1} for ZrH_2), Ti ($\Delta H_h = -124$ kJ·mol H_2^{-1} for TiH_2), and V ($\Delta H_h = -34$ kJ·mol H_2^{-1} for VH_2) [42]. For example, the calculated ΔH_h for TiNi will be -62 kJ·mol H_2^{-1}, which is close to the experimental result of -60 kJ·mol H_2^{-1} [43]. The such-obtained ΔH_h for the constituent phases for P22 are -25.5 (BCC), -49.6 (TiNi), and -46.8 kJ·mol H_2^{-1} (C14). Therefore, the main plateau at higher equilibrium pressure in the PCT isotherm for P22 alloy is attributed to the main BCC phase, and the second plateau, which occurs at lower hydrogen concentration and lower equilibrium pressure, is attributed to the combination of TiNi and C14 secondary phases. The same estimation is performed on alloy P26, and the results are -22.4 (BCC), -25.2 (VNi), -43.8 (TiNi), -41.3 (C14), and -41.7 kJ·mol H_2^{-1} (C15). The first two phases (BCC and VNi) are considered to be storage phases, which are responsible for the plateau in the PCT isotherm, and the other three phases with higher Ni-content are the catalytic phases that do not contribute directly to the hydrogen storage capacity.

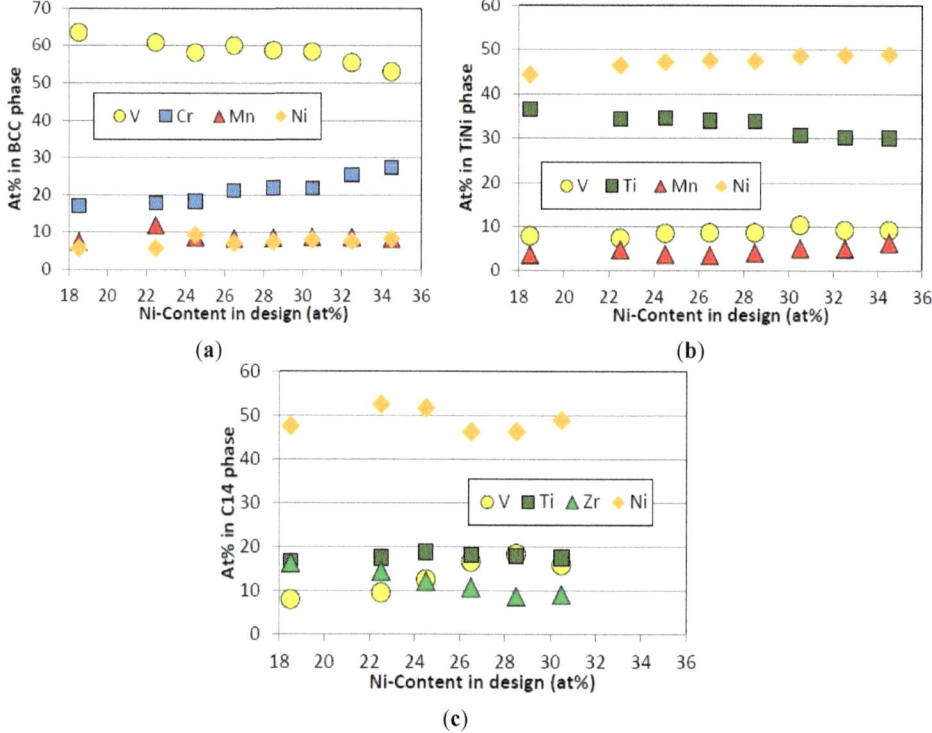

Figure 5. Atomic percentages of key elements in: (**a**) BCC; (**b**) TiNi; and (**c**) C14 phases as functions of Ni-content in design.

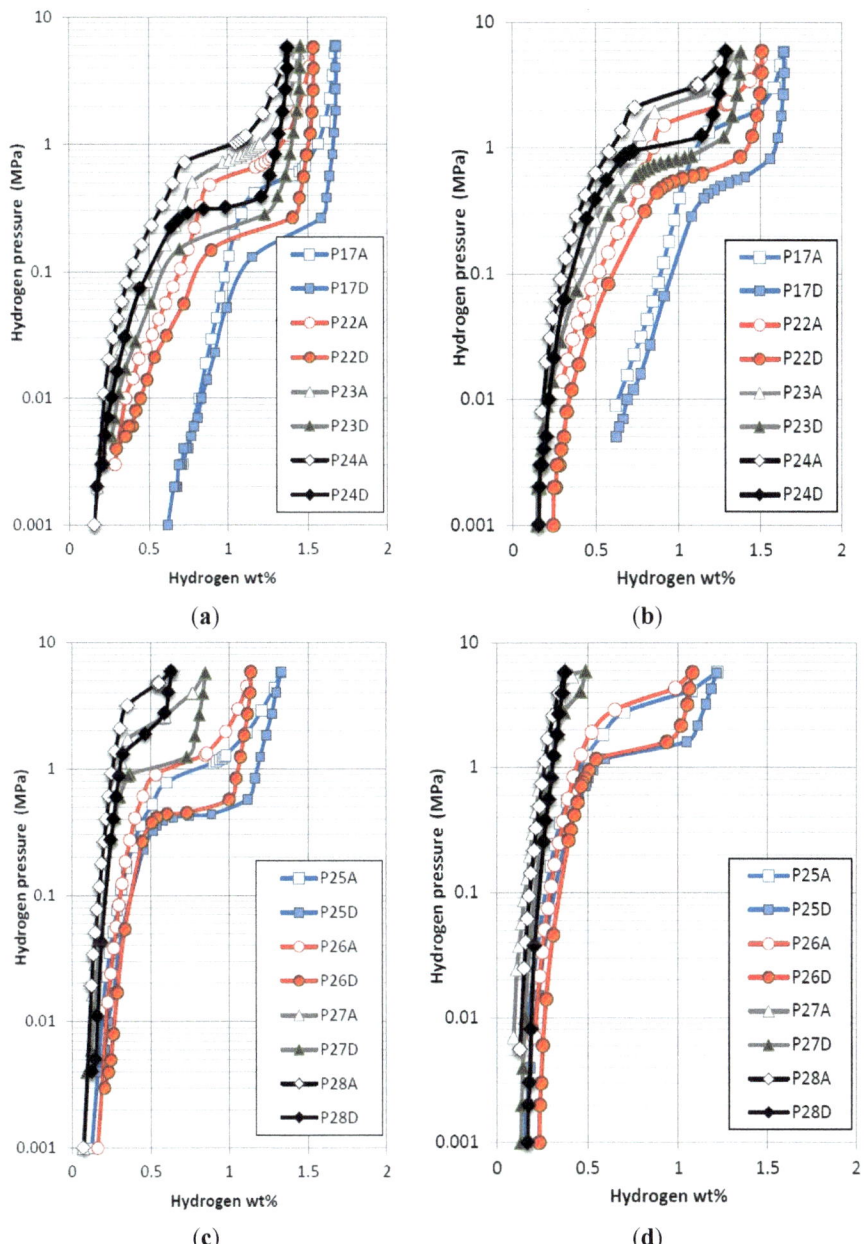

Figure 6. (**a**) 30 °C and (**b**) 60 °C pressure-concentration-temperature (PCT) isotherms of alloys P17, P22, P23, P24; and (**c**) 30 °C and (**d**) 60 °C of alloys P25, P26, P27, and P28. Open and solid symbols are for absorption and desorption curves, respectively.

Table 4. Summary of gaseous phase and thermodynamic properties. The ΔH and ΔS were calculated from the equilibrium pressure plateaus measured at 30, 60, and 90 °C with the error bars of about ±2%.

Alloy No.	Desorption pressure @ 30 °C	Desorption pressure @ 60 °C	Maximum capacity @ 30 °C	Reversible capacity @ 30 °C	PCT hysteresis @ 30 °C	PCT hysteresis @ 60 °C	$-\Delta H$	$-\Delta S$
	(MPa)	(MPa)	(wt%)	(wt%)			(kJ·mol^{-1})	(J·mol^{-1}·K^{-1})
P17	0.17	0.53	1.68	1.01	1.06	1.10	30.9	107
P22	0.21	0.60	1.53	1.24	1.14	1.13	29.9	105
P23	0.22	0.79	1.45	1.25	1.20	1.06	35.3	123
P24	0.32	1.07	1.38	1.20	1.10	0.86	34.2	122
P25	0.45	1.42	1.33	1.16	0.77	0.82	32.1	119
P26	0.43	1.36	1.14	0.93	0.79	0.84	32.4	119
P27	1.06	3.20	0.85	0.75	0.79	0.46	31.0	122
P28	1.90	-	0.63	0.51	0.76	-	-	-

As the Ni-content in the alloy increases, the plateau pressure in the BCC phase increases monotonically (due to the decrease in the BCC unit cell volume), the maximum storage capacity decreases, and the reversible storage first increases (due to the increase in abundance of catalytic C14 phase and the shrinkage in unit cell volumes for BCC, TiNi, and C14 phases) and then decreases (due to the reduction in the maximum capacity). Hysteresis of the PCT isotherm is defined as $\ln(P_a/P_d)$, where P_a and P_d are the absorption and desorption equilibrium pressures at the main plateau (BCC phase), respectively. In general, the hysteresis measured at 30 °C divides the alloys into two groups: alloys with relatively low Ni-content (P17, P22–P24) that show higher hysteresis than those in the other group (P25–P28). PCT hysteresis represents the energy required to elastically distort the lattice at the metal/hydride interface [44]. The first alloys (P17, P22–P24) also have distinctively larger unit cells (Figure 2) and also are free of C15 and σ-VNi phases. The connection between PCT hysteresis and structural properties of these alloys requires further investigation.

In order to study the thermodynamic properties of these alloys, equilibrium pressure in the BCC plateau at 30 °C and 60 °C were used to estimate the ΔH_h and entropy (ΔS) by the equation:

$$\Delta G = \Delta H_h - T\Delta S = RT\ln P \qquad (1)$$

where R is the ideal gas constant and T is the absolute temperature. Results of these calculations are listed in Table 4. The values may not be accurate and can only be used for comparison among these alloys. As the Ni-content in the design increases, both ΔH_h and ΔS decrease first and then increase. The decrease in ΔH_h of the first three alloys (P17, P22 and P23) is due to the reduction in the TiNi phase abundance reducing the proximity effects from the catalytic phases [45]. Further increase in ΔH_h

is due to the shrinking unit cell volume in BCC phase [46]. ΔS is an indication of how far the MH system is from a perfect, ordered state. The theoretical value of ΔS is the entropy of hydrogen gas, which is close to $-135\,\mathrm{J\,mol^{-1}\,K^{-1}}$. Alloy P23 shows the lowest ΔH_h and ΔS, and this indicates that it has the least interaction between the main storage phase and the catalytic phase, acting as a single BCC phase MH alloy.

3.5. Electrochemical Measurement

The electrochemical discharge capacity of each alloy was measured in a flooded-cell configuration against a partially pre-charged $Ni(OH)_2$ positive electrode. No alkaline pretreatment was applied before the half-cell measurement. For the activation behavior study, each sample electrode was charged at a constant current density of $50\,\mathrm{mA\cdot g^{-1}}$ for 10 h and then discharged at a current density of $50\,\mathrm{mA\cdot g^{-1}}$ followed by two pulls at $12\,\mathrm{mA\cdot g^{-1}}$ and $4\,\mathrm{mA\cdot g^{-1}}$. The results of the low-rate (full) capacity and the HRD_{50} (defined as the ratio in capacities obtained between $50\,\mathrm{mA\cdot g^{-1}}$ and $4\,\mathrm{mA\cdot g^{-1}}$) for the first thirteen cycles are plotted in Figure 7. From the comparison, it is easy to see that the activation in low-rate capacity is similar for all alloys (Figure 7a), but the activation in HRD (more related to surface oxidation and pulverization) is easier in alloys with lower Ni-content and becomes more difficult with the increase in Ni-content (Figure 7b). Capacities measured at the 4th cycle are listed in Table 5 together with other important electrochemical and magnetic properties of the alloys in this study. The capacities measured at the 4th cycle are also plotted in Figure 8 with the gaseous phase capacity converted to a theoretical electrochemical capacity by

$$1\ \mathrm{wt\%\ of\ H_2 = 268\ mAh\cdot g^{-1}} \tag{2}$$

As the Ni-content in the design increases, both capacities measured at $50\,\mathrm{mA\cdot g^{-1}}$ and $4\,\mathrm{mA\cdot g^{-1}}$ rates decrease monotonically, which is similar to the decrease in the maximum storage capacities measured in the gaseous phase. In a conventional MH alloy, the electrochemical capacities measured at different rates fall between the boundaries set by the equivalent maximum and reversible hydrogen storage capacities [29,47–50]. In the current study, the electrochemical capacities drop below the lower bound set by the gaseous phase reversible capacity for alloys P24–P28. It is due to the relatively high equilibrium pressure at room temperature (RT) of these alloys (>0.6 MPa) preventing a full charge under the open flooded half-cell configuration. The electrochemical discharge capacities of these alloys should be higher in a sealed cell where the cell pressure can be kept as high as 2.8 MPa.

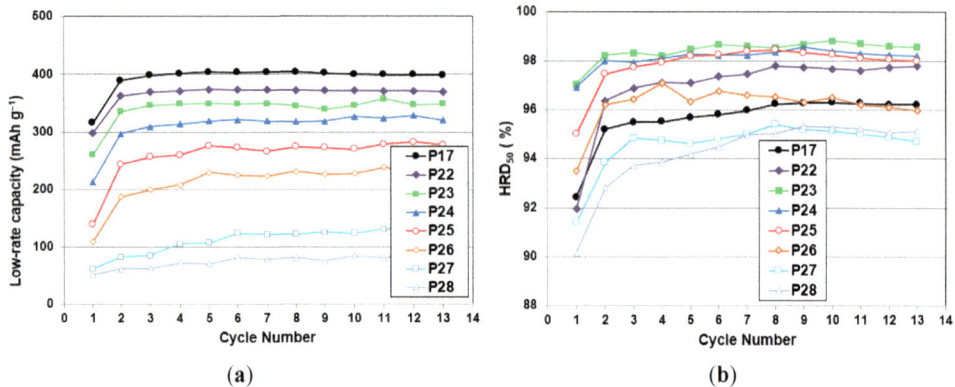

(a) (b)

Figure 7. Activation behavior for the first 13 cycles of (**a**) low-rate capacity and (**b**) HRD_{50} (ratio of capacities at 50 $mA \cdot g^{-1}$ and 4 $mA \cdot g^{-1}$ discharge rates).

Table 5. Summary of electrochemical and magnetic properties. High-rate dischargeability $(HRD)_{200}$ is the ratio of capacities measured at 200 $mA \cdot g^{-1}$ and 4 $mA \cdot g^{-1}$. RT: room temperature.

Alloy No.	4th cycle capacity @ 50 $mA \cdot g^{-1}$	4th cycle capacity @ 4 $mA \cdot g^{-1}$	HRD_{200}	Activation cycle to reach maximum capacity	Diffusion coefficient, D @RT	Exchange current I_o @RT	Ms	$H_{1/2}$	Open circuit voltage
	$(mAh \cdot g^{-1})$	$(mAh \cdot g^{-1})$			$(10^{-10} \cdot cm^2 \cdot s^{-1})$	$(mA \cdot g^{-1})$	$(emu \cdot g^{-1})$	(kOe)	(V)
P17	382.7	400.7	0.815	6	1.69	36.5	0.39	0.11	1.302
P22	360.4	371.0	0.873	5	1.67	25.2	1.40	0.11	1.306
P23	342.6	348.8	0.920	5	2.95	25.5	0.62	0.11	1.326
P24	308.1	314.1	0.939	6	2.85	20.0	0.92	0.10	1.330
P25	253.9	259.2	0.965	5	1.58	18.9	1.03	0.10	1.312
P26	201.6	207.7	0.913	5	1.22	12.8	1.81	0.11	1.288
P27	100.1	105.6	-	6	1.39	12.3	1.42	0.14	1.264
P28	68.0	72.4	-	6	1.46	16.3	1.40	0.12	1.332

Half-cell HRD_{200} of each alloy, defined as the ratio of discharge capacities measured at 200 $mA \cdot g^{-1}$ and 4 $mA \cdot g^{-1}$, is listed in Table 5. As the Ni-content in the alloy increases, HRD_{200} increases in the beginning, maximizes at alloy P25 and then decreases. When the discharge capacities at different discharge rates are compared against a commercially available standard AB_5 MH alloy with a HRD_{200} of 0.958 (Figure 9), only alloy P25 shows better rate-capability, but its capacities are too low. We find that although BCC-TiNi-Laves MH alloys have higher discharge capacities at a lower rate (<200 $mA \cdot g^{-1}$), their high-rate capacities are not as good as in conventional AB_5 MH alloys.

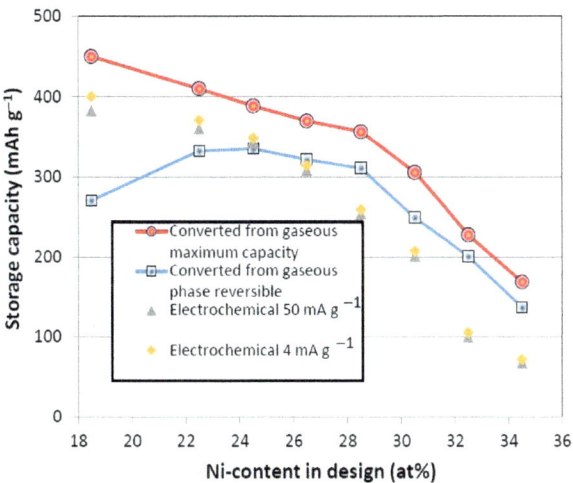

Figure 8. Plot of gaseous phase and electrochemical storage capacities as functions of Ni-content in design.

In order to further study the inferior HRD capability in BCC-TiNi-Laves MH alloys compared to AB_5 MH alloys, both the bulk diffusion coefficient (D) and the surface exchange current (I_o) of the alloys under the current study were measured electrochemically. The details of both parameters' measurements were previously reported [51], and the values are listed in Table 5. As the Ni-content in design increases, the D value increases and then decreases, while I_o decreases. The evolution of the HRD is more similar to the trend in D than in I_o, and therefore we conclude that the high-rate performance in this series of alloys is more related to the bulk diffusion than to the surface reaction. The increase in D initially is due to the shrinking unit cells for BCC, TiNi, and C14 phases (Table 2). In a comparison study, C15 phase shows faster hydrogen bulk diffusion than C14 phase [52]; therefore, the subsequent decrease in D with higher Ni-content is most likely related to the formation of σ-VNi phase. CALPHAD thermodynamic calculations show that hydrogen permeability is reduced with higher Ni-content in V-Ni alloys [53]. Compared to the D value from a typical AB_5 MH alloy (2.55×10^{-10} cm^2·s^{-1} [54]), alloys P23 and P24 have faster bulk hydrogen diffusion. The reason the BCC-TiNi-Laves alloys have lower HRDs despite better bulk diffusion is because of their relatively low I_o values (Table 5) compared to AB_5 MH alloy (43.2 mA·g^{-1}, [54]). When the abundances of the catalytic secondary phases (Laves and TiNi) are plotted against the I_o values as shown in Figure 10, it is clear that while Laves phases (C14 and C15) appear beneficial, TiNi phase appears detrimental to the surface electrochemical reaction. Similar findings regarding TiNi phase hindering the HRD performance of Laves phase based AB_2 MH alloys were reported before [55].

277

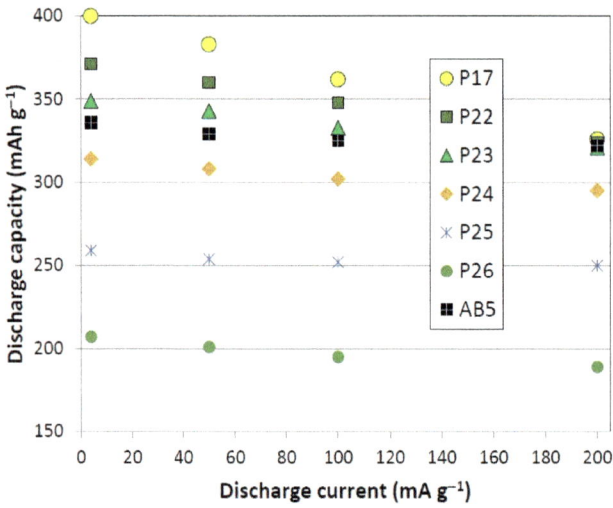

Figure 9. Rate-dependent discharge capacity for alloys P17, P22–P26 and a commercially available AB₅ MH alloy.

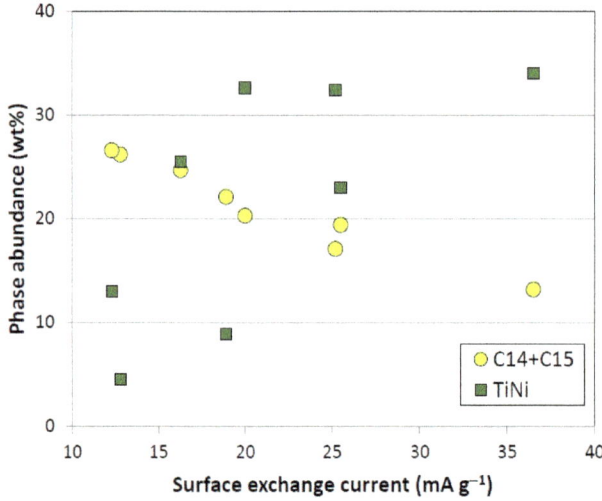

Figure 10. Plots of Laves (C14 + C15) and TiNi phase abundances *vs.* the surface exchange current.

The lower surface reaction currents measured in the current series of alloys were further investigated by magnetic susceptibility measurement. Details of the background and the experimental method are published in [56]. Metallic Ni is an active catalyst for the water splitting and recombination reactions that contribute to the I_o in this electrochemical system, and this technique allows us to obtain the saturated magnetic susceptibility (M_S), a quantification of the amount of surface metallic Ni (the product of preferential oxidation), and the magnetic field strength at one-half of the M_S value ($H_{1/2}$), a measure of the average reciprocal number of Ni atoms in a metallic cluster. M_S and $H_{1/2}$ values for the alloys in this study are listed in Table 5. While the $H_{1/2}$ values of these alloys are very similar, the trend of M_S values with the increase in Ni-content is not obvious. When these values are compared to values from an AB$_5$ MH alloy ($M_S = 0.434$ emu·g^{-1} and $H_{1/2} = 0.17$ kOe [54]), we find that all alloys except P17 have higher M_S values. These BCC-TiNi-Laves alloys have higher weight percentage of metallic Ni in their activated surfaces and should have better HRD [57]. The reason for the lower I_0 values in this current series of alloys may be due to their relatively larger Ni cluster sizes, resulting in less surface area, which is indicated by smaller $H_{1/2}$ values. Systematic transmission electron microscope work is necessary here to continue the investigation of the metallic inclusions, particularly their size and shape, embedded in the activated surface oxide layer of these alloys.

The open circuit voltage (OCV) of the electrochemical cells comprising a negative electrode from each alloy and a Ni(OH)$_2$/NiOOH positive electrode is listed in Table 5. OCV is important for the power density of the cell and is usually proportional to the logarithm of equilibrium pressure, typical of most MH alloys. In the previous study in a series of BCC-C14 alloys, the OCV is inversely proportional to the desorption equilibrium pressure [26]. In the current series of alloys, the plateau pressure increases with the increase in Ni-content, but the OCV first increases and then decreases except for the last alloy P28, which shows very small discharge capacity due to insufficient charging in the open-air configuration. The evolution of OCV in this study is similar to the trends in HRD and maximizes with alloy P24.

4. Conclusions

The Ni-content in a base alloy, P17, was increased with the intention to improve the electrochemical high-rate performance in the negative electrode active material for (Ni/MH) batteries. Although the HRD improved with the increase in Ni-content at the expense of V, the discharge capacity was deteriorated. An alloy, P23, with a composition of Ti$_{15.6}$Zr$_{2.1}$V$_{36}$Cr$_{11.2}$Mn$_{6.9}$Co$_{1.4}$Ni$_{26.5}$Al$_{0.3}$ shows a good balance between capacity and high-rate capability. The increase in Ni-content shifts the Laves phase from C14 into C15 because of the increase in the average electron density in the phase. The increase in Ni-content also promotes a new σ-VNi phase, which has a similar hydrogen-metal bond strength as the main BCC phase. The new phase is associated with reduced bulk diffusion of hydrogen. Further improvement in high-rate capability in this family of alloys should maintain the Ni-level at 22.5 at% to 26.5 at% and increase the content of other C14-promoters, such as Zr and Hf.

Acknowledgments: This work is financially supported by Advanced Research Projects Agency-Energy (ARPA-E) under the robust affordable next generation EV-storage (RANGE) program (DE-AR0000386). The authors thank these colleagues for technical assistance: Simon Ng, Shuli Yan, Shiuan Chang (Wayne State University), Michael A. Fetcenko, Baoquan Huang, Tiejun Meng, David Pawlik, Allen Chan, Ryan J. Blankenship and Su Cronogue (BASF-Ovonic).

Conflicts of Interest: The authors declare no conflict of interest.

References

1. Iba, H.; Akiba, E. Hydrogen absorption and microstructure in BCC alloys with C14-type Lave phase. *J. Jpn. Inst. Met. Mater.* **1994**, *58*, 1225–1232.

2. Akiba, E.; Iba, H. Hydrogen absorption by Laves phase related BCC solid solution. *Intermetallics* **1998**, *6*, 461–470.

3. Enomoto, M. The Cr-Ti-V system. *J. Phase Equilibria* **1992**, *13*, 195–200.

4. Inoue, H.; Arai, S.; Iwakura, C. Crystallographic and electrochemical characterization of $TiV_{4-x}Ni_x$ alloys for nickel-metal hydride batteries. *Electrochim. Acta* **1996**, *41*, 937–939.

5. Yu, X.; Wu, Z.; Xia, B.; Xu, N. A Ti-V-based bcc phase alloy for use as metal hydride electrode with high discharge capacity. *J. Chem. Phys.* **2004**, *121*, 987–990. PubMed]

6. Young, K.-H.; Ouchi, T.; Huang, B.; Nei, J. Structure, hydrogen storage, and electrochemical properties of body-centered-cubic $Ti_{40}V_{30}Cr_{15}Mn_{13}X_2$ alloys (X = B, Si, Mn, Ni, Zr, Nb, Mo, and La). *Batteries* **2015**, *1*, 74–90.

7. Chen, N.; Li, R.; Zhu, Y.; Liu, Y.; Pan, H. Electrochemical hydrogenation and dehydrogenation mechanisms of the Ti-V base multiphase hydrogen storage electrode alloy. *Acta Metall. Sin.* **2004**, *40*, 1200–1204.

8. Iba, H.; Akiba, E. The relation between microstructure and hydrogen absorbing property in Laves phase-solid solution multiphase alloys. *J. Alloy. Compd.* **1995**, *231*, 508–512.

9. Rönnebro, E.; Noréus, D.; Sakai, T.; Tsukahara, M. Structural studies of a new Laves phase alloy (Hf,Ti)(Ni,V)$_2$ and its very stable hydride. *J. Alloy. Compd.* **1995**, *231*, 90–94.

10. Tsukahara, M.; Takahashi, K.; Mishima, T.; Isomura, A.; Sakai, T. V-based solid solution alloys with Laves phase network: Hydrogen absorption properties and microstructure. *J. Alloy. Compd.* **1996**, *236*, 151–155.

11. Qiu, S.; Chu, H.; Zhang, Y.; Sun, D.; Song, X.; Sun, L.; Fen, X. Electrochemical kinetics and its temperature dependence behaviors of $Ti_{0.17}Zr_{0.08}V_{0.35}Cr_{0.10}Ni_{0.30}$ alloy electrode. *J. Alloy. Compd.* **2009**, *471*, 453–456.

12. Iba, H.; Akiba, E. Hydrogen absorption and modulated structure in Ti–V–Mn alloys. *J. Alloy. Compd.* **1997**, *253–254*, 21–24.

13. Shashikala, K.; Banerjee, S.; Kumar, A.; Pai, M.R.; Pillai, C.G.S. Improvement of hydrogen storage properties of TiCrV alloy by Zr substitution for Ti. *Int. J. Hydrog. Energy* **2009**, *34*, 6684–6689.

14. Yang, X.; Li, J.; Zhang, T.; Hu, R.; Xue, X.; Fu, H. Role of defect structure on hydrogenation properties of $Zr_{0.9}Ti_{0.1}V_2$ alloy. *Int. J. Hydrog. Energy* **2011**, *36*, 9318–9323.

15. Huot, J.; Akiba, E.; Ogura, T.; Ishido, Y. Crystal structure, phase abundance and electrode performance of Laves phase compounds $(Zr, A)V_{0.5}Ni_{1.1}Mn_{0.2}Fe_{0.2}$ ($A \equiv Ti$, Nb or Hf). *J. Alloy. Compd.* **1995**, *218*, 101–109.

16. Qiu, S.-J.; Chu, H.-L.; Zhang, J.; Zhang, Y.; Sun, L.-X.; Xu, F.; Sun, D.-L.; Ouyang, L.-Z.; Zhu, M.; Grolier, J.-P.E.; *et al.* Effect of La partial substitution for Zr on the structural and electrochemical properties of $Ti_{0.17}Zr_{0.08-x}La_xV_{0.35}Cr_{0.1}Ni_{0.3}$ ($x = 0$–0.04) electrode alloys. *Int. J. Hydrog. Energy* **2009**, *34*, 7246–7252.

17. Huang, Z.; Cuevas, F.; Liu, X.; Jiang, L.; Wang, S.; Latroche, M.; Du, J. Effects of Si addition on the microstructure and the hydrogen storage properties of $Ti_{26.5}V_{45}Fe_{8.5}Cr_{20}Ce_{0.5}$ BCC solid solution alloys. *Int. J. Hydrog. Energy* **2009**, *34*, 9385–9392.

18. Kamegawa, A.; Shirasaki, K.; Tamura, T.; Kuriiwa, T.; Takamura, H.; Okada, M. Crystal structure and protium absorption properties of Ti–Cr–X alloys. *Mater. Trans.* **2002**, *43*, 470–473.

19. Liu, Y.; Zhang, S.; Li, R.; Gao, M.; Zhong, K.; Miao, H.; Pen, H. Electrochemical performances of the Pd-added Ti-V-based hydrogen storage alloys. *Int. J. Hydrog. Energy* **2008**, *33*, 728–734.

20. Kuriiwa, T.; Tamura, T.; Amemiya, T.; Fuda, T.; Kamegawa, A.; Takamura, H.; Okada, M. New V-based alloys with high protium absorption and desorption capacity. *J. Alloy. Compd.* **1999**, *293–295*, 433–436.

21. Yu, J.; Liu, B.; Cho, K.; Lee, J. The effects of partial substitution of Mn by Cr on the electrochemical cycle life of Ti-Zr-V-Mn-Ni alloy electrodes of a Ni/MH battery. *J. Alloy. Compd.* **1998**, *278*, 283–290.

22. Lee, H.-H.; Lee, K.-Y.; Lee, J.-Y. The Ti-based metal hydride electrode for Ni–MH rechargeable batteries. *J. Alloy. Compd.* **1996**, *239*, 63–70.

23. Kim, J.; Paik, C.; Cho, W.; Cho, B.; Yun, K.; Kim, S. Corrosion behaviour of $Zr_{1-x}Ti_xV_{0.6}Ni_{1.2}M_{0.2}$ (M = Ni, Cr, Mn) AB_2-type metal hydride alloys in alkaline solution. *J. Power Sources* **1998**, *75*, 1–8.

24. Young, K.; Wong, D.; Wang, L. Effect of Ti/Cr content on the microstructures and hydrogen storage properties of Laves phase-related body-centered-cubic solid solution alloys. *J. Alloy. Compd.* **2015**, *622*, 885–893.

25. Young, K.; Ouchi, T.; Nei, J.; Wang, L. Annealing effects on Laves phase-related body-centered-cubic solid solution metal hydride alloys. *J. Alloy. Compd.* **2016**, *654*, 216–225.

26. Young, K.; Ouchi, T.; Nei, J.; Meng, T. Effects of Cr, Zr, V, Mn, Fe, and Co to the hydride properties of Laves phase-related body-centered-cubic solid solution alloys. *J. Power Sources* **2015**, *281*, 164–172.

27. Young, K.; Ouchi, T.; Reichman, B.; Koch, J.; Fetcenko, M.A. Effects of Mo additive on the structure and electrochemical properties of low-temperature AB_5 metal hydride alloys. *J. Alloy. Compd.* **2011**, *509*, 3995–4001.

28. Chai, Y.; Zhao, M. Structure and electrochemical properties of $Ti_{0.25}V_{0.35}Cr_{0.40-x}Ni_x$ ($x = 0.05$–0.40) solid solution alloys. *Int. J. Hydrog. Energy* **2005**, *30*, 279–283.

29. Young, K.; Fetcenko, M.A.; Li, F.; Ouchi, T. Structural, thermodynamic, and electrochemical properties of $Ti_xZr_{1-x}(VNiCrMnCoAl)_2$ C14 Laves phase alloys. *J. Alloy. Compd.* **2008**, *464*, 238–247.

30. Young, K.; Ouchi, T.; Fetcenko, M.A. Roles of Ni, Cr, Mn, Sn, Co, and Al in C14 Laves phase alloys for NiMH battery application. *J. Alloy. Compd.* **2009**, *476*, 774–781.

31. Li, R.; Pan, H.; Gao, M.; Zhu, Y.; Liu, Y.; Jin, Q.; Lei, Y. Structural and electrochemical properties of hydrogen storage alloys $Ti_{0.8}Zr_{0.2}V_{2.7}Mn_{0.5}Cr_{0.8}Ni_x$ (x = 1.50–2.25). *J. Alloy. Compd.* **2004**, *373*, 223–230.

32. Pan, H.; Li, R.; Gao, M.; Liu, Y.; Lei, Y.; Wang, Q. Effects of Ni on the structural and electrochemical properties of Ti-V-based hydrogen storage alloys. *Int. J. Hydrog. Energy* **2006**, *31*, 1188–1195.

33. Pearson, W.; Christian, J. The structure of the σ phase in vanadium-nickel alloys. *Acta Crystallogr.* **1952**, *5*, 157–162.

34. Massalski, T.; Okamoto, H.; Subramanian, P.; Kacprzak, L. *Binary Alloy Phase Diagram*, 2nd ed.; ASM International: Materials Park: Ohio, USA, 1990; p. 2880.

35. Kodentzov, A.A.; Dunaev, S.F.; Slusarenko, E.M. Determination of the phase diagram of the V-Ni-Cr system using diffusion couples and equilibrated alloys. *J. Less Common Met.* **1987**, *135*, 15–24.

36. Yan, Y.; Chen, Y.; Liang, H.; Zhou, X.; Wu, C.; Tao, M.; Pang, L. Hydrogen storage properties of V–Ti–Cr–Fe alloys. *J. Alloy. Compd.* **2008**, *454*, 427–431.

37. Young, K.; Nei, J.; Wong, D.; Wang, L. Structural, hydrogen storage, and electrochemical properties of Laves-phase related body-centered-cubic solid solution metal hydride alloys. *Int. J. Hydrog. Energy* **2014**, *39*, 21489–21499.

38. Johnston, R.L.; Hoffmann, R. Structure-bonding relationships in the Laves phases. *Z. Anorg. Allg. Chem.* **1992**, *616*, 105–120.

39. Nei, J.; Young, K.; Salley, S.O.; Ng, K. Determination of C14/C15 phase abundance in Laves phase alloys. *Mater. Chem. Phys.* **2012**, *136*, 520–527.

40. Young, K.; Fetcenko, M.A.; Li, F.; Ouchi, T.; Koch, J. Effect of vanadium substitution in C14 Laves phase alloys for NiMH battery application. *J. Alloy. Compd.* **2009**, *468*, 482–492.

41. Young, K.; Ouchi, T.; Huang, B.; Reichman, B.; Fetcenko, M.A. Effect of molybdenum content on structure, gaseous storage, and electrochemical properties of C14-predominant AB_2 metal hydride alloys. *J. Power Sources* **2011**, *196*, 8815–8821.

42. Young, K. Metal Hydrides. In *Elsevier Reference Module in Chemistry, Molecular Sciences and Chemical Engineering*; Elsevier B.V.: Waltham, MA, USA, 2013.

43. Griessen, R.; Riesterer, T. Heat of Formation Models. In *Hydrogen in Intermetallic Compounds I*; Schlapbach, L., Ed.; Springer-Verlag: Berlin, Germany, 1988; p. 273.

44. Balasubramaniam, R. Hysteresis in metal-hydrogen systems. *J. Alloy. Compd.* **1997**, *253–254*, 203–206.

45. Wong, D.; Young, K.; Nei, J.; Wang, L.; Ng, K. Effects of Nd-addition on the structural, hydrogen storage, and electrochemical properties of C14 metal hydride alloys. *J. Alloy. Compd.* **2015**, *647*, 507–518.

46. Nakano, H.; Wakao, S. Substitution effect of elements in Zr-based alloys with Laves phase for nickel-hydride battery. *J. Alloy. Compd.* **1995**, *231*, 587–593.

47. Nei, J.; Young, K.; Sally, S.O.; Ng, K. Effects of annealing on $Zr_8Ni_{19}X_2$ (X = Ni, Mg, Al, Sc, V, Mn, Co, Sn, La, and Hf): Hydrogen storage and electrochemical properties. *Int. J. Hydrog. Energy* **2012**, *37*, 8418–8427.

48. Sun, D.; Jiang, J.; Lei, Y.; Liu, W.; Wu, J.; Wang, Q.; Yang, G. Effects of measurement factor on electrochemical capacity of some hydrogen storage alloys. *Mater. Sci. Eng. B* **1995**, *30*, 19–22.

49. Young, K.; Nei, J.; Huang, B.; Ouchi, T.; Fetcenko, M.A. Studies of $Ti_{1.5}Zr_{5.5}V_{0.5}(M_xNi_{1-x})_{9.5}$ (M = Cr, Mn, Fe, Co, Cu, Al): Part 2. Hydrogen storage and electrochemical properties. *J. Alloy. Compd.* **2010**, *501*, 245–254.

50. Young, K.; Chao, B.; Huang, B.; Nei, J. Studies on the hydrogen storage characteristic of $La_{1-x}Ce_x(NiCoMnAlCuSiZr)_{5.7}$ with a B2 secondary phase. *J. Alloy. Compd.* **2014**, *585*, 760–770.

51. Li, F.; Young, K.; Ouchi, T.; Fetcenko, M.A. Annealing effects on structural and electrochemical properties of $(LaPrNdZr)_{0.83}Mg_{0.17}(NiCoAlMn)_{3.3}$ alloy. *J. Alloy. Compd.* **2009**, *471*, 371–377.

52. Young, K.; Ouchi, T.; Huang, B.; Chao, B.; Fetcenko, M.A.; Bendersky, L.A.; Wang, K.; Chiu, C. The correlation of C14/C15 phase abundance and electrochemical properties in the AB_2 alloys. *J. Alloy. Compd.* **2010**, *506*, 841–848.

53. Shim, J.H.; Ko, W.S.; Kim, K.H.; Lee, H.S.; Lee, Y.S.; Suh, J.Y.; Choa, Y.W.; Lee, B.-J. Prediction of hydrogen permeability in V-Al and V-Ni alloys. *J. Membr. Sci.* **2014**, *430*, 234–241.

54. Young, K.; Nei, J. The current status of hydrogen storage alloy development for electrochemical applications. *Materials* **2013**, *6*, 4574–4608.

55. Young, K.; Nei, J.; Ouchi, T.; Fetcenko, M.A. Phase abundances in AB_2 metal hydride alloys and their correlations to various properties. *J. Alloy. Compd.* **2011**, *509*, 2277–2284.

56. Young, K.; Wong, D.; Wang, L.; Nei, J.; Ouchi, T.; Yasuoka, S. Mn in misch-metal based superlattice metal hydride alloy—Part 1 structural, hydrogen storage and electrochemical properties. *J. Power Sources* **2015**, *277*, 426–432.

57. Young, K.; Huang, B.; Regmi, R.K.; Lawes, G.; Liu, Y. Comparisons of metallic clusters imbedded in the surface oxide of AB_2, AB_5, and A_2B_7 alloys. *J. Alloy. Compd.* **2010**, *506*, 831–840.

Studies on the Synergetic Effects in Multi-Phase Metal Hydride Alloys

Kwo-hsiung Young, Taihei Ouchi, Tiejun Meng and Diana F. Wong

Abstract: The electrochemical reactions of multi-phase metal hydride (MH) alloys were studied using a series of Laves phase-related body-centered-cubic (BCC) $Ti_{15.6}Zr_{2.1}V_{43}Cr_{11.2}Mn_{6.9}Co_{1.4}Ni_{18.5}Al_{0.3}X$ (X = V, B, Mg, Y, Zr, Nb, Mo, La, and Nd) alloys. These alloys are composed of BCC (major), TiNi (major), C14 (minor), and Ti_2Ni (minor) phases. The BCC phase was found to be responsible for the visible equilibrium pressure plateau between 0.1 MPa and 1 MPa. The plateaus belonging to the other phases occurred below 0.005 MPa. Due to the synergetic effects of other non-BCC phases, the body-centered-tetragonal (BCT) intermediate step is skipped and the face-centered-cubic (FCC) hydride phase is formed directly. During hydrogenation in both gaseous phase and electrochemistry, the non-BCC phases were first charged to completion, followed by charging of the BCC phase. In the multi-phase system, the side with a higher work function along the grain boundary is believed to be the first region that becomes hydrogenated and will not be fully dehydrided after 8 h in vacuum at 300 °C. While there is a large step at approximately 50% of the maximum hydrogen storage for the equilibrium pressure measured in gaseous phase, the charge/discharge curves measured electrochemically are very smooth, indicating a synergetic effect between BCC and non-BCC phases in the presence of voltage and charge non-neutrality. Compared to the non-BCC phases, the C14 phase benefits while the TiNi phase deteriorates the high-rate dischargeability (*HRD*) of the alloys. These synergetic effects are explained by the preoccupied hydrogen sites on the side of the hydrogen storage phase near the grain boundary.

Reprinted from *Batteries*. Cite as: Young, K.-h.; Ouchi, T.; Meng, T.; Wong, D.F. Studies on the Synergetic Effects in Multi-Phase Metal Hydride Alloys. *Batteries* **2016**, 2, 15.

1. Introduction

The synergetic effects in multi-phase metal hydride (MH) alloys refer to the presence of microsegregated secondary phases occurring in the melted alloys that effectively provide beneficial effects [1]. Some examples of synergetic effects in the gaseous phase and electrochemical environment are summarized in Table 1. In general, the synergetic effects in the gaseous phase hydrogen storage can improve the storage capacity and reversibility, and are characterized by a continuous transition in the pressure-concentration-temperature (PCT) isotherm, from the plateau pressure corresponding to the phase with a stronger metal–hydrogen (M–H) bond strength to the phase with a weaker M–H bond strength (Figure 7b in [2]). The interface region between two phases is considered to be critical for the synergy to take place. Transmission electron microscopy (TEM) studies have demonstrated the interface between the main C14 phase and other secondary phases (C15, Zr_7Ni_{10}, *etc.*) are clean [3,4], and strong crystallographic orientation alignment can be established by electron beam back-scattering diffraction pattern studies [5,6]. Some synergetic effects in the electrochemical environment are similar to those in the gaseous phase, *i.e.*, improvement in capacity and hydrogen absorption/desorption rate. Interestingly, the properties established by the scope of the gaseous phase can be further enhanced by synergetic effects in the electrochemical environment. For example, it has been found that the electrochemical discharge capacities of certain multi-phase MH alloys can be increased substantially by lowering the equivalent plateau pressure (observed through open-circuit voltage (*OCV*) changes during charge/discharge [7]) through synergetic effects [8,9]. Therefore, a systematic study in the difference between the synergetic effects in the two environments is of significant interest.

Table 1. Examples of synergetic effects in multi-phase metal hydride (MH) alloys. GP and EC denote gaseous phase and electrochemistry experiments, respectively. *HRD*: high-rate dischargeability.

Main Phase	Secondary Phase	Main Improvement	Environment	Reference
C14	Zr_7Ni_{10}	Capacity and activation *	EC	[10]
C14/C15	Zr_7Ni_{10}, ZrNi	Capacity and activation	EC	[11]
C15	Zr_7Ni_{10}	*HRD*	EC	[12]
C15	Zr_7Ni_{10}	Capacity	EC	[13]
MgNi	Ti	Cycle stability	EC	[14]
C14/C15	Zr_7Ni_{10} and TiNi	*HRD*	EC	[15]
C14/C15	Zr_7Ni_{10} and Zr_9Ni_{11}	Capacity and *HRD*	EC	[1]
Zr_8Ni_{21}	Zr_7Ni_{10}, Zr_9Ni_{11}	Capacity and *HRD*	EC	[16]
C14	Zr_8Ni_{21}	Activation, bulk diffusion, cycle stability	EC	[17,18]
Zr_7Ni_{10}	C15	*HRD*	EC	[19]
C14/C15	Zr_7Ni_{10} and TiNi	Capacity and reversibility	GP	[20]
C14/C15	Zr_9Ni_{11}	Activation, *HRD*, charge retention, and cycle stability	EC	[21]
C14	Zr_9Ni_{11} and TiNi	Capacity	GP	[21]
C14	Zr_9Ni_{11} and TiNi	Capacity, charge retention, and cycle stability	EC	[21]
C14	Zr_7Ni_{10} and ZrNi	*HRD*	EC	[21]
C14/C15	Zr_7Ni_{10} and TiNi	*HRD*	EC	[22]
Mg	Mg_2Ni	Desorption kinetics	GP	[23]
AB_5	$AlMnNi_2$	Capacity and *HRD*	EC	[24]
BCC, C14	ZrNi	Capacity, activation, and cycle stability	EC	[25]
Zr_2Ni_7	Zr_7Ni_0	Capacity	GP, EC	[26]
Zr_7Ni_0	Zr_8Ni_{21}	Capacity	EC	[26]
Zr_2Ni_7	$ZrNi_3$, $ZrNi_5$, VNi_2, VNi_3	Capacity	EC	[9]
Zr_2Ni_7	$ZrNi_3$ and $ZrNi_5$	Capacity	EC	[8]
$NdNi_5$	Nd_2Ni_7	*HRD*	EC	[27]
$LiBH_4$	Fluorographite	Desorption kinetics	GP	[28]
$Mg(BH_4)_2$	Fluorographite	Desorption kinetics	GP	[29]
MgH_2	In, $TiMn_2$ additives	Desorption kinetics	GP	[30]

* indicates that the authors attributed improvement to micro-cracking at the surface instead of to synergetic effects.

Laves phase-related body-centered-cubic (BCC) solid solution alloys were chosen for this study. By combining the high-capacity main storage BCC phase with catalytic phases, such as C14, TiNi, and Ti_2Ni, one alloy in the family demonstrated a 30% increase in capacity when discharged at a current density of $100\ mA \cdot g^{-1}$, which is adequate for electric vehicle applications [31]. The development of this family of alloys proceeded in stages and the X-ray diffraction (XRD) and PCT analyses demonstrated strong evidence of synergetic effects between the main storage phase and the catalytic phase [32–35]. With a base alloy of P17 ($Ti_{15.6}Zr_{2.1}V_{44}Cr_{11.2}Mn_{6.9}Co_{1.4}Ni_{18.5}Al_{0.3}$), the V/Ni content was adjusted to increase the high-rate dischargeability (*HRD*) [36]. The synergetic effects in an electrochemical environment were studied in detail using this same base alloy, P17, with substitutions of other A-site atoms (B, Mg, Y, Zr, Nb, Mo, La, and Nd) and presented here.

2. Experimental Setup

The alloy samples were prepared using an arc-melting technique. The melting was performed in an Ar environment with an average alloy weight of 12 g. The chemical composition of the ingot, compared to the ratios in the raw materials, was determined with a Varian Liberty 100 inductively coupled plasma-optical emission spectrometer (ICP-OES, Agilent Technologies, Santa Clara, CA, USA). The microstructures of the as-prepared samples were examined by a Philips X'Pert Pro X-ray diffractometer (XRD, Philips, Amsterdam, The Netherlands) and a JEOL-JSM6320F scanning electron microscope (SEM, JEOL, Tokyo, Japan) equipped with energy-dispersive spectroscopy (EDS). The PCT analysis was performed with a Suzuki-Shokan multi-channel PCT (Suzuki Shokan, Tokyo, Japan) system. The MH alloys were compacted on Ni mesh to achieve the negative electrode, and the battery cells were made by using a standard $Ni(OH)_2$-positive electrode and 30 wt% KOH electrolyte. The electrochemical testing of the battery cells was performed at room temperature (RT) using a CTE MCL2 Mini (Chen Tech Electric MFG. Co., Ltd., New Taipei, Taiwan) cell test system.

3. Results

3.1. X-Ray Diffraction Structure Analysis

Eight alloys with the targeted compositions listed in Table 2 were prepared by arc melting. Th base alloy, P17, originated from a composition optimization study [35]. Other alloys were derivatives of this base alloy with 1 at% replacement aimed at the A-site element (B, Mg, Y, Nb, Mo, La, and Nd). The chemical composition of the as-prepared ingot was verified by ICP and shows a large discrepancy only in alloy P39, where the added Mg was not found in the ingot due to loss from evaporation during melting. Mg has a high vapor pressure at temperatures near the melting point of the alloy and low solubility in the AB_2 phase [37,38]. The content of La in the final ingot of P43 was less than 50% of the targeted content due to the high chemical reactivity of La metal with the residual oxygen and the formation of oxide slag during melting. The B/A ratios for the ingots ranged from 4.24 to 4.68, assuming that Ti, Zr, and the additive elements are A-site atoms and the remaining constituent elements are B-site atoms. The XRD patterns of the alloys are shown in Figure 1. The three major peaks detected in all alloys belong to a BCC structure. All the minor peaks can be attributed to a C14 Laves phase. A TiNi phase with a B2 structure (a close derivative of a BCC structure) and a slightly larger lattice constant were separated from the BCC phase using software deconvolution (JADE 9, Christchurch, New Zealand).

Table 2. Design compositions (**bold**) and inductively coupled plasma (ICP) results for alloys in this study in at %.

Alloy	Source	Ti	Zr	V	Cr	Mn	Co	Ni	Al	X	B/A
P17	**Design**	**15.6**	**2.1**	**44**	**11.2**	**6.9**	**1.4**	**18.5**	**0.3**	**0**	**4.65**
	ICP	15.6	2	44.1	11.3	6.4	1.4	18.9	0.3	0	4.68
P38 (X = B)	**Design**	**15.6**	**2.1**	**43**	**11.2**	**6.9**	**1.4**	**18.5**	**0.3**	**1**	**4.35**
	ICP	15.7	2.2	42.7	10.6	6.7	1.4	19.3	0.3	1.1	4.26
P39 (X = Mg)	**Design**	**15.6**	**2.1**	**43**	**11.2**	**6.9**	**1.4**	**18.5**	**0.3**	**1**	**4.35**
	ICP	15.5	2.2	44.5	10.6	7.1	1.4	18.3	0.5	0	4.65
P40 (X = Y)	**Design**	**15.6**	**2.1**	**43**	**11.2**	**6.9**	**1.4**	**18.5**	**0.3**	**1**	**4.35**
	ICP	15.1	2	44.9	11.4	5.4	1.4	18.7	0.3	0.8	4.59
P41 (X = Nb)	**Design**	**15.6**	**2.1**	**43**	**11.2**	**6.9**	**1.4**	**18.5**	**0.3**	**1**	**4.35**
	ICP	15.8	2.1	41.2	11.7	7.4	1.5	19.2	0.4	0.8	4.35
P42 (X = Mo)	**Design**	**15.6**	**2.1**	**43**	**11.2**	**6.9**	**1.4**	**18.5**	**0.3**	**1**	**4.35**
	ICP	15	2	43.1	11	7.7	1.4	18.5	0.4	1	4.56
P43 (X = La)	**Design**	**15.6**	**2.1**	**43**	**11.2**	**6.9**	**1.4**	**18.5**	**0.3**	**1**	**4.35**
	ICP	16.1	2.2	40.9	11.8	7.4	1.5	19.3	0.4	0.4	4.34
P44 (X = Nd)	**Design**	**15.6**	**2.1**	**43**	**11.2**	**6.9**	**1.4**	**18.5**	**0.3**	**1**	**4.35**
	ICP	15.9	2.2	43	9.8	6.8	1.4	19.6	0.3	1	4.24

The lattice constants, crystallite size, and phase abundances calculated from the XRD pattern are listed in Table 3.

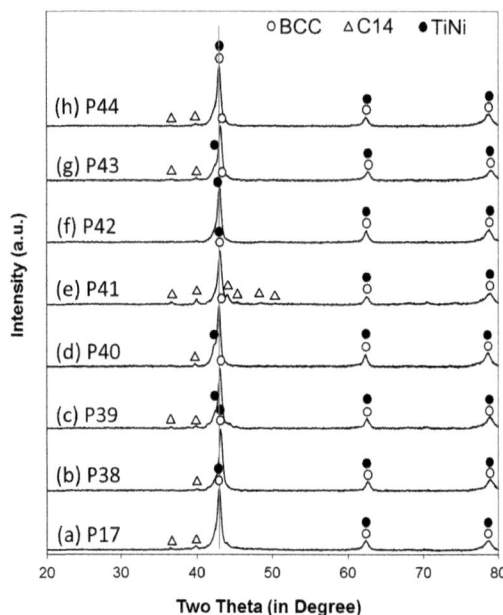

Figure 1. X-ray diffraction (XRD) patterns of alloys (**a**) P17; (**b**) P38; (**c**) P39; (**d**) P40; (**e**) P41; (**f**) P42; (**g**) P43; and (**h**) P44. The vertical line highlights the shifts of the main body-centered-cubic (BCC) and TiNi peaks.

Table 3. Lattice parameters, unit cell volumes, phase abundances, and crystallite sizes of phases derived from XRD analysis. *XS* denotes crystallite size.

Alloys	BCC			C14						TiNi		
	a (Å)	XS (Å)	Abundance (wt%)	a (Å)	c (Å)	c/a	Unit Cell Volume (Å³)	XS (Å)	Abundance (wt%)	a (Å)	XS (Å)	Abundance (wt%)
P17	2.977	171	52.8	4.912	8.010	1.631	167.3	215	13.2	2.993	170	34.0
P38	2.961	181	60.5	4.875	7.998	1.641	164.6	631	2.6	2.974	65	36.9
P39	2.968	260	53.0	4.892	8.019	1.639	166.2	322	7.0	2.982	89	39.9
P40	2.974	343	31.8	4.929	8.030	1.629	169.0	414	6.1	2.986	102	62.1
P41	2.971	187	50.4	4.906	7.991	1.629	166.6	309	20.7	2.986	41	28.9
P42	2.973	221	58.4	4.893	8.013	1.638	166.1	808	0.8	2.987	74	40.8
P43	2.962	262	38.9	4.899	8.003	1.634	166.3	240	9.5	2.976	85	51.6
P44	2.973	228	41.0	4.924	8.007	1.626	168.1	278	6.3	2.988	97	52.7

The lattice constants a from the three constituent phases are plotted against the metallic radii of the additive elements in the Laves-phase alloy [39] in Figure 2a. At a glance, the lattice constants from all three phases follow the same trend: a rapid increase followed by a decrease with increasing radii of the additive. However, the later EDS analysis indicates that most of the additives have zero or limited solubility in these three main phases. Therefore, the changes in lattice constant are tied to other characteristics of this group of alloys. For example, the lattice constants show a reasonably consistent trend of decreasing with increasing Mn-content (with a relatively small radius) in the BCC phase, as shown in Figure 2b. The spread of c/a ratio (a parameter that can be used to estimate the preferential occupation site for foreign atoms [36,40]) in the A-atom substitution (1.626–1.641) is approximately the same as the spread in the B-atom substitution study (1.627–1.644) [35]. No preference for occupation site substitution can be deduced in the current study. From the abundances shown in Table 3, BCC (32–61 wt%) and TiNi (29–62 wt%) are the two major phases, with C14 as the secondary phase. The crystallite of the secondary phase (C14) is larger than those in the main phases (BCC and TiNi), which has previously been observed in BCC-Laves-related alloys [33,36], but which is not common in multi-phase AB_2 MH alloys [41]. It is interesting to observe that there is a correlation between the abundance and the crystallite size in each phase. For example, as the TiNi phase becomes more popular (larger abundance), its crystallite becomes larger. In the case of BCC and C14 phases, the trend is the opposite (Figure 3).

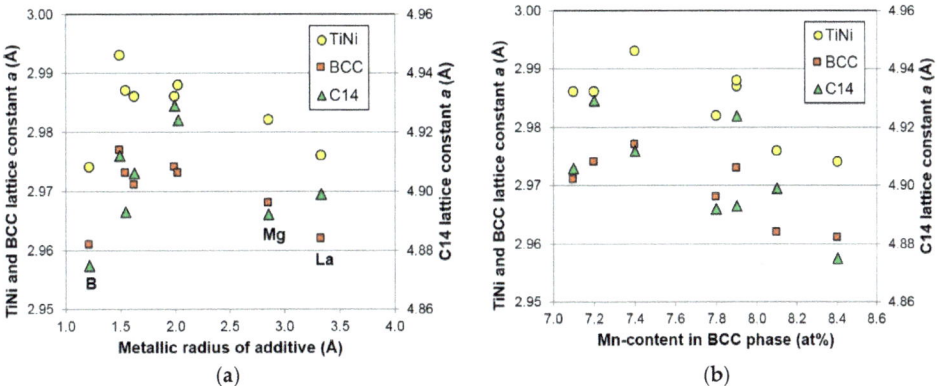

Figure 2. Plots of the lattice constants a for the TiNi, BCC, and C14 phases *vs.* (**a**) the metallic radius of the additives and (**b**) the Mn-content in the BCC phase.

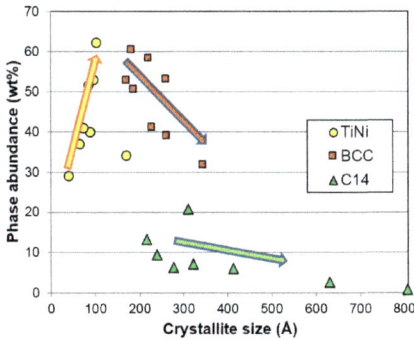

Figure 3. Plots of abundances *vs.* crystallite size for the TiNi, BCC, and C14 phases.

3.2. Scanning Electron Microscope Phase Analysis

The composition of the constituent phase was studied through a combination of SEM and EDS on a few selected areas. A representative SEM-backscattering electron image (BEI) from each alloy is shown in Figure 4, and the corresponding EDS data are listed in Table 4. In contrast to the composition information obtained from ICP, EDS results are far less accurate and representative but are still useful for qualitatively studying the composition in each constituent phase. The two main phases, BCC and TiNi, form separate 3D frameworks that interlace with each other (for a schematic drawing, see Figure 4 in [32]). This microstructure suggests an early formation of a V-rich BCC framework, which pushes the Ni and Ti into grain boundaries and later forms the C14 and TiNi phases. According to an earlier TEM study of the C14-predomintaed MH alloy, the C14 phase is solidified before the formation of the TiNi phase [3,4]. The B/A ratio in each spot was calculated assuming B occupies the A-site in the C14 phase [42] but occupies the A-site in the TiNi and Ti_2Ni phases [34]. All the calculated B/A ratios are slightly above their stoichiometric values (hyper-stoichiometry), *i.e.*, 2.0 (C14), 1.0 (TiNi), and 0.5 (Ti_2Ni). The anti-site defect (Ni in the A-site) is energetically more favorable, compared to the vacancy defect [43], which explains the hyper-stoichiometry. The average electron density (e/a), calculated from the number of conduction electrons in the constituent metals [44] in the C14 phase, is lower than the C14/C15 threshold [45] and thus a C14-structure is expected instead of a C15-structure, except for alloy P17. There may be some of the C15 phase in alloy P17, according to its relatively higher e/a ratio. From the EDS study, we found that only additive Nb participated in all three main phases and only Mo was found in the BCC phase. Boron (B) is too light to be detected by EDS, but definitely exists in the alloy, and Mg is missing completely, which can both be confirmed by ICP. Rare earth elements, such as Y, La, and Nd, promote formation of the AB phase and have undetectable solubilities in the BCC, TiNi, and C14 phases. The Mn-content in the C14 phase has been correlated with the lattice constants of the three phases in Figure 2b. Other correlations are less obvious.

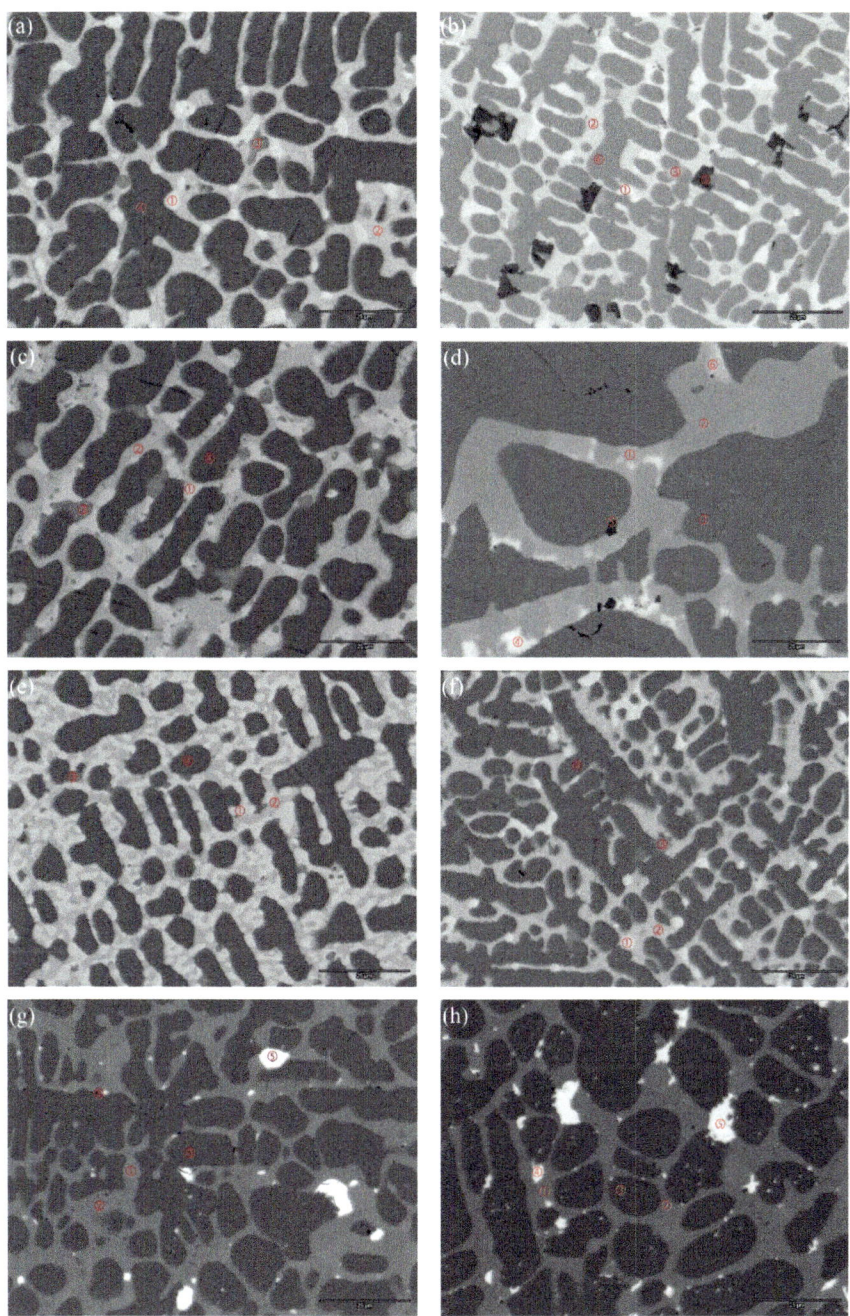

Figure 4. Scanning electron microscope SEM backscattering images from alloys (**a**) P17; (**b**) P38; (**c**) P39; (**d**) P40; (**e**) P41; (**f**) P42; (**g**) P43; and (**h**) P44. The scale bar at the lower right corner represents 25 μm.

Table 4. Energy-dispersive spectroscopy (EDS) composition data from select spots in Figure 4. All numbers are in at%. B/A ratios for the C14 phase were calculated assuming V is in the B-site, while those in the TiNi and VNi phases were calculated assuming V in the A-site. Compositions in the main BCC phase are highlighted in **bold**.

Location	Zr	Ti	V	Cr	Mn	Co	Ni	Al	X	B/A	e/a	Phase
P17-1	16.2	16.3	7.8	4.4	5.9	1.8	47.5	0.1	0.0	2.08	7.28	C14
P17-2	4.7	36.4	7.8	0.8	3.6	2.0	44.4	0.3	0.0	1.04	-	TiNi
P17-3	6.2	48.4	11.7	1.2	2.3	1.8	28.3	0.1	0.0	0.51	-	Ti$_2$Ni
P17-4	**0.1**	**5.5**	**63.5**	**16.9**	**7.4**	**0.8**	**5.6**	**0.2**	**0.0**	**-**	**-**	**BCC**
P38-1	9.9	22.0	20.3	4.7	6.6	2.0	33.9	0.6	0.0	2.13	6.62	C14
P38-2	4.4	37.0	6.4	1.1	2.8	2.7	44.8	0.8	0.0	1.09	-	TiNi
P38-3	5.5	45.4	15.4	2.4	2.8	1.8	26.4	0.3	0.0	0.51	-	Ti$_2$Ni
P38-4	**0.1**	**5.2**	**62.8**	**17.5**	**8.4**	**0.8**	**5.0**	**0.2**	**0.0**	**-**	**-**	**BCC**
P38-5	0.1	13.1	73.4	8.8	3.2	0.2	1.1	0.1	0.0	6.58	-	Oxide
P39-1	9.7	22.8	20.0	4.8	7.1	2.2	32.5	1.0	0.0	2.08	6.56	C14
P39-2	3.8	33.9	11.9	2.4	3.4	2.8	40.6	1.1	0.0	1.01	-	TiNi
P39-3	6.3	48.5	11.4	1.4	2.8	1.8	27.5	0.3	0.0	0.51	-	Ti$_2$Ni
P39-4	**0.1**	**5.7**	**61.7**	**18.5**	**7.8**	**0.8**	**5.1**	**0.3**	**0.0**	**-**	**-**	**BCC**
P40-1	11.8	21.3	20.4	3.3	4.8	1.6	36.2	0.7	0.0	2.02	6.66	C14
P40-2	5.2	37.9	5.5	0.9	2.0	2.8	45.2	0.4	0.0	1.06	-	TiNi
P40-3	**0.1**	**4.6**	**66.2**	**17.1**	**7.2**	**0.8**	**3.7**	**0.2**	**0.0**	**-**	**-**	**BCC**
P40-4	2.0	0.8	1.4	0.5	0.3	0.0	53.4	0.8	40.8	1.22	-	YNi
P40-5	6.6	84.9	5.7	0.5	0.4	0.1	1.7	0.2	0.0	-	-	TiO$_2$
P40-6	0.0	1.4	1.7	0.4	0.0	0.0	2.3	1.3	92.9	-	-	Y
P41-1	6.1	22.9	21.1	5.9	7.4	2.0	31.2	0.5	2.8	2.14	6.54	C14
P41-2	2.8	37.5	7.2	1.4	3.3	2.7	43.8	0.8	0.5	1.08	-	TiNi
P41-3	4.2	47.9	13.3	2.2	2.7	1.8	27.0	0.3	0.6	0.52	-	Ti$_2$Ni
P41-4	**0.1**	**5.2**	**62.9**	**17.7**	**7.1**	**0.7**	**5.7**	**0.2**	**0.5**	**-**	**-**	**BCC**
P42-1	10.3	21.9	20.0	4.4	6.4	2.0	34.3	0.6	0.1	2.10	6.63	C14
P42-2	4.5	37.0	6.3	0.9	2.6	2.7	45.2	0.8	0.0	1.09	-	TiNi
P42-3	6.2	45.5	14.8	2.3	2.5	1.6	26.8	0.3	0.0	0.51	-	Ti$_2$Ni
P42-4	**0.4**	**7.0**	**58.6**	**15.4**	**7.9**	**1.1**	**8.0**	**0.2**	**1.4**	**-**	**-**	**BCC**
P43-1	9.2	23.5	20.6	5.3	7.3	2.1	31.3	0.7	0.0	2.06	6.51	C14
P43-2	4.1	38.5	5.9	1.1	2.5	3.0	44.3	0.7	0.0	1.06	-	TiNi
P43-3	**0.1**	**5.9**	**61.1**	**18.6**	**8.1**	**0.8**	**5.2**	**0.2**	**0.0**	**-**	**-**	**BCC-1**
P43-4	1.1	30.2	45.5	12.7	6.3	0.5	3.3	0.1	0.3	-	-	BCC-2
P43-5	0.2	1.3	2.6	0.8	0.0	0.3	38.3	0.1	56.4	0.65	-	LaNi
P44-1	9.9	23.1	20.8	4.4	6.4	2.1	32.8	0.6	0.0	2.03	6.56	C14
P44-2	4.3	38.4	6.0	0.9	2.5	2.8	44.4	0.7	0.0	1.05	-	TiNi
P44-3	**0.1**	**5.8**	**64.0**	**16.2**		**0.8**	**5.1**	**0.2**	**0.0**	**-**	**-**	**BCC**
P44-4	0.5	2.9	2.9	0.3	0.9	0.1	46.5	0.2	45.7	0.92	-	NdNi
P44-5	0.1	1.8	4.5	0.7	1.2	0	2	0	89.8	-	-	Nd

3.3. Gaseous Phase Characteristics

The characteristics of gaseous phase hydrogen storage for these eight alloys were studied by PCT analysis and the resulting isotherms, obtained at 30 °C and 60 °C, are shown in Figure 5. Similar to other alloys in the same family [32,33,35,36], the PCT isotherm shows only one plateau in the pressure range of our apparatus (0.001–5 MPa). There is at least one more known plateau below 0.001 MPa. The PCT hysteresis from this family of alloys is much larger than those from the AB_2 [46], AB_5 [47], and A_2B_7 [27] MH alloy families. The gaseous phase properties obtained from the PCT analysis are summarized in Table 5. The desorption plateau pressures of the substituted alloys are higher than that in the base alloy (P17), except for Nd-doped P44, which indicates that most of the additives weaken the M–H bond in the hydride. By comparing the capacities of these alloys, we found that while La is beneficial to both maximum and reversible capacities, both B and Nb increase the reversible capacity. The irreversible hydrogen storage capacity (the difference between the maximum and reversible capacities) is proportional to the TiNi phase abundance (shown in Figure 6). The PCT hysteresis originates from the elastic lattice deformation energy needed at the metal (α)-hydride (β) interface during hydrogen absorption [48,49]. The higher PCT hysteresis in this family of alloys suggests an environment which results in difficulty with respect to expansion inside the alloy. The substituted alloys show larger PCT hysteresis at 30 °C, but smaller PCT hysteresis at 60 °C when compared to those from the base alloy P17.

Table 5. Summary of gaseous phase and thermodynamic properties.

Alloy	Desorption Pressure @30 °C (MPa)	Desorption Pressure @60 °C (MPa)	Maximum Capacity @30 °C (wt%)	Reversible Capacity @30 °C (wt%)	PCT Hysteresis @30 °C	PCT Hysteresis @60 °C
P17	0.17	0.53	1.68	1.01	1.06	1.10
P38	0.21	0.65	1.64	1.13	1.14	1.06
P39	0.21	0.71	1.72	1.14	1.16	1.06
P40	0.19	0.60	1.69	0.85	1.22	1.06
P41	0.23	0.76	1.57	1.15	1.13	1.05
P42	0.40	1.29	1.65	1.02	1.13	0.91
P43	0.28	1.05	1.85	1.09	1.23	0.76
P44	0.15	0.47	1.69	0.88	1.27	1.22

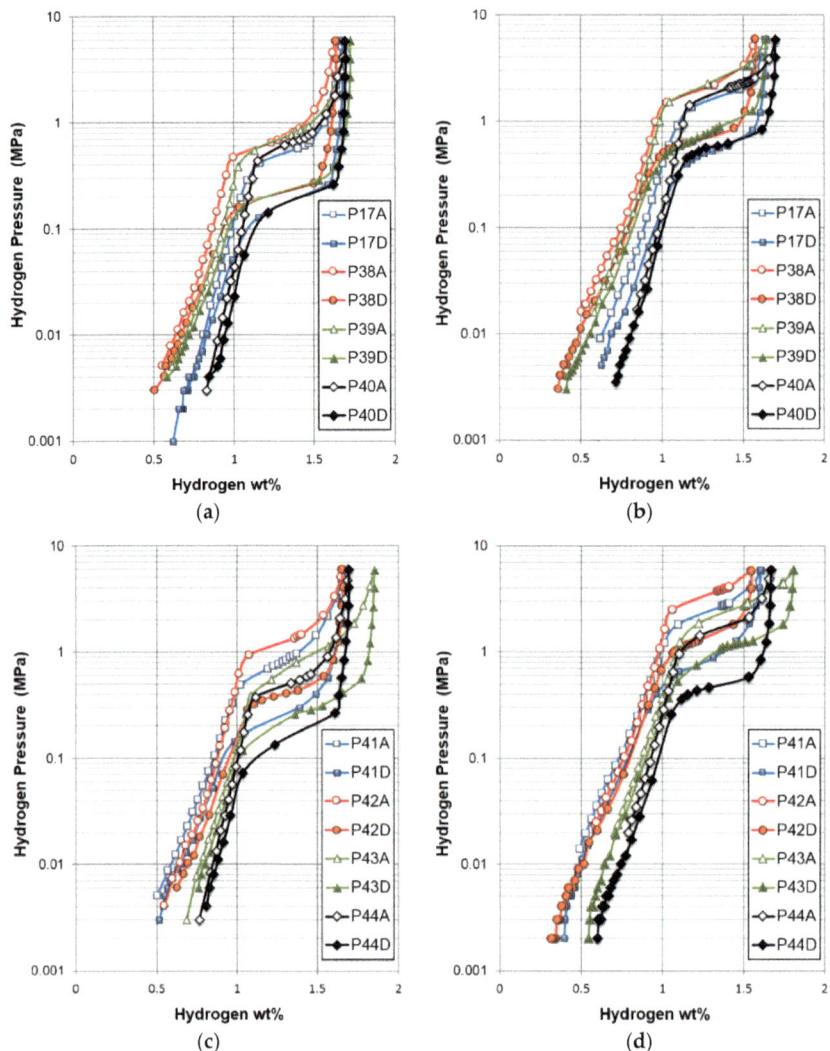

Figure 5. Pressure-concentration-temperature (PCT) isotherms measured at (**a**) 30 °C and (**b**) 60 °C for alloys P17, P38, P39, and P40; and (**c**) 30 °C and (**d**) 60 °C for alloys P41, P42, P43, and P44.

Thermodynamic properties, including changes in enthalpy (ΔH) and entropy (ΔS), were calculated from the equation for free energy (G):

$$\Delta G = \Delta H - T\Delta S = RT\ln P \tag{1}$$

where R is the ideal gas constant and T is the absolute temperature. Based on plateau pressures in the absorption, desorption, and halfway point, three sets of ΔH and ΔS

were calculated and listed in Table 6. Conventionally, the values obtained using the desorption isotherm (ΔH_D and ΔS_D) are adopted because the hysteresis is mainly due to the energy barrier from lattice distortion at the α-β interface during hydrogen absorption. In the current study, the ΔS_A value is more consistent, suggesting that the absorption isotherm may be increasingly unaffected, and at least should be included in the report. As can be seen in Table 6, ΔH_D is always lower than ΔH_A, and $\Delta H_{(A+D)/2}$ falls between the two values. ΔS_D can be higher or lower than ΔS_A, and $\Delta S_{(A+D)/2}$ also falls in the middle. In our previous publication, we adopted an estimation method for ΔH from the atomic percentage and obtained a ΔH of the three hydride-former atoms, Zr ($\Delta H_h = -163$ kJ$\cdot$mol^{-1} H$_2$ for ZrH$_2$), Ti ($\Delta H_h = -124$ kJ$\cdot$mol^{-1} H$_2$ for TiH$_2$), and V ($\Delta H_h = -34$ kJ$\cdot$mol^{-1} H$_2$ for VH$_2$). This estimation is based on the following assumptions. The complete heat of hydride formation of AB$_n$ alloy should be:

$$\Delta H \left(AB_n H_{2m} \right) = \Delta H \left(AH_m \right) + \Delta H \left(B_n H_m \right) - \Delta H \left(AB_n \right) \ [50] \tag{2}$$

However, the ΔH for the alloy AB$_n$ and the non-hydride former atoms are smaller than those in the hydride former atoms (for example, ΔH for VCr$_2$ and NiH$_{0.5}$ are -2 kJ$\cdot$mol^{-1} H$_2$ [51] and -6 kJ$\cdot$mol^{-1} H$_2$ [52], respectively). Therefore, the heat of hydride formation can be estimated by:

$$\Delta H \left(A1_h A2_k A3_l B_n H_2 \right) = h\Delta H \left(A1H_2 \right) + k\Delta H \left(A2H_2 \right) + l\Delta H \left(A3H_2 \right) \tag{3}$$

where $h, k, l,$ and m are the atomic percentages of the hydride former elements A1, A2, and A3 and the non-hyride former element B. According to this calculation, the ΔH for the hydride of the constituent phases of alloy P17 are -28.5 (BCC), -55.4 (TiNi) and -49.3 kJ$\cdot$mol^{-1} H$_2$ (C14). Therefore, the pressure plateaus in the PCT shown in Figure 5 should correspond to the BCC phase, and those from TiNi and C14 phases are below the minimum pressure used in our PCT apparatus (0.001 MPa). This explains why the irrevesible part of the capacity (below 0.001 MPa) has a linear correlation with the abundance of the TiNi phase, as shown in Figure 6.

The ΔHs calculated with Equation (3) and the composition of the BCC phase highlighted in Table 4 for each alloy are listed in the last column of Table 6 and compared to the measured ΔH values in Figure 7a. Except for alloy P42 (Mo), the measured ΔHs, no matter the variation from the absorption (A), desorption isotherms (D) or half-point (half), are lower (more negative) and do not vary much compared to the calculated values. The decrease in the measured ΔH is due to the synergetic effects associated with the other two phases that have a much lower ΔH and will be discussed in the next section of this paper. In addition, the plateau pressure for the PCT and BCC phase lattice constants have been correlated to the calculated ΔH in the BCC phase, and the results are shown in Figure 7b. Except for alloy P42 (Mo), it is logical to connect the larger BCC unit cell volume to both lower plateau pressure and lower (more negative) ΔH, indicating a stronger M–H bond. There is an appreciable amount of Mo in the BCC phase of alloy P42 (1.4 at %). Mo has a relatively large atomic size and contributes to the increase in lattice parameter. However, instead of lowering the ΔH, Mo has a positive contribution (ΔH for $MoH_{0.5} = +10 \text{ kJ} \cdot \text{mol}^{-1}$ H_2 [52]). This explains the abnormal behavior of alloy P42 (Mo) in Figure 7.

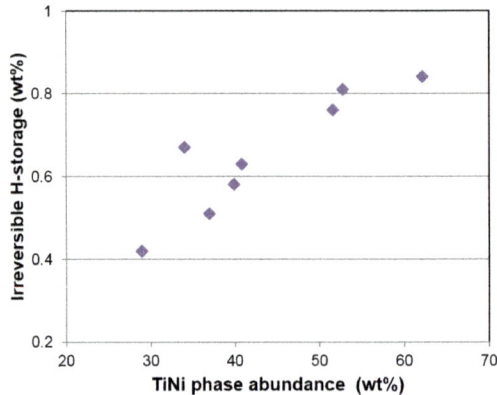

Figure 6. Plot of the amount of irreversible hydrogen storage found in PCT analysis *vs.* the TiNi phase abundance.

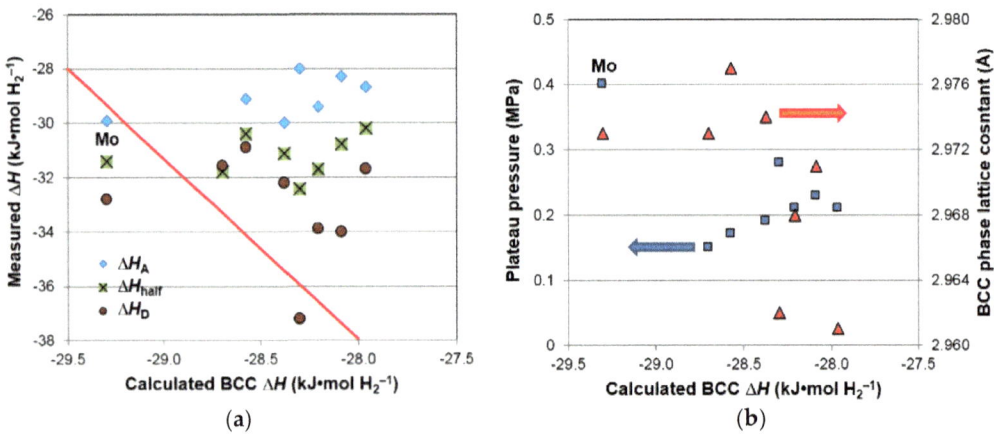

Figure 7. (**a**) The measured ΔH based on the absorption, half-point, and desorption isotherms and (**b**) Plateau pressure and BCC phase lattice constants *vs.* the calculated ΔH based on BCC composition determined by EDS (Table 4) and Equation (3).

Table 6. Thermodynamic properties ΔH and ΔS calculated from the equilibrium pressure plateaus (A: absorption, D: desorption, (A + D)/2:half-point) measured at 30, 45, and 60 °C with error bars of approximately $\pm 3\%$. $-\Delta H_{Calc}$ is the calculated value from the BCC phase composition obtained with SEM-EDS (Table 4) using Equation (3).

Alloy	$-\Delta H_A$ (kJ·mol^{-1})	$-\Delta S_A$ (J·mol^{-1}·K^{-1})	$-\Delta H_D$ (kJ·mol^{-1})	$-\Delta S_D$ (J·mol^{-1}·K^{-1})	$-\Delta H_{(A+D)/2}$ (kJ·mol^{-1})	$-\Delta S_{(A+D)/2}$ (J·mol^{-1}·K^{-1})	$-\Delta H_{Calc}$ (kJ·mol^{-1})
P17	29.2	111	30.9	107	30.4	110	28.6
P38	28.7	111	31.7	111	30.2	111	28.0
P39	29.4	113	33.9	118	31.7	115	28.2
P40	30.0	114	32.2	111	31.1	113	28.4
P41	28.3	111	34.0	119	30.8	114	28.1
P42	29.9	119	32.8	120	31.4	120	29.3
P43	28.0	111	37.2	131	32.4	120	28.3
P44	31.6	118	31.6	108	31.8	113	29.1

3.4. Electrochemical Properties

The electrochemical hydrogen storage properties were determined in a flooded half-cell configuration with a commercially available co-precipitated $Ni_{91}Co_{4.5}Zn_{4.5}$ hydroxide (BASF, Rochester Hills, MI, USA) counter electrode and 30 wt % KOH electrolyte. The pressed electrode was charged with a current density of 100 mA·g^{-1} for 5 h and then discharged with the same current density to a cut-off voltage of 0.9 V with two more pulls at 24 mA·g^{-1} and 8 mA·g^{-1} at the same cut-off voltage. The obtained total capacities and *HRD*s (as defined by the ratio between capacities obtained from 100 mA·g^{-1} and the total discharge capacity) from the first 13 cycles are plotted in Figure 8a,b, respectively, to demonstrate the activation behavior of these alloys. Most of the alloys reach their maximum capacity at Cycle 3 or 4, except for alloys P38 (B) and P40 (Y) which require more cycles to stabilize the capacity. It is easier to achieve *HRD* activation for alloys with higher *HRD* than those with lower *HRD* (alloys P40, P38, and P43). The capacities from 100 mA·g^{-1} and 8 mA·g^{-1} discharge currents and their ratios for Cycle 5 are listed in Table 7. It is obvious that while the increase in the full capacity is only marginal with B (P38) and Nd (P44), the decrease in high-rate capacity and *HRD* is significant. The additives chosen for this study do not significantly improve the electrochemical properties of the base alloy (P17). The gaseous phase capacity is converted to the electrochemical capacity following 1 wt% H_2 = 268 mAh g^{-1} and plotted in Figure 9. As in other MH alloys [53], both the low-rate and high-rate electrochemical capacities fall in between the boundaries set by the maximum and reversible gaseous phase capacities. There is no clear correlation between the gaseous phase maximum storage capacity and the electrochemical low-rate capacity, but there is a strong similarity between the gaseous phase reversible capacity and the electrochemical high-rate capacity, except for the base alloy P17 (V).

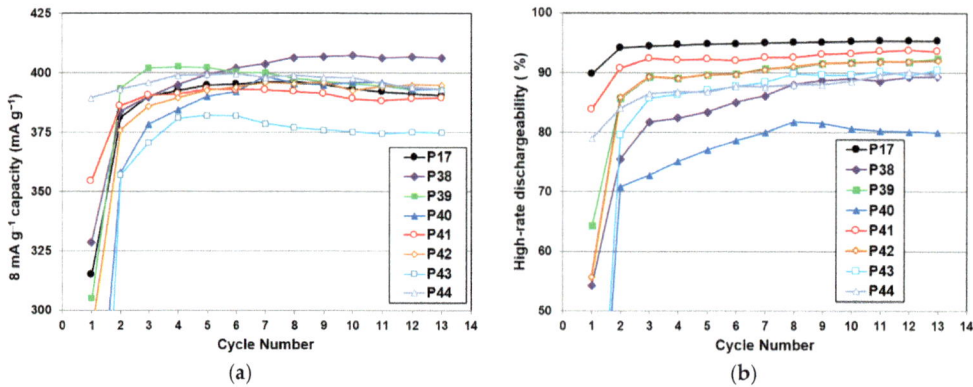

Figure 8. The evolution of (**a**) maximum capacity and (**b**) *HRD* in the first 13 cycles.

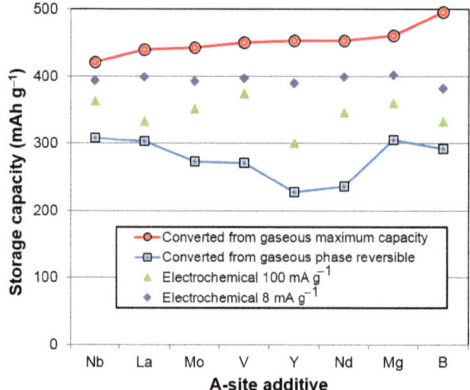

Figure 9. Comparison of capacities obtained from gaseous phase and electrochemical measurements. The gaseous phase capacity is converted by 1 wt% = 268 mAh·g^{-1}.

In order to further study *HRD* performance, the bulk diffusion constant (D) and surface exchange current (I_0) of each sample were measured at RT according to previously described procedures [54] and are listed in Table 7. We found that while D values in the modified alloys are higher than the base alloy, the opposite is true for I_0 values (except for alloy P41 (Nb)). The increase in the degree of disorder, achieved by introducing an additional element, facilitates the diffusion of hydrogen in the bulk; the decrease in the V-content with a high leaching rate in KOH [55] impedes the surface reaction, except for substitutions with Nb, which is in the same column of V and is expected to have a high corrosion rate in KOH solution. It is difficult to attribute the large decrease observed in I_0 to a 1% decrease in V-content, which will be discussed in the next session. The deterioration of *HRD* with various additives is mainly due to the decrease in surface electrochemical reaction activity.

Table 7. Summary of electrochemical properties. Cap_{100} and Cap_8 are discharge capacities measured for discharge currents at 100 mA·g^{-1} and 8 mA·g^{-1}, respectively. HRD_{100} is the ratio of capacities measured at 100 mA·g^{-1} and 8 mA·g^{-1}. RT: room temperature; and OCV: open circuit voltage.

Alloy	Cap_{100} at 5th Cycle (mAh·g^{-1})	Cap_8 at 5th Cycle (mAh·g^{-1})	HRD_{100} (Cap_{100}/Cap_8)	Activation Cycle to Reach Maximum Capacity	Diffusion Coefficient, D @RT (10^{-10} cm^2·s^{-1})	Exchange Current I_o @RT (mA·g^{-1})	OCV (V)
P17	374.0	397.4	0.948	7	1.69	36.5	1.302
P38	332.6	399.0	0.834	10	2.24	22.3	1.303
P39	360.3	402.1	0.896	4	1.81	35.6	1.311
P40	300.2	389.9	0.770	7	1.75	15.0	1.258
P41	362.5	392.9	0.922	5	2.49	37.6	1.319
P42	351.5	392.4	0.896	7	2.27	19.6	1.258
P43	332.6	382.1	0.870	5	2.57	19.2	1.237
P44	346.1	399.2	0.868	4	1.99	15.4	1.266

The OCV was measured at a state-of-charge of 50% and results are listed in Table 7. Theoretically, OCV can be related to the equilibrium pressure in the gaseous phase through the Nernst equation:

$$OCV\ (vs.\ \text{NiOOH at } 0.36\,\text{V}) = 1.294 + 0.029\log p(\text{H}_2)\ [7] \tag{4}$$

However, the plot of the desorption pressure *versus OCV*, as shown in Figure 10, does not follow the trend where higher pressure corresponds to higher OCV, as indicated in Equation (4). However, the OCV in this study seems to correlate closely with the maximum hydrogen storage in the gaseous phase (Figure 10). The alloys with higher OCV show smaller gaseous phase H-storage capacities. Similar phenomena have been observed previously, but were tied to the change in plateau pressure and strength of M–H bonding [56].

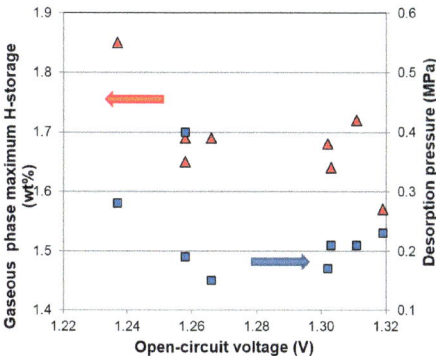

Figure 10. The plots of gaseous phase maximum hydrogen storage capacity and equilibrium plateau pressure from PCT desorption isotherm *vs.* the OCV obtained from the half-cell measurement.

4. Discussion

It is generally believed that in a multi-phase MH system the phase with a lower plateau pressure and more negative ΔH is a main storage phase, which has a larger hydrogen storage capacity (Zr_7Ni_{10}, for example [57]), and the secondary phase with a higher plateau pressure and less negative ΔH is considered to be the catalytic phase (Zr_2Ni_7, for example [57]). In the case of funneling phenomenon, the gaseous phase hydrogen storage is accomplished with the catalytic phase as the necessary funnel to move hydrogen in and out of the alloy [2]. In the case of Laves phase-related BCC MH alloy, the BCC phase is considered to be the main storage phase; although it has a large hydrogen storage capacity, it has limited absorption/desorption kinetics that require an additional catalytic phase to facilitate the hydrogen storage process [58]. Therefore, both C14 and TiNi were considered to be catalytic phases in this family of alloys [32–36]. However, the ΔH values estimated from Equation (3) for the C14 and TiNi phases are much lower than those of the BCC phase, and the visible plateau in the pressure between 0.1 MPa and 1 MPa was assigned to the BCC phase. It is necessary to verify the correctness of such an assignment.

In order to investigate the phase abundance evolution during both gaseous phase and electrochemical hydrogen absorption/desorption, an alloy with a target composition of P17 (Table 2) was reproduced through conventional induction melting. The XRD patterns from the as-prepared samples show an additional Ti_2Ni phase (Figure 11a) which was not observed in the sample prepared by arc melting, due to different cooling rates (Figure 1). The Rietveld refinement indicates percentages of 53.6% BCC, 29.4% TiNi, 10.8% Ti_2Ni, and 6.2% C14 (Table 8). The sample was activated in a 4 MPa hydrogen environment first and then degassed under vacuum for 8 h at 300 °C. XRD analysis at this stage (Figure 11b) shows a shift for TiNi peaks to lower angles, which indicates the presence of some hydrogen remaining in the α-TiNi. The Rietveld refinement shows the β-phase (MH) of C14 and Ti_2Ni. There are still MH remaining in the alloy after 8 h of degassing. In order to understand the cause of the remaining hydride after 8 h degassing, a discussion of two dissimilar metals is necessary. When two metallic phases, M1 and M2, with different work functions (differences between the electron potential in vacuum (E_{VAC}) and the Fermi level (E_F)) are brought into contact (Figure 12b), there will be a small charge transfer that builds a potential (contact potential), preventing electron flow from the metal with a smaller work function (M2) into the metal with a larger work function (M1). During the initial hydrogenation, the neutral hydrogen (proton plus a nearby electron) will reside on the M1 side of the boundary. The extra electron brought by proton will contribute to the conduction-band and raise the Fermi level, as shown in Figure 12c [59]. These protons will stay balanced in equilibrium under vacuum. In the case of MH, the hydrogen storage capability of a hydride former metal is related to its own electron density. For the study in MH from elements, the heat of hydride formation (indicator for M–H bond strength) and the work function for the first row of transition metals (from Sc to Ni) were plotted against their number of 3d electron in Figure 13. As the number of electrons increases, the work function increases due to the increase in the charge of the nucleus, and the host metal starts to resist incorporation of extra electrons brought by the absorbed hydrogen and consequently weakens the MH bond strength (less negative ΔH), with the exception of Mn. Mn has an extraordinarily low work function due to its containing of the maximum number of un-paired electrons (five) and also to a lower ΔH than predicted by the trend. For MH from intermetallic alloys, the situation is the opposite. Comparing the alloys $LaNi_2$ and $LaNi_5$, for example, the former has a lower electron density (lower Ni-content), a lower E_F (as M1 in Figure 13), and a tendency to trap the residual hydrogen near the interface. According to Equation (3), $LaNi_2$ (33% of La) has a stronger MH bond strength compared to that of $LaNi_5$ (16% of La). Therefore, we believe the phase with a stronger MH bond will keep the MH (β-phase) during a total degassing in vacuum. In this study, C14 and Ti_2Ni are deemed to be the phases with the strongest MH bond strength, which agrees with the prediction from Equation (3). These pre-occupied sites at the grain boundary on the side with a higher work function will act as the nucleation center for the β-phase growth (as illustrated in Figure 14). Therefore, the PCT absorption/desorption hysteresis of a single-phase AB_5 MH alloy is always larger than that from a typical multi-phase AB_2 MH alloy [7].

Table 8. Summary of the phase abundance of a P17 alloy prepared by induction melting through various gaseous phases and electrochemical hydrogen absorption/desorption processes.

Stage	BCC		C14		TiNi		Ti$_2$Ni	
	α	β	α	β	α	β	α	β
Pristine	53.6	-	6.2		29.4	-	10.8	
300 °C vacuum	54.0	-	3.0	2.0	30.6	-	4.2	6.2
0.005 MPa	51.7	-	-	4.1	11.3	22.2	3.6	7.1
4 MPa	6.1	46.4	-	3.6	-	34.9	-	9.0
50% SOC	53.4	-	-	5.1	0.7	32.0	-	8.8
100% SOC	29.5	21.1	-	4.7	-	36.5	-	8.2

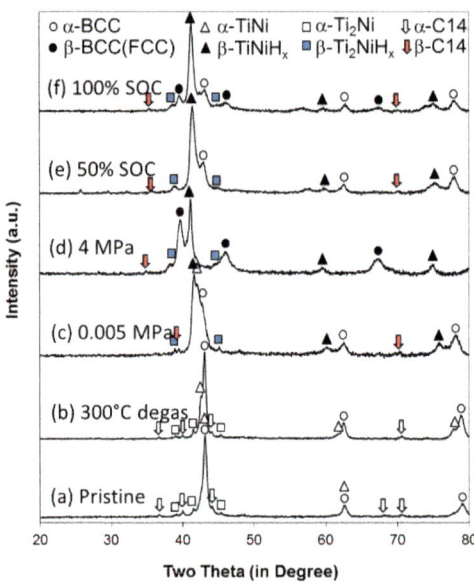

Figure 11. XRD patterns of a P17 alloy prepared by the conventional melt-and-cast process: (**a**) pristine; (**b**) hydrided at 4 MPa H_2 gas and then degassed in vacuum for 8 h at 300 °C; (**c**) withdrawn from an equilibrium state with 0.005 MPa H_2 pressure; and (**d**) pull-out from an equilibrium state with 4 MPa H_2 pressure, and the electrodes made from the same alloy with (**e**) 50% and (**f**) 100% state of charge achieved electrochemically.

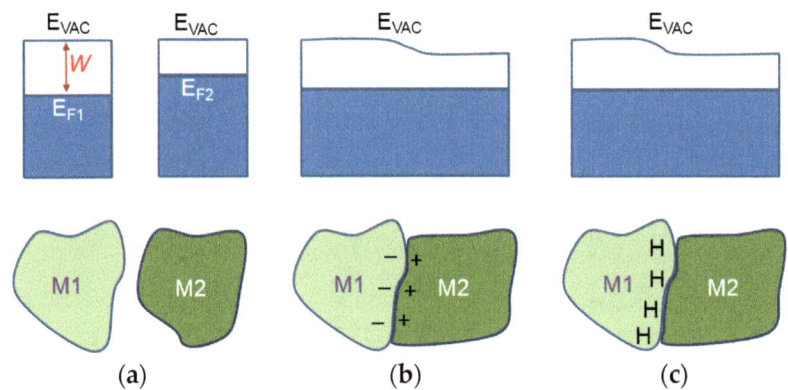

Figure 12. Two MH alloys with different work functions (W): (**a**) before making contact; (**b**) after making contact; and (**c**) after hydrogenation and dehydrogenation.

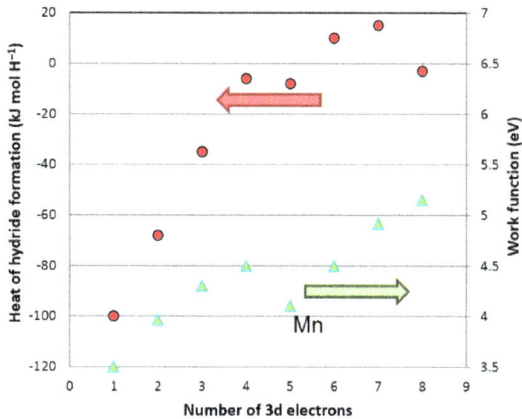

Figure 13. A plot of the heat of hydride formation (data from [52]) and work function (data from [60]) *vs.* the number of 3d electrons for the first row of transition metal elements from Sc to Ni.

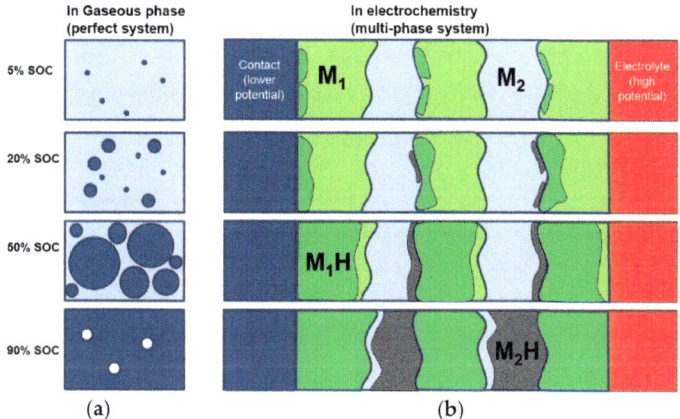

Figure 14. Comparison of hydrogen storage through (**a**) a gaseous phase reaction in a perfect single-component system and (**b**) an electrochemical reaction in a two-phase system with different work functions ($W_{M1} > W_{M2}$) at different states of charge (SOC). In this illustration, the M1 phase has a larger W than M2, and contains a small amount of hydride near the boundary, even at a very low SOC. The MH (β) phase nucleates from a randomly distributed proton (α) in the gaseous phase (**left**) that, during electrochemistry, starts from the remaining MH in the phase with a larger W near the grain boundary.

A sample was removed from the PCT apparatus with an equilibrium hydrogen pressure of 0.005 MPa, a pressure that separates the two plateaus in the PCT isotherms, as shown in Figure 5. From previous studies in BCC MH alloys, we

303

believe the oxide layer formed at the alloy surface during the sudden exposure to air can prevent the loss of hydrogen into atmosphere [61]. Its XRD pattern is shown in Figure 11c, and the Rietveld analysis shows an unchanged BCC phase with 2/3 hydrided TiNi and Ti_2Ni phase (Table 8) and demonstrates that the higher plateau (from 0.1 MPa to 1 MPa) belongs to the BCC phase. Another sample was removed at the maximum hydrogen pressure (4 MPa). The XRD pattern of this sample shows a 90% transition from BCC (M) to FCC (MH) in the BCC phase and complete α- to β-phase transitions for the other three phases.

Besides the phase transition in the gaseous phase hydrogen reaction, we also performed XRD analysis on two pieces of electrode made from the same P17 alloy by induction melting with 50% and 100% state of charge (SOC). The resulting patterns are shown in Figure 11e,f. The Rietveld analysis shows a clear α-BCC phase and almost fully hydrided TiNi, Ti_2Ni, and C14 phases for the 50% SOC sample, while a 40% hydrided BCC (β-BCC) phase with the other phases fully hydrided are seen for the 100% SOC. During charging, the phases with stronger MH bond strength (more negative ΔH) were hydrided first, and BCC was charged last, as in the case of the gaseous phase reaction. The BCC phase was not hydrogenated before completion of hydride formation in other non-BCC phases. The BCC phase was never fully charged in the half-cell testing, due to the nature of the open-to-air configuration used in the electrochemical study. This explains the low electrochemical discharge capacity found in Figure 9, when compared to the gaseous phase maximum capacity.

There are some important differences in hydrogen storage between gaseous phase and electrochemistry, especially when the PCT isotherm (Figure 5) is compared with the electrochemical charge and discharge curves (Figure 15). While the gaseous hydrogen charging occurs through hydrogen gas molecule adsorption and splitting into two hydrogen atoms at the clean surface of metals (free from oxide), the electrochemical hydrogen charging is implemented by an applied voltage, which drives electrons into the negative electrode and leads to water splitting at the surface (which is typically covered with a thin oxide [62]). In PCT, hydrogen absorption can begin as soon as any amount of hydrogen gas is available. However, in the case of an electrochemical environment, a minimal voltage is required to generate enough of a strong electric field to split water into protons and hydroxide ions. Also in PCT, the equilibrium pressure depends only on the concentration of hydrogen in the MH; the electrochemical voltage is mainly determined by the surface reaction and thus a sudden change in the charge voltage profile is not seen when the active storing material switches from one phase into another one (at approximately 50% SOC). The discharge processes between the gaseous phase and electrochemistry are also different. For the gaseous phase, the movement of proton is influenced by diffusion and an equilibrium is reached when the same amount of hydrogen gas leaves and enters the metal. The electrochemical case is far more complicated. During discharge, the electron moves away from the MH alloy into the current collector, which forces the proton to move in the opposite direction, reaching the surface. Unless the surface recombination of either proton-hydroxide or proton-oxygen is very slow (which is unlikely) and there is a large number of protons accumulated at the electrolyte interface, the proton should continue to arrive at the interface to reach charge neutrality. Energy is gained from both the reduction in number of electrons in the metal system and the dehydride process. Therefore, there is no sudden drop in the discharge voltage profile when the system finishes discharging for one phase and starts the discharge for the next phase. This finding is very encouraging with regard to the electrochemical applications of MH alloys. In a multi-phase system, as long as there is an electrochemically active phase, the so called "irreversible" or "no participation" phases found in PCT can be fully utilized in the electrochemical environment. This finding can be used to explain the unpredicted high electrochemical capacity (compared to the gaseous phase capacities) observed in multi-phase systems involving the Zr_2Ni_7 phase [8,9].

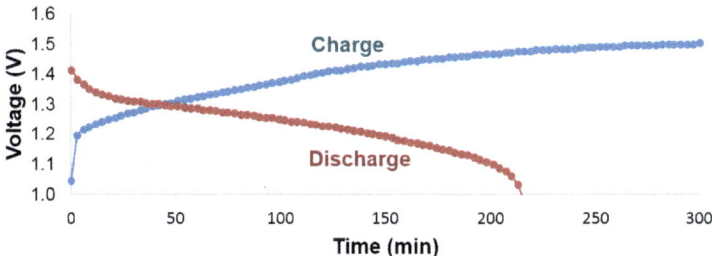

Figure 15. The second cycle electrochemical charge and discharge voltage profiles for an electrode made from alloy P40. The smooth curves seen here are very different from the step-shape PCT isotherm shown in Figure 5.

With the knowledge that the visible plateaus of the PCT isotherm in the 0.1–1 MPa range originate entirely from the phase transition from BCC to FCC, the mechanism of this step, at equilibrium pressure in pure BCC alloys [61,63], is not well understood. Evidently, with synergetic effects through the non-BCC phases, the intermediate body-centered-tetragonal (BCT) phase is skipped. Although the hysteresis of the BCC phase is still very large (compared to all other MH alloy families), it has been significantly reduced through the synergetic effects of the non-BCC phases. In this study, it is clearly observed that, during hydrogenation, the non-BCC phases hydrogenate first and expand the lattice on the non-BCC phase side of the grain boundary. The lattice of the BCC phase is also pre-expanded before hydrogenation by the stress from the hydride on the other side, which reduces the energy barrier needed to expand the BCC lattice, and thus reduces the hysteresis and changes the FCC β-phase directly.

After we identified BCC as the storage phase and the others (C14, TiNi, and Ti$_2$Ni) as the catalytic phase, a correlation study between the properties of the constituent phases and the hydrogen storage characteristics (both in gaseous phase and electrochemistry) was conducted; the results are summarized in Table 9. Those with significant correlation (larger R^2) are highlighted. There are two noticeable correlations in the gaseous phase properties. The first is the reversible capacity, which decreases with the increase in TiNi phase abundance and C14 unit cell volume (Figure 16a). The former has already been explained by the TiNi phase being the largest component in the lower plateau (not seen from the PCT isotherm in Figure 5) and makes the largest contribution to the irreversible part of the gaseous phase hydrogen storage capacity (Figure 6). The source of the latter correlation is less clear. C14, although it has a relatively low ΔH, may be still a catalytic phase. With the decrease in the unit cell volume, hydride from the C14 phase becomes less stable and contributes positively to the reversible capacity of the storage phase (TiNi and BCC). The second noticeable correlation found in the gaseous phase properties is related to the PCT hysteresis—more precisely, the hysteresis of the BCC phase. It has been found that the 30 °C PCT hysteresis increases with increasing TiNi phase abundance and decreasing BCC phase abundance (Figure 16b). Many studies have been conducted on the PCT hysteresis (for a review see [64]). It is generally accepted that the PCT hysteresis is the elastic energy needed for the deformation of the lattice near the α-β boundary. Higher PCT hysteresis in an MH alloy increases the difficulty of hydrogenation. From this aspect, the BCC phase facilitates reaction, while TiNi phase retards the hydrogenation of the BCC phase, which is very difficult to observe from the large value of hysteresis since it involves phase changes from BCC to FCC [6].

Table 9. Table of correlation coefficient (R^2) between hydrogen storage properties and phase component characteristics. Significant correlations are highlighted in **bold**.

Property	Gaseous Phase					Electrochemistry		
	Plateau Pressure @30 °C	Plateau Pressure @60 °C	Maximum Capacity	Reversible Capacity	PCT Hysteresis	High-Rate Discharge Capacity	Full Discharge Capacity	*HRD*
BCC, lattice constant	0.02	0.04	0.14	0.36	0.05	0.07	0.04	0.06
BCC, crystallite size	0.00	0.04	0.21	0.26	0.40	**0.57**	0.15	**0.53**
BCC, abundance	0.11	0.01	0.22	0.41	**0.53**	0.38	0.27	0.28
C14, unit cell volume	0.17	0.08	0.01	**0.72**	0.17	0.09	0.02	0.08
C14, crystallite size	0.31	0.32	0.13	0.00	0.04	0.04	0.00	0.06
C14, abundance	0.10	0.08	0.05	0.09	0.07	0.19	0.02	0.26
TiNi, lattice constant	0.03	0.05	0.13	0.26	0.08	0.17	0.05	0.16
TiNi, crystallite size	0.16	0.15	0.09	0.19	0.06	0.04	0.02	0.04
TiNi, abundance	0.02	0.00	0.31	**0.57**	**0.66**	**0.64**	0.15	**0.59**

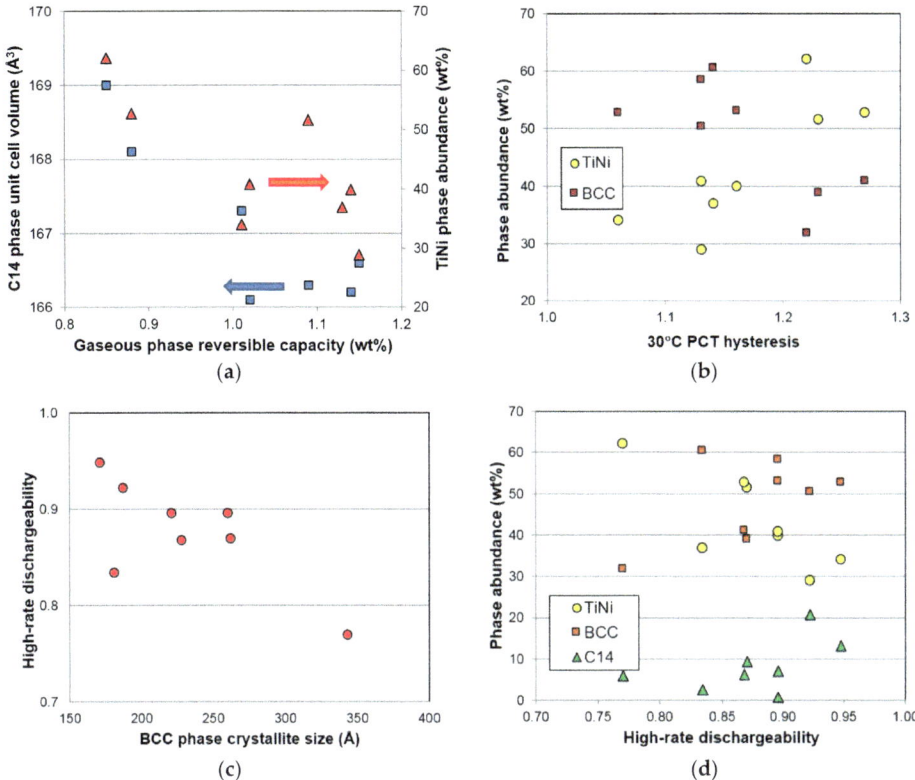

Figure 16. Plots of (**a**) C14 unit cell volume and TiNi phase abundance *vs.* gaseous phase reversible capacity; (**b**) TiNi and BCC phase abundance *vs.* PCT hysteresis measured at 30 °C; (**c**) *HRD vs.* BCC phase crystallite size; and (**d**) TiNi, BCC, and C14 phase abundances *vs. HRD*.

As for correlation to the electrochemical properties, both the BCC crystallite size and the TiNi phase abundance show significant correlation to the high-rate discharge performance. A smaller crystallite in the BCC phase is more desirable, as it gives better high-rate performance (Figure 16c). It is possible that smaller crystallite sizes will generate more grain boundaries (shown in Figure 12c) and facilitate more charge transfer between the two phases. The phase abundances from the three major phases are plotted against *HRD* in Figure 16d. The general trends suggest that the BCC and C14 phases are more beneficial to *HRD*, while the TiNi phase deteriorates it.

5. Conclusions

In this study on the synergetic effects in MH alloys, we challenge the argument that the catalytic phase must have a weaker MH bond strength compared to that in the main storage phase. The C14 phase, with an ostensibly stronger MH bond, contributes to the reversibility in the hydrogen storage of other storage phases. The BCC phase, in a multi-phase system, demonstrates a relatively weak MH bond. On the other hand, the TiNi phase, also with a strong MH bond, hinders the hydrogenation of the BCC phase. The driving force for proton movement in the alloy varies for gaseous phase and electrochemistry. The former occurs by diffusion and the equilibrium with gaseous hydrogen at the clean surface, while the latter proceeds via the electric field and the local distribution of valence electrons. The synergetic effects in a multi-phase MH system can be explained by the pre-occupied hydrogen sites on the side of the metal with a larger work function near the grain boundary.

Acknowledgments: The authors would like to thank the following individuals from BASF-Ovonic for their help: Su Cronogue, Baoquan Huang, Jean Nei, David Pawlik, Allen Chan, and Ryan J. Blankenship.

Author Contributions: Kwo-hsiung Young conceived and designed the experiments; Taihei Ouchi performed the experiments; and Tiejun Meng and Diana Wong analyzed, interpreted the data, and prepared the manuscript.

Conflicts of Interest: The authors declare no conflict of interest.

Abbreviations

BCC	Body-centered-cubic
BCT	Body-centered-tetragonal
FCC	Face-centered-cubic
MH	Metal hydride
PCT	Pressure-centration-temperature
M–H	Metal–hydrogen
XRD	X-ray diffraction
HRD	High-rate dischargeability
ICP-OES	Inductively coupled plasma-optical emission spectrometer
SEM	Scanning electron microscopy
EDS	Energy-dispersive spectroscopy
BEI	Backscattering electrode image
TEM	Transmission electron microscopy
e/a	Average electron density
OCV	Open-circuit voltage
E_{VAC}	Vacuum potential
E_F	Fermi level
W	Work function
SOC	State of charge

References

1. Visintin, A.; Peretti, A.A.; Fruiz, F.; Corso, H.L.; Triaca, W.E. Effect of additional catalytic phases imposed by sintering on the hydrogen absorption behavior of AB_2 type Zr-based alloys. *J. Alloys Compd.* **2007**, *428*, 244–251.

2. Wong, D.F.; Young, K.; Nei, J.; Wang, L.; Ng, K.Y.S. Effects of Nd-addition on the structural, hydrogen storage, and electrochemical properties of C14 metal hydride alloys. *J. Alloys Compd.* **2015**, *647*, 507–518.

3. Boettinger, W.J.; Newbury, D.E.; Wang, K.; Bendersky, L.A.; Chiu, C.; Kattner, U.R.; Young, K.; Chao, B. Examination of multiphase $(Zr,Ti)(V,Cr,Mn,Ni)_2$ Ni-MH electrode alloys: Part I. Dendritic solidification structure. *Metall. Mater. Trans. A* **2010**, *41*, 2033–2047.

4. Bendersky, L.A.; Wang, K.; Boettinger, W.J.; Newbury, D.E.; Young, K.; Chao, B. Examination of multiphase $(Zr,Ti)(V,Cr,Mn,Ni)_2$ Ni-MH electrode alloys: Part II. Solid-state transformation of the interdendritic B_2 phase. *Metall. Mater. Trans. A* **2010**, *41*, 1891–1906.

5. Liu, Y.; Young, K. Microstructure investigation on metal hydride alloys by electron backscatter diffraction technique. *Batteries* **2016**. submitted for publication.

6. Shen, H.; Young, K.; Bendersky, L.A. Clean grain boundary found in C14/bcc multi-phase metal hydride alloys. *Batteries* **2016**. submitted for publication.

7. Mosavati, N.; Young, K.; Meng, T.; Ng, K.Y.S. Electrochemical open-circuit voltage and pressure-concentration-temperature isotherm comparison for metal hydride alloys. *Batteries* **2016**, *2*.

8. Young, M.; Chang, S.; Young, K.; Nei, J. Hydrogen storage properties of $ZrV_xNi_{3.5-x}$ ($x = 0.0–0.9$) metal hydride alloys. *J. Alloys Compd.* **2013**, *580*, S171–S174.

9. Young, K.; Young, M.; Chang, S.; Huang, B. Synergetic effects in electrochemical properties of $ZrV_xNi_{4.5-x}$ ($x = 0.0, 0.1, 0.2, 0.3, 0.4$, and 0.5) metal hydride alloys. *J. Alloys Compd.* **2013**, *560*, 33–41.

10. McCormack, M.; Badding, M.E.; Vyas, B.; Zahurak, S.M.; Murphy, D.W. The role of microcracking in ZrCrNi hydride electrodes. *J. Electrochem. Soc.* **1996**, *143*, L31–L33.

11. Zhang, W.K.; Ma, C.A.; Yang, X.G.; Lei, Y.Q.; Wang, Q.D.; Lu, G.L. Influences of annealing heat treatment on phase structure and electrochemical properties of $Zr(MnVNi)_2$ hydrogen storage alloys. *J. Alloys Compd.* **1999**, *293–295*, 691–697.

12. Zhang, Q.A.; Lei, Y.Q.; Yang, X.G.; Ren, K.; Wang, Q.D. Annealing treatment of AB_2-type hydrogen storage alloys: II. Electrochemical properties. *J. Alloys Compd.* **1999**, *292*, 241–246.

13. Bououdina, M.; Lenain, C.; Aymard, L.; Soubeyroux, J.L.; Fruchart, D. The effects of heat treatments on the microstructure and electrochemical properties of the $ZrCr_{0.7}Ni_{1.3}$ multiphase alloy. *J. Alloys Compd.* **2001**, *327*, 178–184.

14. Rongeat, C.; Roué, L. On the cycle life improvement of amorphous MgNi-based alloy for Ni-MH batteries. *J. Alloys Compd.* **2005**, *404–406*, 679–681.

15. Sun, J.C.; Li, S.; Ji, S.J. The effects of the substitution of Ti and La for Zr in $ZrMn_{0.7}V_{0.2}Co_{0.1}Ni_{1.2}$ hydrogen storage alloys on the phase structure and electrochemical properties. *J. Alloys Compd.* **2007**, *446–447*, 630–634.

16. Ruiz, F.C.; Castro, E.B.; Real, S.G.; Peretti, H.A.; Visintin, A.; Triaca, W.E. Electrochemical characterization of AB_2 alloys used for negative electrodes in Ni/MH batteries. *Int. J. Hydrog. Energy* **2008**, *33*, 3576–3580.

17. Ruiz, F.C.; Peretti, H.A.; Visintin, A.; Real, S.G.; Castro, E.B.; Corso, H.L. Effect of thermal treatment on the electrochemical hydrogen absorption of ZrCrNi alloy. *J. New Mater. Electrochem. Syst.* **2007**, *10*, 249–254.

18. Ruiz, F.C.; Castro, E.B.; Peretti, H.A.; Visintin, A. Study of the different Zr_xNi_y phases of Zr-based AB_2 materials. *Int. J. Hydrog. Energy* **2010**, *35*, 9879–9887.

19. Young, K.; Ouchi, T.; Liu, Y.; Reichman, B.; Mays, W.; Fetcenko, M.A. Structural and electrochemical properties of $Ti_xZr_{7-x}Ni_{10}$. *J. Alloys Compd.* **2009**, *480*, 521–528.

20. Young, K.; Ouchi, T.; Huang, B.; Chao, B.; Fetcenko, M.A.; Bendersky, L.A.; Wang, K.; Chiu, C. The correlation of C14/C15 phase abundance and electrochemical properties in the AB_2 alloys. *J. Alloys Compd.* **2010**, *506*, 841–848.

21. Young, K.; Nei, J.; Ouchi, T.; Fetcenko, M.A. Phase abundances in AB_2 metal hydride alloys and their correlations to various properties. *J. Alloys Compd.* **2011**, *509*, 2277–2284.

22. Young, K.; Chao, B.; Bendersky, L.A.; Wang, K. $Ti_{12.5}Zr_{21}V_{10}Cr_{8.5}Mn_xCo_{1.5}Ni_{46.5-x}$ AB_2-type metal hydride alloys for electrochemical storage application: Part 2. Hydrogen storage and electrochemical properties. *J. Power Sources* **2012**, *218*, 487–494.

23. Pei, L.; Han, S.; Hu, L.; Zhao, X.; Liu, Y. Phase structure and hydrogen storage properties of $LaMg_{3.93}Ni_{0.21}$ alloy. *J. Rare Earths* **2012**, *30*, 534–539.

24. Young, K.; Chao, B.; Huang, B.; Nei, J. Studies on the hydrogen storage characteristic of $La_{1-x}Ce_x(NiCoMnAlCuSiZr)_{5.7}$ with a B2 secondary phase. *J. Alloys Compd.* **2015**, *585*, 760–770.

25. Wang, Y.; Zhao, M. Electrochemical characteristics and synergetic effect of $Ti_{0.10}Zr_{0.15}V_{0.35}Cr_{0.10}Ni_{0.30}$-10 wt.% $LaNi_5$ hydrogen storage composite electrode. *J. Rare Earths* **2012**, *30*, 146–150.

26. Nei, J.; Young, K.; Salley, S.O.; Ng, K.Y.S. Effects of annealing on $Zr_8Ni_{19}X_2$ (X = Ni, Mg, Al, Sc, V, Mn, Co, Sn, La, and Hf): Hydrogen storage and electrochemical properties. *Int. J. Hydrog. Energy* **2012**, *37*, 8414–8427.

27. Young, K.; Ouchi, T.; Huang, B. Effects of various annealing conditions on (Nd, Mg, Zr)(Ni, Al, Co)$_{3.74}$ metal hydride alloys. *J. Power Sources* **2014**, *248*, 147–153.

28. Zhang, L.; Chen, L.; Xiao, X.; Chen, Z.; Wang, S.; Fan, X.; Li, S.; Ge, H.; Wang, Q. Superior dehydrogenation performance of nanoscale lithium borohydride modified with fluorographite. *Int. J. Hydrog. Energy* **2014**, *39*, 896–904.

29. Zhang, L.; Zheng, J.; Chen, L.; Xiao, X.; Qin, T.; Jiang, Y.; Li, S.; Ge, H.; Wang, Q. Remarkable enhancement in dehydrogenation properties of $Mg(BH_4)_2$ modified by the synergetic effect of fluorographite and $LiBH_4$. *Int. J. Hydrog. Energy* **2015**, *40*, 14163–14172.

30. Zhou, C.; Fang, Z.Z.; Sun, P. An experimental survey of additives for improving dehydrogenation properties of magnesium hydride. *J. Power Sources* **2015**, *278*, 38–42.

31. Young, K.; Ng, K.Y.S.; Bendersky, L.A. A Technical Report of the Robust Affordable Next Generation Energy Storage System-BASF Program. *Batteries* **2016**, *2*.

32. Young, K.; Nei, J.; Wong, D.; Wang, L. Structural, hydrogen storage, and electrochemical properties of Laves-phase related body-centered-cubic solid solution metal hydride alloys. *Int. J. Hydrog. Energy* **2014**, *39*, 21489–21499.

33. Young, K.; Wong, D.F.; Wang, L. Effect of Ti/Cr content on the microstructures and hydrogen storage properties of Laves phase-related body-centered-cubic solid solution alloys. *J. Alloys Compd.* **2015**, *622*, 885–893.

34. Young, K.; Ouchi, T.; Nei, J.; Wang, L. Annealing effects on Laves phase-related body-centered-cubic solid solution metal hydride alloys. *J. Alloys Compd.* **2016**, *654*, 216–225.

35. Young, K.; Ouchi, T.; Nei, J.; Meng, T. Effects of Cr, Zr, V, Mn, Fe, and Co to the hydride properties of Laves phase-related body-centered-cubic solid solution alloys. *J. Power Sources* **2015**, *281*, 164–172.

36. Young, K.; Wong, D.F.; Nei, J. Effects of vanadium/nickel contents in Laves phase-related body-centered-cubic solid solution metal hydride alloys. *Batteries* **2015**, *1*, 34–53.

37. Young, K.; Ouchi, T.; Huang, B.; Fetcenko, M.A. Effects of B, Fe, Gd, Mg, and C on the structure, hydrogen storage, and electrochemical properties of vanadium-free AB_2 metal hydride alloy. *J. Alloys Compd.* **2012**, *511*, 242–250.

38. Chang, S.; Young, K.; Ouchi, T.; Meng, T.; Nei, J.; Wu, X. Studies on incorporation of Mg in Zr-based AB_2 metal hydride alloys. *Batteries* **2016**, *2*.

39. Gakkai, N.K. *Hi Kagaku Ryouronteki Kinzoku Kagobutu*; Maruzen: Tokyo, Japan, 1975; p. 296. (In Japanese)

40. Young, K.; Ouchi, T.; Huang, B.; Reichman, B.; Fetcenko, M.A. Effect of molybdenum content on structural, gaseous storage, and electrochemical properties of C14-predominant AB_2 metal hydride alloys. *J. Power Sources* **2011**, *196*, 8815–8821.

41. Young, K.; Wong, D.F.; Nei, J.; Reichman, B. Electrochemical properties of hypo-stoichiometric Y-doped AB_2 metal hydride alloys at ultra-low temperature. *J. Alloys Compd.* **2015**, *643*, 17–27.

42. Young, K.; Fetcenko, M.A.; Li, F.; Ouchi, T.; Koch, J. Effect of vanadium substitution in C14 Laves phase alloys for NiMH battery application. *J. Alloys Compd.* **2009**, *468*, 482–492.

43. Wong, D.F.; Young, K. Density function theory calculation of defects in Zr-Ni intermetallic compounds. *Batteries* **2016**. to be submitted for publication.

44. Johnston, R.L.; Hoffmann, R. Structure-bonding relationships in the Laves phase. *Z. Anorg. Allg. Chem.* **1992**, *616*, 105–120.

45. Nei, J.; Young, K.; Salley, S.O.; Ng, K.Y.S. Determination of C14/C15 phase abundance in Laves phase alloys. *Mater. Chem. Phys.* **2012**, *136*, 520–527.

46. Young, K.; Ouchi, T.; Huang, B.; Reichman, B.; Blankenship, R. Improvement in $-40\,^{\circ}C$ electrochemical properties of AB_2 metal hydride alloy by silicon incorporation. *J. Alloys Compd.* **2013**, *575*, 65–72.

47. Young, K.; Ouchi, T.; Reichman, B.; Koch, J.; Fetcenko, M.A. Effects of Mo additive on the structural and electrochemical properties of low-temperature AB_5 metal hydride alloys. *J. Alloys Compd.* **2011**, *509*, 3995–4001.

48. Schwarz, R.B.; Khachaturyan, A.G. Thermodynamics of open two-phase systems with coherent interfaces: Application to metal–hydrogen systems. *Acta Mater.* **2006**, *54*, 313–323.

49. Lexcellent, Ch.; Gondor, G. Analysis of hydride formation for hydrogen storage: Pressure-composition isotherm curves modeling. *Intermetallic* **2007**, *15*, 934–944.

50. Osumi, Y. *Suiso Kyuzou Goukin–Sono Bussei to Ouyou*; Agne Gijutsu Center Inc.: Tokyo, Japan, 1993; p. 57. (In Japanese)

51. De Boer, F.R.; Boom, R.; Mattens, W.C.M.; Miedema, A.R.; Niessen, A.K. *Cohesion in Metal Transition Metal Alloys*; North-Holland Physics Publishing: Amsterdam, The Netherlands, 1989; Volume 1, p. 150.

52. Griessen, R.; Riesterer, T. Heat of Formation Models. In *Hydrogen in Intermetallic Compounds I Electronic, Thermodynamic, and Crystallographic Properties, Preparation*; Schlapbach, L., Ed.; Springer-Verlag: Berlin, Germany, 1988; p. 267.

53. Li, F.; Young, K.; Ouchi, T.; Fetcenko, M.A. Annealing effects on structural and electrochemical properties of $(LaPrNdZr)_{0.83}Mg_{0.17}(NiCoAlMn)_{3.3}$ alloy. *J. Alloys Compd.* **2009**, *471*, 371–377.

54. Young, K.; Fetcenko, M.A.; Li, F.; Ouchi, T. Structural, thermodynamic, and electrochemical properties of $Ti_xZr_{1-x}(VNiCrMnCoAl)_2$ C14 Laves phase alloys. *J. Alloys Compd.* **2008**, *464*, 238–247.

55. Young, K.; Huang, B.; Regmi, R.K.; Lawes, G.; Liu, Y. Comparisons of metallic clusters imbedded in the surface oxide of AB_2, AB_5, and A_2B_7 alloys. *J. Alloys Compd.* **2010**, *506*, 831–840.

56. Wang, L.; Young, K.; Meng, T.; Ouchi, T.; Yasuoka, S. Partial substitution of cobalt for nickel in mixed rare earth metal based superlattice hydrogen absorbing alloy—Part 1 structural, hydrogen storage and electrochemical properties. *J. Alloys Compd.* **2016**, *660*, 407–415.

57. Joubert, J.M.; Latroche, M.; Percheron-Guégan, A. Hydrogen absorption properties of several intermetallic compounds of the Zr-Ni system. *J. Alloys Compd.* **1995**, *31*, 494–497.

58. Qiu, S.; Chu, H.; Zhang, Y.; Sun, D.; Song, X.; Sun, L.; Xu, F. Electrochemical kinetics and its temperature dependence behaviors of $Ti_{0.17}Zr_{0.08}V_{0.35}Cr_{0.10}Ni_{0.30}$ alloy electrode. *J. Alloys Compd.* **2009**, *471*, 453–456.

59. Gupta, M.; Schlapbach, L. Electronic Properties. In *Hydrogen in Intermetallic Compounds I: Electronic, Thermodynamic, and Crystallographic Properties, Preparation*; Schlapbach, L., Ed.; Springer-Verlag: Berlin, Germany, 1988; pp. 139–216.

60. Drummond, T.J. *Work Functions of the Transition Metals and Metal Silicides*; No. SAND99-0391J. Sandia National Labs.: Albuquerque, NM, USA; Livermore, CA, USA. Available online: http://www.osti.gov/scitech/biblio/3597/ (accessed on 20 February 2016).

61. Young, K.; Fetcenko, M.A.; Ouchi, T.; Im, J.; Ovshinsky, S.R.; Li, F.; Reinhout, M. Hydrogen Storage Materials Having Excellent Kinetics, Capacity, and Cycle Stability. U.S. Patent 7,344,676, 18 March 2008.

62. Young, K.; Chao, B.; Liu, Y.; Nei, J. Microstructures of the oxides on the activated AB_2 and AB_5 metal hydride alloys surface. *J. Alloys Compd.* **2014**, *606*, 97–104.

63. Young, K.; Ouchi, T.; Huang, B.; Nei, J. Structure, hydrogen storage, and electrochemical properties of body-centered-cubic $Ti_{40}V_{30}Cr_{15}Mn_{13}X_2$ alloys (X = B, Si, Mn, Ni, Zr, Nb, Mo, and La). *Batteries* **2015**, *1*, 74–90.

64. Young, K.; Ouchi, T.; Fetcenko, M.A. Pressure-composition-temperature hysteresis in C14 Laves phase alloys: Part 1. Simple ternary alloys. *J. Alloys Compd.* **2009**, *480*, 428–433.

First-Principles Point Defect Models for Zr_7Ni_{10} and Zr_2Ni_7 Phases

Diana F. Wong, Kwo-Hsiung Young, Taihei Ouchi and K. Y. Simon Ng

Abstract: Synergetic effects in multi-phased AB_2 Laves-phase-based metal hydride (MH) alloys enable the access of high hydrogen storage secondary phases, despite the lower absorption/desorption kinetics found in nickel/metal hydride (Ni/MH) batteries. Alloy design strategies to further tune the electrochemical properties of these secondary phases include the use of additives and processing techniques to manipulate the chemical nature and the microstructure of these materials. It is also of particular interest to observe the engineering of constitutional point defects and how they may affect electrochemical properties and performance. The Zr_7Ni_{10} phase appears particularly prone to point defects, and we use density functional theory (DFT) calculations coupled with a statistical mechanics model to study the theoretical point defects. The Zr_2Ni_7 phase appears less prone to point defects, and we use the Zr_2Ni_7 point defect model, as well as experimental lattice parameters, with Zr_7Ni_{10} phases from X-ray diffraction (XRD) as points of comparison. The point defect models indicate that anti-site defects tend to form in the Zr_7Ni_{10} phase, and that these defects form more easily in the Zr_7Ni_{10} phase than the Zr_2Ni_7 phase, as expected.

Reprinted from *Batteries*. Cite as: Wong, D.F.; Young, K.-H.; Ouchi, T.; Ng, K.Y.S. First-Principles Point Defect Models for Zr_7Ni_{10} and Zr_2Ni_7 Phases. *Batteries* **2016**, 2, 23.

1. Introduction

Nickel/metal hydride (Ni/MH) batteries utilizing multi-phased AB_2 Laves-phase-based metal hydride (MH) alloy active materials leverage the synergetic effects between secondary phases and the main Laves phases to allow access to the high hydrogen storage of the secondary phases, such as Zr_7Ni_{10}, despite their lower absorption/desorption kinetics [1–5]. The Zr_2Ni_7 phase, on the other hand, has excellent absorption/desorption kinetics, but poor hydrogen storage capacity [4,6]. Modifiers to the Zr-Ni alloys, including Ti in $Ti_xZr_{7-x}Ni_{10}$, have shown improvements in diffusion to help the kinetics [5], and V in $ZrV_xNi_{3.5-x}$ has shown improvements in capacity [6]. In addition to modifiers, constitutional defect structures in the alloys may also affect mechanical and electrochemical properties. Defects, including vacancies, can act to trap hydrogen and inhibit the transport of hydrogen through the alloy [7]. In addition, other defects, such as anti-sites, can promote lower atomic packing ratios, which can improve the cycling capability due to a higher propensity to deform rather than to crack [8,9]. The ability to tune the ratio between hydride formers, such as Zr, and hydride modifiers, such as Ni, as well as to add other modifiers while maintaining the structure of the alloys is an important feature for designing battery materials targeting a specific application, and can strongly affect battery performance properties [10–13]. We have constructed point defect models for the Zr_7Ni_{10} and Zr_2Ni_7 phases from first-principle calculations to lay the groundwork for examining how structure and defects can affect properties that can be linked to Ni/MH battery performance, adding another dimension to consider in the design of negative active materials. Some comparison to experimental lattice parameters show a consistency with predicted defected structures and c/b lattice ratio trends.

2. Computational Details

Point defect models for intermetallics have been constructed and studied in the literature for binary compounds, such as ZrNi [14], FeAl [15,16], NiAl [17], and $NiAl_3$ [18]. We apply similar techniques and assumptions to construct the point defect models for the Zr_7Ni_{10} and Zr_2Ni_7 phases (each with Zr-rich, stoichiometric, and Ni-rich cases); statistical mechanics use defect formation energies and the formation of defect combinations while preserving the targeted phase composition to provide defect compositions that minimize the potential energy of the system, and density functional theory (DFT) calculations supply the defect formation energies needed in the statistical mechanics model. The details in constructing the point defect models are described in Appendix. The *fsolve* function from the SciPy library for Python was used to solve the equations generated for vacancies and anti-site defects for each sublattice, defined by the equivalent atom sites in Zr_7Ni_{10} and Zr_2Ni_7 phases (e.g., 4*a* sites comprise a single sublattice). We neglected dumbbell and interstitial defects, due to the limited space available near each of the sites. $LaNi_5$ is known to form dumbbell interstitials due to the hexagon-shaped holes above and below La on the $z = 0$ plane [19]. Ni–Ni distances across the hexagon vertices measure ~5.1 Å across with Ni–Ni bond lengths of ~2.5 Å. Holes or spaces near the Zr local environment in Zr_7Ni_{10} and Zr_2Ni_7 phases are typically pentagon- or square-shaped, with vertices measuring 3.8–4.7 Å and Ni-Ni bond lengths measuring ~2.6–2.8 Å. Defect concentrations and effective formation energies were calculated at a reference temperature of 1000 °C, a temperature near the melting points for the Zr_7Ni_{10} and Zr_2Ni_7 systems [20]. When precision limitations were encountered in Python (typically at low temperature, which results in extremely low defect concentrations), the logarithmic terms containing defect concentration variables were analyzed and dropped when low concentration assumptions were valid.

Electronic structure calculations were performed using the plane-wave-based DFT code implemented in Quantum ESPRESSO (Version 5.1.0, Quantum ESPRESSO

Foundation, London, UK) [21] and ultra-soft pseudopotentials from the Garrity–Bennett–Rabe–Vanderbilt (GBRV) Pseudopotential Library (Piscataway, NJ, USA) [22]. The exchange–correlation potential applied the Perdew–Burke–Ernzerhof (PBE) version of the generalized gradient approximation [23]. The recommended plane-wave cutoff energy of 40 Ry and charge-density cutoff energy of 120 Ry allowed convergence within 1×10^{-5} Ry/atom of the energy. A Methfessel–Paxton smearing width of 0.02 Ry with a Monkhorst–Pack k-point grid that yields 100–200 k-points also met convergence criteria with reasonable speed [24,25]. Spin-polarization was not included in the calculations, as differences in energies and stresses were low (<0.5 kbar) when included in the base equilibrium structures.

Cell structural optimizations for a given composition and structure were conducted by a variable cell relaxation calculation that minimizes forces and stresses within the cell. Symmetry of the cell was conserved during the structural optimizations based on a starting input structure. Non-defected structures preserved the symmetry based on starting experimental values. Using the optimized non-defected structure as a base, we removed or substituted atoms to create starting input values for the defected structures. The cell is considered optimized when forces were below 1×10^{-3} Ry/Å, the minimum energy converged at below 1×10^{-5} Ry/atom, and stresses converged within 0.5 kbar.

The Zr_7Ni_{10} phase has an orthorhombic structure and can sometimes occur as a metastable tetragonal phase [5,26,27]. It was originally reported to have space group symmetry $Aba2$ [28,29] and has since been revised to have space group symmetry $Cmca$ [30]. Its crystal structure is shown in Figure 1a, rendered using VESTA Graphical Software (Version 3.2.1, Momma and Izumi, Tsukuba, Japan) [31]. It contains 68 atoms per unit cell (Z = 4 formula units) with experimental parameters a = 12.381 Å, b = 9.185 Å, and c = 9.221 Å [30]. The unit cell contains four equivalent Zr atoms in the $4a$, $8d$, $8e$, and $8f_1$ positions and three equivalent Ni atoms in the $8f_2$, $16g_1$, and $16g_2$ positions. A supercell was not constructed due to the inherent large size of the cell. The structure with the space group $Aba2$ was evaluated for the calculation and allowed to relax, yielding the energy of the non-defected structure. Point defects involving all equivalent sites of the optimized non-defected structure were also evaluated, obtaining the energy of each defected structure after relaxation.

The Zr_2Ni_7 phase is monoclinic, and its crystal structure is shown in Figure 1b. It contains 36 atoms per unit cell (Z = 4) with experimental parameters a = 4.698 Å, b = 8.235 Å, c = 12.193 Å and β = 95.83° [32]. It contains two equivalent Zr atoms in the $4i_1$ and $4i_2$ positions and four equivalent Ni atoms in the $4i_3$, $8j_1$, $8j_2$, and $8j_3$ positions. A $2 \times 1 \times 1$ supercell containing 72 atoms was constructed to reduce interactions for the defects in adjacent cells. Point defects involving all equivalent sites were evaluated to obtain the energy of each defected structure after relaxation. The structures for $ZrNi_5$ (cubic $AuBe_5$ structure [33]), Zr_8Ni_{21} (triclinic Hf_8Ni_{21}

structure [34]), and ZrNi (orthorhombic CrB structure [35]) were also optimized to calculate the energies that define the tie lines for the formation energy diagrams.

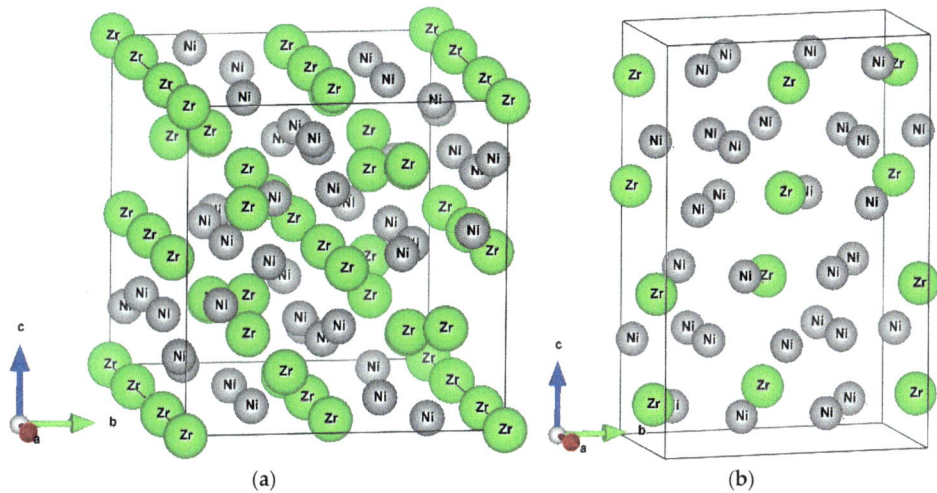

(a)   (b)

Figure 1. Unit-cell crystal structures for (**a**) Zr_7Ni_{10} and (**b**) Zr_2Ni_7.

The point defect model is developed for stoichiometric and near stoichiometric compositions in which assumptions for low defect concentrations are valid and interactions between defects are considered negligible. Defect concentrations beyond ~1% concentration would require a new model that considers defect interactions and the dependence of defect formation energy parameters on defect concentration.

3. Experimental Setup

Lattice parameters shrink with vacancy-predominant defects, and grow with certain anti-site defects (e.g., Zr → Ni anti-site defects). Experimental lattice parameters offer some evidence of the consistency of the Zr_7Ni_{10} point defect model. An arc melting process under continuous argon flow with a non-consumable tungsten electrode and water-cooled copper tray was used to prepare Zr-rich and Ni-rich Zr_7Ni_{10} ingot samples. Before each arc melt, a piece of sacrificial titanium was repeatedly melted and cooled for several cycles to reduce the residual oxygen concentration in the system. Similarly, each 5 g ingot was turned over, melted, and cooled several times to ensure uniformity in the chemical composition. A Rigaku Miniflex X-ray diffractometer (XRD, Rigaku, Tokyo, Japan) was used to study the phase component. JEOL-JSM32C and JEOL-JSM6400 scanning electron microscopes (SEM, JEOL, Tokyo, Japan) with energy dispersive spectroscopy (EDS) were applied to investigating the phase distribution and composition.

4. Results and Discussion

4.1. Theoretical Point Defects in Zr_7Ni_{10}

DFT calculation of the Zr_7Ni_{10} phase in the space group *Aba*2 results in a relaxed structure with a space group *Cmca* symmetry. The calculated lattice parameters are shown in Table 1, with comparative experimental lattice parameters. The lattice parameters show reasonable agreement with the experiment results, although the calculated structure appears to converge to a near tetragonal unit cell. This near tetragonal structure differs from the tetragonal meta-stable phase observed experimentally after hydrogenation–dehydrogenation of Zr_7Ni_{10} alloys; the dehydrogenated Zr_7Ni_{10} alloy has a smaller lattice parameter ($a = b = 6.496$ Å) with different symmetry [5,26,27]. Constitutional point defects may play a role in the different structures that were observed and calculated for the Zr_7Ni_{10} phase, for example, by introducing distortion to the lattice that favors the orthorhombic unit cell. Defects and distortion to the crystal lattice may promote diffusion kinetics for improved rate performance and access to hydrogen storage capacity if energy barriers between hydrogen sites are lowered with the defects/distortion. We examine the relative ease with which defects can form by first calculating the DFT defect formation energies for the Zr_7Ni_{10} phase.

Table 1. Zr_7Ni_{10} unit-cell lattice parameters.

Parameter	This Work	References [28,29]	Reference [30]
Space group	*Cmca*	*Aba*2	*Cmca*
a (Å)	12.419	12.386	12.381
b (Å)	9.179	9.156	9.185
c (Å)	9.180	9.211	9.221
Unit-cell volume (Å^3)	1046.5	1044.6	1048.6

4.1.1. Density Functional Theory Defect Formation Energies

Ground state DFT formation energies (at $T = 0$ K) for Zr_7Ni_{10} phase and the theoretical Ni $\rightarrow$ Zr anti-site, Zr vacancy, Ni vacancy, and Zr $\rightarrow$ Ni anti-site defects are plotted in Figure 2 with the tie lines to the neighboring compounds Zr_8Ni_{21} and ZrNi. The formation energies were calculated using following the equation:

$$\Delta H_f = \frac{E\left(Zr_nNi_m\right) - \frac{n}{2}E\left(Zr_2\right) - \frac{m}{4}E\left(Ni_4\right)}{n + m} \qquad (1)$$

where $E(Zr_nNi_m)$ is the energy of the Zr-Ni compound in the structure of interest, $E(Zr_2)$ is the energy of Zr in hexagonal close-packed structure, and $E(Ni_4)$ is the energy of Ni in a face-centered-cubic structure. Tie lines connect the ground state DFT formation energies of neighboring equilibrium structures (e.g., Zr_7Ni_{10} and Zr_8Ni_{21} phases), and represent the ground state energies of phase mixtures based on the overall Zr content (as opposed to a single phase with the same Zr content containing vacancy/anti-site defects). All of the defect energies at 0 K lie above the stoichiometric compound, as well as above the tie lines, indicating energy is required for the defects to form, and that the defects are in competition with formation of phases or mixtures of phases. There can be a large difference in formation energies between the investigated sublattice sites, and each type of defect appears to prefer different sites. Out of the Zr sublattices, the Ni → Zr anti-site defect appears to preferentially form on the $8e$ sublattice, while the Zr vacancy defect appears to prefer the $4a$ sublattice. For the Ni sublattices, the Ni vacancy defect appears to preferentially form on one of the $16g$ sublattices, while the Zr → Ni anti-site defect appears to preferentially form on the $8f_2$ sublattice. Note that defect formation energies on the $16g_1$ and $16g_2$ sublattices are similar and appear indistinguishable in Figure 2.

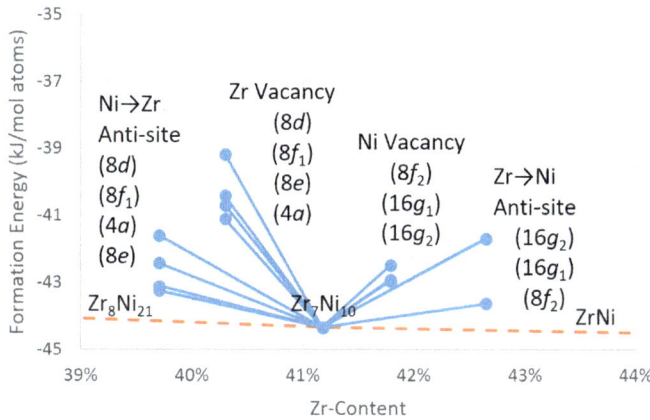

Figure 2. Density functional theory (DFT) formation energies for Zr_7Ni_{10} and its point defects with tie lines.

The model considers the effects of the heat of mixing, entropy, vibrational energy, and defect interaction on the free energy of each of the possible phases to be negligible, and a more comprehensive study of the point defect models could consider the effect of defect interactions. However, the DFT formation energies indicate that, out of the defects considered, the $8e$ Ni → Zr anti-site and the $8f_2$ Zr→Ni anti-site defects are the most stable of the point defects for the Zr_7Ni_{10} phase. However, since the Zr_7Ni_{10} phase is an ordered, binary compound, a mixture of all point defects is necessarily generated in order to maintain bulk homogeneity and required stoichiometry, and this is addressed by the statistical mechanics model (see Appendix).

4.1.2. Theoretical Effective Defect Formation Model

The theoretical defect concentrations for Zr_7Ni_{10} phase at 1000 °C calculated from the statistical mechanics model using defect energy parameters derived from DFT calculations are plotted as functions of Zr-content for stoichiometric and theoretical off-stoichiometric compositions in Figure 3. Concentrations of defects for each of the different site sublattices were generated and combined to give a total concentration for a defect on the atomic sublattice. The A atoms represent Zr and the B atoms represent Ni in the binary statistical mechanics model, where c_v^{Zr} is the concentration of vacancies of the Zr sublattices, c_v^{Ni} is the concentration of vacancies of the Ni sublattices, c_{Ni}^{Zr} is the concentration of total Ni → Zr anti-site defects, and c_{Zr}^{Ni} is the concentration of total Zr → Ni anti-site defects.

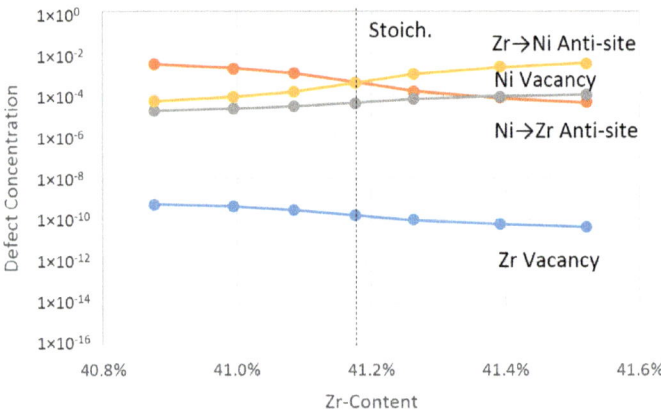

Figure 3. Theoretical defect concentrations for Zr_7Ni_{10} at 1000 °C.

The stoichiometric Zr_7Ni_{10} phase at 1000 °C is dominated by Ni → Zr and Zr → Ni anti-site defects in an approximately 1:1 ratio. Zr-rich compositions promote more defects of the Ni sublattices, particularly Zr → Ni anti-site defects. Ni → Zr anti-site defects dominate the analysis for Ni-rich compositions, and while the concentration of Zr vacancies increases, it does not approach the concentrations of the other defects. Each of the defects in the sublattices generated an effective formation energy, and the weighted average by concentration of sublattice sites was used to calculate the theoretical effective formation energies of the atomic point defects. The theoretical effective formation energies for the respective point defects at 1000 °C are tabulated in Table 2 and show the relative ease at which the defects can form. The effective formation energies for the defects are consistent with the defect concentration trends observed in Figure 3, with Zr → Ni anti-site defects showing the lowest effective formation energy at stoichiometry and in the Zr-rich state.

Table 2. Theoretical effective formation energies for point defects in Zr_7Ni_{10} at 1000 °C.

Defects	Ni-Rich	Stoichiometric	Zr-Rich
$\Delta H_v{}^{Zr}$ (eV)	2.12	2.22	2.33
$\Delta H_v{}^{Ni}$ (eV)	1.10	1.03	0.95
$\Delta H_{Ni}{}^{Zr}$ (eV)	0.49	0.66	0.86
$\Delta H_{Zr}{}^{Ni}$ (eV)	0.80	0.63	0.44

The non-defected and defected unit-cell lattice parameters based on the structure that yielded the lowest DFT formation energy are tabulated in Table 3. Comparison of these calculated parameters to experimental lattice parameters, as measured by XRD, may offer some insight into the point defect structures that may occur in the Zr_7Ni_{10} phase.

Table 3. Unit-cell lattice parameters for non-defected and defected Zr_7Ni_{10}.

Parameters	Perfect Crystal	Zr Vacancy (4a)	Ni → Zr Anti-Site (8e)	Ni Vacancy (16g_2)	Zr→Ni Anti-Site (8f_2)
a (Å)	12.418	12.334	12.365	12.379	12.437
b (Å)	9.179	9.152	9.135	9.172	9.224
c (Å)	9.180	9.139	9.152	9.162	9.230
Unit-cell volume (Å³)	1046.4	1031.6	1033.7	1040.3	1058.8

We have noted that the calculated crystal structure for the stoichiometric Zr_7Ni_{10} phase based on parameters given in literature appears to converge towards a tetragonal unit cell. Experimental lattice parameters, however, show c/b lattice parameter ratios of 1.004–1.006 (~0.5% difference from the calculated value). Calculated anti-site defected lattice parameters also show c/b ratios >1, while vacancy defected lattice parameters show ratios <1. The point defect model for the Zr_7Ni_{10} phase presented here predicts higher concentrations of anti-site defects for stoichiometric and off-stoichiometric compositions. Curiously, the anti-site defected unit-cell aspect ratio trends appear more consistent to the experimental literature results than for the perfect crystal case (0.2% difference vs. 0.5% difference, respectively). This is despite experimental conditions that include an annealing treatment at a high temperature for a long period of time (one month) to obtain equilibrium phases [30]. We also note that the equilibrium phase diagram for the Ni-Zr system has been revised to correct the solubility window for the Zr_7Ni_{10} phase, originally reporting a Zr-content range of 41.1 at% to 43.5 at%, to a maximum threshold of 41.5 at% [20,27,36]. The relative difficulty in removing defects (or conversely, the relative ease of forming defects) in the Zr_7Ni_{10} phase could have led to the construction of the larger solubility window, which supports the case for the presence of anti-site defects at stoichiometric compositions, but does not reconcile the observation that the defects do not appear to be the most thermodynamically favorable structures, based on the assumptions made in these calculations. Possible future work can re-examine the assumptions involved in calculating the free energies of the non-defected and defected Zr_7Ni_{10} phase, taking into account effects of temperature on vibrational and entropic contributions. Supercells with both $Zr \rightarrow Ni$ and $Ni \rightarrow Zr$ anti-site defects for stoichiometric compositions can be examined for possible interactive configurations and optimized for comparison.

In considering the types of point defects for the Zr_7Ni_{10} phase that may promote better electrochemical performance in Ni/MH batteries, the $Zr \rightarrow Ni$ anti-site defects that are prevalent in Zr-rich phases offer a slightly larger unit-cell volume for hydrogen storage, lowering the plateau pressure and stabilizing the hydride and hydrogen storage sites. Conversely, the $Ni \rightarrow Zr$ anti-site defects that are prevalent in Ni-rich phases slightly reduce the unit-cell volume, destabilizing the hydride and hydrogen storage sites. $Zr \rightarrow Ni$ anti-site defects may also further locally trap hydrogen, with Zr strongly binding to hydrogen (heats of hydride formation are -163 kJ$\cdot$mol^{-1} for metallic Zr [37] and -50 kJ$\cdot$mol^{-1} for Zr_7Ni_{10} [5]). However, the effects of these defects on hydrogen storage capacity and reversibility appear incremental compared to the direct effects of structure and the hydrogen binding energy related to the ratio of hydride formers (Zr) and modifiers (Ni) in the alloy composition, which would need to be considered for comparing properties across different phases, such as the Zr_7Ni_{10} and Zr_2Ni_7 phases. For example, the effect of hydrogen binding with higher Zr content can be seen in the reversibility performance (Zr_7Ni_{10} phase has a Zr/Ni ratio of 0.70 with 77% reversible capacity and an initial capacity of 1.01 hydrogen to metal (H/M), while Zr_2Ni_7 phase has a Zr/Ni ratio of 0.29 with 100% reversible capacity and an initial capacity of 0.29 H/M [4]; the point defect models predict defect concentrations of 10^{-5}–10^{-3} at stoichiometry for local site effects). Within a single phase, however, insights may be gleaned for future alloy design, particularly since synergetic effect studies show that Zr_7Ni_{10} is a good catalytic phase, and improving its thermodynamic properties can further improve the performance of multi-phase negative electrode active materials [1]. One such strategy is to try to maintain the Zr_7Ni_{10} crystal structure, while substituting elements that have less of an affinity for hydrogen (Ti has a heat of hydrogen formation of -124 kJ$\cdot$mol^{-1} [37]) while balancing lattice shrinkage with combination of anti-site defects and other additives. The point defect models offer a first step towards laying the groundwork for understanding the role of constitutional defects as it relates to alloy design and optimization of electrochemical properties for Ni/MH batteries.

Investigation of absorption–desorption or diffusion kinetics, through the calculation of the activation energy by methods such as nudged elastic band, can be of interest. Future work involving additives can also offer further tuning of the defect structures; addition of Ti has experimentally improved hydrogen reversibility, as well as the kinetics needed for high-rate discharge [5].

4.2. Theoretical Point Defects in Zr_2Ni_7

The Zr_2Ni_7 phase was studied to provide a comparison to the Zr_7Ni_{10} phase in terms of the types of defects formed and the ease of defect formation. The DFT structure optimization calculation for the Zr_2Ni_7 phase is consistent with the monoclinic $C2/m$ symmetry and structure reported in literature. The calculated lattice parameters are shown in Table 4 with experimental comparative lattice parameters. The lattice parameters show reasonable agreement with the experimental results. The DFT defect formation energies for the Zr_2Ni_7 phase are calculated based on this optimized calculated structure.

Table 4. Zr_2Ni_7 unit-cell lattice parameters.

Lattice Parameter	This Work	From Reference [32]
a (Å)	4.677	4.698
b (Å)	8.239	8.235
c (Å)	12.176	12.193
β (°)	95.20	95.83

4.2.1. Density Functional Theory Formation Energies

Ground state DFT formation energies for the Zr_2Ni_7 phase and its theoretical Ni → Zr anti-site, Zr vacancy, Ni vacancy, and Zr → Ni anti-site defects are plotted in Figure 4, with tie lines to neighboring compounds $ZrNi_5$ and Zr_8Ni_{21}. All defect energies at 0 K lie above that of the stoichiometric compounds, as well as the tie lines. However, the Ni → Zr anti-site defect energy lies considerably higher than the stoichiometric compound and the $ZrNi_5$-Zr_2Ni_7 mixture tie line, indicating a possible shift in the dominating defects found in a Ni-rich compound. The Zr → Ni anti-site defect energy is the lowest of the defects for the Zr_2Ni_7 phase, followed by the Ni vacancy defect energy. There is a small difference in defect energy between the $4i$ and $8j$ sites for the Ni vacancy defect, but there is a larger difference between the sites of the Zr → Ni anti-site defect. In general, point defects appear to form preferentially in the $8j$ sublattices. Note that defect formation energies for several sets of sublattices are similar and appear indistinguishable in Figure 4 (i.e., $4i_1$ and $4i_2$, $4i_3$ and $8j_3$, and $8j_1$ and $8j_2$).

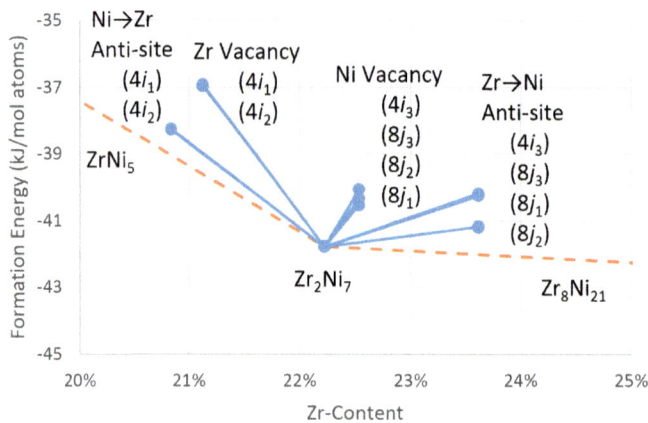

Figure 4. Formation energies for Zr_2Ni_7 and its point defects with tie lines.

4.2.2. Theoretical Effective Defect Formation Model

The theoretical defect concentrations for the Zr_2Ni_7 phase at 1000 °C are plotted as functions of Zr-content in Figure 5. Stoichiometric Zr_2Ni_7 phase at 1000 °C is dominated by Ni vacancy defects. Zr → Ni anti-site defects overtake the formation of Ni vacancy defects in Zr-rich compositions, making Ni → Zr anti-site defects the predominant defect for Ni-rich compositions. Zr vacancy defects also increase on the Ni-rich side, but again, the concentration is overall relatively low.

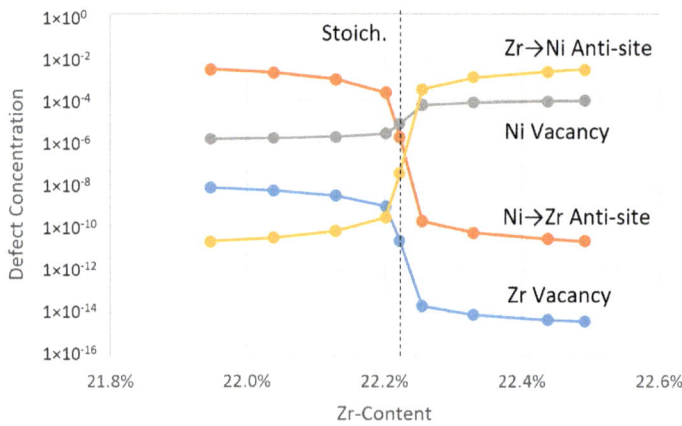

Figure 5. Theoretical defect concentrations for Zr_2Ni_7 at 1000 °C.

The weighted theoretical effective formation energies for the respective point defects at 1000 °C are tabulated in Table 5 and define the relative ease for which a particular defect can form. The effective formation energies for the defects are consistent with the defect concentration trends observed in Figure 5, with Ni vacancies showing the lowest effective formation energy at stoichiometric ratios. By comparing the point defect models between the Zr_7Ni_{10} and Zr_2Ni_7 phases, the data suggests that it is easier to form defects in the Zr_7Ni_{10} phase than the Zr_2Ni_7 phase, which is consistent with our hypothesis.

Table 5. Theoretical effective formation energies for point defects in Zr_2Ni_7 at 1000 °C.

Defects	Ni-Rich	Stoichiometric	Zr-Rich
ΔH_v^{Zr} (eV)	1.89	2.52	3.47
ΔH_v^{Ni} (eV)	1.42	1.24	0.97
ΔH_{Ni}^{Zr} (eV)	0.48	1.29	2.52
ΔH_{Zr}^{Ni} (eV)	2.60	1.79	0.56

In considering the types of point defects within the Zr_2Ni_7 phase that may promote better electrochemical performance in Ni/MH batteries, we again express caution for comparing different phases, but within the Zr_2Ni_7 phase, $Zr \rightarrow Ni$ anti-site defects that are prevalent in Zr-rich phases offer a slightly larger unit-cell volume for hydrogen storage at the expense of possible hydrogen trapping, due to local Zr concentrations. Conversely, $Ni \rightarrow Zr$ anti-site defects that are prevalent in Ni-rich phases slightly reduce the unit-cell volume and, thereby, offer less space to accommodate hydrogen storage.

4.3. Experimental Zr-Rich and Ni-Rich Zr_7Ni_{10} Phase Analysis

The Zr-rich and Ni-rich Zr_7Ni_{10} alloy samples were used to compare their experimental lattice parameters to defected lattice parameters, as predicted by the model and calculated from DFT. No annealing treatment was applied to the samples to preserve meta-stable, off-stoichiometric phases. Representative SEM images of the Zr-rich sample are shown in Figure 6a,b, and their phase compositions, represented by differing contrasts in the images, are analyzed by EDS and listed in Table 6. The stoichiometric Zr_7Ni_{10} phase has a B/A ratio of 1.43 (where B = Ni, Al, and Mn and A = Zr), and the SEM-EDS analysis shows that the sample is mainly composed of a Zr-rich phase and smaller distributions of a Ni-rich phase, with a small presence of the Zr_2Ni phase. There is a small amount of Mn contamination in this sample, which may affect the interpretation of the results. However, previous first-principle calculations investigating substitutions in $LaNi_5$ and $TiMn_2$ alloys suggest that Mn and Ni substitutions do not alter the alloys' hydrogen stability sites, which offers an indication that the small amount of observed contamination may also have little effect [38]. A representative SEM image of the Ni-rich sample is shown in Figure 6c, with the phase composition listed similarly in Table 6. The analysis shows that the second sample is composed mainly of Ni-rich phases with the minor presence of the Zr_2Ni phase. Contamination was not detected in this particular sample.

Table 6. Energy dispersive spectroscopy (EDS) phase compositions of select areas for Zr-rich and Ni-rich Zr_7Ni_{10} alloy samples depicted in Figure 6.

Sample	Area	Zr	Ni	Al	Mn	B/A	Phase
Zr-rich Zr_7Ni_{10}	1	42.6	57.3	0.0	0.1	1.35	Zr_7Ni_{10}
	2	35.6	64.4	0.0	0.1	1.81	Zr_7Ni_{10}
	3	38.4	59.4	0.0	2.2	1.60	Zr_7Ni_{10}
	4	66.0	33.9	0.0	0.1	0.52	Zr_2Ni
Ni-rich Zr_7Ni_{10}	1	40.4	59.6	0.0	0.0	1.48	Zr_7Ni_{10}
	2	39.9	60.1	0.0	0.0	1.51	Zr_7Ni_{10}
	3	40.3	59.7	0.0	0.0	1.48	Zr_7Ni_{10}
	4	40.2	59.8	0.0	0.0	1.49	Zr_7Ni_{10}
	5	68.0	32.0	0.0	0.0	0.47	Zr_2Ni
	6	59.9	40.1	0.0	0.0	0.67	Zr_2Ni

Experimental lattice parameters for the Zr-rich and Ni-rich Zr_7Ni_{10} samples were obtained from XRD measurement, and are listed in Table 7. In comparison to the theoretical and experimental lattice parameters for the stoichiometric Zr_7Ni_{10} phase, both off-stoichiometric samples show unit cells that are smaller than expected. However, the Zr-rich sample shows a larger unit cell than the Ni-rich samples, which is consistent with expectations. The c/b ratios are also >1, which is consistent with the anti-site defected structures predicted by the point defect model.

Table 7. Experimental lattice parameters for Zr-rich and Ni-rich Zr_7Ni_{10} samples.

Parameter	Zr-Rich Zr_7Ni_{10}	Ni-Rich Zr_7Ni_{10}
a (Å)	12.365	12.356
b (Å)	9.172	9.162
c (Å)	9.208	9.194
Unit-cell volume (Å^3)	1044.2	1040.7

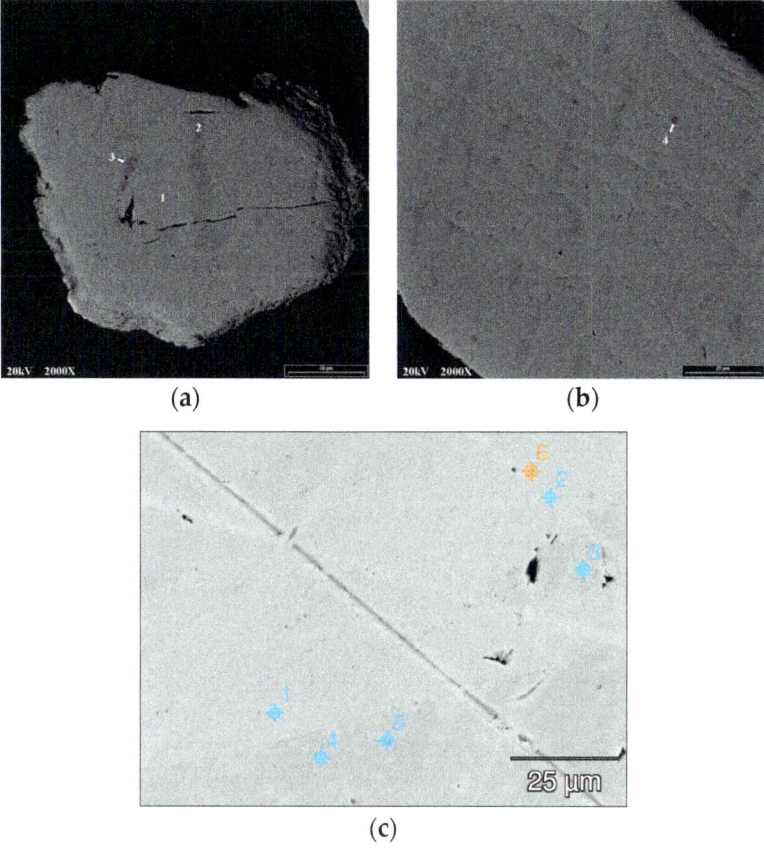

(a)   (b)

(c)

Figure 6. Scanning electron microscopy (SEM) images of a Zr-rich Zr_7Ni_{10} alloy sample showing representative phase distributions of (**a**) mainly Zr-rich phase and a smaller distribution of Ni-rich phases with (**b**) the presence of some amount of the Zr_2Ni phase; and (**c**) a Ni-rich Zr_7Ni_{10} alloy sample consisting of mainly Ni-rich phase and the presence of some amount of the Zr_2Ni phase. The bar on the right lower corner of the Zr-rich samples indicates a length of 20 μm. Chemical compositions associated with the numbered areas are listed in Table 6.

5. Conclusions

Defect models for the Zr_7Ni_{10} and Zr_2Ni_7 intermetallic phases were calculated from first-principles using DFT and statistical mechanics. Zr-Ni based active negative electrode materials in Ni/MH batteries offer performance benefits such as improved capacity through synergetic phase effects, and the understanding and engineering of point defects can contribute to further improvement in these materials through the framework that is laid by the defect models. DFT calculations confirmed that the Zr_7Ni_{10} phase structure has space group *Cmca* symmetry in the ground state.

The point defect model indicates that at stoichiometry, the Zr_7Ni_{10} phase tends to form $Zr \rightarrow Ni$ and $Ni \rightarrow Zr$ anti-site defects in a 1:1 ratio, while Zr_2Ni_7 tends to form Ni vacancy defects. Zr vacancies appear almost negligible in both Zr_7Ni_{10} and Zr_2Ni_7 compounds. $Zr \rightarrow Ni$ anti-site defects are the most prevalent defects in Zr-rich compositions in both compounds, and $Ni \rightarrow Zr$ anti-site defects are the most prevalent defects in Ni-rich compositions. In general, it is easier to form defects in the Zr_7Ni_{10} phase than the Zr_2Ni_7 phase. Comparison to experimental lattice parameters for the Zr_7Ni_{10} phase from XRD measurements show some discrepancy (~0.5% difference in c/b ratios) with regard to the orthorhombic versus tetragonal unit cells predicted by first principle calculations, which warrants further re-examination of the typical assumptions, as well as supercells that consider the distribution and interaction of defects in the calculations. Despite this discrepancy, the experimental lattice trends for off-stoichiometric phases (c/b ratios > 1) appear to be consistent with the anti-site defects and the c/b ratios predicted by the model. The point defect models can be used to further investigate absorption–desorption or diffusion kinetics through nudged elastic band DFT calculations, offering further tuning of alloy design and the optimization of electrochemical properties for Ni/MH batteries.

Acknowledgments: The authors would like to thank the following individuals from BASF-Ovonic for their help: Jean Nei, David Pawlik, Alan Chan, Ryan J. Blankenship, and Su Cronogue.

Author Contributions: Kwo-Hsiung Young and Diana F. Wong conceived and designed the experiments; Diana F. Wong and Taihei Ouchi performed the experiments; Kwo-Hsiung Young, Diana F. Wong, Taihei Ouchi, and K. Y. Simon Ng analyzed the data; Diana F. Wong prepared the manuscript.

Conflicts of Interest: The authors declare no conflict of interest.

Appendix

Defect models for Zr_7Ni_{10} and Zr_2Ni_7 intermetallic phases are developed using statistical mechanics within the generalized grand canonical μPT formalism for an ordered, binary system A_nB_m, where Zr is an A-site atom and Ni is a B-site atom. The model is similar to the ones developed for NiAl [17] and $NiAl_3$ [18] systems, but is extended to account for a variable number of sublattices for each type of atom. These statistical mechanics-based models address the issue of non-homogeneity that can occur if we apply a monatomic crystal defect model wherein the formation energy of a vacancy defect is described by removing an atom and inserting it at a surface site; hypothetically by analogy in a binary system, A-site vacancies in an $A_{0.5}B_{0.5}$ crystal would generate a surface of strictly A atoms, introducing non-homogeneity into the system [16,39]. To maintain homogeneity and composition of the bulk, such A-site vacancies are then necessarily accompanied by B-site vacancies, $A \rightarrow B$ anti-site defects, or a mixture of the defects. The statistical mechanics model considers the different possible local defect configurations possible, minimizing the system potential to obtain equilibrium concentrations of each type of defect.

The system is described on a unit cell basis, so that $n = 28$ and $m = 40$ for Zr_7Ni_{10} phase and $n = 8$ and $m = 28$ for Zr_2Ni_7 phase. The generalized grand canonical formalism prescribes the independent variable temperature T, pressure P, and chemical potential μ_A and μ_B. The generalized grand canonical potential J' is defined as:

$$J' = U + PV - TS - \mu_A N_A - \mu_B N_B \tag{A1}$$

where U is the internal energy of the system, V is the volume, S is the entropy, N_A is the number of A atoms in the system, and N_B is the number of B atoms in the system. The equilibrium concentrations of each type of point defect are found by minimizing the potential with respect to the concentration c_i^ν or particle number N_i^ν of each type of defect i (vacancy or anti-site) on the sublattice ν. Each system contains M total possible lattice sites, which may be divided into sublattices α for A-site atoms and β for B-site atoms. The sublattices can be further subdivided into $\alpha_1, \alpha_2, \ldots, \alpha_k$, and $\beta_1, \beta_2, \ldots, \beta_l$, for k equivalent A atoms and l equivalent B atoms. For example in the Zr_7Ni_{10} phase, we assign $4a$, $8d$, $8e$, and $8f_1$ sites for the Zr atoms to sublattices $\alpha_1, \alpha_2, \alpha_3$, and α_4, respectively, and $8f_2$, $16g_1$, and $16g_2$ sites for the Ni atoms to sublattices β_1, β_2, and β_3, respectively. M is then also the sum of the lattice sites for each of the sublattices:

$$M = \sum_k M^{\alpha_k} + \sum_l M^{\beta_l} \tag{A2}$$

where M^{α_k} is the total number of lattice sites for the α_k sublattice and M^{β_l} is the total number of lattice sites for the β_l sublattice. The number of A and B atoms N_A and N_B in the system that can occupy the lattice sites is described by the accounting of the lattice sites and the number of defects:

$$N_A = \sum_k M^{\alpha_k} - \sum_k N_v^{\alpha_k} - \sum_k N_B^{\alpha_k} + \sum_l N_A^{\beta_l} \tag{A3}$$

$$N_B = \sum_l M^{\beta_l} - \sum_l N_v^{\beta_l} - \sum_l N_A^{\beta_l} + \sum_k N_B^{\alpha_k} \tag{A4}$$

where $N_v^{\alpha_k}$ is the number of vacancy defects on the α_k sublattice, $N_B^{\alpha_k}$ is the number of B $\rightarrow$ A antisite defects on the α_k sublattice, $N_v^{\beta_l}$ is the number of vacancy defects on the β_l sublattice, and $N_A^{\beta_l}$ is the number of A$\rightarrow$B antisite defects on the β_l sublattice.

The internal energy U can be described as a sum of the ground state energy of the non-defected system and the defect formation energies:

$$U = M\varepsilon_0 + \sum_k N_v^{\alpha_k}\varepsilon_v^{\alpha_k} + N_B^{\alpha_k}\varepsilon_B^{\alpha_k} + \sum_l N_v^{\beta_l}\varepsilon_v^{\beta_l} + N_A^{\beta_l}\varepsilon_A^{\beta_l} \tag{A5}$$

329

The ground state energy per atom ε_0 and the defect formation energy parameters ε_i^ν are calculated from DFT using a supercell of N total lattice sites, such that

$$\varepsilon_0 = \frac{1}{N} E(N, 0) \tag{A6}$$

$$\varepsilon_v^{\alpha_k, \beta_l} = E\left(N - 1, 1^{\alpha_k, \beta_l}\right) - E(N, 0) \tag{A7}$$

$$\varepsilon_A^{\beta_l} = E\left(N, A^{\beta_l}\right) - E(N, 0) \tag{A8}$$

$$\varepsilon_B^{\alpha_k} = E(N, B^{\alpha_k}) - E(N, 0) \tag{A9}$$

where $E(N, 0)$ is the ground state energy of the supercell of size N with zero defects, $E(N - 1, 1^{\alpha_k, \beta_l})$ is the ground state energy of the supercell of size $N - 1$ with one vacancy on the respective sublattice, $E(N, A^{\beta_l})$ is the ground state energy of the supercell of size N with one A atom on the β_l sublattice, and $E(N, B^{\alpha_k})$ is the ground state energy of the supercell of size N with one B atom on the α_k sublattice. For low defect concentrations and cases near stoichiometry, the defects are assumed to be non-interacting so that ε_i^ν does not change with concentration.

The volume of the system can also be written as a sum of its equilibrium volume in the non-defected state and the changes in the volume due to its defects:

$$V = M v_0 + \sum_k N_v^{\alpha_k} v_v^{\alpha_k} + N_B^{\alpha_k} v_B^{\alpha_k} + \sum_l N_v^{\beta_l} v_v^{\beta_l} + N_A^{\beta_l} v_A^{\beta_l} \tag{A10}$$

where v_0 is the unit-cell volume per atom and v_i^ν is the defect volume formation parameter for defect i on sublattice ν, defined similarly as in Equations (A6)–(A9).

The entropy S is the sum of the equilibrium entropy in the non-defected state, the defect formation entropies, and the configurational entropy:

$$\begin{aligned}
S = M s_0 &+ \sum_k N_v^{\alpha_k} s_v^{\alpha_k} + N_B^{\alpha_k} s_B^{\alpha_k} + \sum_l N_v^{\beta_l} s_v^{\beta_l} + N_A^{\beta_l} s_A^{\beta_l} \\
&+ k_B \ln \left(\prod_k \frac{M^{\alpha_k}!}{N_v^{\alpha_k}! N_B^{\alpha_k}! \left(M^{\alpha_k} - N_v^{\alpha_k} - N_B^{\alpha_k}\right)!} \right. \\
&\qquad \left. \cdot \prod_l \frac{M^{\beta_l}!}{N_v^{\beta_l}! N_A^{\beta_l}! \left(M^{\beta_l} - N_v^{\beta_l} - N_B^{\beta_l}\right)!} \right)
\end{aligned} \tag{A11}$$

where s_0 is the entropy per atom in the non-defected structure and s_i^ν is the defect entropy formation parameter for defect i on sublattice ν, defined similarly as in Equations (A6)–(A9). Concentrations for each type of defect c_i^ν can be derived by substituting Equations (A3)–(A5), (A10) and (A11) into Equation (A1) and minimizing the potential with respect to N_i^v, such that:

$$c_v^{\alpha_k} = \frac{N_v^{\alpha_k}}{M} = \frac{M^{\alpha_k}}{M} \frac{e^{s_v^{\alpha_k}/k_B}e^{-(\varepsilon_v^{\alpha_k}+\mu_A+Pv_v^{\alpha_k})/k_BT}}{1+e^{s_v^{\alpha_k}/k_B}e^{-(\varepsilon_v^{\alpha_k}+\mu_A+Pv_v^{\alpha_k})/k_BT}+e^{s_B^{\alpha_k}/k_B}e^{-(\varepsilon_B^{\alpha_k}+\mu_A-\mu_B+Pv_B^{\alpha_k})/k_BT}}$$ (A12)

$$c_B^{\alpha_k} = \frac{N_B^{\alpha_k}}{M} = \frac{M^{\alpha_k}}{M} \frac{e^{s_B^{\alpha_k}/k_B}e^{-(\varepsilon_B^{\alpha_k}+\mu_A-\mu_B+Pv_B^{\alpha_k})/k_BT}}{1+e^{s_v^{\alpha_k}/k_B}e^{-(\varepsilon_v^{\alpha_k}+\mu_A+Pv_v^{\alpha_k})/k_BT}+e^{s_B^{\alpha_k}/k_B}e^{-(\varepsilon_B^{\alpha_k}+\mu_A-\mu_B+Pv_B^{\alpha_k})/k_BT}}$$ (A13)

Similar expressions are derived for $c_v^{\beta_l}$ and $c_A^{\beta_l}$.

The chemical potentials μ_A and μ_B are Lagrangian parameters that can be determined from the thermodynamic relation $J' = 0$ for the generalized grand canonical potential [16]. From this relation, we combine Equations (A1), (A5) and (A10)–(A13) and their analogous expressions to obtain:

$$
\begin{aligned}
\varepsilon_0 + Pv_0 - Ts_0 \quad = & \sum_k \frac{M^{\alpha_k}}{M}\mu_A + \sum_l \frac{M^{\beta_l}}{M}\mu_B \\
& - \sum_k \frac{M^{\alpha_k}}{M}k_BT\ln\left(1 - \frac{M}{M^{\alpha_k}}c_v^{\alpha_k} - \frac{M}{M^{\alpha_k}}c_B^{\alpha_k}\right) \\
& - \sum_l \frac{M^{\beta_l}}{M}k_BT\ln\left(1 - \frac{M}{M^{\beta_l}}c_v^{\beta_l} - \frac{M}{M^{\beta_l}}c_A^{\beta_l}\right)
\end{aligned}
$$ (A14)

A final expression describing the composition of the compound preserves the composition of the compound of the system of interest:

$$\frac{N_A}{N_B} = \frac{\sum_k M^{\alpha_k} - \sum_k N_B^{\alpha_k} - \sum_k N_v^{\alpha_k} + \sum_l N_A^{\beta_l}}{\sum_l M^{\beta_l} - \sum_l N_A^{\beta_l} - \sum_l N_v^{\beta_l} + \sum_k N_B^{\alpha_k}} = \frac{x}{1-x}$$ (A15)

where x is the atomic fraction of A atoms of the system A_xB_{1-x}.

The effective formation energy of the atomic defects is defined as:

$$\Delta H_i^v = -k_B\frac{\partial c_i^v}{\partial\left(\frac{1}{T}\right)}$$ (A16)

which simplifies to

$$\Delta H_v^{\alpha_k} = \varepsilon_v^{\alpha_k} + \mu_A$$ (A17)

$$\Delta H_B^{\alpha_k} = \varepsilon_B^{\alpha_k} + \mu_A - \mu_B$$ (A18)

and similarly for $\Delta H_v^{\beta_l}$ and $\Delta H_A^{\beta_l}$.

Within this statistical mechanics model for a binary intermetallic compound, the contributions from PV and entropy are much smaller than the contributions from the internal energy, and, therefore, Pv_0, v_i^v, s_0 and s_i^v are typically considered negligible for room temperature and pressure.

Abbreviations

c_i^ν	Concentration of defect i on the sublattice ν
c_v^{Ni}	Concentration of vacancies on the Ni sublattices
c_v^{Zr}	Concentration of vacancies on the Zr sublattices
c_{Zr}^{Ni}	Concentration of Zr anti-site atoms in the Ni sublattices
c_{Ni}^{Zr}	Concentration of Ni anti-site atoms in the Zr sublattices
DFT	Density functional theory
EDS	Energy dispersive spectroscopy
J'	Generalized grand canonical potential
k_B	Boltzmann constant
M	Number of lattice sites
N_A	Number of A atoms
N_B	Number of B atoms
N_i^ν	Number of defect i on the sublattice ν
MH	Metal hydride
Ni/MH	Nickel/metal hydride
P	Pressure
s_0	Entropy per atom for non-defected structure
s_i^ν	Defect entropy formation parameter for defect i on sublattice ν
SEM	Scanning electron microscopy
T	Temperature
U	Internal energy
v_0	Unit-cell volume per atom for non-defected structure
v_i^ν	Defect volume formation parameter for defect i on sublattice ν
XRD	X-ray diffraction
Z	Number of formula units
α_k	kth sublattice for A-site atoms
β_l	lth sublattice for B-site atoms
ΔH_f	Energy of formation
ΔH_v^{Ni}	Effective defect formation energy for vacancies in the Ni sublattices
ΔH_v^{Zr}	Effective defect formation energy for vacancies in the Zr sublattices
ΔH_{Zr}^{Ni}	Effective defect formation energy for Zr anti-site atoms in the Ni sublattices
ΔH_{Ni}^{Zr}	Effective defect formation energy for Ni anti-site atoms in the Zr sublattices
ε_0	Energy per atom for non-defected structure
ε_i^ν	Defect energy formation parameter for defect i on sublattice ν
μ_A	Chemical potential for A atom
μ_B	Chemical potential for B atom

References

1. Joubert, J.M.; Latroche, M.; Percheron-Guégan, A.; Bouet, J. Improvement of the electrochemical activity of Zr–Ni–Cr laves phase hydride electrodes by secondary phase precipitation. *J. Alloy. Compd.* **1996**, *240*, 219–228.

2. Young, K.; Nei, J.; Ouchi, T.; Fetcenko, M. Phase abundances in AB_2 metal hydride alloys and their correlations to various properties. *J. Alloy. Compd.* **2011**, *509*, 2277–2284.

3. Young, K.; Ouchi, T.; Fetcenko, M.A.; Mays, W.; Reichman, B. Structural and electrochemical properties of $Ti_{1.5}Zr_{5.5}V_xNi_{10-x}$. *Int. J. Hydrog. Energy* **2009**, *34*, 8695–8706.

4. Joubert, J.M.; Latroche, M.; Percheron-Guégan, A. Hydrogen absorption properties of several intermetallic compounds of the Zr Ni system. *J. Alloy. Compd.* **1995**, *231*, 494–497.

5. Young, K.; Ouchi, T.; Liu, Y.; Reichman, B.; Mays, W.; Fetcenko, M. Structural and electrochemical properties of $Ti_xZr_{7-x}Ni_{10}$. *J. Alloy. Compd.* **2009**, *480*, 521–528.

6. Young, M.; Chang, S.; Young, K.; Nei, J. Hydrogen storage properties of $ZrV_xNi_{3.5-x}$ (x = 0.0–0.9) metal hydride alloys. *J. Alloy. Compd.* **2013**, *580*, S171–S174.

7. Myers, S.M.; Baskes, M.I.; Birnbaum, H.K.; Corbett, J.W.; DeLeo, G.G.; Estreicher, S.K.; Haller, E.E.; Jena, P.; Johnson, N.M.; Kirchheim, R.; et al. Hydrogen interactions with defects in crystalline solids. *Rev. Mod. Phys.* **1992**, *64*, 559–617.

8. Chen, K.C.; Peterson, E.J.; Thoma, D.J. $HfCo_2$ laves phase intermetallics—Part I: Solubility limits and defect mechanisms. *Intermetallics* **2001**, *9*, 771–783.

9. Chen, K.C.; Chu, F.; Kotula, P.G.; Thoma, D. $HfCo_2$ laves phase intermetallics—Part II: Elastic and mechanical properties as a function of composition. *Intermetallics* **2001**, *9*, 785–798.

10. Ovshinsky, S.R.; Fetcenko, M.A.; Ross, J. A nickel metal hydride battery for electric vehicles. *Science* **1993**, *260*, 176–181.

11. Züttel, A. Materials for hydrogen storage. *Mater. Today* **2003**, *6*, 24–33.

12. Young, K. Stoichiometry in inter-Metallic Compounds for Hydrogen Storage Applications. In *Stoichiometry and Materials Science—When Numbers Matter*; Innocenti, A., Kamarulzaman, N., Eds.; InTech: Rijeka, Croatia, 2012; pp. 147–172.

13. Stein, F.; Palm, M.; Sauthoff, G. Structure and stability of Laves phases. Part I. Critical assessment of factors controlling Laves phase stability. *Intermetallics* **2004**, *12*, 713–720.

14. Moura, C.S.; Motta, A.T.; Lam, N.Q.; Amaral, L. Atomistic simulations of point defects in zrni intermetallic compounds. *Nucl. Instrum. Methods Phys. Res. Sect. B Beam Interact. Mater. At.* **2001**, *180*, 257–264.

15. Mayer, J.; Elsasser, C.; Fahnle, M. Concentrations of atomic defects in $B2\text{-}Fe_xAl_{1-x}$—An ab-initio study. *Phys. Status Solidi B Basic Res.* **1995**, *191*, 283–298.

16. Mayer, J.; Fahnle, M. On the meaning of effective formation energies, entropies and volumes for atomic defects in ordered compounds. *Acta Mater.* **1997**, *45*, 2207–2211.

17. Meyer, B.; Fähnle, M. Atomic defects in the ordered compound B2-NiAl: A combination of ab initio electron theory and statistical mechanics. *Phys. Rev. B* **1999**, *59*, 6072–6082.

18. Rasamny, M.; Weinert, M.; Fernando, G.; Watson, R. Electronic structure and thermodynamics of defects in $NiAl_3$. *Phys. Rev. B* **2001**, *64*.

19. Notten, P.; Latroche, M.; Percheron-Guégan, A. The influence of Mn on the crystallography and electrochemistry of nonstoichiometric AB_5-type hydride-forming compounds. *J. Electrochem. Soc.* **1999**, *146*, 3181–3189.

20. Okamoto, H. Ni-Zr (nickel-zirconium). *J. Phase Equilibria Diffus.* **2007**, *28*, 409–409.

21. Giannozzi, P.; Baroni, S.; Bonini, N.; Calandra, M.; Car, R.; Cavazzoni, C.; Ceresoli, D.; Chiarotti, G.L.; Cococcioni, M.; Dabo, I. Quantum Espresso: A modular and open-source software project for quantum simulations of materials. *J. Phys. Condens. Matter* **2009**, *21*, 395–502.

22. Garrity, K.F.; Bennett, J.W.; Rabe, K.M.; Vanderbilt, D. Pseudopotentials for high-throughput DFT calculations. *Comput. Mater. Sci.* **2014**, *81*, 446–452.

23. Perdew, J.P.; Burke, K.; Ernzerhof, M. Generalized gradient approximation made simple. *Phys. Rev. Lett.* **1996**, *77*, 3865–3868.

24. Methfessel, M.; Paxton, A.T. High-precision sampling for brillouin-zone integration in metals. *Phys. Rev. B* **1989**, *40*, 3616–3621.

25. Monkhorst, H.J.; Pack, J.D. Special points for brillouin-zone integrations. *Phys. Rev. B* **1976**, *13*, 5188–5192.

26. Takeshita, H.T.; Fujiwara, N.; Oishi, T.; Noréus, D.; Takeichi, N.; Kuriyama, N. Another unusual phenomenon for Zr_7Ni_{10}: Structural change in hydrogen solid solution and its conditions. *J. Alloy. Compd.* **2003**, *360*, 250–255.

27. Takeshita, H.; Kondo, S.; Miyamura, H.; Takeichi, N.; Kuriyama, N.; Oishi, T. Re-examination of Zr_7Ni_{10} single-phase region. *J. Alloy. Compd.* **2004**, *376*, 268–274.

28. Kirkpatrick, M.; Smith, J.; Larsen, W. Structures of the intermediate phases $Ni_{10}Zr_7$ and $Ni_{10}Hf_7$. *Acta Crystallogr.* **1962**, *15*, 894–903.

29. Glimois, J.L.; Forey, P.; Feron, J.; Becle, C. Structural investigations of the pseudo-binary compounds $Ni_{10-x}Cu_xZr_7$. *J. Less Common Met.* **1981**, *78*, 45–50.

30. Joubert, J.M.; Cerný, R.; Yvon, K.; Latroche, M.; Percheron-Guégan, A. Zirconium–nickel, Zr_7Ni_{10}: Space group revision for the stoichiometric phase. *Acta Crystallogr. Sect. C Cryst. Struct. Commun.* **1997**, *53*, 1536–1538.

31. Momma, K.; Izumi, F. *VESTA 3* for three-dimensional visualization of crystal, volumetric and morphology data. *J. Appl. Crystallogr.* **2011**, *44*, 1272–1276.

32. Eshelman, F.R.; Smith, J.F. The structure of Zr_2Ni_7. *Acta Crystallogr. Sect. B Struct. Crystallogr. Cryst. Chem.* **1972**, *28*, 1594–1600.

33. Villars, P. Ni_5Zr ($ZrNi_5$) Crystal Structure. In *Material Phases Data System (MPDS)*; Springer-Verlag GmbH: Heidelberg, Germany, 2014.

34. Joubert, J.M.; Černý, R.; Yvon, K.; Latroche, M.; Percheron-Guégan, A. Refinement of the crystal structure of zirconium nickel, Zr_8Ni_{21}. *Z. Kristallogr. New Cryst. Struct.* **1998**, *213*, 227–228.

35. Villars, P. NiZr (ZrNi) Crystal Structure. In *Material Phases Data System (MPDS)*; Springer-Verlag GmbH: Heidelberg, Germany, 2014.

36. Nash, P.; Jayanth, C.S. The Ni-Zr (nickel-zirconium) system. *Bull. Alloy Phase Diagr.* **1984**, *5*, 144–148.

37. Young, K. Metal Hydrides. In *Elsevier Reference Module in Chemistry, Molecular Sciences and Chemical Engineering*; Elsevier B.V.: Waltham, MA, USA, 2013.

38. Wong, D.F.; Young, K.; Ng, K.Y.S. First-principles study of structure, initial lattice expansion, and pressure-composition-temperature hysteresis for substituted LaNi$_5$ and TiMn$_2$ alloys. *Model. Simul. Mater. Sci. Eng.* 2016, submitted for publication.

39. Schott, V.; Fähnle, M. Concentration of atomic defects in ordered compounds: Canonical and grandcanonical formalism. *Phys. Status Solidi B* **1997**, *204*, 617–624.

New Type of Alkaline Rechargeable Battery—Ni-Ni Battery

Lixin Wang, Kwo-Hsiung Young and Hao-Ting Shen

Abstract: The feasibility of utilizing disordered Ni-based metal hydroxide, as both the anode and the cathode materials, in alkaline rechargeable batteries was validated for the first time. Co and Mn were introduced into the hexagonal $Ni(OH)_2$ crystal structure to create disorder and defects that resulted in a conductivity increase. The highest discharge capacity of $55.6 \ mAh \cdot g^{-1}$ was obtained using a commercial Li-ion cathode precursor, specifically NCM111 hydroxide, as anode material in the Ni-Ni battery. Charge/discharge curves, cyclic voltammetry (CV), X-ray diffraction (XRD) analysis, scanning electron microscopy (SEM), transmission electron microscopy (TEM), X-ray energy dispersive spectroscopy (EDS) analysis, and electron energy loss spectroscopy (EELS) were used to study the capacity degradation mechanism, and the segregation of Ni, Co, and Mn hydroxides in the mixed hydroxide. Further optimization of composition and control in micro-segregation are needed to increase the discharge capacity closer to the theoretical value, $578 \ mAh \cdot g^{-1}$.

Reprinted from *Batteries*. Cite as: Wang, L.; Young, K.-H.; Shen, H.-T. New Type of Alkaline Rechargeable Battery—Ni-Ni Battery. *Batteries* **2016**, 2, 16.

1. Introduction

The first $Ni(OH)_2$-based alkaline rechargeable batteries, Ni-Fe and Ni-Cd, were patented by Thomas A. Edison in 1901 and 1902, respectively [1,2]. Since then, many works have been done to improve the performance of rechargeable batteries [3–5]. The half-cell reactions at positive and negative electrodes and the full cell reaction are shown in Equations (1)–(3), respectively (where M_t is a transition metal, for example Fe or Cd).

$$Ni(OH)_2 + OH^- \rightleftharpoons NiOOH + H_2O + e^- \text{ (forward : charge, reverse : discharge)} \quad (1)$$

$$M_t(OH)_2 + 2e^- \rightleftharpoons M_t + 2OH^- \text{ (forward : charge, reverse : discharge)} \quad (2)$$

$$M_t(OH)_2 + 2Ni(OH)_2 \rightleftharpoons M_t + 2NiOOH + 2H_2O \text{ (forward : charge, reverse : discharge)} \quad (3)$$

Basically, the negative electrode uses the transformation of a transition metal between the +2 and 0 oxidation states during charge/discharge operation. During the last two decades, the Ni-TM (transition metal) batteries extended into Ni-Zn and Ni-Co systems. More recently, even a Ni-Mn rechargeable system has been proposed [6]. The comparisons between various Ni-TMs have been summarized in Table 1. The common features of these rechargeable batteries are low-cost and wide working temperature ranges. Although the reaction in Equation (2) involves a two-electron transfer (resulting in a very high theoretical capacity, see Table 2), the utilization of active material is not sufficient. As a result, practical Ni-TM cells often occur with low gravimetric energy and, thus, are not suitable for mobile and automobile applications. In the late 1980s, a new type of alkaline rechargeable battery, nickel/metal hydride (Ni/MH), which uses a metal hydride (MH) alloy as the negative electrode (anode) active material [7] was commercialized by Matsushita, Sanyo, Toshiba, Yuasa, and Ovonic [8]. The new anode half-cell reaction and full-cell reactions are shown in Equations (4) and (5), respectively, where M_h is one of the MH alloys.

$$M_h + H_2O + e^- \rightleftharpoons M_hH + OH^- \text{ (forward : charge, reverse : discharge)} \quad (4)$$

$$M_h + Ni(OH)_2 \rightleftharpoons M_hH + NiOOH \text{ (forward : charge, reverse : discharge)} \quad (5)$$

Table 1. Comparisons of Ni-TM alkaline rechargeable battery. TM: transition metal.

Battery	Anode (Charge/Discharge)	Pros	Cons	Commercial Product	Energy Density
Ni-Mn	Mn/Mn(OH)$_2$	• Low cost • High voltage (2.2 V)	Only theoretical	No	130–190 Wh· kg^{-1} [6]
Ni-Fe	Fe/Fe(OH)$_2$	• Low cost • Long cycle life	• Low energy density • Low power	Yes	30 Wh· kg^{-1} [9]
Ni-Co	Co/Co(OH)$_2$	• High capacity	High cost	No	Projected to be 165 Wh· kg^{-1} [10,11]
Ni-Zn	Zn/Zn(OH)$_2$	• Low cost • Higher voltage (1.5 V)	Cycle life still has room to improve	Yes	65-120 Wh· kg^{-1} [12]
Ni-Cd	Cd/Cd(OH)$_2$	• Low cost • Long cycle life • High power at low temperature	• Toxic • Low energy density	Yes	40-60 Wh· kg^{-1} [13]

Table 2. Properties of hydroxides as negative electrode (anode) candidates in Ni-TM battery systems. All data are from reference [14] unless otherwise cited.

Hydroxide	Formula Weight	Theoretical Capacity (mAh· g^{-1})	Density (g· cm^{-3})	Solubility in Cold Water (g· 100· cm^{-3})	$E_0(M_t)$ in Equation (7) [15] (V)	M–O Bond Strength (kJ· mol^{-1}· M)
Mn(OH)$_2$	88.94	603	3.258	0.0002	−0.163	402.9
Fe(OH)$_2$	89.55	597	3.4	0.00015	0.493	390.4
Co(OH)$_2$	92.93	577	3.597	0.00032	0.659	384.5
Ni(OH)$_2$	92.69	578	4.15	0.013	0.648	382.0
Zn(OH)$_2$	99.39	539	3.258	0.0002	0.034	180 [16]
Cd(OH)$_2$	146.41	366	4.79	0.00025	0.583	235.6

With the higher capacity found in the MH alloy ($330 \, \text{mAh} \cdot \text{g}^{-1}$ and $400 \, \text{mAh} \cdot \text{g}^{-1}$ for AB_5 and AB_2 MH alloys, respectively), Ni/MH soon controlled the consumer market and was used to power the first commercially built electric vehicle (EV-1) of modern times by General Motors [17]. The history of battery for electrical vehicles, the EV-1 and its immediate predecessors and successors were reviewed by Matthé and Eberle [18]. Later on, Li-ion batteries entered the consumer market and dominated it with a higher energy density compared to that in Ni/MH battery. However, Ni/MH batteries still power more than 10 million hybrid electric vehicles (mainly the Prius, made by Toyota [19]), due to its high power, longevity, and excellent abuse tolerance. With the focus on rechargeable batteries in the consumer market switching to a stationary market, the battery industry is facing two major challenges, specifically cost and cycle stability, in replacing the currently used lead-acid battery [20,21]. The Ni-TM system happens to fall in the right direction for the stationary market. Additionally, by comparing the half-cell reactions in Equations (2) and (4), the Ni-TM batteries demonstrate an advantage over Ni/MH batteries at very low temperatures by not diluting the electrolyte with water generated from the discharge process. A few anode material candidates for the Ni-TM system are compared in Table 2. Although the theoretical capacities obtained from these materials are very high (due to two-electron transfer), the utilization is limited by the poor conductivity of the hydroxide. The electrochemical reaction, Equation (2), can be re-written into a standard form as Equation (6), with the equilibrium potential given by Equation (7) [15]:

$$HM_tO_2^- + 3H^+ + 2e^- \rightleftharpoons M_t + 2H_2O \tag{6}$$

$$E_o = E_o(M_t) - 0.00886pH + 0.0295\log\left(HM_tO_3^-\right) \tag{7}$$

Lower E_o values correspond to metals with lower oxidation energy (easier to be oxidized). From this table, it is easy to see that both Ni and Co have a potential similar to Cd and can have highly reversibility redox reactions occur in the voltage range of interest. Significant research has been devoted into Ni-Co batteries and results have been summarized in a review article [22]. In the past, a few methodologies were developed to improve the utilization (reversibility of Equation (2)), including: Reducing crystallite size [22], alloying with B [23], Si [24], P [25], and S [26] to increase the degree of disorder (DOD [27]), mixing silica [28], nitride [29], carbon nanotube [30], and CMK-3 (an ordered mesoporous form of carbon) [31], and forming Co_3O_4 nanowires [32].

Ni, with a similar oxidation potential and metal–oxygen (M–O) bond strength (Table 2) but a much lower raw material cost compared to Co, is a rational choice for a Ni-TM rechargeable system. The challenge is the same as $Co(OH)_2$: $Ni(OH)_2$ has a very poor electronic conductance. Intrinsic pure $Ni(OH)_2$ is a good insulator. In order to increase the conductivity, both composition [33–35] and structural [36] modifications are necessary and have been successfully developed. It is interesting to see the use of the disordered $Ni(OH)_2$ as the anode material for the Ni-Ni alkaline battery. With the funding from a U.S. Department of Energy sponsored Robust Affordable Next Generation EV (RANGE) program [37], we are able to investigate the feasibility of such a Ni-Ni battery and present the results in this paper.

2. Experimental Setup

The anode materials were prepared by a co-precipitation method in a continuous stirring tank reactor (CSTR) [7,38], where the sulfate salts of the nickel and/or manganese and cobalt were dissolved in deionized water. A suitable amount of solution was gradually pumped into the reactor in accordance with the desired mole ratio between the three elements. The pH value of the mixture was maintained between 10.5 and 12 by pumping sodium hydroxide solution (30 wt %) at a specific flow rate. Stirring rate (800 rpm), reaction temperature (60 °C), pH value, salt concentration, and residence time are well controlled and experimentally varied to achieve a desirable particle size and morphology.

Carbon black and polyvinylidene fluoride (PVDF) were used in the negative electrode to enhance conductivity and electrode integrity. To construct the negative electrode, 100 mg of active material (hydroxide from one, two, or three transition metals) were stir-mixed thoroughly with carbon black and PVDF in a weight ratio of 3:2:1. A 0.5×0.5 inch^2 nickel mesh was used as a substrate and current collector with a nickel mesh tab leading out of the square substrate for testing connections. The mixture of carbon black, PVDF, and metal hydroxides was evenly pressed onto both sides of the nickel mesh by a hydraulic press under 300 MPa for 5 s to form the anode for this experiment. The positive electrode was a sintered type of $Ni_{0.9}Co_{0.1}$ active material, which is commonly used in Ni-Cd batteries. The positive and negative electrodes were sandwiched together with a polypropylene/polyethylene separator in a flooded half-cell configuration. The capacity at the positive electrode was significantly more than that at the negative electrode, resulting in a negative limited design. Electrochemical testing was performed with an Arbin electrochemical testing station (Arbin Instrument, College Station, TX, USA). Cyclic voltammetry (CV) was obtained using a Gamry Potentiostats (Gamry Instruments Inc., Warminster, PA, USA). The particle size distribution of the powder was measured with a Microtrac-SRA 150 (Microtrac, Montgomeryville, PA, USA). X-ray diffraction (XRD) analysis was performed with a Philips X'Pert Pro X-ray diffractometer (Philips, Amsterdam, The Netherlands) and the generated patterns were fitted, and peaks indexed, by Jade 9 software (Jade Software Corp. Ltd., Christchurch, New Zealand). A JEOL-JSM6320F scanning electron microscope (SEM, JEOL, Tokyo, Japan) with energy dispersive spectroscopy (EDS) was applied in investigating the phase distributions and compositions of the powders. A FEI Titan 80–300 (scanning) transmission electron microscope (TEM/STEM, Hillsboro, OR, USA) was employed to study the microstructure of the alloy samples. For TEM characterization, mechanical polishing was used to thin samples, followed by ion milling.

3. Results

3.1. Electrochemical Measurements

Seven hydroxides, two elements (Ni and Mn), three binaries (Ni-Co, Ni-Zn, and Ni-Mn), and two ternaries (both based on Ni-Co-Mn) were prepared by the CSTR process. The ternary hydroxides are the precursor material for the cathode materials used in Li-ion rechargeable batteries, and herein is used as an anode material for a Ni-Ni battery. Their composition and observed discharge capacities at a discharge current of 5 mA$\cdot$g^{-1} from Cycles 1 and 5 are listed in Table 3. All the elemental and binary hydroxides show zero, or close to zero, capacity and both ternary hydroxide, NCM111 hydroxide (Ni$_{0.33}$Co$_{0.33}$Mn$_{0.33}$) and NCM424 hydroxide (Ni$_{0.4}$Co$_{0.2}$Mn$_{0.4}$), show the best capacities at approximately 20 mAh$\cdot$g^{-1} (Figure 1). It is interesting to find that the capacity improves with increasingly DOD through addition of more ingredients in the co-precipitation process. In principle, the idea of using a disordered hydroxide during the CSTR process is validated, but composition and process optimizations are necessary for a further improvement in capacity.

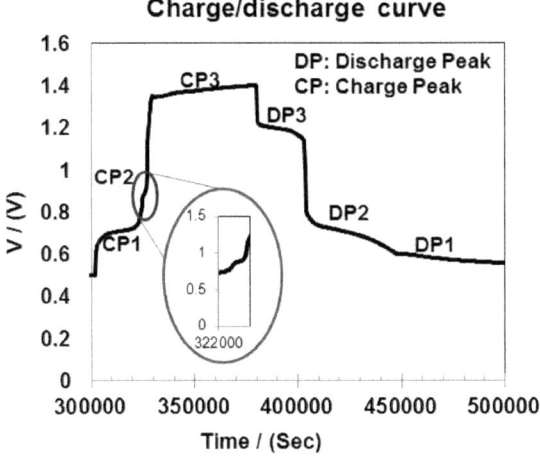

Figure 1. First cycle charge and discharge curves with a NCM111 hydroxide (Ni$_{0.33}$Co$_{0.33}$Mn$_{0.33}$) negative electrode.

Table 3. Discharge capacity in $mAh \cdot g^{-1}$ obtained at a discharge current of $5\ mA \cdot g^{-1}$.

M in M(OH)$_2$	Ni	Mn	Ni$_{0.91}$Co$_{0.09}$	Ni$_{0.91}$Zn$_{0.09}$	Ni$_{0.91}$Mn$_{0.09}$	Ni$_{0.33}$Co$_{0.33}$Mn$_{0.33}$	Ni$_{0.4}$Co$_{0.2}$Mn$_{0.4}$
First cycle	1.0	0.0	0.0	0.0	0.0	17.2	14.5
Fifth cycle	0.7	0.0	0.0	0.0	0.0	20.7	20.1

The electrochemical properties of the NCM111 hydroxide were further tested with a charge rate of $5\ mA \cdot g^{-1}$ and discharge rate of $0.4\ mA \cdot g^{-1}$. The charge and discharge voltage curves during the first cycle are plotted in Figure 1. The first charge plateau is observed at 0.71 V (*versus* Ni(OH)$_2$ cathode), corresponding to a discharge plateau at 0.68 V. A second plateau occurs at 1.38 V, with a corresponding discharge plateau at 1.19 V. Discharge voltage efficiency is defined as discharge voltage divided by the charge voltage, which is read at the middle point of the voltage plateau. The first charge reaction has a discharge voltage efficiency of 96%, compared to 86% for the second redox reaction. The highest discharge capacity of the NCM111 hydroxide is 55.6 $mAh \cdot g^{-1}$ and was observed during the first cycle (Figure 2). Degradations in both capacity and utilization (ratio of charge out *versus* charge in) with cycling can be easy seen in Figure 3. The cause of these degradations at a relatively low rate will be reported in the following sessions.

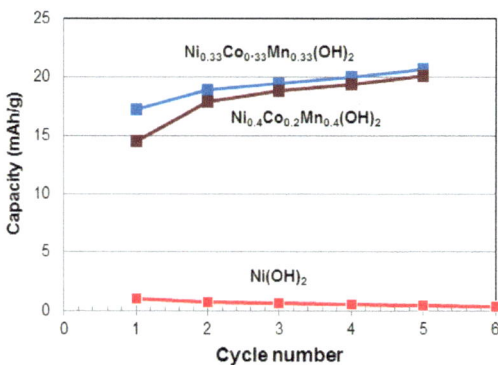

Figure 2. Discharge capacity obtained from Ni(OH)$_2$, NCM111 hydroxide, and NCM424 hydroxide (Ni$_{0.4}$Co$_{0.2}$Mn$_{0.4}$) negative electrodes with a discharge current of $5\ mA \cdot g^{-1}$.

The CV curves of the NCM111 hydroxide anode are plotted in Figure 4. The same cell configuration was used in the CV testing as in the charge/discharge testing, except that an Hg/HgO reference electrode was employed. Corresponding to the charge/discharge plateau at 1.38 V (*versus* cathode) and 1.19 V (*versus* cathode) from the charge and discharge curves, two peaks at -1.0 V (*versus* Hg/HgO) and -0.8 V (*versus* Hg/HgO) further verified the existence of one of the two redox reactions in Ni-Ni batteries. Similarly, the charge/discharge peaks at −0.37 V (*versus* Hg/HgO) and −0.2 V (*versus* Hg/HgO) correspond to the peaks at 0.71 V (*versus* cathode) and 0.68 V (*versus* cathode). The additional redox peaks in the CV curve at close to 0.0 V (*versus* Hg/HgO) is due to the oxidation of water into oxygen in the electrolyte during overcharge. The reduction of the residue oxygen on the surface was lowered during charging, causing a wide peak at −0.21 V (*versus* Hg/HgO).

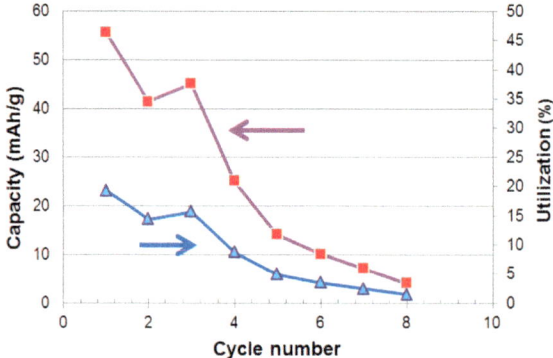

Figure 3. Discharge capacity and utilization obtained from a NCM111 hydroxide negative electrode with a discharge current of 0.4 mA· g^{-1}.

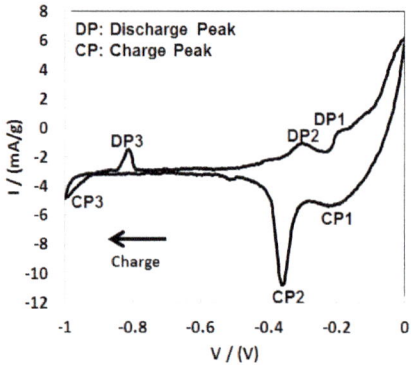

Figure 4. Cyclic voltammetry (CV) curve for a NCM111 hydroxide negative electrode.

3.2. X-ray Diffraction Analysis

The crystal structure of the NCM111 hydroxide anode at the charged state, the 1st discharge plateau, and the discharged state were characterized by XRD analysis and the resulting patterns are shown in Figure 5. The lattice constants of the hexagonal hydroxide phase were obtained through the curve fitting function offered in Jade 9.0 Software and the results are listed in Table 4. The charge or discharge process of the cell was terminated at the desired discharge state before it was disassembled. The anode samples were soaked in deionized water overnight and rinsed thoroughly to remove any KOH electrolyte residue. The anodes were then completely dried in air under room temperature before XRD analysis. However, some peaks from residue KOH (with a slightly larger lattice constant when comparing to pure KOH, due to some water content) are still present in Figure 5b.

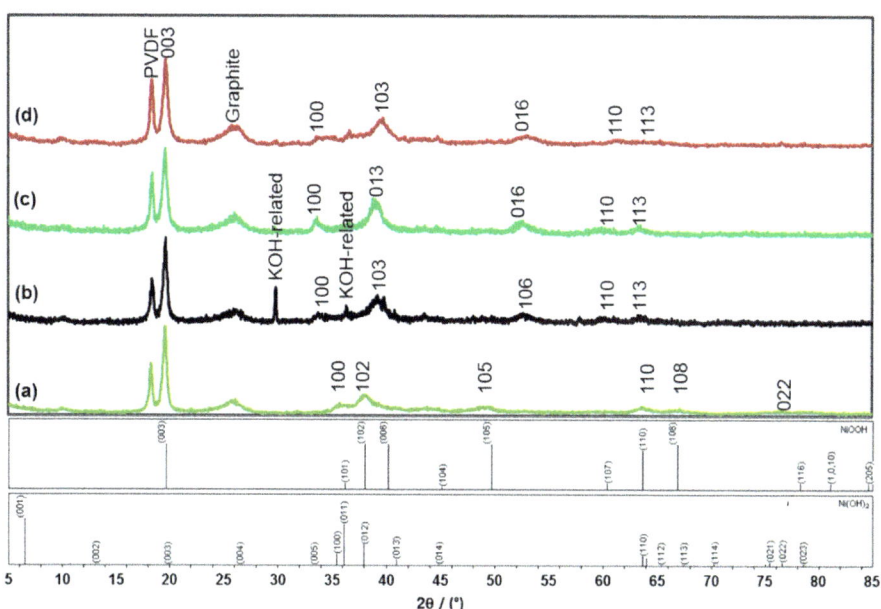

Figure 5. X-ray diffraction (XRD) patterns of: (**a**) a pristine NCM111 hydroxide powder; (**b**) a NCM111 hydroxide electrode at the first voltage plateau; (**c**) a charged NCM111 hydroxide fresh electrode; and (**d**) a NCM111 hydroxide electrode at a discharged state. Peak positions of a NiOOH and a Ni(OH)$_2$ structures from Powder Diffraction File (PDF) database [39] 00-006-0075 and 00-059-0462, respectively, are shown in the bottom for comparison.

Table 4. Lattice constants a and c of hydroxide in NCM111 hydroxide at different stages obtained from the XRD patterns shown in Figure 5.

Lattice Constant (Å)	Pristine	At the First Discharge Voltage Plateau	After First Charge	After Discharge
a	2.92	3.09	3.11	3.02
c	13.44	13.82	13.87	13.85

The XRD patterns of the NCM111 hydroxide anode at different discharge state can be fitted with a hexagonal unit cell (as the original Ni(OH)$_2$). Compared to that

343

in the fresh sample (Figure 5a), the (100) peak shows shifts from 35.73° to 33.59° and 33.46° for those in the first plateau (Figure 5b) and charged state (Figure 5c), respectively. Accordingly, the lattice parameter, a, increases from 2.92 Å to 3.09 Å and 3.11 Å, respectively. The lattice parameter, c, fitted by Jade 9, also showed a 2.8% increase, from 13.44 Å to 13.82 Å. The changes in both lattice parameters, a and c, indicate unit cell expansion during the charge process. During the charge phase, Ni was reduced to a lower state and extracted from the NCM111 hydroxide unit cell, leaving Co and Mn with a larger radius supporting the original hexagonal crystal structure. The extraction of small radius Ni from the basal plane may be the reason for the 5.8% increase in lattice parameter a. Comparably, the 2.8% decrease in c might be due to the extraction of a hydroxide group from between the basal planes, which caused partial collapse of this layer. The separation of Ni from the NCM111 hydroxide was also detected by EDS analysis, as demonstrated in the next section.

At a discharged state, the 100 peak shifted back to 33.69° (Figure 5d), corresponding to a decrease in a to 3.02 Å, which is smaller than observed in the charged state and the first plateau, but not fully reverted to the fresh state before cycling. Compared to the decreases in a, the parameter c decreased less during the discharge. The major changes in parameter a and the almost same changes in c indicate that the reaction sites are majorly involved in the basal plane during the charging, causing an expansion in the a–b planes and a slightly expansion in the c direction. During discharge, a and c recovered to 3.02 Å and 13.85 Å, but did not achieve the values seen in the original fresh anode state. The extraction in the c direction was insignificant after the initial charge, so that NCM111 hydroxide can still maintain a hexagonal structure with Co and Mn supporting the basal planes.

3.3. Scanning Electron Microscopy/Energy Dispersive Spectroscopy Characterizations

As shown in Figure 6a, the diameter range of pristine NCM111 hydroxide particles varies from several microns to as large as 30 μm. The surface of the NCM111 hydroxide particles is covered by thin flakes vertical to the particle surface and approximately 100 nm thick with diameters around 500 nm. After cycling, the flakes change their orientation from perpendicular to the surface (Figure 6b) to parallel with the surface (Figure 6d), indicating a regrowth of hydroxide during cycling.

The EDS mappings of the NCM hydroxide anode before and after cycling are shown in Figure 7a,b, respectively. The nickel, cobalt, and manganese are distributed evenly in the sample before cycling. However, the nickel distribution of the NCM hydroxide anode after cycling differs from the distribution of cobalt and manganese. The two areas are indicated by red and blue circles that show the differences between Ni and Co/Mn after the cycling. This difference can be explained by the segregation of $Ni(OH)_2$ out of the hexagonal structure of NCM111 hydroxide material during the cycling, as observed in the EDS map. This is in agreement with the XRD pattern, that the lattice constant *a* increases during the charge process and decreases during discharge, while *c* less noticeable changes. This segregation of Ni is the main cause for the capacity and utilization degradations discussed in Section 3.1. The Ni segregation is more severe during slower rate discharges and needs to be controlled to sustain the initial capacity. Approaches for such segregation-prevention may include implantation of grain growth inhibitor [40], control of oxygen impurities [41], addition of elements with different solubilities among phases [42], addition of rare earth elements [43], addition of other transition metals with larger ionic radii, introduction of a pulse charge/discharge method to interrupt the formation of $Ni(OH)_2$ isolated grain, surface modification with surfactant to make it more difficult to precipitate into larger $Ni(OH)_2$ grains, a new combination of alkali [44] and salt [45] used in the electrolyte, and improvements in the co-precipitation process to evenly distribute the cation in the hydroxide.

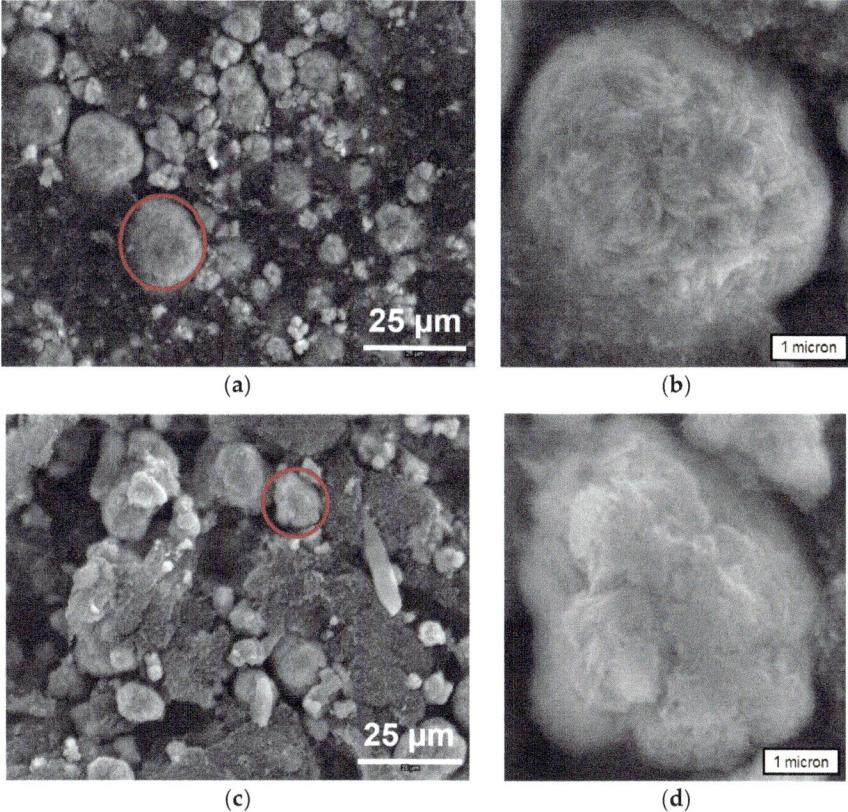

Figure 6. Scanning electron microscopy (SEM) micrographs of the NCM111 hydroxide negative material (**a,b**) before and (**c,d**) after cycling. Circled areas in (**a,c**) have been examined at higher magnification, as shown in (**b,d**).

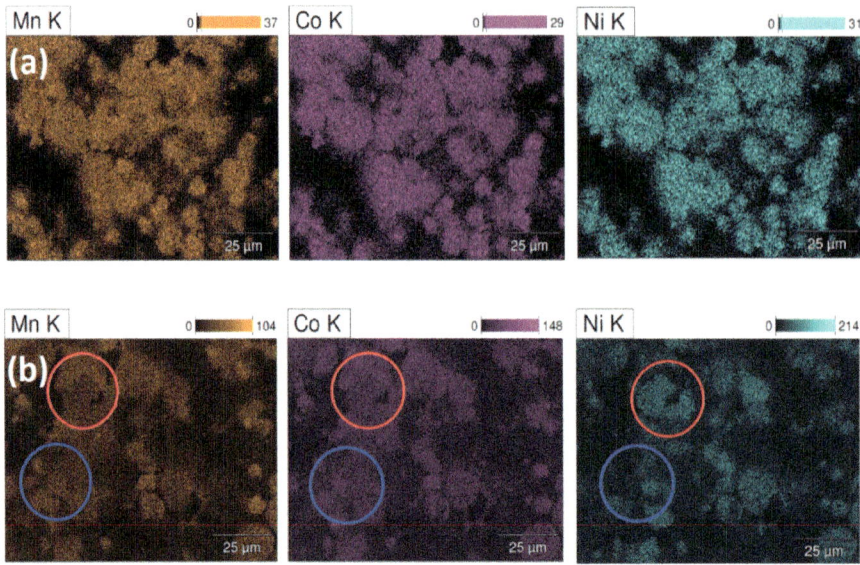

Figure 7. Energy dispersive spectroscopy (EDS) mappings of a NCM111 hydroxide anode (**a**) before and (**b**) after cycling.

3.4. Transmission Electron Microscopy Characterizations

To locally study the distribution change of Ni, Co, and Mn after cycling, scanning TEM was performed. As shown in Figure 8a, the EDS spectrum was collected from seven evenly distributed spots along the line, with an electron beam spot size of 1 nm. The relative percent of these three elements are plotted in Figure 8b. The plot clearly shows that while the distribution of Co is generally uniform, Ni content is higher in the center of the particle, and Mn content is higher at the surface. Compared to the uniform elemental distribution in fresh samples, such an elemental distribution change should be attributed to micro-segregation that occurs during cycling.

Electron energy loss spectroscopy (EELS) data was also obtained by scanning the indicated region in Figure 9a, which is consistent with EDS analysis results. The phase segregation with cycling in the NCM111 hydroxide sample forms an isolated $Ni(OH)_2$ region which is known to be electrochemically inactive and deteriorates the discharge capacity.

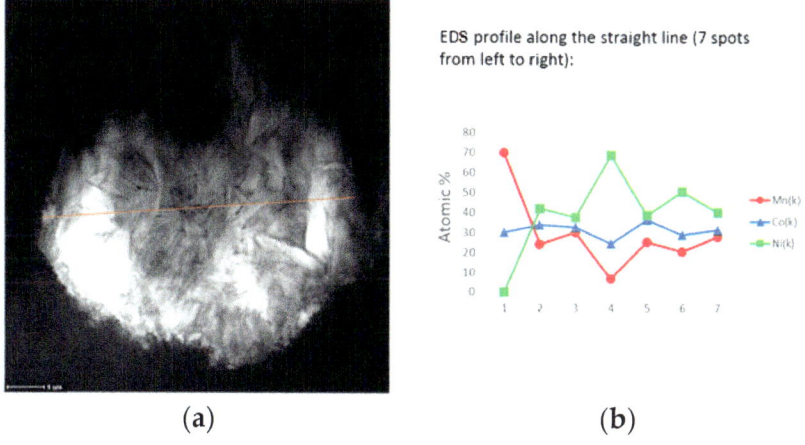

Figure 8. (**a**) Transmission electron microscopy (TEM) dark field image of a cross-section of a NCM111 hydroxide after 10 electrochemical cycles and (**b**) the EDS composition profile of Ni, Co, and Mn along the orange straight line shown in (**a**).

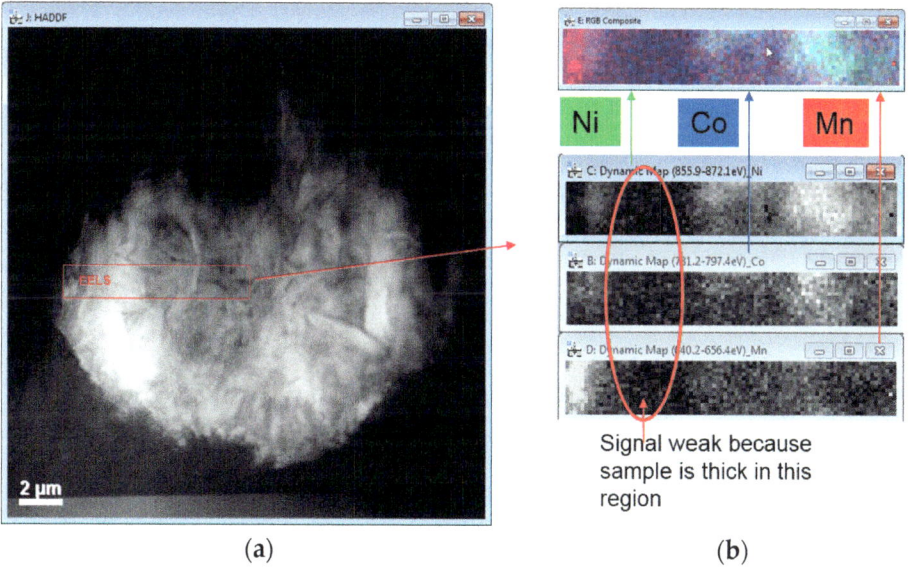

Figure 9. (**a**) TEM dark field image of a cross-section of a NCM111 hydroxide after 10 electrochemical cycles and (**b**) electron energy loss spectroscopy (EELS) mappings of Ni, Co, and Ni in the rectangle area highlighted in (**a**).

347

3.5. Discussion

In this work, NCM111 shows the highest degree of compositional disorder and also the highest discharge capacity. However, the high cost of Co prevents the commercialization of this material unless other strong incentives can be obtained. Considering the low voltage of the aqueous system compared to the Li-ion rival using the NCM111 as the cathode, it will be tough for NCM111 Ni-Ni alkaline battery to compete with Li-ion battery unless a capacity close to the theoretical value ($ca.$ 580 mAh$\cdot$g^{-1}) can be realized. When the Li-industries are moving from NCM111 to NCM424, NCM523, even NCM811 [46,47] to reduce the cost and increase the capacity, such high-Ni precursor is certainly a good candidate for Ni-Ni alkaline battery testing. Other inexpensive transition metals (Cr, Fe, Cu, and Zn) or even non-transition metals (Li, Al, S, and Mg) should also be tested as components to further increase the DOD and cycle stability.

4. Conclusions

A novel ternary NCM111 hydroxide (NiCoMn(OH)$_6$) material was demonstrated for the first time to act as an anode material for alkaline rechargeable batteries. A capacity of 55.6 mAh$\cdot$g^{-1} was achieved with a cycle life of 8, without further optimizing. Charge/discharge testing and CV testing verified the existence of a two-step reaction. The chemical reactions have been preliminarily suggested to be two-step reactions within which the Ni(OH)$_2$ is reduced to NiOH at the first plateau and then to Ni at the second plateau. During the cycling, Ni and Ni hydroxide aggregated, while Co and Mn remained at the same location, which led to the loss of disorder in the NCM111 hydroxide structure after cycling, as revealed by XRD, SEM, and TEM analysis. This reduction in DOD is considered to be the major reason of capacity loss over the course of cycling and necessitates immediate attention for future research.

Acknowledgments: This work was financially supported by US DOE ARPA-E under the RANGE program (DE-AR0000386).

Author Contributions: Lixin Wang conceived and performed the experiments, and wrote the paper; Kwo-Hsiung Young helped in data interpretation and paper preparation; Hao-Ting Shen performed the TEM analysis.

Conflicts of Interest: The authors declare no conflict of interest.

348

Abbreviations

TM	Transition metal
Ni/MH	Nickel/metal hydride
MH	Metal hydride
EV	Electric vehicle
DOD	Degree of disorder
M–O	Metal–oxygen
RANGE	Robust Affordable Next Generation EV
CSTR	Continuous stirring tank reactor
rpm	Revolution per minute
PVDF	Polyvinylidene fluoride
CV	Cyclic voltammetry
XRD	X-Ray diffraction
SEM	Scanning electron microscopy
EDS	X-Ray energy dispersive spectroscopy
DP	Discharge peak
CP	Charge peak
TEM	Transmission electron microscopy
PDF	Powder diffraction file
EELS	Electron energy loss spectroscope

References

1. Edison, T.A. Reversible Galvanic Battery. U.S. Patent 678,722, 16 July 1901.
2. Edison, T.A. Reversible Galvanic Battery. U.S. Patent 692,507, 4 February 1902.
3. Shukla, A.K.; Venugopalan, S.; Hariprakash, B. Nickel-based rechargeable batteries. *J. Power Sources* **2001**, *100*, 125–148.
4. Morioka, Y.; Narukawa, S.; Itou, T. State-of-the-art of alkaline rechargeable batteries. *J. Power Sources* **2001**, *100*, 107–116.
5. Tarascon, J.M. Key challenges in future Li-battery research. *Philos. Trans. R. Soc. A* **2010**, *368*, 3227–3241.
6. Carmichael, C. Making Economical, Green, High-Energy Nickel-Manganese (NiMn) Batteries. Available online: http://www.saers.com/recorder/craig/TurquoiseEnergy/BatteryMaking/BatteryMaking.html (accessed on 1 April 2016).
7. Chang, S.; Young, K.; Nei, J.; Fierro, C. Reviews on the U.S. patents regarding nickel/metal hydride batteries. *Batteries* **2016**, *2*.
8. Ouchi, T.; Young, K.; Moghe, D. Reviews on the Japanese patent applications regarding nickel/metal hydride batteries. *Batteries* **2016**, *2*.
9. Jiang, W.; Wu, Y.; Cheng, Y.; Wang, L. Industrial application of nickel-iron battery and its recent research progress. *Chin. J. Appl. Chem.* **2014**, *31*, 749–756.

10. Gao, X.P.; Yao, S.M.; Yan, T.Y.; Zhou, Z. Alkaline rechargeable Ni/Co batteries: Cobalt hydroxides as negative electrode materials. *Energy Environ. Sci.* **2009**, *2*, 502–505.

11. Gao, X.P.; Yang, H.X. Multi-electron reaction materials for high energy density batteries. *Energy Environ. Sci.* **2010**, *3*, 174–189.

12. PowerGenix Batteries. The Nickel-Zinc Battery. Available online: http://www.powergenix.com/the-nickel-zinc-battery/ (accessed on 6 April 2016).

13. Wikipedia Webpage. Nickel-Cadmium Battery. Available online: https://en.wikipedia.org/wiki/Nickel%E2%80%93cadmium_battery/ (accessed on 6 April 2016).

14. Lide, D.R. *CRC Handbook of Chemistry and Physics*, 74th ed.; CRC Press Inc.: Boca Raton, FA, USA, 1993.

15. Pourbaix, M. *Atlas of Electrochemical Equilibrium in Aqueous Solutions*; National Association of Corrosion Engineers: Houston, TX, USA, 1974.

16. Nimmermark, A.; Ohrstrom, L.; Reedijk, J. Metal-ligand bond lengths and strengths: Are they correlated? A detailed CSD analysis. *Z. Krist. Cryst. Mater.* **2013**, *228*, 311–317.

17. Wikipedia Webpage. General Motors EV1. Available online: https://en.wikipedia.org/wiki/General_Motors_EV1 (accessed on 17 May 2016).

18. Matthé, R.; Eberle, U. The Voltec System: Energy Storage and Electric Propulsion. Available online: https://www.selidori.com/tech/scarica.php?id_doc=1090 (accessed on 17 May 2016).

19. Tanoue, K.; Yanagihara, H.; Kusumi, H. Hybrid Is A Key Technology for Future Automobiles. In *Hydrogen Technology*; Léon, A., Ed.; Springer: Berlin, Germany, 2008; pp. 235–272.

20. Zelinsky, M.; Koch, J.; Fetcenko, M. Heat Tolerant NiMH Batteries for Stationary Power. Available online: https://www.battcon.com/PapersFinal2010/ZelinskyPaper2010Final_12.pdf (accessed on 28 March 2016).

21. Zelinsky, M.; Koch, J. Batteries and Heat—A Recipe for Success? Available online: https://www.battcon.com/PapersFinal2013/16-Mike%20Zelinsky%20-%20Batteries%20and%20Heat.pdf (accessed on 28 March 2016).

22. Zhao, X.; Ma, L.; Shen, X. Co-based anode materials for alkaline rechargeable Ni/Co batteries: A review. *J. Mater. Chem.* **2012**, *22*, 277–285.

23. Liu, Y.; Wang, Y.; Xiao, L.; Song, D.; Wang, Y.; Jiao, L.; Yuan, H. Structure and electrochemical behaviors of a series of Co-B alloys. *Electrochim. Acta* **2008**, *53*, 2265–2271.

24. Wang, Y.; Lee, J.M.; Wang, X. An investigation of the origin of the electrochemical hydrogen storage capacities of the ball-milled Co–Si composites. *Int. J. Hydrog. Energy* **2010**, *35*, 1669–1673.

25. Cao, Y.; Zhou, W.; Li, X.; Ai, X.; Gao, X.; Yang, H. Electrochemical hydrogen storage behaviors of ultrafine Co-P particles prepared by direct ball-milling method. *Electrochim. Acta* **2006**, *51*, 4285–4290.

26. Wang, Q.; Jiao, L.; Du, H.; Peng, W.; Liu, S.; Wang, Y.; Yuan, H. Electrochemical hydrogen storage property of Co-S alloy prepared by ball-milling method. *Int. J. Hydrog. Energy* **2010**, *35*, 8357–8362.

27. Sapru, K.; Reichman, B.; Reger, A.; Ovshinsky, S.R. Rechargeable Battery and Electrode Used Therein. U.S. Patent 4,623,597, 18 November 1986.

28. Han, Y.; Wang, Y.; Wang, Y.; Jiao, L.; Yuan, H. Characterization of CoB-silica nanochains hydrogen storage composite prepared by in-situ reduction. *Int. J. Hydrog. Energy* **2010**, *35*, 8177–8181.

29. Yao, S.M.; Xi, K.; Li, G.R.; Gao, X. Preparation and electrochemical properties of Co-Si_3N_4 nanocomposites. *J. Power Sources* **2008**, *184*, 657–662.

30. Du, H.; Jiao, L.; Wang, Q.; Peng, W.; Song, D.; Wang, Y.; Yuan, H. Structure and electrochemical properties of ball-milled Co-carbon nanotube composites as negative electrode material of alkaline rechargeable batteries. *J. Power Sources* **2011**, *196*, 5751–5755.

31. Li, L.; Xu, Y.; An, C.; Wang, Y.; Jiao, L.; Yuan, H. Enhanced electrochemical properties of Co/CMK-3 composite as negative material for alkaline secondary battery. *J. Power Sources* **2013**, *238*, 117–122.

32. Xu, Y.; Wang, X.; An, C.; Wang, Y.; Jiao, L.; Yuan, H. Effect of the length and surface area on electrochemical performance of cobalt oxide nanowires for alkaline secondary battery application. *J. Power Sources* **2014**, *272*, 703–710.

33. Fierro, C.; Fetcenko, M.A.; Young, K.; Ovshinsky, S.R.; Sommers, B.; Harrison, C. Nickel Hydroxide Positive Electrode Material Exhibiting Improved Conductivity and Engineered Activation Energy. U.S. Patent 6,228,535, 8 May 2001.

34. Fierro, C.; Fetcenko, M.A.; Young, K.; Ovshinsky, S.R.; Sommers, B.; Harrison, C. Nickel Hydroxide Positive Electrode Material Exhibiting Improved Conductivity and Engineered Activation Energy. U.S. Patent 6,447,953, 10 September 2002.

35. Ovshinsky, S.R.; Corrigan, D.; Venkatesan, S.; Young, R.; Fierro, C.; Fetcenko, M.A. Chemically and Compositionally Modified Solid Solution Disordered Multiphase Nickel Hydroxide Positive Electrode for Alkaline Rechargeable Electrochemical Cells. U.S. Patent 5,348,822, 20 September 1994.

36. Wong, D.F.; Young, K.; Wang, L.; Nei, J.; Ng, K.Y.S. Evolution of stacking faults in substituted nickel hydroxide spherical powders. *J. Alloys Compd.* **2016**, submitted.

37. Young, K.; Ng, K.Y.S.; Bendersky, L.A. A technical report of the robust affordable next generation energy storage system-BASF program. *Batteries* **2016**, *2*.

38. Fierro, C.; Zallen, A.; Koch, J.; Fetcenko, M.A. The influence of nickel-hydroxide composition and microstructure on the high-temperature performance of nickel metal hydride batteries. *J. Electrochem. Soc.* **2006**, *153*, A492–A496.

39. *Powder Diffraction File (PDF) Database*; International Centre for Diffraction Data: Newtown Square, PA, USA, 2011.

40. Siozios, A.; Zoubos, H.; Pliatsikas, N.; Koutsogeorgis, D.C.; Vourlias, G.; Pavlidou, E.; Cranton, W.; Patsalas, P. Growth an annealing strategies to control the microstructure of AlN:Ag nanocomposite films for plasmonic applications. *Surf. Coat. Technol.* **2014**, *255*, 28–36.

41. Ishijima, Y.; Kannari, S.; Kurishita, H.; Hasegawa, M.; Hiraoka, Y.; Takida, T.; Takebe, K. Processing of fine-grained W materials without detrimental phases and their mechanical properties ta 200–432 K. *Mater. Sci. Eng. A* **2008**, *473*, 7–15.

42. Wang, M.; Du, J.; Deng, Q.; Tian, Z.; Zhu, J. The effect of phosphorus on the microstructure and mechanical properties of ATI 718Plus alloy. *Mater. Sci. Eng. A* **2015**, *626*, 382–389.

43. Fu, H.; Xiao, Q.; Li, Y. A study of the microstructures and properties of Fe-V-W-Mo alloy modified by rare earth. *Mater. Sci. Eng. A* **2005**, *395*, 281–287.

44. Nei, J.; Young, K.; Rotarov, D. Studies in Mg-Ni based metal hydride electrode with electrolytes composed of various hydroxides. *Batteries* **2016**, submitted.

45. Yan, S.; Young, K.; Ng, K.Y.S. Effects of salt additives to the KOH electrolyte used in Ni/MH batteries. *Batteries* **2015**, *1*, 54–73.

46. BASF Catalysts. NCM Cathode Materials. Available online: http://www.catalysts.basf.com/p02/USWeb-Internet/catalysts/en/content/microsites/catalysts/prods-inds/batt-mats/NCM (accessed on 17 May 2016).

47. Fetcenko, M.A. BASF-ANL Collaboration on NCM Cathode Materials. Available online: http://www.energy.gov/sites/prod/files/2014/11/f19/Fetcenko%20-%20Industry%20Partners%20Panel_0.pdf (accessed on 17 May 2016).

Failure Mechanisms of Nickel/Metal Hydride Batteries with Cobalt-Substituted Superlattice Hydrogen-Absorbing Alloy Anodes at 50 °C

Tiejun Meng, Kwo-hsiung Young, John Koch, Taihei Ouchi and Shigekazu Yasuoka

Abstract: The incorporation of a small amount of Co in the A_2B_7 superlattice hydrogen absorbing alloy (HAA) can benefit its electrochemical cycle life performance at both room temperature (RT) and 50 °C. The electrochemical properties of the Co-substituted A_2B_7 and the failure mechanisms of cells using such alloys cycled at RT have been reported previously. In this paper, the failure mechanisms of the same alloys cycled at 50 °C are reported. Compared to that at RT, the trend of the cycle life at 50 °C versus the Co content in the Co-substituted A_2B_7 HAAs is similar, but the cycle life is significantly shorter. Failure analysis of the cells at 50 °C was performed using X-ray diffraction (XRD), scanning electron microscopy (SEM), X-ray energy dispersive spectroscopy (EDS), and inductively coupled plasma (ICP) analysis. It was found that the elevated temperature accelerates electrolyte dry-out and the deterioration (both pulverization and oxidation) of the A_2B_7 negative electrode, which are major causes of cell failure when cycling at 50 °C. Cells from HAA with higher Co-content also showed micro-shortage in the separator from the debris of the corrosion of the negative electrode.

Reprinted from *Batteries*. Cite as: Meng, T.; Young, K.-h.; Koch, J.; Ouchi, T.; Yasuoka, S. Failure Mechanisms of Nickel/Metal Hydride Batteries with Cobalt-Substituted Superlattice Hydrogen-Absorbing Alloy Anodes at 50 °C. *Batteries* **2016**, 2, 20.

1. Introduction

Depending on their particular high energy density and long cycle life, lithium-ion batteries have become the dominant battery type in the consumer and electric vehicle market. However, high cost, limited temperature range, and safety concerns have caused inconvenience and restricted the application of lithium-ion batteries in many fields, including uninterruptible power supply for home appliances, vending machines, cell phone towers, and other on-grid or off-grid stationary energy storage applications. In comparison, the nickel/metal hydride (Ni/MH) battery is a low cost alternative with high safety, long cycle life, excellent rate capabilities, and exceptional high-temperature capabilities for stationary applications [1,2]. The continuous research and development in Ni/MH batteries has enabled continuous use in both consumer type and hybrid electric vehicles [3].

The Ni/MH type battery is capable of operating at high temperatures, up to 85 °C [4], while maintaining a long cycle life [5]. These exceptional high-temperature capabilities make the Ni/MH batteries ideal candidates in extreme environments. In comparison, the performance of commercial lithium-ion batteries deteriorates dramatically at above 70 °C due to electrolyte decomposition [6]. Furthermore, the breakdown of the solid electrolyte interphase starts at 80 °C [7,8] and may result in a thermal runaway [9], which eventually causes fire and explosion [10]. Most rechargeable batteries, including Ni/MH batteries, are optimized to operate at room temperature (RT), and their performances degrade with the increase of temperature. To further improve the high-temperature performance of Ni/MH batteries, their degradation mechanisms at elevated temperatures must be clarified. There are two concerns for Ni/MH applications at high-temperature (>40 °C): charge acceptance and capacity loss. The former has been addressed with composition modifications for the active material used in the positive electrode [11]. The later can be further classified into losses in the positive and negative electrodes. The capacity loss in the positive electrode at high temperature was reviewed in a recent article [12], while the capacity loss in the negative electrode remains to be thoroughly investigated.

There are basically three types of hydrogen absorbing alloy (HAA) that are used as the negative electrode active materials in commercial Ni/MH batteries: misch-metal based AB_5 (mainly $CaCu_5$ structure), Laves phase AB_2 (mainly C14 and C15 structures), and A_2B_7 superlattice (mainly Nd_2Ni_7 and $PrNi_3$ structures). The capacity deteriorations in AB_5 and AB_2 HAAs were identified to be due to surface passivation of $La(OH)_3$ [13] and surface corrosion which leaches out soluble ions [14], respectively. However, studies on the failure mechanism of A_2B_7 superlattice HAAs are limited. In the past, we have examined Nd-based [15], La-based [16], and La, Pr, Nd-based [17,18] superlattice HAAs and found that these materials share the same failure mechanism: a passivated rare earth hydroxide surface. Both the Mg-incorporation and the higher rare earth content contribute to the higher rate of surface passivation of superlattice HAAs compared to that of AB_5 HAAs.

It is well accepted that B-site substitution can effectively improve the electrochemical performance of HAAs (AB_5, AB_2, and A_2B_7) [19–22]. Previously, we have reported on the electrochemical properties of two series of Mn- and Co-substituted superlattice HAAs as the active materials in the negative electrodes in the sealed cells [23,24], and analyzed their failure mechanisms in sealed full cells at RT [23,24]. These results have shown that both Mn- and Co- substitutions can improve low-temperature cell performance, but not the high-temperature performance [23,24]. The failure mechanisms of cells with superlattice HAAs at high temperatures have not yet been fully understood and require clarification for

354

further improvement of the cell performance at high temperature. In this work, we focused on sealed full cells made with Co-substituted superlattice HAAs and studied the failure mechanisms of the cells cycled at 50 °C.

2. Experimental Setup

The metal hydride (MH) alloys used in this study include five superlattice A_2B_7 HAAs (C1–C5) and one La-rich AB_5 HAA (AB5) as a reference. The compositions of the five A_2B_7 HAAs are $Mm_{0.83}Mg_{0.17}Ni_{3.14-x}Al_{0.17}Co_x$ (x = 0, 0.1, 0.2, 0.3, and 0.4), with variations corresponding to the Co amount of 0 (C1), 2.3 (C2), 4.7 (C3), 7.0 (C4), and 9.3 (C5) at%. These alloys were made by Japan Metals and Chemicals Co. (JMC, Tokyo, Japan). Herein, Mm is the mixed rare earth metal with a composition of 19.6 wt% lanthanum (La), 40.2 wt% praseodymium (Pr), and 40.2 wt% neodymium (Nd). The designed formula for the five A_2B_7 HAAs is $AB_{3.31}$. Details on the composition of the Co-substituted A_2B_7 HAAs can be found in Table 1 of [24]. The La-rich AB_5 HAA has a composition of $La_{10.5}Ce_{4.3}Pr_{0.5}Nd_{1.4}Ni_{60.0}Co_{12.7}Mn_{5.9}Al_{4.7}$ and was supplied by Eutectix (Troy, MI, USA). The same alloys have been used in our previous study of the Co-substituted superlattice A_2B_7 HAAs [18,24]. The powders of the six different alloys were dry-compacted onto nickel mesh current collectors as negative electrodes. Pairing such negative electrodes with the positive electrodes made with 94.1 wt% CoOOH-coated $Ni_{0.85}Co_{0.12}Zn_{0.03}(OH)_2$, 4.9 wt% Co, and 1 wt% Y_2O_3 on nickel foam substrates, C-size Ni/MH cells were assembled for electrochemical testing. The separator used was Scimat 700/79 acrylic acid grafted polypropylene/polyethylene, which was supplied by Freudenberg (Weinheim, Germany) and the electrolyte was a NaOH electrolyte (26.8 wt%) with LiOH (1.5 wt%) as an additive for improved high-temperature performance [25]. The designed negative-to-positive capacity ratio was kept at 2.0 in order to balance the over-charge and over-discharge reservoirs [15]. For convenience, C1–C5 and AB5 are used to represent not only the HAA alloys C1–C5 and AB5, but also the cells with corresponding HAA alloys.

The sealed cells were tested using a Maccor Battery Cycler (Tulsa, OK, USA) in a Blue M Oven (TPS Thermal Power Solutions, White Deer, PA, USA) with a set temperature of 50 °C. After the formation process in which the newly sealed cells were charged/discharged at C/3 for three cycles with cut-off voltages of 1.45 V/1 V, they were tested at 50 °C using a 0.5C charge/discharge rate with a discharge cut-off voltage of 0.9 V. The cut-off of the charging process was triggered when the cells reached 105% state-of-charge, based on the initial discharge capacity. The cycle test ended when the capacity dropped below 70% of the initial discharge capacity.

The cells were disassembled after the cycle test and their failure modes were studied by X-ray diffraction (XRD), scanning electron microscopy (SEM), energy dispersive spectroscopy (EDS), and inductively coupled plasma (ICP). The electrolyte residues in the cycled electrodes were removed in a Soxhlet extractor (Thermo

Fisher Scientific, Walthan, MA, USA). XRD analysis in the 2θ range of 10°–80° was performed using Cu Kα radiation (λ = 1.5406 Å) on a Philips X'Pert Pro X-ray diffractometer (Amsterdam, The Netherlands) to study the crystal structure of the positive and negative electrodes after cycling. The powders were scraped from the cycled electrodes and sifted into the well of a glass sample holder for XRD. The working voltage of the X-ray generator is 30 kV and current is 15 mA. Scanning was carried out with the angular step size of 0.02° and counting time of 5 s. SEM was carried out using a JEOL-JSM6320F microscope (Tokyo, Japan), with EDS capabilities, to study the morphology and composition of the negative electrodes after cycling. Another JEOL JSM7100 field-emission SEM with EDS was used to obtain the elemental mapping across the battery cross-section (positive electrode/separator/negative electrode). After Soxhlet extraction and drying in nitrogen, the cycled cells were vacuum-impregnated in epoxy and cured overnight. Later, they were ground with SiC paper (180, 320 and 600 grits) and polished with 6 μm and 1 μm diamond compound. SEM analysis was carried out with a 20 kV beam accelerating voltage, 300 pA current and 15 mm working distance. A Thermo iCAP 7400 ICP system (Thermo Fisher Scientific) was used to determine the composition of the Soxhlet extraction solution. The parameters used in ICP measurement were: RF power—1150 W, exposure time—10 s, nebulizer gas flow—0.55 L·min^{-1} and three replicates.

3. Results and Discussion

High-temperature charge/discharge tests were performed on the C-size cells at a fixed temperature, 50 °C, with a rate of 0.5C for charge and discharge, until the capacity dropped below 70% of the initial capacity, which is deemed as the end of cycle life. Later, the cells were disassembled and the electrodes and separators were characterized by XRD, SEM/EDS, and ICP to study the failure mechanisms of the cells cycled at 50 °C.

3.1. Cycle Life

The cycle life versus Co content at 50 °C and RT (20 °C) for samples C1–C5 and AB5 are shown in Figure 1a, and the evolution of capacity as a function of cycle number at 50 °C is shown in Figure 1b (adapted from [24]). The blue dotted line and red dashed line indicate the cycle life of AB5 at RT and 50 °C, respectively. For each material, three identical cells were fabricated and tested, and the results show that the error in cycle life measurement is no more than 10% in this study. To show the best performance for each material, the data points in Figure 1 are from the cells with the longest cycle life. At 50 °C, the cycle life for all samples is dramatically lower than those cycled at RT. The maximum cycle life for the 50 °C series is 255 cycles for both C2 and C3, compared to 775 and 990 cycles for C2 and C3, respectively, at RT. Similar to the trend for the RT series, at 50 °C the cycle life increases with incorporation of Co and then decreases with further increases in Co content. The AB$_5$ sample exhibited a longer cycle life at 50 °C than all five superlattice A$_2$B$_7$ HAA samples.

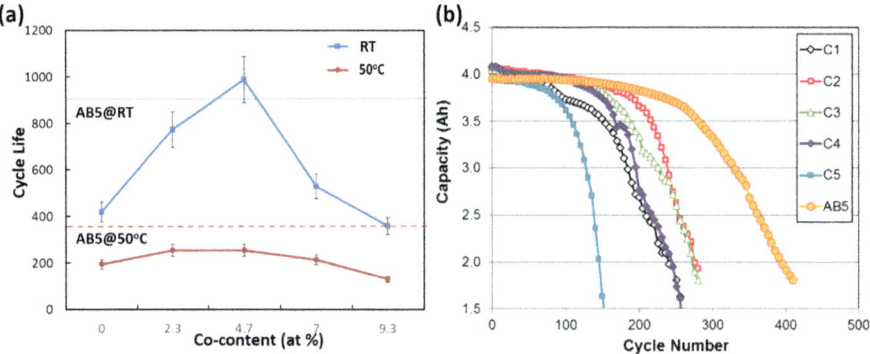

Figure 1. (**a**) Cycle life (the number of cycles when the discharge capacity drops to 70% of the original capacity) as a function of Co-content in the superlattice A$_2$B$_7$ hydrogen absorbing alloys (HAAs) at room temperature (RT, 20 °C) and 50 °C. The blue dotted line and red dashed line denote the cycle life for AB$_5$ at RT and 50 °C, respectively. (**b**) Capacity vs. cycle number for cells with HAAs C1–C5 and AB5 at 50 °C (adapted from [24]).

3.2. X-Ray Diffraction Structure Analysis

The XRD patterns from the cycled positive and negative electrodes are presented in Figure 2a,b, respectively. In Figure 2a, the cycled positive electrodes are found to be comprised mainly of β-Ni(OH)$_2$ (JCPDS PDF No. 00-014-0117) [26] with a small amount of CoO (JCPDS PDF No. 00-043-1004) [27] and metallic Ni (JCPDS PDF No. 00-004-0850) [28]. Metallic Ni should originate from the debris of the Ni foam that occurs when scraping the powder off the electrode. CoO is the oxidation product of the Co powder, which was added into the positive electrode as an additive in order to form the CoOOH conduction network. There is no distinguishable peak of α-Ni(OH)$_2$ or γ-NiOOH in the XRD patterns. The existence of α-Ni(OH)$_2$ or γ-NiOOH often results from Al-contamination from the negative electrode [29,30] or overcharge [31–34], which may cause electrode pulverization and shorten its cycle life. In addition, no obvious differences were observed for the cycled positive electrodes for all the samples.

On the other side of the system, there are obvious signs that the negative electrodes were heavily oxidized, and peak intensities of the oxidation products are prominent for all the cycled negative electrodes. There are two main types of observed oxidation products: rare earth hydroxides and Ni(OH)$_2$. The rare earth hydroxides include La(OH)$_3$, Nd(OH)$_3$, and Pr(OH)$_3$, which have very similar

2θ-diffraction angles and their peaks overlap to form broadened peaks in the XRD patterns. Figure 3 shows the XRD patterns of the negative electrodes cycled at 50 °C and RT at the end of cycle life, in comparison with those obtained from pristine HAA alloys. As indicated by the XRD patterns, the relative peak heights of the oxidation products increase dramatically when the test temperature increases to 50 °C from RT, especially for C1, which exhibits the least oxidation among all the samples at RT.

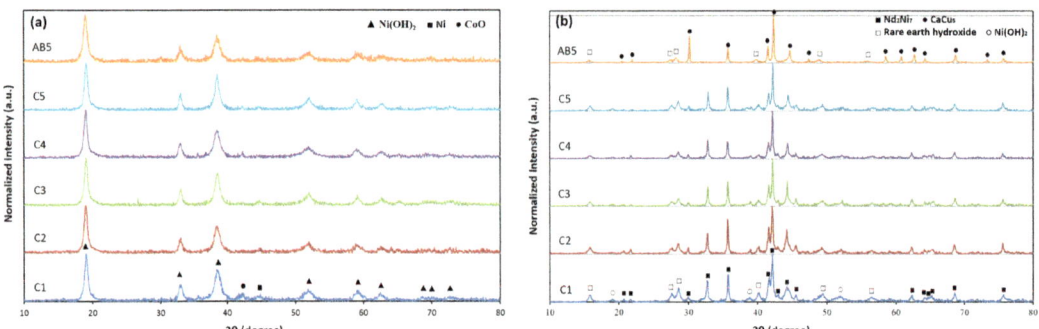

Figure 2. X-ray diffraction (XRD) patterns of the cycled (**a**) positive and (**b**) negative electrodes for cells with HAAs C1–C5 and AB5 at the end of cycle life.

3.3. Scanning Electron Microscopy/Energy Dispersive Spectroscopy Analysis

The SEM backscattered electron images (BEIs) of the cycled positive electrodes are shown in Figure 4 and the chemical compositions at different spots on the surface of the electrode (electrode/electrolyte interface) to the bulk of the electrode, measured by EDS, are listed in Table 1. No pulverization was observed for all the samples, despite a certain degree of swelling, particularly for AB5. It is well known that Al can stabilize α-Ni(OH)$_2$ phase, which causes a large volume expansion and will eventually lead to pulverization and capacity degradation. Al is detected in all the positive electrodes in this study. Since Al exists only in the negative electrode, where HAA is corroded, Al leaches out during cycling and migrates to the positive electrode. Also, the amount of Al found in the positive electrode may be an indication of the extent of corrosion in the negative electrode. Herein, AB5 has the highest amount of Al (6.2–7.7 at%) in the positive electrode and the most severe swelling of Ni(OH)$_2$, as indicated by the blurry Ni(OH)$_2$ grain boundaries in the SEM BEI micrograph. C5 has a higher amount of Al (2.8–4.1 at%) in the cycled positive electrode, compared to C1–C4. Al diffuses deep into the bulk of the positive electrode, with a slight decline in at% from Spot 1 (surface) to Spot 4 (about 50–75 μm from the surface), though it is expected that the diffusion length of Al is much larger than observed. Besides Al, Mg and Mn were also observed in the cycled positive electrodes with A$_2$B$_7$ and AB$_5$ HAAs, respectively. Mg accumulates at the surface region of the cycled positive electrodes of the cells C1–C5 and does not diffuse far into the bulk, while a small amount of Mn (1.6 at%) gathers mainly at the surface region of the cycled positive electrodes of the cell AB5.

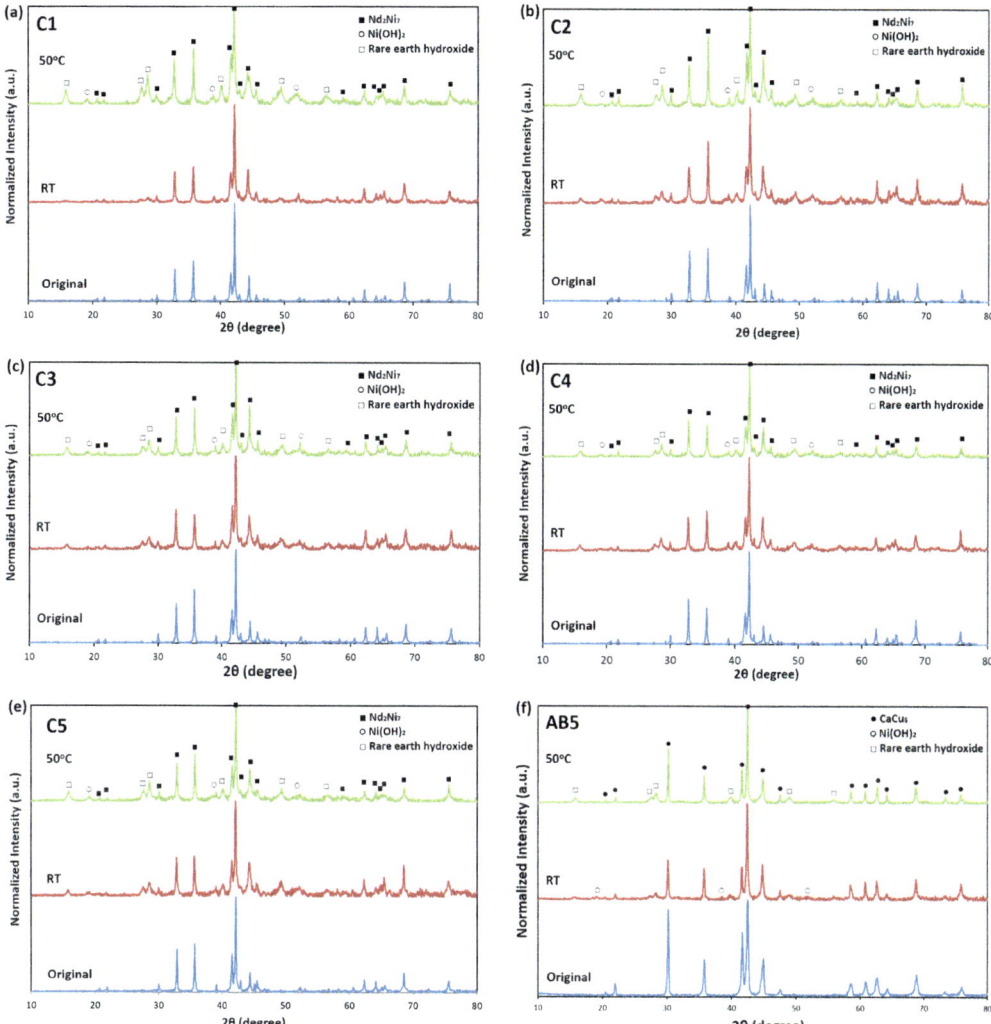

Figure 3. Comparison of the XRD patterns of the negative electrodes before cycling (original) and after cycling at RT (20 °C) and 50 °C containing HAAs (**a**) C1; (**b**) C2; (**c**) C3; (**d**) C4; (**e**) C5; and (**f**) AB5.

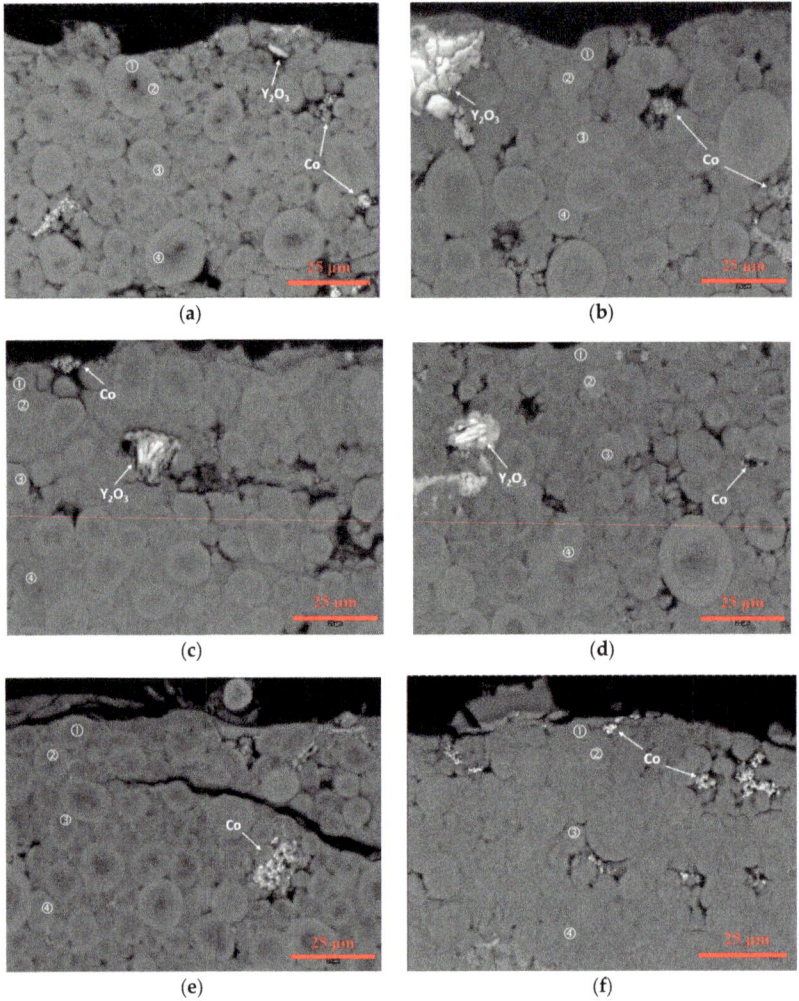

Figure 4. Scanning electron microscopy (SEM) backscattered electron image (BEI) micrographs of the positive electrodes at the end of cycle life at 50 °C for (**a**) C1; (**b**) C2; (**c**) C3; (**d**) C4; (**e**) C5; and (**f**) AB5.

The SEM BEIs of the cycled negative electrodes are shown in Figure 5 and their chemical compositions at different spots, measured by EDS, are listed in Table 2. The spot EDS results indicate that the bright regions (Spots 1 and 2) on the BEI micrographs correspond to the un-oxidized A_2B_7 phase (the small amount of oxygen may originate from sample preparation and transfer for SEM), while the dark regions (Spots 3 and 4) correspond to the oxidized alloy surfaces and are characterized by a high oxygen content (high oxygen-to-metal ratio, O/M) and much lower content of

Al and/or Mg (Mn) for C1–C5 (AB5). All the samples show a substantial degree of oxidation. The Co-free C1 exhibits a more severe oxidation than the low Co-content C2 and C3, which is consistent with the XRD analysis discussed above and could be one of the main causes of why C1 has a shorter cycle life than C2 and C3. Although the oxidation for C1, C2, and C3 are obvious, the pulverization for these three samples is not as severe as the high Co-content C4 and C5. Except for oxidation and pulverization, in C5, there are regions with very high O content (C5-4), which passivate the active surfaces and result in capacity loss. The secondary phases, other than A_2B_7, for C5 may have caused the most severe oxidation and pulverization among all the samples, which leads to a much shorter cycle life. Oxidation and pulverization are also observed for AB5 after cycling. Similar to A_2B_7 HAAs, there is substantially less Al and Mn in the oxidized regions of the AB_5 HAA. Compared to C1–C5, the oxidation and pulverization of AB_5 HAA is not severe.

Table 1. Chemical compositions in at% of selected spots (as shown in Figure 4) determined by energy dispersive spectroscopy (EDS) in the positive electrodes at the end of cycle life. ND denotes non-detectable.

Location	Ni	Co	Zn	Y	Al	Mg	Mn
C1-1	78.2	16	2.0	ND	2.5	1.3	ND
C1-2	77.3	15.7	2.4	ND	3.4	1.2	ND
C1-3	78.4	17.0	2.1	ND	2.5	ND	ND
C1-4	77.5	18.0	1.9	ND	2.6	ND	ND
C2-1	78.1	14.4	2.1	0.4	3.6	1.3	ND
C2-2	78.0	16.6	2.4	0.3	2.7	ND	ND
C2-3	78.2	16.8	2.3	0.2	2.4	ND	ND
C2-4	78.8	16.1	2.4	0.4	2.4	ND	ND
C3-1	75.1	17.3	1.8	0.2	2.9	2.6	ND
C3-2	74.6	18.0	2.2	0.3	2.7	2.1	ND
C3-3	76.4	18.4	2.4	0.3	2.5	ND	ND
C3-4	76.8	18.2	2.4	0.2	2.3	0.1	ND
C4-1	77.2	15.2	2.1	0.2	2.8	2.5	ND
C4-2	78.1	15.0	1.9	0.3	2.5	2.2	ND
C4-3	78.3	16.8	2.4	0.3	2.2	ND	ND
C4-4	78.4	16.4	2.5	0.2	2.5	ND	ND
C5-1	78.4	12.4	1.7	ND	4.1	3.4	ND
C5-2	79.3	13.2	1.7	ND	3.2	2.6	ND
C5-3	80.2	13.1	2.2	0.2	3.6	0.9	ND
C5-4	81.5	13.7	1.9	0.1	2.8	ND	ND
AB5-1	74.0	14.9	1.7	0.2	7.6	ND	1.6
AB5-2	77.5	12.4	1.8	ND	7.7	ND	0.5
AB5-3	75.9	15.0	2.2	ND	6.8	ND	0.2
AB5-4	77.0	14.8	2.0	0.1	6.2	ND	ND

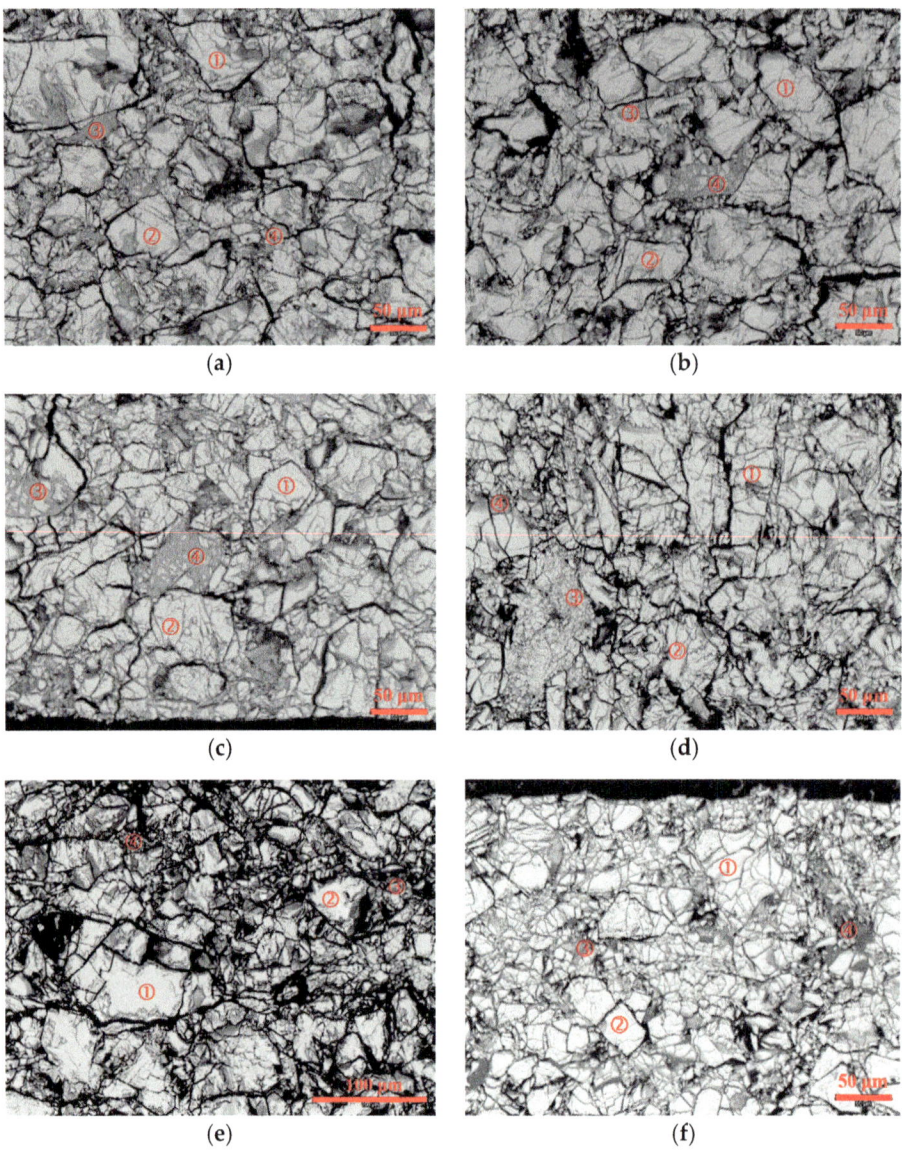

Figure 5. SEM BEI micrographs of the negative electrodes at the end of cycle life at 50 °C for HAAs (**a**) C1; (**b**) C2; (**c**) C3; (**d**) C4; (**e**) C5; and (**f**) AB5.

Table 2. Chemical compositions in at% of selected spots (as shown in Figure 5) determined by EDS in the negative electrodes at the end of cycle life. B/A is the ratio of the sum of the atomic percentages of A-site atoms (La, Ce, Pr, Nd, and Mg) over that of the B-site atoms (Ni, Co, Al, and Mn). O/M denotes the ratio of oxygen to total metallic content. ND denotes non-detectable.

Location	La	Ce	Pr	Nd	Ni	Co	Al	Mg	Mn	O	B/A	O/M
C1-1	4.1	ND	7.9	7.8	73.1	ND	4.2	2.9	ND	ND	3.41	ND
C1-2	3.9	ND	7.9	7.8	7.6	ND	4.0	2.9	ND	ND	3.45	ND
C1-3	10.2	ND	0.3	2.0	64.0	ND	0.2	ND	ND	23.4	5.14	0.31
C1-4	2.3	ND	3.5	3.5	64.1	ND	0.2	0.2	ND	26.2	6.77	0.36
C2-1	3.6	ND	7.2	7.1	66.6	2.3	3.3	2.8	ND	7.1	3.49	0.08
C2-2	3.4	ND	7.6	7.4	66.4	2.1	3.3	2.6	ND	7.2	3.42	0.08
C2-3	2.0	ND	4.9	5.3	63.3	2.7	0.4	0.5	ND	20.9	5.23	0.26
C2-4	2.2	ND	4.8	5.4	59.3	2.1	ND	0.3	ND	25.9	4.83	0.35
C3-1	3.3	ND	7.5	7.5	62.9	4.4	3.0	3.0	ND	8.3	3.30	0.09
C3-2	3.4	ND	7.3	7.2	62.8	4.4	3.4	2.8	ND	8.8	3.41	0.10
C3-3	2.7	ND	6.3	6.1	52.5	3.7	0.7	1.1	ND	26.9	3.51	0.37
C3-4	2.8	ND	5.6	5.8	51.3	4.2	0.6	0.6	ND	29.2	3.79	0.41
C4-1	3.4	ND	7.4	7.2	61.7	6.9	3.2	2.9	ND	7.2	3.44	0.08
C4-2	3.7	ND	7.1	7.0	59.5	6.6	4.0	3.0	ND	9.1	3.37	0.10
C4-3	3.3	ND	7.1	6.9	58.0	6.0	2.6	2.9	ND	13.2	3.30	0.15
C4-4	2.2	ND	5.5	5.7	57.8	7.1	3.1	0.5	ND	18.1	4.89	0.22
C5-1	3.2	ND	7.1	7.3	59.0	8.8	3.2	3.2	ND	8.1	3.41	0.09
C5-2	4.5	ND	7.2	6.9	58.3	8.8	3.8	2.7	ND	7.8	3.33	0.08
C5-3	2.7	ND	4.8	4.6	60.4	7.2	2.6	1.8	ND	15.8	5.05	0.19
C5-4	7.5	ND	14.4	14.8	19.0	2.1	2.1	35.0	ND	36.5	0.58	0.58
AB5-1	9.8	4.1	0.7	0.7	57.3	12.4	4.0	ND	6.5	3.6	5.24	0.04
AB5-2	9.7	4.2	0.6	0.6	58.9	12.5	4.2	ND	5.1	3.3	5.34	0.03
AB5-3	8.0	3.1	0.3	0.3	44.8	10.5	0.5	ND	0.7	30.5	4.83	0.44
AB5-4	1.0	0.4	0.2	0.2	1.9	0.5	0.6	ND	57.6	37.6	33.67	0.60

The above XRD and SEM/EDS results indicate that the positive electrodes remain in good shape at the end of cycle life. No pulverization or severe swelling was observed, even though Al migrated from the negative electrode and diffuses into the bulk of the positive electrode. Thus, it is unlikely that the failure of the cells results from the degradation of the positive electrodes. On the negative electrode side, a higher degree of oxidation is observed at 50 °C for all samples, compared to cells cycled at RT, which may be a major cause of the battery failure at 50 °C. In particular, for C4 and C5, failure is closely related to the severe pulverization of the negative electrodes.

In addition to the positive and negative electrodes, the micro-shortage caused by the conducting or semiconducting deposits in the separator also contributes to the capacity loss as a result of self-discharge [35–37] and is examined in this work. SEM was used to study the cross-section of the positive electrode/separator/negative electrode sandwich structure at the end of cycle life. The BEI micrographs are presented in Figure 6a–d. For all the samples, the positive electrodes do not exhibit any sign of pulverization, which is consistent with our SEM observations on the surfaces of the positive electrodes (Figure 4). For Cell C1 (Figure 6a), the separator and the two electrode/separator interfaces are clean, despite some scattered debris from SEM-sample preparation (loose particles from negative electrode). The EDS elemental mapping does not show any micro-shortage network forming in the separator by any element of this battery system. In comparison with the other samples, the Co- and Ni-mapping micrographs, observed using EDS, are selected and shown in Figure 6b. It is clear that no Co- or Ni- micro-shortage network formed in the separator for Cell C1. However, the negative electrode from the same cell showed severe pulverization, which was not seen from the surface SEM study (Figure 5a). Since the positive electrode remains intact after cycling, the failure of Cell C1 is attributed to the oxidation and pulverization of the negative electrode. For Cell C3 (Figure 6c), it is obvious that the separator loses electrolyte, leading to dry-out, as seen from fact that the positive electrode materials are squeezed into the separator. Although the separator is squeezed, no clear micro-shortage networks were observed from the Co and Ni EDS mapping (Figure 6d). Thus, the failure of cell C3 is ascribed to electrolyte dry-out. For Cell C5 (Figure 6e), again the separator is embedded with debris, which strongly indicates electrolyte dry-out. In addition, EDS mapping shows a clear Co and Ni micro-shortage network (Figure 6f) across the separator. The formation of such a network is a strong indication that the negative electrode was severely corroded and Co and Ni leached out, which formed deposits on the separator surface and decreased the charge/discharge depth and cycle life. The heavy corrosion of HAA C5 can also be observed in Figure 5e, and may be related to its high Co-content, which warrants further investigation. Similar to Cell C5, a Mn-rich micro-shortage network (Figure 6h, reported before in AB5 MH alloy [38]) is observed for cell AB5, which is believed to be the main cause of failure, considering that both electrodes were in good condition. The deposits of Mn in the separator from the leach-out of Mn from the HAA AB5, as can be seen from the decreased Mn content in certain regions of the negative electrode (Table 2), decrease the cell capacity. Pulverization can be seen from the HAA AB5, which also contributes to the capacity loss.

From the analysis above, it can be concluded that, at 50 °C, the failure of the cells using Co-substituted superlattice HAAs can be mainly attributed to issues related to the separator and negative electrode, rather than the positive electrode, which differs from what is seen during RT cycling, where swelling/pulverization of the positive electrode also plays a role in cell failure [18]. Elevated temperature accelerates the electrolyte dry-out and the corrosion of HAA, which results in electrode pulverization, as well as the formation of a micro-shortage network created from leach-out elements (Co, Ni, Mn) from HAA. Cells with the low Co-content

HAAs (C2 and C3) exhibit better cycle life performance compared to those with the Co-free (C1) and high Co-content HAAs (C4 and C5), which is attributed to their low degree of oxidation and pulverization properties. As a comparison, the failure of cell AB5 is attributed to the pulverization of the negative electrode and formation of a Mn-containing micro-shortage network in the separator.

The high-temperature behaviors of AB_5 MH alloys have been extensively studied. Begum and his coworkers reported a decrease in capacity of a standard misch-metal based AB_5 MH alloy at higher temperature and attributed it to the "instability" of the alloy [39]. Lin and his coworkers found the cycle stability and discharge capacity of a Pr-Nd-free AB_5 MH alloy degrade when the temperature rises to 80 °C due to pulverization of the alloy and disintegration of the electrode [40]. Khaldi and his coworkers reported that both the corrosion current and potential of a La-only AB_5 MH alloy increase with the increase of temperature [41]. Zhou and his coworkers found the needle-shaped corrosion product using SEM and correlated it to the capacity degradation at high temperature for a La-rich AB_5 MH alloy [42]. The higher corrosion rate of Mg-containing A_2B_7 versus the Mg-free A_2B_7 due to the formation of $Mg(OH)_2$ on the surface was also reported before [43]. Our findings in the failure mechanisms for AB_5 and superlattice MH alloys are consistent with these reported results, but are from various analytic measurements conducted from the sealed cells, instead of HAA itself.

As for the possible solution to improve the cycle stability at an elevated temperature, Shangguan and his coworkers proposed a $NaBO_2$ [44] and $NaWO_4$ [45] added NaOH electrolyte to improve the capacity at 70 °C by increasing the oxygen evolution potential, but the capacity degradation remained unchanged. Adding Al [46] or Fe [47] to a Pr-Nd-free AB_5 MH alloy improves its anti-corrosion capability at 60 °C as reported by researchers in Sichuan University [46]. Adding Al into the La-only AB_5 alloy, reported by Balogun et al. [48], also increases its anti-corrosion capability at 50°C. Li and his coworkers also demonstrated the benefit of adding Al to increase the stability of misch-metal based AB_5 MH alloy at 60 °C [49]. We also developed a surface coating of Y, Si-containing compound to stabilize the high-temperature cycle performance of AB_5 and Mg-containing superlattice MH alloys, and further studies are ongoing.

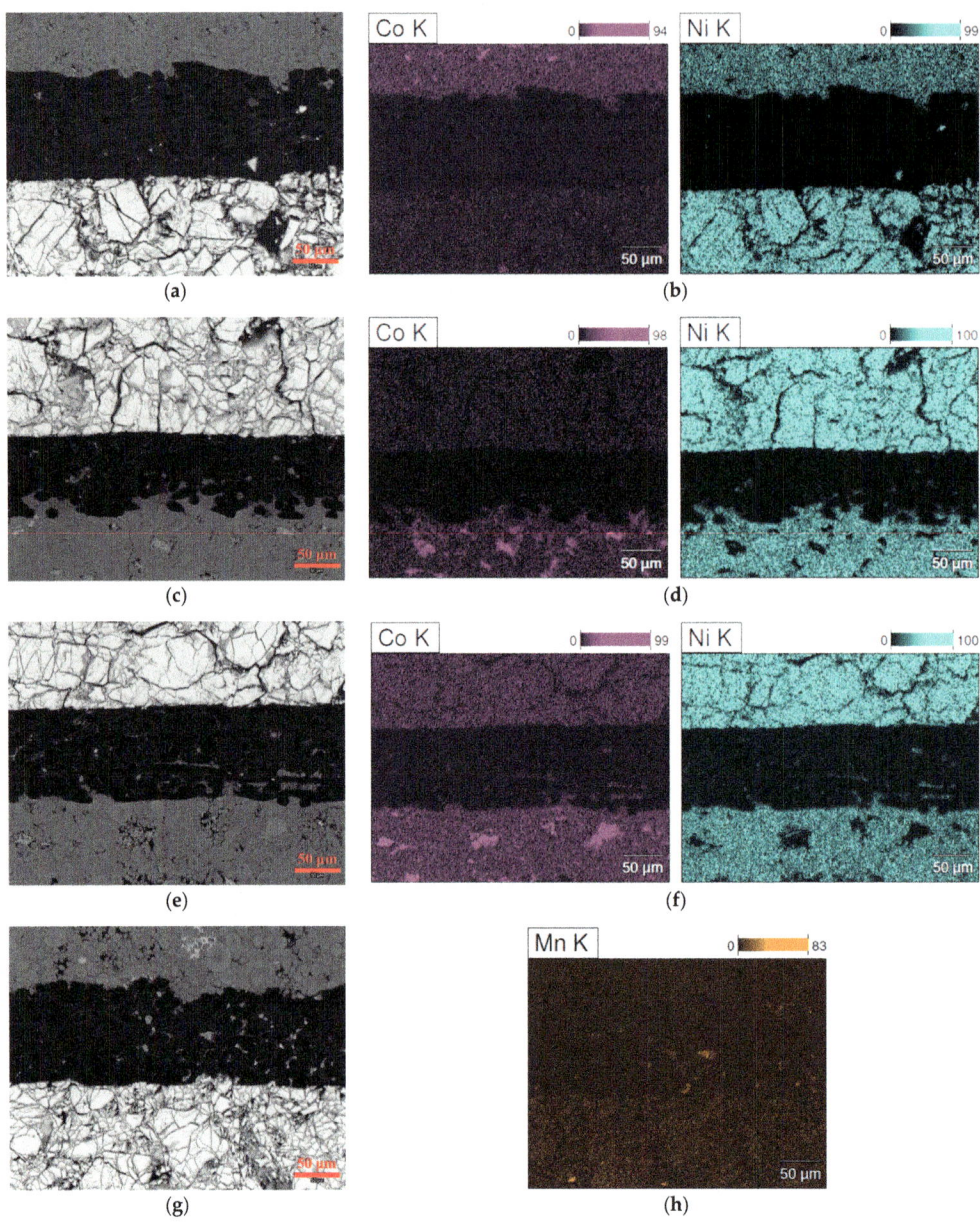

Figure 6. SEM BEI micrographs and selected EDS elemental mappings of the cross-sections of cells (**a**,**b**) C1; (**c**,**d**) C3; (**e**,**f**) C5; and (**g**,**h**) AB5.

4. Conclusions

The failure mechanisms of cells using Co-substituted A_2B_7 superlattice HAAs cycled at 50 °C were assessed. The positive electrodes remained in good condition at the end of cycle life for all samples, as the XRD patterns showed no signs of α-Ni(OH)$_2$ or γ-NiOOH formation and SEM exhibited no signs of pulverization or severe swelling. Thus, the positive electrode was excluded as a cause of battery failure at 50 °C. For the cells with Co-substituted A_2B_7 HAAs, the inferior performance at 50 °C can be attributed to accelerated dry-out of electrolyte and deterioration of the negative electrode. In these cases, the negative electrodes of the cycled cells at 50 °C exhibit much more severe oxidation than those at RT, as indicated by XRD. At 50 °C, the incorporation of a small amount of Co (C2 and C3) in A_2B_7 leads to mitigated oxidation/pulverization and, therefore, longer cycle life compared to the Co-free C1. However, further increases in Co-content leads to more severe oxidation/pulverization and the formation of micro-shortage networks in the separator, which has a negative effect on cycle life.

Acknowledgments: The authors would like to thank the following from BASF-Ovonic for technical assistance: Alan Chen, Ryan Blankenship, Nathan English, and Su Cronogue.

Author Contributions: Tiejun Meng designed the experiments and analyzed test results. Kwo-hsiung Young interpreted the data. John Koch performed the cell disassembly. Taihei Ouchi performed the XRD analysis. Shigekazu Yasuoka designed and obtained the test samples.

Conflicts of Interest: The authors declare no conflict of interest.

Abbreviations

HAA	Hydrogen absorbing alloy
Ni/MH	Nickel/metal hydride
MH	Metal hydride
XRD	X-ray diffraction
SEM	Scanning electron microscopy
EDS	Energy dispersive spectroscopy
ICP	Inductively coupled plasma
BEI	Backscattered electron image
O/M	Oxygen to metal ratio

References

1. Zelinsky, M.; Koch, J.; Fetcenko, M. Heat Tolerant NiMH Batteries for Stationary Power. Available online: www.battcon.com/PapersFinal2010/ZelinskyPaper2010Final_12.pdf (accessed on 28 March 2016).

2.	Zelinsky, M.; Koch, J. Batteries and Heat—A Recipe for Success? Available online: www.battcon.com/PapersFinal2013/16-Mike%20Zelinsky%20-%20Batteries%20and%20Heat.pdf (accessed on 28 March 2016).
3.	Young, K.; Ng, K.Y.S.; Bendersky, L. A Technical Report of the Robust Affordable Next Generation Energy Storage System-BASF Program. *Batteries* **2016**, *2*.
4.	VARTA Microbattery GmbH. Powerful High Temperature NiMH Batteries −20/+85 °C, 2008. Available online: http://www.varta-microbattery.com/applications/mb_data/documents/product_information/PI20081111_Electronica_Ni-MH_High_Temp_en.pdf (accessed on 28 March 2016).
5.	Kai, T.; Ishida, J.; Yasuoka, S.; Takeno, K. The Effect of Nickel-Metal Hydride Battery's Characteristics with Structure of the Alloy. In Proceedings of the 54th Battery Symposium in Japan, Osaka, Japan, 7–9 October 2013; p. 210.
6.	Li, W.; Campion, C.; Lucht, B.L.; Ravdel, B.; DiCarlo, J.; Abraham, K.M. Additives for stabilizing $LiPF_6$-based electrolytes against thermal decomposition. *J. Electrochem. Soc.* **2005**, *152*, A1361–A1365.
7.	Lee, H.H.; Wan, C.C.; Wang, Y.Y. Thermal stability of the solid electrolyte interface on carbon electrodes of lithium batteries. *J. Electrochem. Soc.* **2004**, *151*, A542–A547.
8.	Yang, H.; Bang, H.; Amine, K.; Prakash, J. Investigations of the exothermic reactions of natural graphite anode for Li-ion batteries during thermal runaway. *J. Electrochem. Soc.* **2005**, *152*, A73–A79.
9.	Hammami, A.; Raymond, N.; Armand, M. Lithium-ion batteries: Runaway risk of forming toxic compounds. *Nature* **2003**, *424*, 635–636.
10.	Wang, Q.; Ping, P.; Zhao, X.; Chu, G.; Sun, J.; Chen, C. Thermal runaway caused fire and explosion of lithium ion battery. *J. Power Sources* **2012**, *208*, 210–224.
11.	Fierro, C.; Zallen, A.; Koch, J.; Fetcenko, M.A. The influence of nickel-hydroxide composition and microstructure on the high-temperature performance of nickel metal hydride batteries. *J. Electrochem. Soc.* **2006**, *153*, A492–A496.
12.	Young, K.; Yasuoka, S. Capacity degradation mechanisms in nickel/metal hydride batteries. *Batteries* **2016**, *2*.
13.	Young, K.; Chao, B.; Liu, Y.; Nei, J. Microstructures of the oxides on the activated AB_2 and AB_5 metal hydride alloys surface. *J. Alloys Compd.* **2014**, *616*, 97–104.
14.	Young, K.; Huang, B.; Regmi, R.K.; Lawes, G.; Liu, Y. Comparisons of metallic clusters imbedded in the surface oxide of AB_2, AB_5, and A_2B_7 alloys. *J. Alloys Compd.* **2010**, *506*, 831–840.
15.	Young, K.; Wu, A.; Qiu, Z.; Tan, J.; Mays, W. Effects of H_2O_2 addition to the cell balance and self-discharge of Ni/MH batteries with AB_5 and A_2B_7 alloys. *Int. J. Hydrog. Energy* **2012**, *37*, 9882–9891.
16.	Zhou, X.; Young, K.; West, J.; Regalado, J.; Cherisol, K. Degradation mechanisms of high-energy bipolar nickel metal hydride battery with AB_5 and A_2B_7 alloys. *J. Alloys Compd.* **2013**, *580*, S373–S377.

17. Young, K.; Wong, D.F.; Wang, L.; Nei, J.; Ouchi, T.; Yasuoka, S. Mn in misch-metal based superlattice metal hydride alloy—Part 2 Ni/MH battery performance and failure mechanism. *J. Power Sources* **2015**, *277*, 433–442.

18. Wang, L.; Young, K.; Meng, T.; English, N.; Yasuoka, S. Partial substitution of cobalt for nickel in mixed rare earth metal based superlattice hydrogen absorbing alloy—Part 2 Battery performance and failure mechanism. *J. Alloys Compd.* **2016**, *664*, 417–427.

19. Young, K.; Nei, J. The current status of hydrogen storage alloy development for electrochemical applications. *Materials* **2013**, *6*, 4574–4608.

20. Zhang, Y.; Zhao, D.; Li, B.; Zhao, X.; Wu, Z.; Wang, X. Microstructures and electrochemical characteristics of the $La_{0.75}Mg_{0.25}Ni_{2.5}M_x$ (M = Ni, Co; $x = 0$–1.0) hydrogen storage alloys. *Int. J. Hydrog. Energy* **2008**, *33*, 1868–1875.

21. Reilly, J.J.; Adzic, G.D.; Johnson, J.R.; Vogt, T.; Mukerjee, S.; Mcbreen, J. The correlation between composition and electrochemical properties of metal hydride electrodes. *J. Alloys Compd.* **1999**, *293–295*, 569–582.

22. Liu, J.; Yang, Y.; Li, Y.; Yu, P.; He, Y.; Shao, H. Comparative study of $LaNi_{4.7}M_{0.3}$ (M = Ni, Co, Mn, Al) by powder microelectrode technique. *Int. J. Hydrog. Energy* **2007**, *32*, 1905–1910.

23. Young, K.; Wong, D.F.; Wang, L.; Nei, J.; Ouchi, T.; Yasuoka, S. Mn in misch-metal based superlattice metal hydride alloy—Part 1 Structural, hydrogen storage and electrochemical properties. *J. Power Sources* **2015**, *277*, 426–432.

24. Wang, L.; Young, K.; Meng, T.; Ouchi, T.; Yasuoka, S. Partial substitution of cobalt for nickel in mixed rare earth metal based superlattice hydrogen absorbing alloy—Part 1 Structural, hydrogen storage and electrochemical properties. *J. Alloys Compd.* **2016**, *660*, 407–415.

25. Nei, J.; Young, K.; Rotarov, D. Studies on MgNi-based metal hydride electrode with aqueous electrolytes composed of various hydroxides. *Batteries* **2016**, *2*.

26. JCPDS-International Centre for Diffraction Data®. *Powder Diffraction File (PDF) No. 00-014-0117*; JCPDS: Newtown Square, PA, USA, 2015.

27. JCPDS-International Centre for Diffraction Data®. *Powder Diffraction File (PDF) No. 00-043-1004*; JCPDS: Newtown Square, PA, USA, 2015.

28. JCPDS-International Centre for Diffraction Data®. *Powder Diffraction File (PDF) No. 00-004-0850*; JCPDS: Newtown Square, PA, USA, 2015.

29. Liu, B.; Wang, X.Y.; Yuan, H.T.; Zhang, Y.S.; Song, D.Y.; Zhou, X.Y. Physical and electrochemical characteristics of aluminum-substituted nickel hydroxide. *J. Appl. Electrochem.* **1999**, *29*, 855–860.

30. Liu, B.; Yuan, H.; Zhang, T. Impedance of Al-substituted α-nickel hydroxide electrodes. *Int. J. Hydrog. Energy* **2004**, *29*, 453–458.

31. Bode, H.; Dehmelt, K.; Witte, J. Zur kenntnis der nickelhydroxidelektrode—I. Über das nickel (II)-hydroxidhydrat. *Electrochim. Acta* **1966**, *11*, 1079–1087. (In German)

32. Oliva, P.; Leonardi, J.; Laurent, J.F.; Delmas, C.; Braconnier, J.J.; Figlarz, M.; Fievet, F.; de Guibert, A. Review of the structure and the electrochemistry of nickel hydroxides and oxy-hydroxides. *J. Power Sources* **1982**, *274*, 8, 229–255.

33. Van der Ven, A.; Morgan, D.; Meng, Y.S.; Ceder, G. Phase stability of nickel hydroxides and oxyhydroxides. *J. Electrochem. Soc.* **2006**, *153*, A210–A215.

34. Miao, C.; Zhu, Y.; Huang, L.; Zhao, T. The relationship between structural stability and electrochemical performance of multi-element doped alpha nickel hydroxide. *J. Power Sources* **2015**, *274*, 186–193.

35. Guo, H.; Qiao, Y.; Zhang, H. Fast determination of micro short circuit in sintered MH-Ni battery. *Chin. J. Power Sources* **2010**, *34*, 608–609.

36. Shinyama, K.; Magari, Y.; Kumagae, K.; Nakamura, H.; Nohma, T.; Takee, M.; Ishiwa, K. Deterioration mechanism of nickel metal-hydride batteries for hybrid electric vehicles. *J. Power Sources* **2005**, *141*, 193–197.

37. Zhu, W.H.; Zhu, Y.; Tatarchuk, B.J. Self-discharge characteristics and performance degradation of Ni-MH batteries for storage applications. *Int. J. Hydrog. Energy* **2014**, *39*, 19789–19798.

38. Shinyama, K.; Magari, Y.; Akita, H.; Kumagae, K.; Nakamura, H.; Matsuta, S.; Nohma, T.; Takee, M.; Ishiwa, K. Investigation into the deterioration in storage characteristics of nickel metal-hydride batteries during cycling. *J. Power Sources* **2005**, *143*, 265–269.

39. Begum, S.N.; Muralidharan, V.S.; Basha, C.A. Electrochemical investigations and characterization of a metal hydride alloy (MmNi$_{3.6}$Al$_{0.4}$Co$_{0.7}$Mn$_{0.3}$) for nickel metal hydride batteries. *J. Alloys Compd.* **2009**, *467*, 124–129.

40. Lin, J.; Cheng, Y.; Liang, F.; Sun, L.; Yin, D.; Wu, Y.; Wang, L. High temperature performance of La$_{0.6}$Ce$_{0.4}$Ni$_{3.45}$Co$_{0.75}$Mn$_{0.7}$Al$_{0.1}$ hydrogen storage alloy for nickel/metal hydride batteries. *Int. J. Hydrog. Energy* **2014**, *39*, 13231–13239.

41. Khaldi, C.; Boussami, S.; Tliha, M.; Azizi, S.; Fenineche, N.; El-kedim, O.; Mathlouthi, H.; Lamloumi, J. The effect of the temperature on the electrochemical properties of the hydrogen storage alloy for nickel-metal hydride accumulators. *J. Alloys Compd.* **2013**, *574*, 59–66.

42. Zhou, H.; Wang, P.; Wang, Z.; Zou, R.; Ni, C. Influence of temperature on self-discharge and high-rate discharge characteristics of La-rich AB$_5$-based MH alloy electrode. *Mater. Sci. Forum* **2010**, *654–656*, 2835–2838.

43. Monnier, J.; Chen, H.; Joiret, S.; Bourgon, J.; Latroche, M. Identification of a new pseudo-binary hydroxide during calendar corrosion of (La, Mg)$_2$Ni$_7$-type hydrogen storage alloys for nickel-metal hydride batteries. *J. Power Sources* **2014**, *266*, 162–169.

44. Shangguan, E.; Wang, J.; Li, J.; Dan, G.; Chang, Z.; Yuan, X.; Wang, H. Enhancement of the high-temperature performance of advanced nickel-metal hydride batteries with NaOH electrolyte containing NaBO$_2$. *Int. J. Hydrog. Energy* **2013**, *38*, 10616–10624.

45. Shangguan, E.; Li, J.; Chang, Z.; Tang, H.; Li, B.; Yuan, X.; Wang, H. Sodium tungstate as electrolyte additive to improve high-temperature performance of nickel-metal hydride batteries. *Int. J. Hydrog. Energy* **2013**, *38*, 5153–5158.

46. Zhou, W.; Ma, Z.; Wu, C.; Zhu, D.; Huang, L.; Chen, Y. The mechanism of suppressing capacity degradation of high-Al AB$_5$-type hydrogen storage alloys at 60 $^{\circ}$C. *Int. J. Hydrog. Energy* **2016**, *41*, 1801–1810.

47. Chao, D.; Zhong, C.; Ma, Z.; Yang, F.; Wu, Y.; Zhu, D.; Wu, C.; Chen, Y. Improvement in high-temperature performance of Co-free high-Fe AB_5-type hydrogen storage alloys. *Int. J. Hydrog. Energy* **2012**, *37*, 12375–12383.

48. Balogun, M.; Wang, Z.; Huang, H.; Yao, Q.; Deng, J.; Zhou, H. Effect of high and low temperature on the electrochemical performance of $LaNi_{4.4-x}Co_{0.3}Mn_{0.3}Al_x$ hydrogen storage alloys. *J. Alloys Compd.* **2013**, *579*, 438–443.

49. Li, Z.; Rei, Y.; Han, F. Influence of small amount of addition elements on electrochemical and high temperature performance of La-riched RE-based hydrogen storage materials. *Chin. J. Rare Metals* **2002**, *26*, 47–50.

Chapter 3:
Electrolyte

Studies on MgNi-Based Metal Hydride Electrode with Aqueous Electrolytes Composed of Various Hydroxides

Jean Nei, Kwo-Hsiung Young and Damian Rotarov

Abstract: Compositions of MgNi-based amorphous-monocrystalline thin films produced by radio frequency (RF) sputtering with a varying composition target have been optimized. The composition $Mg_{52}Ni_{39}Co_3Mn_6$ is identified to possess the highest initial discharge capacity of 640 mAh·g^{-1} with a 50 mA·g^{-1} discharge current density. Reproduction in bulk form of $Mg_{52}Ni_{39}Co_3Mn_6$ alloy composition was prepared through a combination of melt spinning (MS) and mechanical alloying (MA), shows a sponge-like microstructure with >95% amorphous content, and is chosen as the metal hydride (MH) alloy for a sequence of electrolyte experiments with various hydroxides including LiOH, NaOH, KOH, RbOH, CsOH, and $(C_2H_5)_4N(OH)$. The electrolyte conductivity is found to be closely related to cation size in the hydroxide compound used as 1 M additive to the 4 M KOH aqueous solution. The degradation performance of $Mg_{52}Ni_{39}Co_3Mn_6$ alloy through cycling demonstrates a strong correlation with the redox potential of the cation in the alkali hydroxide compound used as 1 M additive to the 5 M KOH aqueous solution. NaOH, CsOH, and $(C_2H_5)_4N(OH)$ additions are found to achieve a good balance between corrosion and conductivity performances.

Reprinted from *Batteries*. Cite as: Nei, J.; Young, K.-H.; Rotarov, D. Studies on MgNi-Based Metal Hydride Electrode with Aqueous Electrolytes Composed of Various Hydroxides. *Batteries* **2016**, *2*, 27.

1. Introduction

Mg-based metal hydride (MH) alloys are very attractive for various hydrogen storage applications due to Mg's abundance, low cost, light weight, and the availability of many intermetallic compounds. In particular, two groups of MgNi-based MH alloys have been proposed for use as the negative electrode active material in nickel/metal hydride (Ni/MH) battery: amorphous MgNi and crystalline Mg_2Ni. The former has a higher initial discharge capacity but a much lower cycle stability due to its amorphous nature, which makes protection layer formation very difficult compared to that in the latter [1]. Research into these two alloy families were reviewed in 2013 [2]. More recently, additives such as B [3], Ti [4,5], Pt [4], Pd [4], Nd [6], Cr [7], AB$_5$ [8], La [9], Co [10], nano-sized Ni [11], Li [12], and Cu [5,13] were added to the bulk or surface of Mg_2Ni-based MH alloys to improve the capacity, cycle stability, and high-rate dischargeability (HRD). For MgNi-based MH alloys, TiO_2 addition [14], Ni coating [15], Mn substitution [16], and Nb substitution [17] were experimented for electrochemical property enhancements. Alloy modifications including annealing, addition of anti-corrosion agents, and composition modification in both the surface and bulk, control of charge input, and alloy particle size selection have been previously proposed to improve the cycle stability of MgNi-based electrodes [18–21], but reducing the corrosive nature of electrolyte has never been investigated. Advantages of the current method that uses 30 wt% KOH aqueous solution as the electrolyte are low cost, high ionic conductivity, low toxicity, and low freezing point. However, KOH aqueous solution at such concentration rapidly reacts with amorphous MgNi MH alloy. Therefore, an adjustment in electrolyte composition is necessary to balance some of the high ionic conductivity with a reduction in corrosive nature of electrolyte. Salt additives in KOH aqueous solution were previously investigated as the modified electrolytes in our laboratory, and the results have been reported [22], leaving cations in various hydroxides as the main topic in this paper.

Many studies investigating the alkali cation species and electrolyte concentrations in Ni/MH battery were previously conducted and are summarized in Table 1. However, none of these studies focused on amorphous MgNi MH alloy, which has the highest capacity but is also the most susceptible to electrolyte oxidation. Furthermore, while there is only a limited number of soluble inorganic hydroxides, several organic hydroxides are available. The cations of organic hydroxides are much larger than those in inorganic hydroxides, and their effects on electrochemical performances as the electrolyte additives is a subject of acute interest. Therefore, a study of inorganic and organic hydroxides in electrolyte was launched in the U.S. Department of Energy (DOE)-sponsored Robust Affordable Next Generation EV-storage (RANGE) program [23], and the obtained results are summarized in this paper. We will discuss MgNi-based thin film composition optimization, bulk MgNi-based alloy powder process optimization, electrolyte conductivity, and, finally, MgNi-based electrode degradation in electrolytes composed of various hydroxides.

2. Experimental Setup

Thin film was prepared using the radio frequency (RF) sputtering technique with an MRC Model 8667 RF Sputtering System (Material Research Corporation, Gilbert, AZ, USA). Bulk powder was prepared by a combination of melt spinning (MS) and mechanical alloying (MA) processes, which were conducted using a homebuilt system with a 2 kg induction furnace and a 20″ diameter rotating copper wheel (rotation speed up to 100 rpm) and a SC-10 attritor (Union Process, Akron, OH, USA), respectively. Microstructure was examined via a Philips X'Pert Pro X-ray diffractometer (XRD, Philips, Amsterdam, The Netherlands) and a CM200/FEG transmission electron microscope (TEM, Philips). Electrochemical testing was performed with a CTE MCL2 Mini cell test system (Chen Tech Electric MFG. Co., Ltd., New Taipei, Taiwan). Approximately 70 mg of AR3 ($Mg_{52}Ni_{39}Co_3Mn_6$) alloy powder, made by MS + MA, was compacted onto an expanded nickel substrate using a 10 ton press to form the negative working electrode, which was approximately 0.2 mm in thickness, without any binder. Different concentrations of KOH aqueous solutions (up to 6 M) and various levels of $X(OH)_y$ (X = Li, Na, Rb, Cs, tetraethylammonium (tEA, $(C_2H_5)_4N$), Mg, Ca, Sr, or Ba, y = 1 or 2) replacements/additions in KOH aqueous solutions (with the total concentration of OH^- up to 6.5 M) were prepared and used as the electrolyte. Two halves of a sintered $Ni(OH)_2$-positive electrode, each 1 cm^2 in area and 1.5 mm in thickness, were connected by a nickel tab strip and used as the counter electrode. A piece of grafted polypropylene/polyethylene separator was folded in half twice and used to sandwich the negative electrode, so that two layers of separator were on each side of the negative electrode. Next, the wrapped negative electrode was again sandwiched with the two halves of positive electrode. The electrode assembly was placed into a plastic sleeve, which was then slid into an acrylic cell holder. The sleeve was filled with the electrolyte using a pipette. After absorbing the electrolyte for five minutes, the sleeve was refilled with the electrolyte so that the half-cell is in a flooded cell configuration. For the discharge capacity measurement, the half-cell was first charged at a current density of 100 mA·g^{-1} for 5 h and then discharged at a current density of 100 mA·g^{-1} until a cut-off voltage of 0.9 V was reached. Then, the cell was discharged at a current density of 24 mA·g^{-1} until a cut-off voltage of 0.9 V was reached, and finally discharged at a current density of 8 mA·g^{-1} until a cut-off voltage of 0.9 V was reached. There were 10 cycles of the charge/discharge procedure performed for each half-cell. Full discharge capacity is the sum of capacities measured at 100, 24, and 8 mA·g^{-1} for each cycle. Electrolyte conductivity was measured with an YSI Model 3200 Conductivity Meter (YSI Incorporated, Yellow Springs, OH, USA).

Table 1. Summary of previous electrolyte studies on alkali cations in a chronological order. LT, RT, and HT denote low temperature (<0 °C), room temperature, and high temperature (>40 °C), respectively. A and B in the alloy formula represent misch metals (La, Ce, Pr, and/or Nd) and transition metals (Ni, Co, Mn, and/or Al), respectively. MH: metal hydride; and HRD: high-rate dischargeability.

Tested Electrode	Hydroxide	Main Findings	Reference
Ni(OH)$_2$	From LiOH to CsOH	• LiOH increases HT charge acceptance. • RbOH and CsOH increase LT charge acceptance.	[24]
Ni(OH)$_2$	From LiOH to CsOH	• CsOH and a low concentration of RbOH do not promote α-NiOOH.	[25]
Ni(OH)$_2$	From LiOH to CsOH	• Highest capacities obtained from LiOH and NaOH at HT, KOH at RT, and RbOH and CsOH at LT.	[26]
Not specified	KOH	• Conductivities of KOH at various concentrations and temperatures were reported.	[27]
AB$_5$	8.7 M KOH-0.5 M NaOH-0.7 M LiOH	• Corrosion behavior was reported.	[28]
(Ti,Zr)B$_2$	8.5 M mixture of LiOH, NaOH, KOH	• Corrosion behavior was reported.	[29]
Zircaloy (ZrSn)	From LiOH to CsOH	• Relative corrosion rates: LiOH > NaOH > KOH > RbOH > CsOH.	[30]
Ni(OH)$_2$	LiOH additive	• LiOH increases HT capacity and discharge voltage.	[31]
AB$_5$	Mixture of LiOH, NaOH, KOH	• Best results observed with a mixture of 78% KOH, 20% NaOH, and 2% LiOH.	[32]
AB$_5$	Mixture of LiOH, NaOH, KOH	• NaOH improves HT. • Best HT result observed with a mixture of 36% KOH, 43% NaOH, and 6% LiOH.	[33]
AB$_5$	KOH, NaOH	• NaOH reduces the corrosion rate of AB$_5$ MH alloy.	[34]
AB$_5$	From LiOH to CsOH	• Best LT electrolyte is 6.2 M KOH + 1.2 M LiOH. • RbOH and CsOH help LT performance.	[35]

378

Table 1. *Cont.*

Tested Electrode	Hydroxide	Main Findings	Reference
AB_5	Mixture of LiOH, NaOH, KOH	• Suggestion of a ternary mixture to improve LT performance.	[36]
AB_5	Mixture of LiOH, NaOH, KOH	• Highest capacity at $-18\ °C$ is obtained from mixture of 3.6 M KOH, 3.38 M NaOH, and 0.14 M LiOH.	[37]
$LaNi_5$	KOH, NaOH	• KOH has better performance at LT compared to NaOH.	[38,39]
$LaNi_5$	KOH, NaOH	• NaOH is good for HT but bad for LT.	[40]
LaB_5	1 M KOH and 8 M KOH	• 1 M KOH has higher capacity and lower corrosion.	[41]
$(Ti,Zr)Ni$	6 M KOH and 8 M KOH	• 8 M KOH has higher capacity and corrosion.	[42]
AB_5	KOH, NaOH	• KOH is good for LT, and NaOH is good for HT.	[43]
$LaCrO_3$	5.6–12.5 M KOH	• Higher concentrations of KOH increase capacity.	[44]
AB_5	LiOH, KOH	• 26% KOH has a much better cycle life than 31% KOH. • LiOH prevents $Ni(OH)_2$ grain growth and increases charge efficiency in the positive electrode.	[45]
LaB_5	2, 4, 6, 8 M KOH	• 6 M KOH and 8 M KOH correspond to the highest capacity and HRD.	[46]
$(Ti,Zr)B_2$	2, 4, 6, 8 M KOH	• 4 M KOH corresponds to the best cycle stability. • 8 M KOH corresponds to the highest capacity.	[47]
AB_5	From LiOH to CsOH	• KOH demonstrates the highest capacity. • Highest corrosion rates are found with RbOH.	[48]
AB_5	LiOH, KOH	• KOH demonstrate better activation and lower corrosion.	[49]

3. Results

The current study begins with MgNi-based alloy composition optimization, which is performed with thin film deposition using RF sputtering technique. After identifying the compositions that demonstrated the highest electrochemical capacities in the thin film form, bulk powder process consisting of MS + MA was adopted to reproduce the optimized compositions. Once the electrochemical performances of alloys in the bulk form were confirmed, the most suitable alloy composition in the bulk form was then selected for the investigation on modified electrolytes composed of various hydroxides. The abovementioned developments are discussed in detail in the following sections.

3.1. MgNi-Based Thin Film Prepared by Radio Frequency Sputtering

Thin film deposition was performed in a multi-target RF sputtering unit with a modified target and sample holder (Figure 1). The deposition rate was approximately $0.5\ \mu m \cdot h^{-1}$, and the resulting average film thickness was approximately 1 μm. In each run, 10–25 nickel substrates (1 cm $\times$ 1 cm) were deposited by thin films with different compositions, which were determined by energy-dispersive X-ray spectroscopy from the silicon witness samples placed among the nickel substrates. Results from the binary composition optimization are shown in Table 2. Discharge capacity maximizes at the composition $Mg_{52}Ni_{48}$ (AR1), which is used as the basis for ternary Mg-Ni-Co and then quaternary Mg-Ni-Co-Mn composition optimizations (Figure 2). Among all the compositions in the ternary and quaternary matrices, AR2 ($Mg_{52.1}Ni_{45.1}Co_{2.8}$) and AR3 ($Mg_{52}Ni_{39}Co_3Mn_6$) demonstrate the highest discharge capacities, respectively (Table 3). Furthermore, the cycle stabilities of several RF-sputtered thin films are shown in Figure 3, where three films with different chemical positions show similar capacity degradation in the first 20 cycles. More specifically, two or three activation cycles are needed for the thin film electrodes; once the electrodes are fully activated, steady capacity degradation is observed until around cycle 10; capacity remains stable thereafter. According to the XRD analysis, formation of $Mg(OH)_2$ on the surface impedes further electrochemical reaction and is the major source for capacity degradation (approximately 1.2% per cycle).

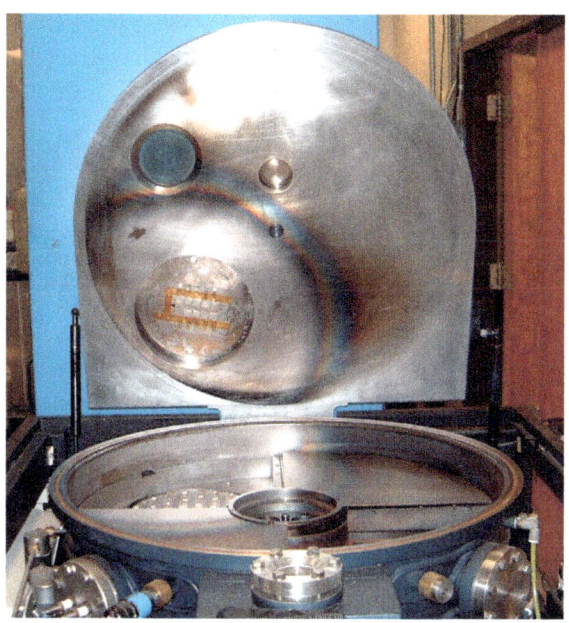

Figure 1. An radio frequency (RF)-sputtering unit with an 8″ nickel target decorated with different amounts of modifying elements (**bottom**) and an array of sandblasted nickel substrates with silicon witness substrates for composition measurements (**top**). The substrates are taped to a copper plate with circulating ice water or liquid nitrogen.

Table 2. Maximum discharge capacities of Mg-Ni binary thin films deposited on ice water-cooled nickel substrates with 20 mA·g^{-1} charge and discharge current densities. Mg$_{52}$Ni$_{48}$ (AR1) is identified as the composition with the highest discharge capacity. Microstructures and phase abundances were obtained from the X-ray diffraction (XRD) patterns in reference [50].

Alloy Composition	Discharge Capacity (mAh·g^{-1})	Microstructure
Mg$_{48}$Ni$_{52}$	302	30% microcrystalline + 70% amorphous
Mg$_{52}$Ni$_{48}$	327	20% microcrystalline + 80% amorphous
Mg$_{57}$Ni$_{43}$	260	10% polycrystalline + 90% amorphous
Mg$_{61}$Ni$_{39}$	75	20% polycrystalline + 80% amorphous
Mg$_{65}$Ni$_{35}$	20	30% polycrystalline + 70% amorphous

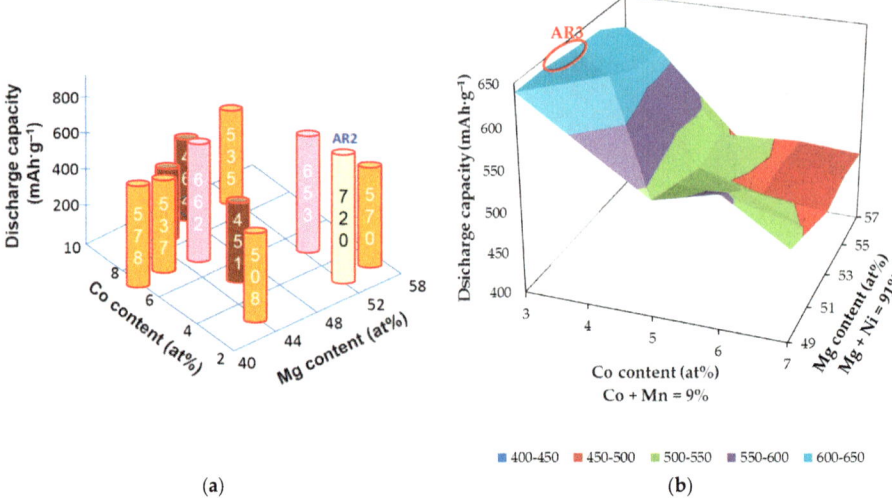

(a)   (b)

Figure 2. Maximum discharge capacities of (**a**) Mg-Ni-Co ternary and (**b**) Mg-Ni-Co-Mn quaternary thin films deposited on liquid nitrogen-cooled nickel substrates with 50 mA·g^{-1} charge and discharge current densities. AR2 (Mg$_{52.1}$Ni$_{45.1}$Co$_{2.8}$) and AR3 (Mg$_{52}$Ni$_{39}$Co$_3$Mn$_6$) are identified as compositions corresponding to the highest discharge capacities.

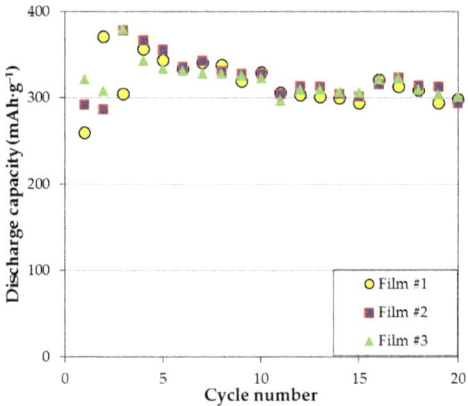

Figure 3. Cycling performances of Mg-Ni-Co thin film: #1 (Mg$_{41.9}$Ni$_{50.9}$Co$_{7.2}$ deposited on liquid nitrogen-cooled nickel substrate); #2 (Mg$_{44.8}$Ni$_{46.8}$Co$_{8.4}$ deposited on liquid nitrogen-cooled nickel substrate); and #3 (Mg$_{37.7}$Ni$_{54.6}$Co$_{7.7}$ deposited on ice water-cooled nickel substrate) with 100 mA·g^{-1} charge and discharge current densities. All films show similar cycling behavior. Two or three cycles are needed for activation, and once the electrodes are fully activated, steady capacity degradation is observed until around cycle 10. Capacity maintains stability from cycle 10 to 20.

Table 3. Evolution of an MgNi-based MH alloy formula with increasing degree of disorder (number of constituent elements). Bulk powder was produced by melt spinning (MS) + mechanical alloying MA.

Formula Name	Composition	Thin Film Discharge Capacity (mAh·g^{-1})	Bulk Powder Discharge Capacity (mAh·g^{-1})	Power	Cycle Life
AR1	$Mg_{52}Ni_{48}$	327	-	Low	Low
AR2	$Mg_{52.1}Ni_{45.1}Co_{2.8}$	662	-	Low	Low
AR3	$Mg_{52}Ni_{39}Co_3Mn_6$	639	791	Medium	Low
AR4	$Mg_{51.5}Ni_{37}Co_6Mn_4Fe_{1.5}$	823	472	Medium	Improved
AR5	$Mg_{50}Ni_{40}Co_6Mn_3Zr_1$	592	456	Good	Good

3.2. MgNi-Based Alloy Powder Prepared by Melt Spinning + Mechanical Alloying

According to the Mg-Ni binary phase diagram, AR2 and AR3 compositions cannot be reproduced by the conventional melt-and-cast method [51]. Therefore, we used a powder fabrication method combining the MS and MA techniques. The former (see Figure 4a for the MS system) produces ribbons with closely packed polycrystalline Mg_2Ni and $MgNi_2$ phases, and the latter (see Figure 4b for the MA attritor system) produces powder with an amorphous MgNi phase and therefore reproduces the microstructure and electrochemical capacity of the thin film work.

 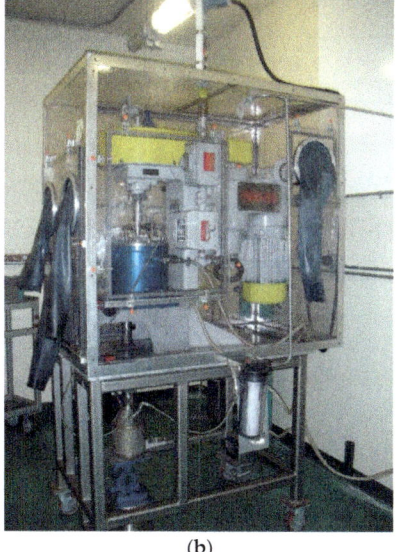

(a)   (b)

Figure 4. Photos of (**a**) a home-built MS system with a 2 kg induction furnace (**top**) and a rotating copper wheel (**center**) and (**b**) an attritor enclosed in a glovebox filled with argon gas. Powder is collected in a canister (**bottom**) that can be separated from the system by a butterfly valve.

Compared to the method of using MA with raw material in elemental powder form alone, incorporation of MS in the alloy fabrication process can reduce the MA time from 72 h to 12 h and lower the material cost. The polycrystalline and amorphous natures of MS ribbon before and after the MA process, respectively, are revealed by XRD, as shown in Figure 5. The cycle stability of the AR3 powder made by MS + MA, shown in Figure 6, is much more severe than that of its thin film counterpart. More specifically, degradation of the bulk AR3 powder occurs at approximately 10% per cycle in the first 10 cycles. While the addition of 1 wt% Si or Zr in the MA process does not affect the cycling performance, the addition of 1 wt% Fe reduces the capacity degradation to around 4% per cycle in the first 10 cycles. Furthermore, although the initial capacity is decreased with the addition of Fe to AR3 powder, the maintained capacity at higher cycle number is much higher than the original or modified AR3 powder with other additions. Addition of Fe also shows significant improvement in cycle stability in our thin film results.

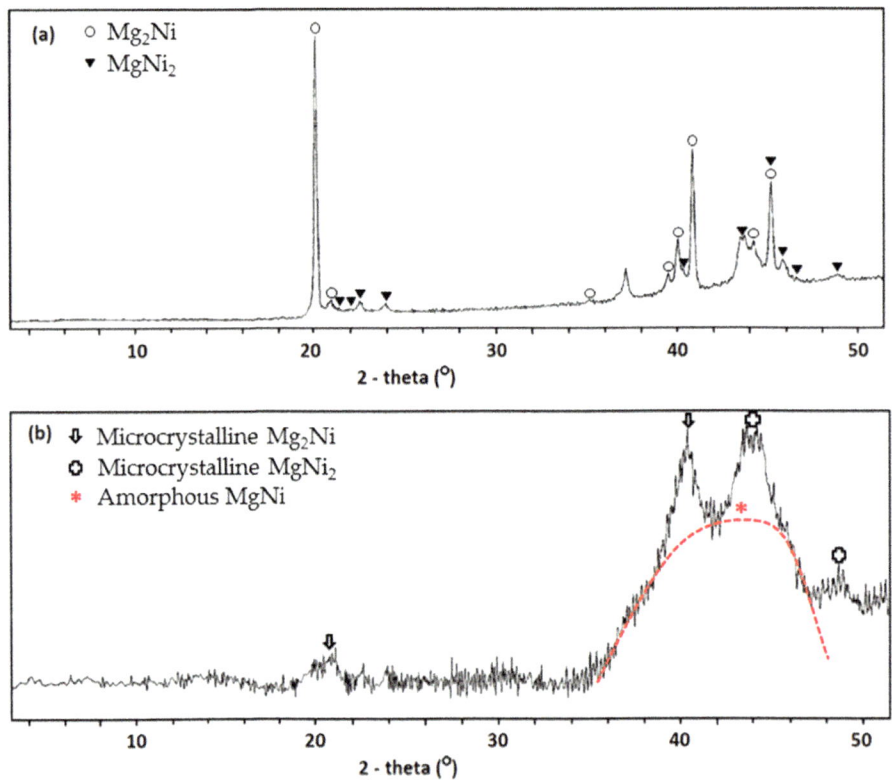

Figure 5. XRD patterns from MS AR3 ribbon samples (**a**) before and (**b**) after MA. With MA, the microstructure changes from polycrystalline Mg_2Ni and $MgNi_2$ phases to amorphous MgNi and microcrystalline Mg_2Ni and $MgNi_2$ phases.

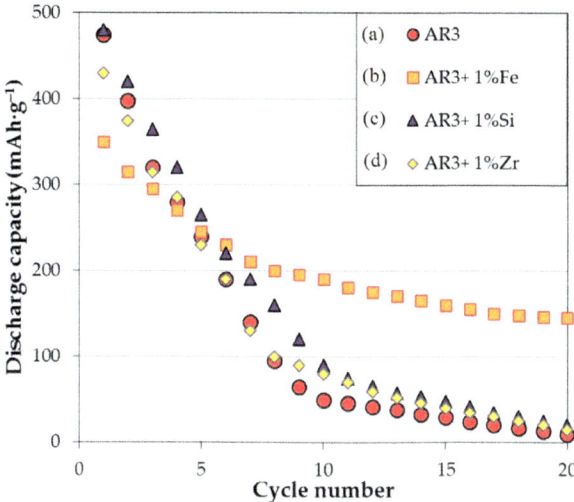

Figure 6. Cycling performances of MS + MA AR3 powders with (**a**) no additive; (**b**) 1 wt% Fe; (**c**) 1 wt% Si; and (**d**) 1 wt% Zr MA for 3 h with 100 mA·g^{-1} charge and discharge current densities. Cycle stability in the bulk form is much worse compared to that in the thin film form. While the addition of Si or Zr in the MA process does not affect the cycling performance, the addition of Fe reduces the capacity degradation. Although the initial capacity is decreased with the addition of Fe, the maintained capacity at higher cycle number is much higher than the original or modified AR3 powder with other additions.

Figure 7 shows TEM micrographs of the AR3 powder sample made by MS + MA, and it reveals that the material density is not uniform. Large areas with lower density (brighter contrast) are found alongside the scroll-type denser regions (darker contrast). This type of microstructure cannot form the protective oxide layer that prevents further oxidation found in other MH alloys [52] and consequently causes the more severe degradation observed in the bulk form compared to that the thin film form. Therefore, in order to magnify the effect of electrolyte composition modification, the AR3 powder made by MS + MA is the perfect candidate for the current study on corrosion of MH alloy in electrolytes composed of various hydroxides, despite the fact that we have developed a series of new compositions based on the quaternary AR3 composition, new surface treatments, coatings, binder/surfactant additions, and other methods to improve the cycle stability of MgNi-based MH alloys.

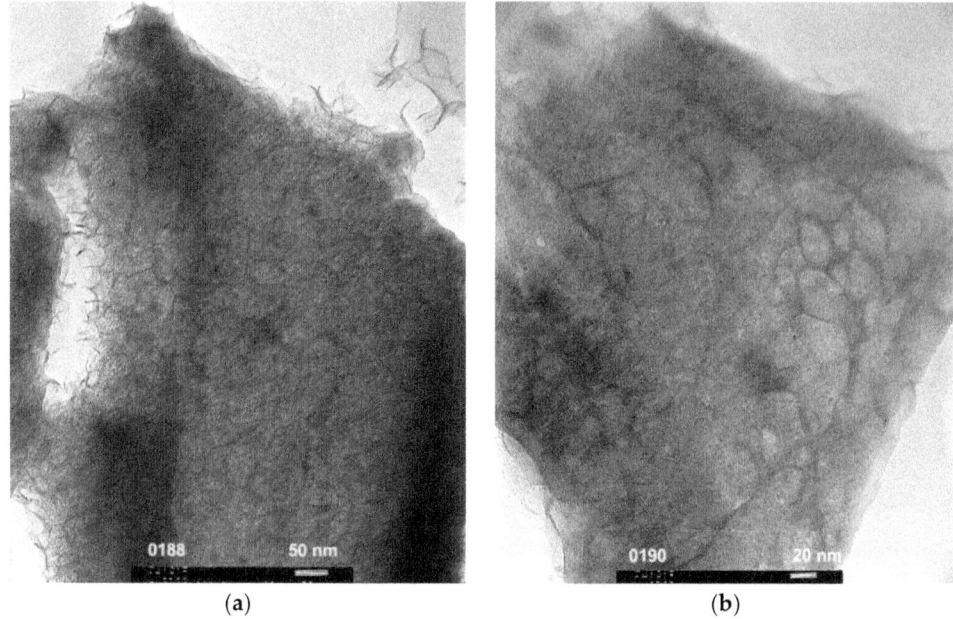

(a)   (b)

Figure 7. Transmission electron microscope (TEM) micrographs at (**a**) ×100,000 and (**b**) ×200,000 magnifications of the AR3 powder made by MS + MA at different magnifications. The cellular structures seen in the micrographs are from the cross-section of a sponge-scroll type of microstructure. Areas with darker contrast represent the scroll-type denser regions.

3.3. Conductivities of Various Hydroxide Electrolytes

Room temperature conductivities of different concentrations of KOH and other hydroxide aqueous solutions were measured and then normalized to that of 6 M KOH aqueous solution, and the results are summarized in Table A1 in Appendix. Due to its low solubility, no effect on conductivity is observed with the addition of $Mg(OH)_2$, $Ca(OH)_2$, $Sr(OH)_2$, or $Ba(OH)_2$ in 6 M KOH aqueous solution. Moreover, as a representative example, the electrolyte conductivities are plotted against the total OH^- concentration in solution for 1 M LiOH, NaOH, RbOH, and CsOH, and 0.6 M and 1.2 M tEAOH (tetraethylammonium hydroxide) in various concentrations of KOH aqueous solutions in Figure 8. As the total OH^- concentration increases, the electrolyte conductivity increases. Also, at any fixed KOH concentration, the electrolyte conductivity increases in agreement with the following trend: tEAOH < LiOH < NaOH < RbOH < CsOH addition. However, none of the conductivities of mixed hydroxide electrolytes exceeds that of pure KOH aqueous solution. According

386

to Stokes' Law, as the object size increases, it experiences more friction or drag when moving through fluid, as indicated in the equation for a spherical particle object:

$$\zeta = 6\pi\eta r \tag{1}$$

where ζ is the drag coefficient of object, η is the fluid viscosity, and r is the object radius. The drag coefficient also presents in the Einstein relation:

$$D = \frac{k_B T}{\zeta} \tag{2}$$

where D is the diffusion coefficient, k_B is the Boltzmann's constant, and T is the absolute temperature. Thus, the diffusion coefficient can be directly related to the size of object moving in fluid by combining Equations (1) and (2), which gives the Stokes-Einstein-Sutherland equation:

$$D = \frac{k_B T}{6\pi\eta r} \tag{3}$$

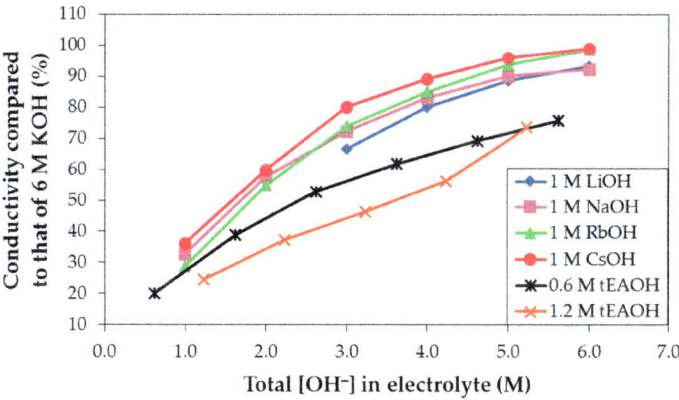

Figure 8. Plot of conductivities normalized to that of 6 M KOH aqueous solution vs. the total OH^- concentration in solution for various hydroxide aqueous solutions. The KOH concentration is the difference between the total OH^- concentration and additive concentration. As the total OH^- concentration increases, the electrolyte conductivity increases. Also, at any fixed KOH concentration, the electrolyte conductivity increases in the order of tetraethylammonium hydroxide (tEAOH) < LiOH < NaOH < RbOH < CsOH addition.

Furthermore, molar conductivity of each ionic species in solution can be linked to the diffusion coefficient by the Nernst-Einstein equation:

$$\Lambda^0_{m,i} = \left(\frac{F^2}{RT}\right) z_i^2 D_i \tag{4}$$

where $\Lambda^0_{m,i}$ is the limiting molar conductivity (molar conductivity at the limit of infinite dilution) of ionic species, F is the Faraday constant, R is the gas constant, z_i is the charge number of ionic species, and D_i is the diffusion coefficient of ionic species. In the case of hydroxide solution, electrolyte conductivity is determined by the combination of contributions from both the cation and anion species, which can be expressed by the Kohlrausch's law of independent migration of ions:

$$\Lambda^0_m = \nu_+\Lambda^0_{m,+} + \nu_-\Lambda^0_{m,-} \tag{5}$$

where Λ^0_m, $\Lambda^0_{m,+}$, and $\Lambda^0_{m,-}$ are the limiting molar conductivities of electrolyte, cation, and anion, respectively, and ν_+ and ν_- are the stoichiometric coefficients of cation and anion, respectively. Since the anion species in the current study is unchanged, the observed differences in overall electrolyte conductivity are caused by the various cation species. As the cation size increases, the diffusion coefficient decreases, and the limiting molar conductivity consequently decreases. Among all the cation species in the current study, tEA$^+$ is the largest and results in the lowest conductivity upon tEAOH addition. However, this trend is not observed with the smaller alkali cations. When the cation size is very small, the cation has higher charge density and therefore shows a higher tendency to attract water molecules. In other words, a smaller cation has a larger amount of surrounding water, which makes its transport in solution more difficult (higher drag coefficient and lower diffusion constant) and contributes to the observed lower electrolyte conductivity. The normalized electrolyte conductivities of alkali hydroxides at the same level of addition in 4 M KOH aqueous solution are plotted against the cation radii in Figure 9, and a linear dependency with a very high correlation factor (R^2) is clearly demonstrated. In addition, adding hydroxides with other alkali cations in KOH aqueous solution reduces the electrolyte conductivity slightly compared to pure KOH aqueous solution (5 M KOH aqueous solution has approximately the same conductivity as 6 M KOH aqueous solution at room temperature, which has the normalized conductivity value of 100%), indicating that a pure electrolyte system may have higher electrolyte conductivity than a heterogeneous electrolyte system at the same OH$^-$ solution concentration. Such a phenomenon is interesting to observe and deserves further investigation of, for example, the effect of interaction between cations in a heterogeneous hydroxide electrolyte system on conductivity.

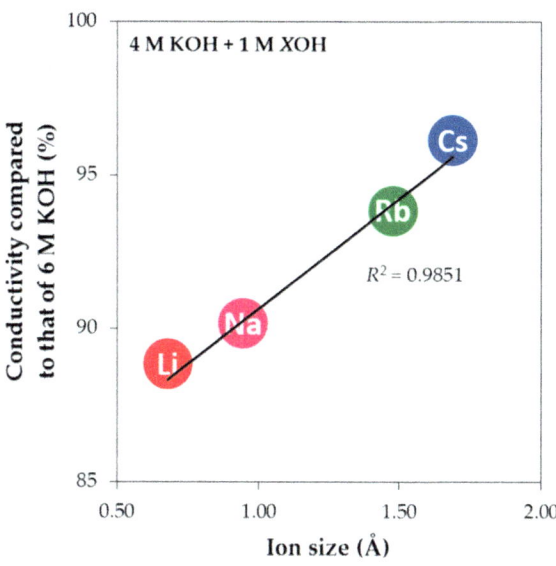

Figure 9. Plot of the conductivities of 4 M KOH + 1 M XOH aqueous solutions normalized to that of 6 M KOH aqueous solution vs. the cation sizes of alkali elements Xs.

3.4. Corrosion Performances in Various Hydroxide Electrolytes

The AR3 powder made by MS + MA is used as the negative electrode material for the corrosion study in various hydroxide aqueous solutions. Before studying the relative corrosion strengths of different electrolytes, an evaluation of capacity degradation must be established. We use three hypothetic electrolytes (Electrolytes 1, 2, and 3) in Figure 10 as the representatives to explain and describe the differences in discharge characteristics in various hydroxide aqueous solutions observed in the current study. Looking closely at the first cycle discharge voltage profiles from half-cells with the three different electrolytes (Figure 10a), the largest initial discharge capacity is observed with Electrolyte 1. The difference in discharge voltage curve length indicates that Electrolyte 1 has the strongest activation power compared to Electrolytes 2 and 3. In other words, Electrolyte 1 activates the largest portion of MH alloy within the same initial charging period as in Electrolytes 2 and 3, which is the main reason for its superior initial capacity performance. However, higher activation power also corresponds to stronger corrosion strength, increasing in the order of Electrolyte 3 < Electrolyte 2 < Electrolyte 1. The discharge capacities are plotted against the cycle number in Figure 10b, and the degradation in Electrolyte 1 is worse (higher capacity loss per cycle) compared to that in Electrolyte 3. Therefore, in order to separate the contribution of activation to degradation, degradation calculation in this study is defined as:

$$Degradation\ in\ capacity\ loss\ \%\ per\ cycle = \frac{Cap_{high} - Cap_{low}}{\left(n_{high} - n_{low}\right) Cap_{high}} \times 100\% \quad (6)$$

where Cap_{high} and Cap_{low} are the highest and lowest capacities throughout cycling (Cap_{high} is usually the initial capacity in the current study), respectively, and n_{high} and n_{low} are the corresponding cycle numbers. It is worth noting that a positive correlation is found between the activation power and electrolyte conductivity in the current study, i.e., electrolyte conductivity increases in the order of Electrolyte 3 < Electrolyte 2 < Electrolyte 1. The highest conductivity of Electrolyte 1 (or the lowest internal resistance, as observed from its highest voltage plateau in Figure 10a) is also a contributing factor for its highest initial capacity, albeit minor compared to the influence of activation power. Therefore, the contribution of conductivity on capacity is neglected in the degradation calculation.

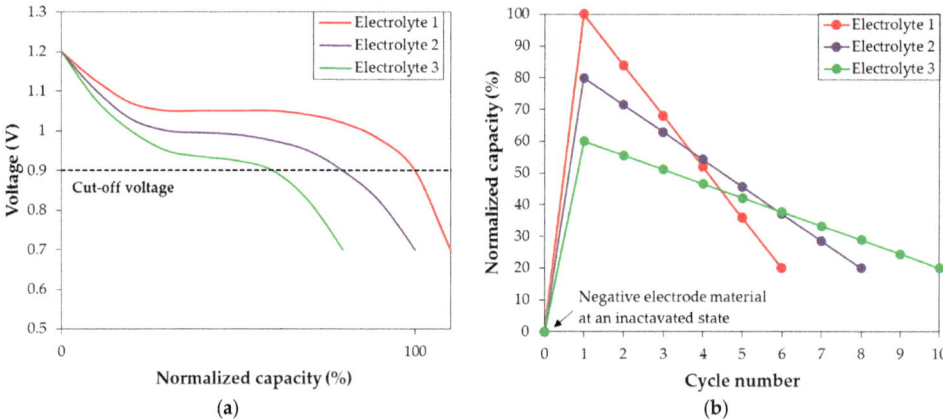

Figure 10. Schematics of (**a**) the first cycle discharge voltage profiles and (**b**) capacity cycle data from three hypothetical electrolytes with different activation powers (Electrolyte 3 < Electrolyte 2 < Electrolyte 1). Electrolyte 1 with the fastest activation shows the highest plateau voltage (lowest impedance) and consequently the highest initial capacity. However, activation power is strongly associated with corrosion strength, so Electrolyte 1 also exhibits the highest degradation rate.

Room temperature degradation of different concentrations of KOH and/or other hydroxide aqueous solutions were measured, and the results are summarized in Table A2 in Appendix. Due to its low solubility, no effect on degradation is observed with the addition of $Mg(OH)_2$, $Ca(OH)_2$, $Sr(OH)_2$, or $Ba(OH)_2$ in 6 M

KOH aqueous solution. For the pure KOH aqueous solution systems at various concentrations, the initial capacities and degradations normalized to those of 6 M KOH aqueous solution are plotted against the total OH^- concentration in solution in Figure 11a,b, respectively. Even after the attempt to eliminate the contribution of activation on degradation, a strong similarity between the initial capacity and degradation performances is still present, meaning that activation and corrosion are very closely related and cannot be completely separated by our calculation for KOH-based electrolytes. Such correlation can also be seen in mixed electrolytes. For example, a resemblance between the initial capacity (Figure 12a) and degradation (Figure 12b) performances in various electrolyte mixtures of KOH and LiOH can be observed. It is also interesting to find that higher OH^- concentrations in electrolytes do not necessarily correspond to higher corrosion.

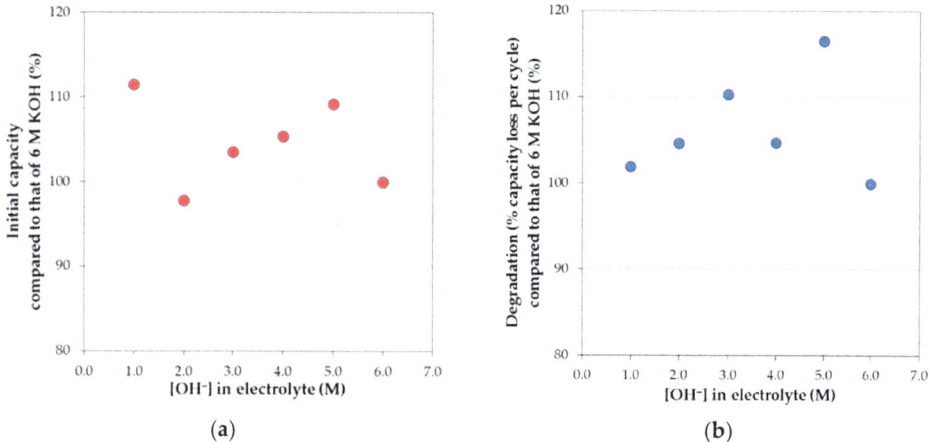

Figure 11. Plots of (**a**) the initial capacities and (**b**) degradation performances normalized to those of 6 M KOH aqueous solution vs. the total OH^- concentration in solution for several pure KOH aqueous solution systems.

As a representative example, the degradation performances compared to that of 6 M KOH aqueous solution are plotted against the total OH^- concentration in solution for 1 M LiOH, NaOH, RbOH, and CsOH, and 0.6 M tEAOH in 5 M KOH aqueous solution in Figure 13. With regard to lowering the corrosion strength of pure KOH aqueous solution, both NaOH and tEAOH additions are beneficial while CsOH, RbOH, and LiOH additions worsen the degradation behavior. More specifically, the degradation performance improves in the trend of LiOH < RbOH < CsOH < KOH < tEAOH < NaOH addition. Since the anion species in the current study is unchanged, the differences in degradation are caused by the various cation species. The trend in reactivity series of alkali elements was considered to be responsible for the observed degradation trend, which increases in the order of Cs < Rb < K < Na < Li. Interestingly, it is clear that the two trends do not correlate very well. Determination of reactivity relies on the element characteristics alone, so the trend in reactivity series is similar to another qualitative measure: ionization energy. However, the element's reaction partner also affects the reactivity of reaction. For example, for the reaction of an alkali element in water, the element's characteristics and interaction with water, such as enthalpy of sublimation ($M_{(s)} \rightarrow M_{(g)}$), ionization energy ($M_{(g)} \rightarrow M_{(g)}^+ + e^-$), and enthalpy of dissolution ($M_{(g)}^+ \rightarrow M_{(aq)}^+$), must be taken into consideration in predicting how reactive the reaction is. Therefore, the qualitative electrochemical series is used to correlate with the degradation trend. The degradation performances of alkali hydroxides at the same level of addition in 5 M KOH aqueous solution are plotted against the standard redox potential in Figure 14, resulting in a linear dependency with a high R^2. Due to the lowest and highest redox potentials of Li and Na, respectively, the addition of LiOH appears to be the most corrosive while the addition of NaOH retards degradation. Consequently, NaOH is used as an electrolyte supplement for HT applications, where corrosion is much more aggressive, and LiOH is added in electrolyte to increase electrolyte activity for the LT discharge performance in Ni/MH battery [53].

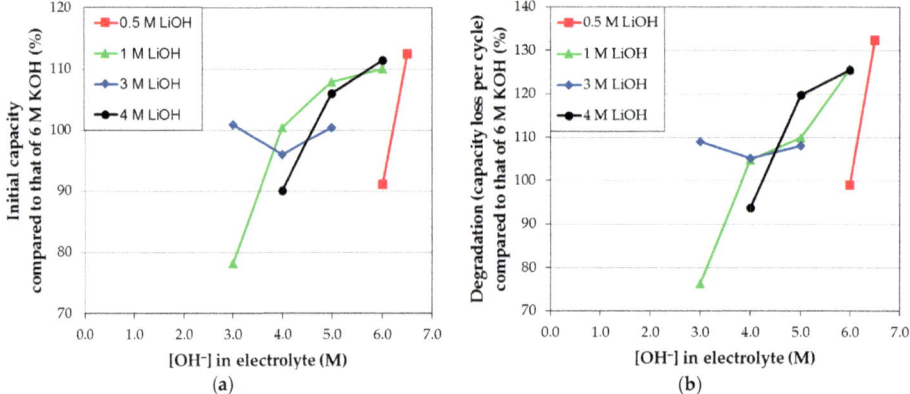

Figure 12. Plots of (**a**) the initial capacities and (**b**) degradation performances normalized to those of 6 M KOH aqueous solution vs. the total OH⁻ concentration in solution for various electrolyte mixtures of KOH and LiOH. The trends in activation power (indicated by the initial capacity) and degradation are similar: higher total OH⁻ concentration in solution corresponds to higher activation power and degradation. However, an exception is observed with the KOH-free solution with lower OH⁻ concentration (the leftmost data point of the 3 M LiOH series), which unexpectedly shows a higher activation power and a higher degradation compared to the mixed solutions in the same series. Such abnormality was found in many similar cases in this study (see data in Table A2 in Appendix).

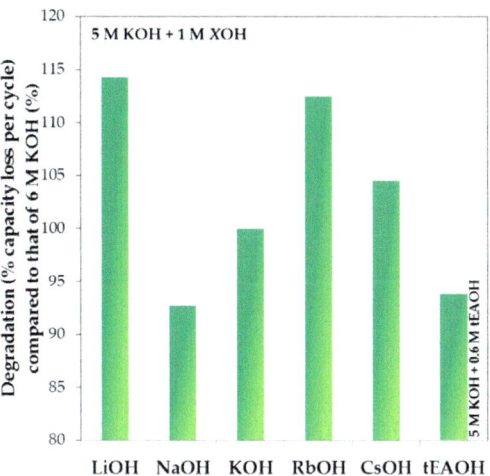

Figure 13. An example from the electrolyte degradation study. The KOH concentration is 5 M, and the concentrations of additions are 1 M for LiOH, NaOH, RbOH, and CsOH, and 0.6 M for tEAOH.

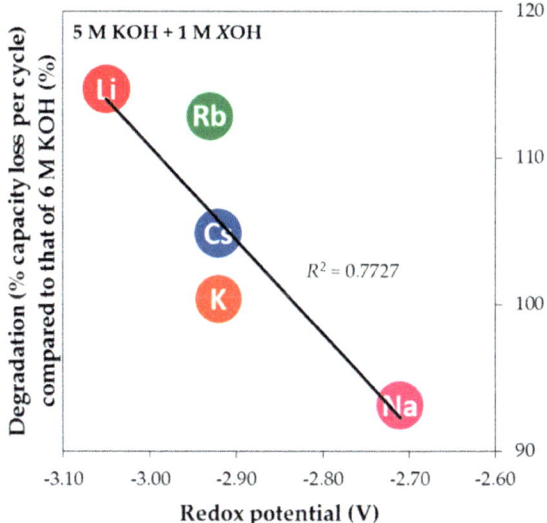

Figure 14. Plot of the degradation performances of 5 M KOH + 1 M XOH aqueous solutions normalized to that of 6 M KOH aqueous solution vs. the standard redox potentials of alkali elements Xs.

The entire data sets from Appendix are plotted in Figure 15. Most of the data plot to Quadrant III (with lower degradation but also lower conductivity). Those points in Quadrant II show lower conductivity and higher degradation and represent

393

the most unsuitable electrolytes. From this chart, we also find that while NaOH and CsOH additions can reduce the corrosive nature of an electrolyte without sacrificing excessive amounts of conductivity, tEAOH is also a good candidate if electrolyte conductivity is not a major concern for the application (e.g., high energy Ni/MH battery without strict power requirements). We plan to continue this electrolyte study with additions of other organic hydroxides and amphoteric hydroxides/oxides, such as quaternary ammonium hydroxides, bis(ethylenediamine) copper(II) hydroxide, choline base solution, tetrabutylphosphonium hydroxide, hexamethonium hydroxide, and zinc oxide/hydroxide, in order to find possible candidates that fall in Quadrant IV.

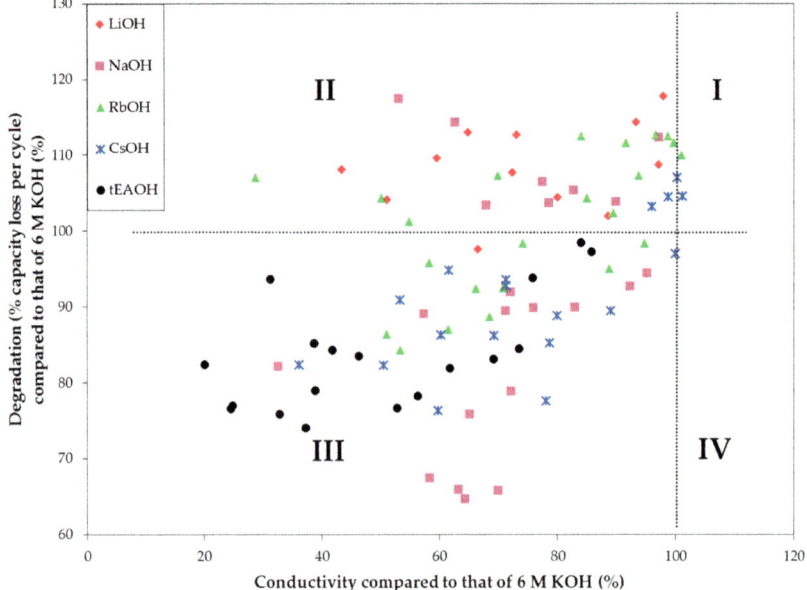

Figure 15. Plot of the degradation performances vs. the conductivities for various hydroxide aqueous solutions. Both quantities are normalized to those of 6 M KOH aqueous solution. Vertical and horizontal lines corresponding to the values from standard 6 M KOH aqueous solution divide the plot into Quadrants I, II, III, and IV. While the solutions in Quadrant II are less desirable (lower conductivity and higher degradation), those in Quadrant III represent a compromise between conductivity and degradation (lower conductivity and lower degradation). Future research will focus on finding solutions in Quadrant IV. Data originate from Appendix.

4. Conclusions

Using a quaternary MgNiCoMn MH alloy powder prepared by MS followed by MA, we have shown a clear map of the balancing of conductivity for a reduction in corrosive nature of electrolyte and lower capacity degradation in KOH-based electrolytes. Among all modifying hydroxides, NaOH, CsOH, and $(C_2H_5)_4N(OH)$ show the best conductivity/degradation results. Combining the results from this study on various hydroxides with other methods of improving negative electrode cyclability, such as salt/surfactant additions, surface coatings/treatments, and composition optimizations, the use of high-capacity MgNi-based MH alloy in commercial Ni/MH battery is possible.

Acknowledgments: This work was financially supported by U.S. DOE ARPA-E under the RANGE program (DE-AR0000386). Authors thank colleagues from BASF-Ovonic: Jun In and Jonathan Tao for alloy preparation, Alan Chan, Ryan Blackenship and David Pawlik for analytic work, Taihei Ouchi, Baoquan Huang and Su Cronogue for technical assistance, and Benjamin Reichman and Benjamin Chao for fruitful discussions.

Author Contributions: Jean Nei designed the experiments and analyzed the data; Kwo-Hsiung Young prepared the manuscript and Damian Rotarov performed the experiments.

Conflicts of Interest: The authors declare no conflict of interest.

Abbreviations

MH	Metal hydride
Ni/MH	Nickel/metal hydride
HRD	High-rate dischargeability
LT	Low temperature
RT	Room temperature
HT	High temperature
DOE	U.S. Department of Energy
RANGE	Robust affordable next generation EV-storage
RF	Radio frequency
MS	Melt spinning
MA	Mechanical alloying
XRD	X-ray diffractometer
TEM	Transmission electron microscope
tEAOH	Tetraethylammonium hydroxide
R^2	Correlation factor

Appendix A.

Table A1. Room temperature conductivities of electrolytes with various concentrations of KOH and other hydroxides normalized to that of 6 M KOH aqueous solution. Electrolyte with conductivity value higher than 100% implies that it is more conductive compared to the standard electrolyte (6 M KOH aqueous solution).

Content	KOH (M)	$X(OH)_y$ (M)	$[OH^-]$ (M)	Conductivity (%)	Content	KOH (M)	$X(OH)_y$ (M)	$[OH^-]$ (M)	Conductivity (%)
	0.00	3.00	3.00	43.43		0.00	1.00	1.00	32.70
	0.00	4.00	4.00	51.15		0.00	2.00	2.00	53.15
	1.00	3.00	4.00	59.61		0.00	3.00	3.00	62.71
	1.00	4.00	5.00	64.96		0.00	4.00	4.00	65.23
	2.00	1.00	3.00	66.64		0.00	5.00	5.00	63.33
$X = Li, y = 1$	2.00	3.00	5.00	72.41		0.00	6.00	6.00	58.47
	2.00	4.00	6.00	73.06		1.00	1.00	2.00	57.48
	3.00	1.00	4.00	80.02		1.00	2.00	3.00	67.98
	4.00	1.00	5.00	88.65		1.00	3.00	4.00	72.19
	5.00	1.00	6.00	93.29		1.00	4.00	5.00	70.02
	5.50	0.50	6.00	97.23	$X = Na, y = 1$	1.00	5.00	6.00	64.41
	6.00	0.50	6.50	98.03		2.00	1.00	3.00	72.13
	0.00	1.00	1.00	28.79		2.00	2.00	4.00	77.44
	0.00	2.00	2.00	50.18		2.00	3.00	5.00	75.93
	0.00	3.00	3.00	66.16		2.00	4.00	6.00	71.31
	0.00	4.00	4.00	51.08		3.00	1.00	4.00	82.89
	0.00	5.00	5.00	58.26		3.00	2.00	5.00	82.71
	0.00	6.00	6.00	89.38		3.00	3.00	6.00	78.60
	1.00	1.00	2.00	54.83		4.00	1.00	5.00	89.96
	1.00	2.00	3.00	69.90		5.00	1.00	6.00	92.27
	1.00	3.00	4.00	53.39		5.50	0.50	6.00	95.16
	1.00	4.00	5.00	61.53		6.00	0.50	6.50	97.21
$X = Rb, y = 1$	1.00	5.00	6.00	68.47		0.00	1.00	1.00	36.18
	2.00	1.00	3.00	74.04		0.00	3.00	3.00	78.65
	2.00	2.00	4.00	83.90		0.00	4.00	4.00	78.10
	2.00	3.00	5.00	88.69		0.00	5.00	5.00	53.39
	2.00	4.00	6.00	70.78		0.00	6.00	6.00	100.31
	3.00	1.00	4.00	84.95		1.00	1.00	2.00	59.81
	3.00	2.00	5.00	91.47		1.00	2.00	3.00	50.59
	3.00	3.00	6.00	94.65		1.00	4.00	5.00	61.59
	4.00	1.00	5.00	93.68	$X = Cs, y = 1$	1.00	5.00	6.00	71.33
	4.00	2.00	6.00	96.59		2.00	1.00	3.00	79.92
	5.00	1.00	6.00	98.66		2.00	3.00	5.00	60.30
	5.50	0.50	6.00	99.68		2.00	4.00	6.00	71.26
	6.00	0.50	6.50	100.93		3.00	1.00	4.00	88.99
	0.00	0.62	0.62	20.13		3.00	3.00	6.00	69.31
	0.00	1.24	1.24	24.55		4.00	1.00	5.00	95.97
	0.00	1.86	1.86	24.83		5.00	1.00	6.00	98.78
	1.00	0.62	1.62	38.87		5.50	0.50	6.00	99.95
	1.00	1.24	2.24	37.29		6.00	0.50	6.50	101.21
$X = tEA, y = 1$	1.00	1.86	2.86	32.94					
	2.00	0.62	2.62	52.90	$X = Mg, y = 2$	6.00	0.0130	6.0259	99.58
	2.00	1.24	3.24	46.27					
	2.00	1.86	3.86	38.71					
	2.00	2.48	4.48	31.31					
	3.00	0.62	3.62	61.80	$X = Ca, y = 2$	6.00	0.0180	6.0360	98.81
	3.00	1.24	4.24	56.29					

Table A1. *Cont.*

Content	KOH (M)	$X(OH)_y$ (M)	$[OH^-]$ (M)	Conductivity (%)	Content	KOH (M)	$X(OH)_y$ (M)	$[OH^-]$ (M)	Conductivity (%)
	3.00	1.86	4.86	41.75					
	4.00	0.62	4.62	69.22	$X = Sr, y = 2$	6.00	0.0004	6.0008	98.73
	4.00	1.24	5.24	73.48					
	5.00	0.62	5.62	75.74					
	5.50	0.31	5.81	83.96	$X = Ba, y = 2$	6.00	0.0005	6.0011	98.74
	6.00	0.31	6.31	85.68					

Table A2. Room temperature degradation performances (defined as percentage capacity loss per cycle) of electrolytes with various concentrations of KOH and other hydroxides normalized to that of 6 M KOH aqueous solution. Electrolyte with degradation value higher than 100% implies that it is more corrosive compared to the standard electrolyte (6 M KOH aqueous solution).

Content	KOH (M)	$X(OH)_y$ (M)	$[OH^-]$ (M)	Degradation (%)	Content	KOH (M)	$X(OH)_y$ (M)	$[OH^-]$ (M)	Degradation (%)
	1.00	0.00	1.00	101.92		0.00	1.00	1.00	82.11
	2.00	0.00	2.00	104.64		0.00	2.00	2.00	117.41
	3.00	0.00	3.00	110.26		0.00	3.00	3.00	114.34
	4.00	0.00	4.00	104.69		0.00	4.00	4.00	75.89
	5.00	0.00	5.00	116.52		0.00	5.00	5.00	65.91
	5.50	0.00	5.50	83.76		0.00	6.00	6.00	67.41
	6.00	0.00	6.00	100.00		1.00	1.00	2.00	89.07
	0.00	3.00	3.00	108.06		1.00	2.00	3.00	103.36
$X = Li, y = 1$	0.00	4.00	4.00	104.05		1.00	3.00	4.00	78.91
	1.00	3.00	4.00	109.58		1.00	4.00	5.00	65.76
	1.00	4.00	5.00	113.00	$X = Na, y = 1$	1.00	5.00	6.00	64.64
	2.00	1.00	3.00	97.59		2.00	1.00	3.00	91.91
	2.00	3.00	5.00	107.62		2.00	2.00	4.00	106.48
	2.00	4.00	6.00	112.67		2.00	3.00	5.00	89.88
	3.00	1.00	4.00	104.37		2.00	4.00	6.00	89.46
	4.00	1.00	5.00	101.92		3.00	1.00	4.00	89.90
	5.00	1.00	6.00	114.28		3.00	2.00	5.00	105.34
	5.50	0.50	6.00	108.68		3.00	3.00	6.00	103.66
	6.00	0.50	6.50	117.71		4.00	1.00	5.00	103.81
	0.00	1.00	1.00	107.00		5.00	1.00	6.00	92.73
	0.00	2.00	2.00	104.29		5.50	0.50	6.00	94.45
	0.00	3.00	3.00	92.36		6.00	0.50	6.50	112.35
	0.00	4.00	4.00	86.38		0.00	1.00	1.00	82.36
	0.00	5.00	5.00	95.83		0.00	2.00	2.00	81.80
	0.00	6.00	6.00	102.33		0.00	3.00	3.00	85.26
	1.00	1.00	2.00	101.22		0.00	4.00	4.00	77.61
	1.00	2.00	3.00	107.25		0.00	5.00	5.00	90.85
	1.00	3.00	4.00	84.30		0.00	6.00	6.00	107.00
	1.00	4.00	5.00	86.97		1.00	1.00	2.00	76.35
$X = Rb, y = 1$	1.00	5.00	6.00	88.66		1.00	2.00	3.00	79.92
	2.00	1.00	3.00	98.37		1.00	3.00	4.00	82.31
	2.00	2.00	4.00	112.52		1.00	4.00	5.00	94.88
	2.00	3.00	5.00	95.00		1.00	5.00	6.00	92.78
	2.00	4.00	6.00	92.46	$X = Cs, y = 1$	2.00	1.00	3.00	88.79
	3.00	1.00	4.00	104.29		2.00	2.00	4.00	77.35
	3.00	2.00	5.00	111.65		2.00	3.00	5.00	86.31
	3.00	3.00	6.00	98.37		2.00	4.00	6.00	93.55

Table A2. Cont.

Content	KOH (M)	$X(OH)_y$ (M)	$[OH^-]$ (M)	Degradation (%)
	4.00	1.00	5.00	107.25
	4.00	2.00	6.00	112.63
	5.00	1.00	6.00	112.52
	5.50	0.50	6.00	111.65
	6.00	0.50	6.50	109.97
	0.00	0.62	0.62	82.42
	0.00	1.24	1.24	76.59
	0.00	1.86	1.86	76.97
	0.00	2.48	2.48	85.49
	1.00	0.62	1.62	78.96
	1.00	1.24	2.24	74.06
	1.00	1.86	2.86	75.89
	1.00	2.48	3.48	75.15
	2.00	0.62	2.62	76.65
X = tEA, y = 1	2.00	1.24	3.24	83.52
	2.00	1.86	3.86	85.17
	2.00	2.48	4.48	93.69
	3.00	0.62	3.62	81.94
	3.00	1.24	4.24	78.27
	3.00	1.86	4.86	84.33
	4.00	0.62	4.62	83.07
	4.00	1.24	5.24	84.43
	5.00	0.62	5.62	93.81
	5.50	0.31	5.81	98.44
	6.00	0.31	6.31	97.26

Content	KOH (M)	$X(OH)_y$ (M)	$[OH^-]$ (M)	Degradation (%)
	3.00	1.00	4.00	89.42
	3.00	2.00	5.00	81.53
	3.00	3.00	6.00	86.18
	4.00	1.00	5.00	103.22
	4.00	2.00	6.00	76.46
	5.00	1.00	6.00	104.51
	5.50	0.50	6.00	97.03
	6.00	0.50	6.50	104.57
X = Mg, y = 2	6.00	0.0130	6.0259	98.38
X = Ca, y = 2	6.00	0.0180	6.0360	98.29
X = Sr, y = 2	6.00	0.0004	6.0008	96.13
X = Ba, y = 2	6.00	0.0005	6.0011	97.90

References

1. Liu, D.; Zhu, Y.; Li, L. Effect of surface oxidation on the hydriding and dehydriding of Mg$_2$Ni alloy produced by hydriding combustion synthesis. *J. Mater. Sci.* **2007**, *42*, 9725–9729.

2. Young, K.; Nei, J. The current status of hydrogen storage alloy development for electrochemical applications. *Materials* **2013**, *6*, 4574–4608.

3. Redzeb, M.; Zlatanova, Z.; Spassov, T. Influence of boron on the hydriding of nanocrystalline Mg$_2$Ni. *Intermetallics* **2013**, *34*, 63–68.

4. Nikkuni, F.R.; Santos, S.F.; Ticianelli, E.A. Microstructures and electrochemical properties of Mg$_{49}$Ti$_6$Ni$_{(45-x)}$M$_x$ (M = Pd and Pt) alloy electrodes. *Int. J. Energy Res.* **2013**, *37*, 706–712.

5. Zhang, Z.; Elkedim, O.; Balcerzak, M.; Jurczyk, M. Structural and electrochemical hydrogen storage properties of MgTiNi$_x$ (x = 0.1, 0.5, 1, 2) alloys prepared by ball milling. *Int. J. Hydrog. Energy* **2016**, *41*, 11761–11766.

6. Zhang, Y.; Li, C.; Cai, Y.; Hu, F.; Liu, Z.; Guo, S. Highly improved electrochemical hydrogen storage performances of the Nd-Cu-added Mg$_2$Ni-type alloys by melt spinning. *J. Alloys Compd.* **2014**, *584*, 81–86.

7. Wang, Y.T.; Wan, C.B.; Wang, R.L.; Meng, X.H.; Huang, M.F.; Ju, X. Effect of Cr substitution by Ni on the cycling stability of Mg$_2$Ni alloy using EXAFS. *Int. J. Hydrog. Energy* **2014**, *39*, 14858–14867.

8. Pu, Z.; Zhu, Y.; Zhu, J.; Yuan, J.; Zhang, J.; Chen, W.; Fang, J.; Li, L. Kinetics and electrochemical characteristics of Mg_2NiH_{4-x} wt.% $MmNi_{3.8}Co_{0.75}Mn_{0.4}Al_{0.2}$ (x = 5, 10, 20, 40) composites for Ni-MH battery. *Int. J. Hydrog. Energy* **2014**, *39*, 3887–3894.

9. Hou, X.; Hu, R.; Zhang, T.; Kou, H.; Song, W.; Li, J. Microstructure and electrochemical hydrogenation/dehydrogenation performance of melt-spun La-doped Mg_2Ni alloys. *Mater. Charact.* **2015**, *106*, 163–174.

10. Verbovytskyy, Y.; Zhang, J.; Cuevas, F.; Paul-Boncour, V.; Zavaliy, I. Synthesis and properties of the $Mg_2Ni_{0.5}Cu_{0.5}H_{4.4}$ hydride. *J. Alloys Compd.* **2015**, *645*, S408–S411.

11. Li, M.; Zhu, Y.; Yang, C.; Zhang, J.; Chen, W. Enhanced electrochemical hydrogen storage properties of Mg_2NiH_4 by coating with nano-nickel. *Int. J. Hydrog. Energy* **2015**, *40*, 13949–13956.

12. Shang, J.; Ouyang, Z.; Liu, K.; Xing, C.; Liu, W.; Wang, L. Effect of Li atom infiltration by the way of electro-osmosis on electrochemical properties of amorphous $Mg_{62}Ni_{27}La_8$ alloy used as negative electrode materials for the nickel-metal hydride secondary batteries. *J. Non Cryst. Solids* **2015**, *415*, 30–35.

13. Shao, H.; Li, X. Effect of nanostructure and partial substitution on gas absorption and electrochemical properties in Mg_2Ni-based alloys. *J. Alloys Compd.* **2016**, *667*, 191–197.

14. Shahcheraghi, A.; Dehghani, F.; Raeissi, K.; Saatchi, A.; Enayati, M.H. Effects of TiO_2 additive on electrochemical hydrogen storage properties of nanocrystalline/amorphous Mg_2Ni intermetallic alloy. *Iran. J. Mater. Sci. Eng.* **2013**, *10*, 1–9.

15. Ohara, R.; Lan, C.; Hwang, C. Electrochemical and structural characterization of electroless nickel coating on Mg_2Ni hydrogen storage alloy. *J. Alloys Compd.* **2013**, *580*, S368–S372.

16. Haghighat-Shishavan, S.; Kashani-Bozorg, S.F. Nano-crystalline $Mg_{(2-x)}Mn_xNi$ compounds synthesized by mechanical alloys: Microstructure and electrochemistry. *J. Ultrafine Grained Nanostruct. Mater.* **2014**, *47*, 43–49.

17. Venkateswari, A.; Nithya, C.; Kumaran, S. Electrochemical behaviour of $Mg_{67}Ni_{(33-x)}Nb_x$ (x = 0, 1, 2, and 4) alloy synthesized by high energy ball milling. *Proc. Mater. Sci.* **2014**, *5*, 679–687.

18. Rongeat, C.; Roué, L. On the cycle life improvement of amorphous MgNi-based alloy for Ni-MH batteries. *J. Alloys Compd.* **2005**, *404–406*, 679–681.

19. Rongeat, C.; Grosjean, M.-H.; Ruggeri, S.; Dehmas, M.; Bourlot, S.; Marcotte, S.; Roué, L. Evaluation of different approaches for improving the cycle life of MgNi-based electrodes for Ni-MH batteries. *J. Power Sources* **2006**, *158*, 747–753.

20. Wang, J.; Na, E.; Wu, F. Study on the improvement of cycle life of magnesium-based hydrogen storage alloy. *J. Wuhan Univ. Technol.* **2006**, *28*, 371–373.

21. Kong, F.; Yan, H.; Xiong, W.; Li, B.; Li, J. Investigation on the cycling stability of Mg-based hydrogen storage electrode improved by anti-corrosion method. *Chin. Rare Earths* **2006**, *27*, 39–42.

22. Yan, S.; Young, K.; Ng, K.Y.S. Effects of salt additives to the KOH electrolyte used in Ni/MH batteries. *Batteries* **2015**, *1*, 54–73.

23. Young, K.; Ng, K.Y.S.; Bendersky, L.A. A technical report of the robust affordable next generation energy storage system-BASF program. *Batteries* **2016**, *2*.

24. Rubin, E.J.; Baboian, R. A correlation of the solution properties and the electrochemical behavior of the nickel hydroxide electrode in binary aqueous alkali hydroxides. *J. Electrochem. Soc.* **1971**, *118*, 428–433.

25. Barnard, R.; Randell, C.F.; Tye, F.L. Studies concerning changes nickel hydroxide electrodes. IV. Reversible potentials in LiOH, NaOH, RbOH and CdOH. *J. Appl. Electrochem.* **1981**, *11*, 517–523.

26. Oliva, P.; Leonardi, J.; Laurent, J.F.; Delmas, C.; Braconnier, J.J.; Figlarz, M.; Fievet, F.; Guibert, A. Review of the structure and the electrochemistry of nickel hydroxides and oxy-hydroxides. *J. Power Sources* **1982**, *8*, 229–255.

27. See, D.M.; White, R.E. Temperature and concentration dependence of the specific conductivity of concentrated solutions of potassium hydroxide. *J. Chem. Eng. Data* **1997**, *42*, 1266–1268.

28. Leblanc, P.; Jordy, C.; Knosp, B.; Blanchard, P. Mechanism of alloy corrosion and consequences on sealed nickel—metal hydride battery performance. *J. Electrochem. Soc.* **1998**, *145*, 860–863.

29. Knosp, B.; Vallet, L.; Blamchard, P. Performance of an AB$_2$ alloy in sealed Ni-MH batteries for electric vehicles: Qualification of corrosion rate and consequences on the battery performance. *J. Alloys Compd.* **1999**, *293–295*, 770–774.

30. Jeong, Y.H.; Kim, H.G.; Jung, Y.H.; Ruhmann, H. *Effect of LiOH, NaOH and KOH on Corrosion and Oxide Microstructure of Zr-Based Alloys*; International Atomic Energy Agency: Wien, Austria, 1999.

31. Liu, J.; Wang, D.; Liu, S.; Feng, X. Improving high temperature performance of MH/Ni battery by orthogonal design. *Battery Bimon.* **2003**, *33*, 218–220.

32. Hou, X.L.; Na, J.M.; Han, D.M.; Zhao, J.F. Preparation and performance of high-rated A-type MH-Ni batteries. *Chin. J. Appl. Chem.* **2004**, *21*, 1169–1173. (In Chinese)

33. Lv, J.; Liu, X.; Zhang, J.; Fan, L.; Wang, L.; Zhang, Z. Studies on high-power nickel-metal hydride battery. *Chin. J. Power Sources* **2005**, *29*, 826–830.

34. Li, X.; Dong, H.; Zhang, A.; Wei, Y. Electrochemical impedance and cyclic voltammetry characterization of a metal hydride electrode in alkaline electrolytes. *J. Alloys Compd.* **2006**, *426*, 93–96.

35. Park, C.; Shim, J.; Jang, M.; Park, C.; Choi, J. Influences of various electrolytes on the low-temperature characteristics of Ni-MH secondary battery. *Trans. Korean Hydrog. New Energy Soc.* **2007**, *18*, 284–291.

36. Chen, R.; Li, L.; Wu, F.; Qiu, X.; Chen, S. Effects of low temperature on performance of hydrogen-storage alloys and electrolyte. *Min. Metall. Eng.* **2007**, *27*, 44–46.

37. Yang, D.C.; Park, C.N.; Park, C.J.; Choi, J.; Sim, J.S.; Jang, M.H. Design of additives and electrolyte for optimization of electrode characteristics of Ni-MH secondary battery at room and low temperatures. *Trans. Korean Hydrog. New Energy Soc.* **2007**, *18*, 365–373.

38. Zhang, X.; Chen, Y.; Tao, M.; Wu, C. Effect of electrolyte concentration on low-temperature electrochemical properties of LaNi$_5$ alloy electrode at 233 K. *J. Rare Earths* **2008**, *26*, 402–405.

39. Zhang, X.; Chen, Y.; Tao, M.; Wu, C. Effect of electrolyte on the low-temperature electrochemical properties of LaNi$_5$ alloy electrode at 253 K. *Rare Metal. Mater. Eng.* **2008**, *37*, 2012–2015. (In Chinese)

40. Pei, L.; Yi, S.; He, Y.; Chen, Q. Effect of electrolyte formula on the self-discharge properties of nickel-metal hydride batteries. *J. Guangdong Univ. Technol.* **2008**, *25*, 10–12.

41. Khaldi, C.; Mathlouthi, H.; Lamloumi, J. A comparative study of 1 M and 8 M KOH electrolyte concentrations used in Ni-MH batteries. *J. Alloys Compd.* **2009**, *469*, 464–471.

42. Guiose, B.; Cuevas, F.; Décamps, B.; Leroy, E.; Percheron-Guégan, A. Microstructural analysis of the aging of pseudo-binary (Ti,Zr)Ni intermetallic compounds as negative electrodes of Ni-MH batteries. *Electrochim. Acta* **2009**, *54*, 2781–2789.

43. Qiu, Z.; Wu, A. Study on wide temperature characteristics of Ni-MH battery. *J. South China Norm. Univ.* **2009**, *S1*, 79–81. (In Chinese)

44. Song, M.; Chen, Y.; Tao, M.; Wu, C.; Zhu, D.; Yang, H. Some factors affecting the electrochemical performances of LaCrO$_3$ as negative electrodes for Ni/MH batteries. *Electrochim. Acta* **2010**, *55*, 3103–3108.

45. Ma, H.; Cheng, F.; Chen, J. Nickel-Metal Hydride (Ni-MH) Rechargeable Battery. In *Electrochemical Technologies for Energy Storage and Conversion*; Zhang, J., Zhang, L., Liu, H., Sun, A., Liu, R., Eds.; John Wiley & Sons, Inc.: New York, NY, USA, 2011; p. 204.

46. Ruiz, F.C.; Martínez, P.S.; Castro, E.B.; Humana, R.; Peretti, H.A.; Visintin, A. Effect of electrolyte concentration on the electrochemical properties of an AB$_5$-type alloy for Ni/MH batteries. *Int. J. Hydrog. Energy* **2013**, *38*, 240–245.

47. Martínez, P.S.; Ruiz, F.C.; Visintin, A. Influence of different electrolyte concentrations on the performance of an AB$_2$-type alloy. *J. Electrochem. Soc.* **2014**, *161*, A326–A329.

48. Karwowska, M.; Jaron, T.; Fijalkowski, K.J.; Leszczynski, P.J.; Rogulski, Z.; Czerwinski, A. Influence of electrolyte composition and temperature on behavior of AB$_5$ hydrogen storage alloy used as negative electrode in Ni-MH batteries. *J. Power Sources* **2014**, *263*, 304–309.

49. Giza, K. Influence of electrolyte on capacity and corrosion resistance of anode material used in Ni-MH cells. *Ochr. Koroz.* **2016**, *59*, 167–169.

50. Young, K. Stoichiometry in Inter-Metallic Compounds for Hydrogen Storage Applications. In *Stoichiometry and Materials Science—When Numbers Matter*; Innocenti, A., Kamarulzaman, N., Eds.; InTech: Rijeka, Croatia, 2012.

51. Nayeb-Hashemi, A.A.; Clark, J.B. Mg-Ni (Magnesium-Nickel). In *Binary Alloy Phase Diagram*, 2nd ed.; Massalski, T.B., Okamoto, H., Subramanian, P.R., Kacprzak, L., Eds.; ASM International: Geauga County, OH, USA, 1990; Volume 3, pp. 2529–2530.

52. Young, K.; Chao, B.; Liu, Y.; Nei, J. Microstructures of the oxides on the activated AB$_2$ and AB$_5$ metal hydride alloys surface. *J. Alloys Compd.* **2014**, *606*, 97–104.

53. Fetcenko, M.; Koch, J.; Zelinsky, M. Nickel-Metal Hydride Batteries and Nickel-Zinc Batteries for Hybrid Electric Vehicles and Battery Electric Vehicles. In *Advances in Battery Technologies for Electric Vehicles*; Scrosati, B., Garche, J., Tillmet, W., Eds.; Woodhead Publishing Ltd.: Cambridge, UK, 2015; pp. 107–108.

Effects of Salt Additives to the KOH Electrolyte Used in Ni/MH Batteries

Suli Yan, Kwo-Hsiung Young and K.Y. Simon Ng

Abstract: KOH-based electrolytes with different salt additives were investigated to reduce their corrosive nature toward Mg/Ni metal hydride alloys used as negative electrodes in nickel metal hydride (Ni/MH) batteries. Alkaline metal halide salts and oxyacid salts were studied as additives to the traditional KOH electrolyte with concentrations varying from 0.005 M to 1.77 M. Effects of the cations and anions of the additives on charge/discharge performance are discussed. The reduction potential of alkaline cations and radii of halogen anions were correlated with initial capacity and degradation of the metal hydride alloy. A synergistic effect between KOH and some oxyacid salt additives was observed and greatly influenced by the nature of the salt additives. It was suggested that both the formation of a solid film over the metal hydride surface and the promotion of proton transfer in the additives containing electrolytes led to a decreased degradation of the electrodes and an increased discharge capacity. 12 salt additives, $NaC_2H_3O_2$, $KC_2H_3O_2$, K_2CO_3, Rb_2CO_3, Cs_2CO_3, K_3PO_4, Na_2WO_4, Rb_2SO_4, Cs_2SO_4, NaF, KF, and KBr, were found to increase the corrosion resistance of the MgNi-based metal hydride alloy.

Reprinted from *Batteries*. Cite as: Yan, S.; Young, K.-H.; Ng, K.Y.S. Effects of Salt Additives to the KOH Electrolyte Used in Ni/MH Batteries. *Batteries* **2015**, *1*, 54–73.

1. Introduction

Nickel metal hydride (Ni/MH) batteries are currently one of the most widely used energy storage devices with many applications including hybrid electric vehicles, vacuum cleaners, electric toys, power tools, cordless phones, *etc.* [1,2]. They have many advantages such as having a high specific power, excellent tolerance to abuse, long cycle life, wide temperature operation range, and low self-discharge. Facing competition from Li-ion batteries, scientists continue to improve the various properties of Ni/MH batteries. One of the key areas demanding immediate attention is the gravimetric energy density.

A basic Ni/MH battery consists of an assembly of a highly conductive electrolyte, a metal hydride (MH) negative electrode, and a nickel hydroxide positive electrode. The cell reactions on the electrodes are shown in following equations:

$$\text{Negative electrode: } M + H_2O + e^- \rightleftharpoons OH^- + MH \tag{1}$$

$$\text{Positive electrode: } Ni(OH)_2 + OH^- \rightleftharpoons NiOOH + H_2O + e^- \tag{2}$$

The most widely used electrolytes are pure KOH aqueous solutions with concentrations in the range of 20–36 wt% because these solutions are highly conductive and can efficiently convey protons in an electric field. AB_5 and A_2B_7 type MH alloys with electrocapacity in the range of 300–350 mAh·g^{-1} have been commercialized as the negative electrodes for a long time. For the positive electrode, β-Ni(OH)$_2$ co-precipitated with Co and Zn with an energy density of about 240 mAh·g^{-1} is the most commonly used material.

During charge, a negative voltage (with respect to the counter electrode) is applied to the MH electrode, and electrons enter the metal through the current collector to neutralize the protons produced from the splitting of water that occurs at the metal/electrolyte interface. During discharge, protons in the MH leave the surface and recombine with OH$^-$ in the alkaline electrolyte to form H$_2$O and the resulting charge neutrality pushes the electrons out of the MH through the current collector, performing electrical work in the outer circuitry. Due to their demanding applications in Ni/MH batteries, MH alloys are required to possess a high capacity and moderate hydride stability. Iwakura *et al.* [3] and Geng *et al.* [4] tested the pressure-composition-temperature (P-C-T) isotherm curves of MH alloys, and correlated the hydrogen content with electrochemical potential through the Nernst equation. It was found that higher hydrogen content of alloys lead to higher electrochemical capacity. A wide set of hydrogen storage electrode alloys, other than AB_5 and A_2B_7 type, have been reported to contain high hydrogen content and high electrochemical capacity. Among them, MgNi-based amorphous and nanocrystalline MH alloys are reported to have higher capacity than those in AB_5 and A_2B_7 [5,6]. For example, Young and Nei [7] reported that the theoretical capacity of MgNi metal alloys is about 1080 mAh·g^{-1} and Anik *et al.* [8] reported that the practical electrocapacity values of MgNi and Mg$_{0.9}$M$_{0.1}$Ni (M = B, Ti, and Zr) type alloys are up to 495 mAh·g^{-1} and 508 mAh·g^{-1}, respectively. The theoretical capacities of Mg$_{80}$Sc$_{20}$, Mg$_{80}$Ti$_{20}$, Mg$_{80}$V$_{20}$, and Mg$_{80}$Cr$_{20}$ reach 1790, 1750, 1700, and 1270 mAh· g^{-1}, respectively. High hydrogen storage capacity, light weight, and the low cost of Mg type alloys make them one of the most promising commercially viable materials for high capacity Ni/MH batteries.

However, as many researchers have stated, MgNi-based electrodes are still far from commercial application because of the rapid decay of electrochemical capacity. Rongeat *et al.* [9] reported a 70% decay in capacity after 20 charge/discharge cycles. A lot of effort has been put into increasing the corrosion resistance of MgNi-based alloys to KOH electrolytes, mostly through studying various types of substitutions in the MgNi formula, such as replacements of the A- (by rare earth, transition, or other metals [10]) and B-sites (by transition metals [11,12]), different fabrication procedures [13], and surface treatment [14].

It is well known that the use of electrolyte additives is one of the most economic and effective methods to improve the performance of electrodes as they do not affect the gravimetric and volumetric energy density of the battery. However, there are only a few studies of the modification of traditional KOH electrolytes to improve the cycle life of Ni/MH batteries. For example, Danczuk *et al.* [15] and Karwowska *et al.* [16] studied the effects of alkaline cation additives in KOH electrolyte to AB$_5$ type alloys; Shangguan *et al.* [17] found that Na$_2$WO$_4$ additives could increase the high temperature performance of Ni electrodes; Vaidyanathan *et al.* [18] reported that KSiO$_4$ and KNO$_3$ additives could increase the end-of-charge voltage for Ni/MH batteries. Until now, very few studies for electrolyte additives' effects to MgNi-based Ni/MH batteries have been reported. In this study, a systematic investigation on the effect of salt additives to the performance of traditional KOH electrolyte to an Mg$_{52}$Ni$_{39}$Co$_3$Mn$_6$ alloy prepared by the melt-spin and consequently mechanical alloying is performed. The experiment is accomplished by partially replacing KOH in the traditional electrolyte with salt additives while keeping the total ionic concentrations the same; thus, a better understanding of the effect of the anion and cation substitute of KOH electrolyte in Ni/MH battery can be obtained. The ultimate objective of this study is to develop a novel salt containing electrolyte formulation, which is as highly conductive as the traditional KOH electrolyte, but less corrosive to MgNi-based negative electrodes. With this new electrolyte, MgNi-based NiMH battery is expected to have the same, or even higher, electrochemical capacity and improved capacity stability compared to the traditional 30 wt% KOH electrolyte. The mechanism for proton transfer and improved corrosion resistance is also briefly discussed in this work.

2. Experimental

2.1. Materials and Electrolyte Matrix

The MgNi alloy (Mg$_{52}$Ni$_{39}$Co$_3$Mn$_6$) was provided by BASF/Battery Materials-Ovonic, Rochester Hills, MI, USA. The alloy was prepared by melt-spin followed by mechanical alloying. Details of the alloy preparation were described elsewhere [19]. Sintered β-Ni(OH)$_2$ was also obtained from BASF/Battery Materials-Ovonic. The matrix examined on the effects of salt additives in aqueous KOH solution consisted of the following: (1) alkaline salts, including LiCl, NaCl, KCl, RbCl, and CsCl (with salt concentrations at 0.44, 0.88, 1.33, and 1.77 M each and KOH concentrations change accordingly as 6.33, 5.88, 5.44, and 5.00 M, respectively, with the total ionic concentration kept at 6.77 M); (2) halogen containing salts including KF, KCl, KBr, and KI (with salt concentrations the same as previously mentioned); (3) oxyacid containing salts, including LiNO$_3$, NaNO$_3$, KNO$_3$, RbNO$_3$, CsNO$_3$, K$_2$CO$_3$, Na$_2$CO$_3$, Cs$_2$CO$_3$, Rb$_2$CO$_3$, K$_3$PO$_4$, KIO$_4$, Li$_2$SO$_4$, K$_2$SO$_4$, Na$_2$SO$_4$, Rb$_2$SO$_4$,

Cs_2SO_4, $LiCHO_2$, $NaCHO_2$, $KCHO_2$, $CsCHO_2$, Na_2WO_4, $NaC_2H_3O_2$, and $KC_2H_3O_2$ (with salt concentration at 0.44 M and 6.33 M KOH each for $LiNO_3$, $NaNO_3$, KNO_3, $RbNO_3$,$CsNO_3$, KIO_4, $LiCHO_2$, $NaCHO_2$, $KCHO_2$, $CsCHO_2$, $NaC_2H_3O_2$, and $KC_2H_3O_2$; 0.29 M and 6.33 M KOH each for K_2CO_3, Na_2CO_3, Cs_2CO_3, Rb_2CO_3, Na_2WO_4; 0.005 M and 6.33 M KOH each for Li_2SO_4, K_2SO_4, Na_2SO_4, Rb_2SO_4, and Cs_2SO_4; and 0.22 M salt and 6.33 M KOH for K_3PO_4); (4) other salts, including NaF, LiBr, and NaBr with salt concentration at 0.44 M and 6.33 M KOH each. All salt additives were purchased from Sigma-Aldrich Company (St. Louis, MO, USA). Except for the $SO_4{}^{2-}$ containing electrolytes, the total ion concentration of all electrolytes is kept at 13.54 M.

2.2. Measurements and Calculations

Electrochemical charge/discharge cycling tests were performed with an Arbin BT2000 battery tester (Arbin Instrument, College Station, TX, USA) at room temperature. For the test cells, the positive electrode was the sintered β-$Ni(OH)_2$ and the negative electrode was made from directly dry-compacting the MgNi-based $Mg_{52}Ni_{39}Co_3Mn_6$ alloy onto the Ni-expanded metal without using any binder. A hydrophilic nonwoven polyolefin was used as the separator. The cell was charged at 100 mA·g^{-1} for 5 h, and discharged first at 100 mA·g^{-1} to reach a cut-off voltage fixed at 0.9 V. This is followed by resting for 30 s and the voltage will be recovered to a higher value, and then discharged at 24 mA·g^{-1} to reach a cut-off voltage fixed at 0.9 V. The process is repeated by a final discharge at 8 mA·g^{-1} to 0.9 V. The conductivity of the electrolyte was analyzed by an YSI model 3200 conductivity meter with a probe produced by Traceable VWR Inc. (West Chester, PA, USA). Cell is considered failed when its capacity drops below 70% of the initial capacity before cycling.

Three parameters: conductivity, discharge capacity, and degradation, were measured and compared to those with traditional 30% KOH electrolyte. Each test was repeated three times. Degradation was determined as the percent capacity loss per cycle within the initial 10 cycles, as shown in the following:

$$Degradation\ \% = \frac{\dfrac{Cap_{high} - Cap_{low}}{(n_{high} - n_{low}) \times Cap_{high}}}{\dfrac{Cap_{0,high} - Cap_{0,low}}{(n_{0,high} - n_{0,low}) \times Cap_{0,high}}} \times 100\% \tag{3}$$

where:

Cap_{high} is the highest value of discharge capacity in the initial 10 cycles;

Cap_{low} is the lowest value of discharge capacity in the initial 10 cycles;

n_{high} is the cycle number of the highest discharge capacity in the initial 10 cycles;

n_{low} is the cycle number of the lowest discharge capacity in the initial 10 cycles; Subscript $_0$ is for 6.77 M (30%) KOH electrolyte.

3. Results and Discussion

3.1. Electrochemical Performance of MgNi-Based NiMH with 30% KOH Electrolyte

The cycle stability and capacity/degradation as functions of KOH concentration of negative electrode composed of $Mg_{52}Ni_{39}Co_3Mn_6$ MH alloy in a traditional 30% KOH electrolyte are shown in Figures 1 and 2, respectively. This composition is selected due to its weak resistance to corrosion in traditional alkaline electrolyte, which allows for a better comparison of the effectiveness of additives. Other MgNi MH alloy compositions with high capacities and better cycle stability in 30% KOH are also available but were not used in this work. Figure 1 depicts the discharge capacities of $Mg_{52}Ni_{39}Co_3Mn_6$ in the first 10 cycles at five different KOH concentrations: 6.77, 6.33, 5.88, 5.44, and 5.00 M. It shows a rapid decay of discharge capacity in the first four cycles for all KOH concentrations and the capacity loss is more than 80% after 10 cycles. This result is consistent with other reported MgNi-based MH alloys [9]. Figure 2 shows the normalized capacity and degradation as functions of KOH concentration. A low KOH concentration corresponds to a low degradation; however, a low discharge capacity as well. Therefore, it is necessary to balance the degradation and discharge capacity of negative electrode by a matrix of salt additives.

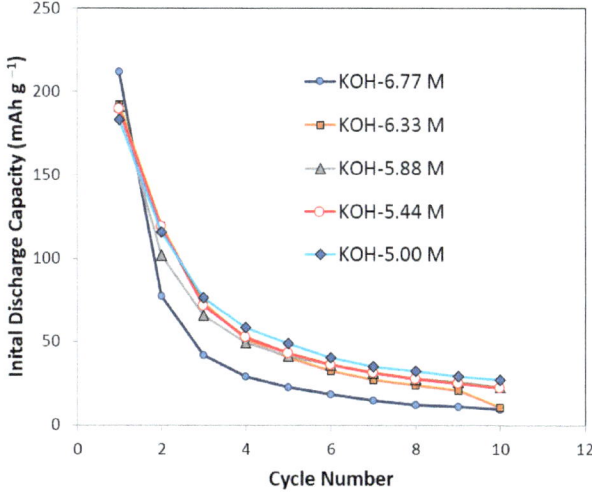

Figure 1. Initial discharge capacity of pure KOH solutions in initial 10 cycles. KOH concentrations are 6.77 M, 6.33 M, 5.88 M, 5.33 M, and 5.00 M separately. It shows a decay of discharge capacity for more than 80% after 10 cycles' running for all the pure KOH electrolytes.

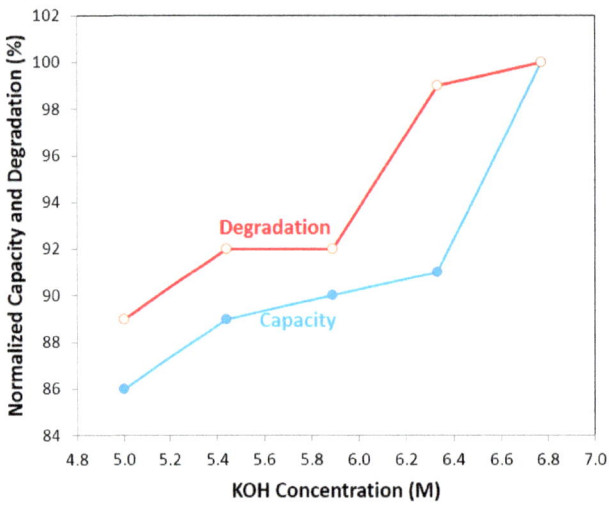

Figure 2. Capacity and degradation of pure KOH electrolytes *vs.* KOH concentration. It shows at low KOH concentration, electrolyte has low degradation but low discharge capacity; at high KOH concentration, electrolyte has a high discharge capacity but high degradation.

3.2. Electrochemical Performance of Pure Salt Solutions

The conductivity, capacity, and degradation characteristics of KOH and various pure salt solutions are shown in Table 1. For a 6.77 M KOH electrolyte, the cells were cycled for more than 20 cycles with the total charging/discharging time of more than 130 h before cell failure. However, most of the cells with pure salt solutions as electrolyte totally lost activity within the first few cycles. For example, the cells with a 1.77 M LiCl solution lost all electrochemical activity only after five cycles. The cells were disassembled after the test and both the MgNi-based MH alloy and nickel hydroxide electrodes were found to be heavily corroded. Table 1 also shows that the conductivities of pure salt solutions are less than 20% of the conductivity of 6.77 M KOH electrolytes. Low conductivities could cause a low electronic efficiency for Ni/MH battery systems. Most of the pure salt solutions have a higher degradation compared to KOH, as shown in Table 1. While some salt solutions such as KF and RbCl resulted in a lower degradation, their capacities decreased significantly, which can be attributed to low electronic efficiencies. Table 1 indicates that all of the pure salt solutions were more corrosive to the electrodes than the standard KOH solution and are therefore unsuitable as electrolytes for Ni/MH systems.

Table 1. Test time, test cycles, capacity, degradation and conductivity of pure 6.77 M KOH solution, pure 1.77 M salt solutions. The salts include KF, KCl, KBr, KI, LiCl, NaCl, KCl, RbCl, and CsCl. Note that the high corrosion could be observed in pure salt solutions.

Salt	Concentration in M	Test time before cell failed in h	Test cycles before cell failed	Capacity % [1]	Degradation % [2]	Conductivity % [3]
KOH	6.77	>130	>20	100.0	100.0	100.0
KF	1.77	105	20	12.3	49.4	12.9
KCl	1.77	73	8	79.7	136.5	14.6
KBr	1.77	56	4	167.4	309.9	12.9
KI	1.77	128	20	16.3	n/a	15.0
LiCl	1.77	56	5	84.7	185.4	10.3
NaCl	1.77	100	11	100.0	121.2	12.0
KCl	1.77	73	8	79.7	136.5	14.6
RbCl	1.77	70	9	82.8	88.8	15.0
CsCl	1.77	42	4	79.4	105.0	14.2

[1]: test error for capacity is 10%; [2]: test error for degradation is 10%; [3]: test error for conductivity is 5%.

3.3. Electrochemical Performance of Mixtures of Salt and KOH Solutions

3.3.1. Effect of Salt Additives on Conductivity

Traditionally, the concentration of KOH used in commercial Ni/MH batteries varies from 4.1 M to 8.5 M. Thus, the conductivities of KOH electrolytes with concentrations of 4.15, 4.65, 5.15, 5.67, 6.21, 6.77, 7.33, 7.91, and 8.52 M were tested. If we normalize the conductivities to the conductivity of 6.77 M KOH electrolyte as 100%, the conductivities of each of the electrolytes varied from 88% to 100%, as shown in Table 2.

Table 2. Conductivity of traditional KOH electrolyte with concentration varies from 20 wt% to 36 wt%. Note the conductivities vary from 88% to 100%.

Sample number	KOH concentration in M	Normalized conductivity in % [1]
1	4.15	88.2
2	4.64	92.6
3	5.15	95.6
4	5.67	98.5
5	6.21	98.5
6	6.77	100.0
7	7.33	95.6
8	7.91	92.6
9	8.52	89.7

[1]: test error for normalized conductivity is 5%.

Electrolytes containing 6.33 M KOH and 32 types of salt additives were prepared and their conductivities were tested and shown in Table 3. It was found that the conductivities of all the additive-containing electrolytes vary from 88% to 100%. It shows that the salt containing electrolytes can be as conductive as traditional KOH electrolytes in some cases (KIO_4 and KNO_3).

Table 3. Conductivity and concentration of additive containing electrolyte. Note the conductivities vary from 88% to 100%.

Salt	Concentration in M	Conductivity in % [1]	Salt	Concentration in M	Conductivity in % [1]	Salt	Concentration in M	Conductivity in % [1]
LiCl	0.44	97.4	Na_2WO_4	0.29	88.3	Li_2SO_4	0.005	90.7
NaCl	0.44	97.4	KIO_4	0.44	100.0	Na_2SO_4	0.005	91.4
KCl	0.44	97.0	$LiNO_3$	0.44	97.7	K_2SO_4	0.005	89.0
RbCl	0.44	96.3	$NaNO_3$	0.44	97.7	Rb_2SO_4	0.005	89.3
CsCl	0.44	96.3	$RbNO_3$	0.44	95.5	Cs_2SO_4	0.005	88.7
KF	0.44	96.8	KNO_3	0.44	100	$LiCHO_2$	0.44	92.4
KBr	0.44	98.3	K_2CO_3	0.29	99.6	$KCHO_2$	0.44	93.6
KI	0.44	96.1	Cs_2CO_3	0.29	97.0	$CsCHO_2$	0.44	94.1
LiBr	0.44	93.8	Rb_2CO_3	0.29	98.5	$KC_2H_3O_2$	0.44	89.4
NaBr	0.44	94.5	Na_2CO_3	0.29	89.0	$NaC_2H_3O_2$	0.44	90.5
NaF	0.44	96.9	K_3PO_4	0.22	99.2	-	-	-

[1]: test error for conductivity is 5%.

3.3.2. Effect of Salt Additives on Discharge Capacity and Degradation

Figure 3 shows the effects of the salt additive containing electrolytes on the discharge capacity and degradation. Data are normalized to the degradation and discharge capacity of 6.77 M KOH. Figure 3 can be divided into four different areas depending on the degradation and capacity.

Area 1: normalized degradation <100% and normalized capacity >100%. The electrolytes in this area are the most desirable since they offer lower degradations with improved capacities. It includes $NaC_2H_3O_2$ (0.44 M and 6.33 M KOH solution), $KC_2H_3O_2$ (0.44 M and 6.33 M KOH solution), K_2CO_3 (0.29 M and 6.33 M KOH, 0.59 M and 5.88 M KOH, 0.89 M and 5.44 M KOH solutions), Rb_2CO_3 (0.29 M and 6.33 M KOH solution), Cs_2CO_3 (0.29 M and 6.33 M KOH solution), K_3PO_4 (0.22 M and 6.33 M KOH, 0.44 M and 5.88 M KOH, 0.66 M and 5.44 M KOH, 0.88 M and 5 M KOH solutions), NaF (0.44 M and 6.33 M KOH solution), KF (0.44 M and 6.33 M KOH, 0.88 M and 5.88 M KOH solutions), NaBr (0.44 M and 6.33 M KOH solution), Rb_2SO_4 (0.005 M and 6.33 M KOH solution), Cs_2SO_4 (0.005 M and 6.33 M KOH solution), KBr (0.44 M and 6.33 M KOH solution), LiBr (0.44 M and 6.33 M KOH solution), and RbCl (0.88 M and 5.88 M KOH solution).

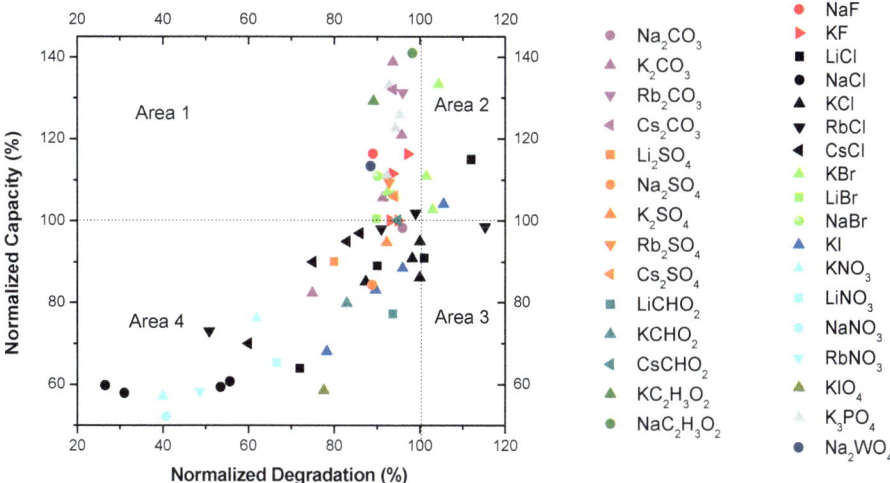

Figure 3. Screen test results of discharge capacity and degradation for the electrolytes containing various salt additives.

Area 2: normalized degradation >100% and normalized capacity >100%. This area contains electrolytes with chemical activity which increase both the capacity and the degradation of the MgNi-based MH alloy. KBr (0.88 M and 5.88 M KOH, 1.33 M and 5.44 M KOH, 1.77 M and 5 M KOH solutions), LiCl (0.44 M and 6.33 M KOH solution), and KI (0.44 M and 6.33 M KOH solution) are examples.

Area 3: normalized degradation >100% and normalized capacity <100%. It includes RbCl (1.33 M and 5.44 M KOH solution), LiCl (0.88 M and 5.88 M KOH solution), and KCl (0.44 M and 6.33 M KOH, 1.77 M and 5 M KOH solutions) and is the worst scenario with low capacities and high degradations.

Area 4: normalized degradation <100% and normalized capacity <100%. It contains additives with reduced chemical activity resulting in reduced capacity and degradation such as NaCl (0.44 M and 6.33 M KOH, 0.88 M and 5.88 M KOH, 1.33 M and 5.44 M KOH, 1.77 M and 5 M KOH solutions), KNO_3 (0.44 M and 6.33 M KOH, 0.88 M and 5.88 M KOH solutions), $RbNO_3$ (0.44 M and 6.33 M KOH solution), $LiNO_3$ (0.44 M and 6.33 M KOH solution), $NaNO_3$ (0.44 M and 6.33 M KOH solution), RbCl (0.44 M and 6.33 M KOH, 1.77 M and 5 M KOH solutions), LiCl (1.33 M and 5.44 M KOH, 1.77 M and 5 M KOH solutions), KI (0.88 M and 5.88 M KOH, 1.33 M and 5.44 M KOH, 1.77 M and 5 M KOH solutions), K_2CO_3 (1.77 M and 5 M KOH solution), $KCHO_2$ (0.44 M and 6.33 M KOH solution), $LiCHO_2$ (0.44 M and 6.33 M KOH solution), KCl (0.88 M and 5.88 M KOH, 1.33 M and 5.44 M KOH solutions), CsCl (0.44 M and 6.33 M KOH, 0.88 M and 5.88 M KOH, 1.33 M and 5.44 M KOH, 1.77 M and 5 M KOH solutions), K_2SO_4 (0.44 M and 6.33 M KOH solution), Na_2SO_4 (0.44 M and 6.33 M KOH solution), Na_2CO_3 (0.44 M and 6.33 M KOH solution), Li_2SO_4 (0.005 M and 6.33 M KOH solution), and $CsCHO_2$ (0.44 M and 6.33 M KOH solution).

In order to gain a better understanding of the effects of the anions and cations of additive salts on the normalized discharge capacity and electrode degradation, the data in Figure 3 was further analyzed based on the anions and cations.

3.3.3. Performance of Alkaline Cations Containing Electrolytes

Figures 4 and 5 illustrate the effects of alkaline cations (Li, Na, K, Rb, and Cs) on normalized degradation and discharge capacity, respectively, at 0.44 M of salt (with 6.33 M KOH), 0.88 M of salt (with 5.88 M KOH), 1.33 M of salt (with 5.44 M KOH), and 1.77 M of salt (with 5 M KOH). Figure 4 shows that degradation values vary from 26% to 115% with the addition of alkaline salts to traditional KOH electrolytes. CsCl and NaCl containing electrolytes decrease significantly in degradation compared to that in a pure KOH solution. At a low salt concentration (0.44 M salt with 6.33 M KOH), KCl containing electrolyte has a similar degradation value to KOH; LiCl containing electrolyte has an increased degradation, which is unfavorable; and RbCl electrolyte has a decreased degradation. At a high salt concentration (1.77 M salt with 5 M KOH), all electrolytes have a decreased degradation. Figure 5 illustrates that, except for NaCl containing electrolytes, the discharge capacity for electrolytes with a low salt concentration (0.44 M salt with 6.33 M KOH) is increased, while the discharge capacity for electrolytes with a high salt concentration (1.77 M salt with 5 M KOH) is greatly decreased. Figures 4 and 5 show that the concentration of alkaline salt in electrolytes has a significant influence on the degradation and discharge capacity. A linear correlation between degradation and reduction potential of alkaline ions was found for electrolytes containing a low concentration of salt, shown in Figure 6. It suggests that degradation is influenced by the reducibility of the salt additive. There is no apparent relationship between capacity and reduction potential of alkaline metals. There is also no apparent relationship between degradation and reduction potential in the electrolytes with high salt concentration. It is supposed that the high concentration of salt would cause the effect of electrostatic interaction among ions to become stronger and influence the degradation performance.

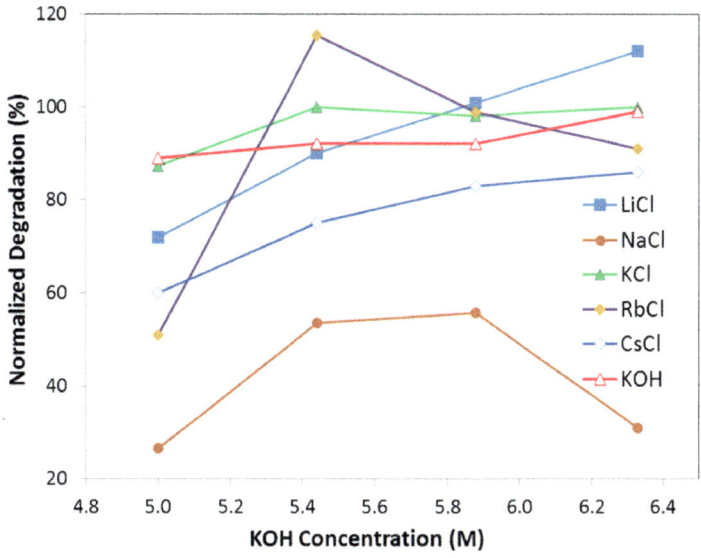

Figure 4. Degradation of electrolyte with 6.33 M KOH (and 0.44 M salts), 5.88 M KOH (and 0.88 M salts), 5.44 M KOH (and 1.33 M salts), 5.00 M KOH (and 1.77 M salts). Salts are LiCl, NaCl, KCl, RbCl, and CsCl, respectively.

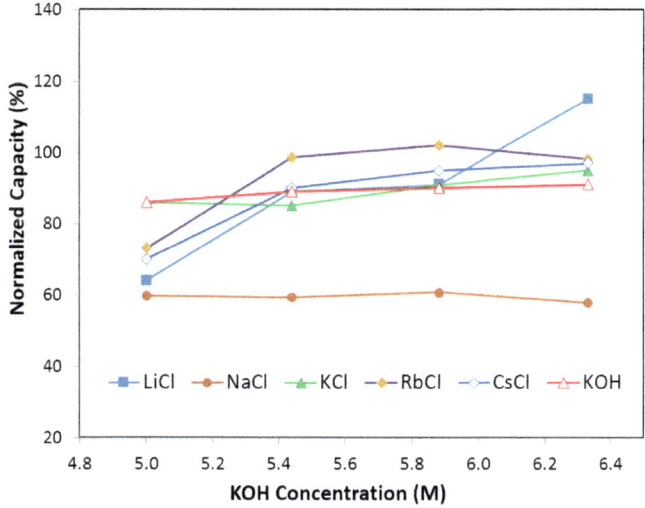

Figure 5. Capacity of electrolyte with 6.33 M KOH (and 0.44 M salts), 5.88 M KOH (and 0.88 M salts), 5.44 M KOH (and 1.33 M salts), 5.00 M KOH (and 1.77 M salts). Salts are LiCl, NaCl, KCl, RbCl, and CsCl, respectively.

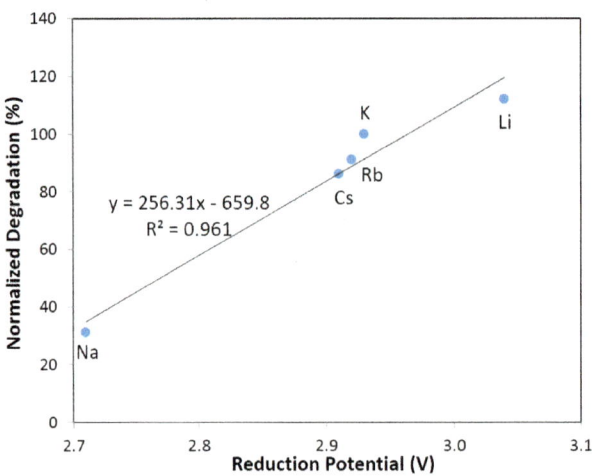

Figure 6. Linear fit of degradation with reduction potential of alkaline ions (Li$^+$, Na$^+$, K$^+$, Rb$^+$ and Cs$^+$).

3.3.4. Performance of Halogen Containing Electrolytes

Figures 7 and 8 illustrate the effects of anions (F, Cl, Br and I) on normalized degradation and discharge capacity, respectively, with 0.44 M of salt (with 6.33 M KOH), 0.88 M of salt (with 5.88 M KOH), 1.33 M of salt (with 5.44 M KOH), and 1.77 M of salt (with 5 M KOH). Figure 7 illustrates that the normalized degradation varies from 78% to 105%, which suggests that halogen ions' effects on degradation is not as strong as those in alkaline ions'. Figure 7 also indicates that at low (0.44 M) and high salt concentrations (1.77 M), the additives slightly increase the degradation value except for the electrolyte with 0.44 M KF and 6.33 M KOH. Figure 8 illustrates that discharge capacity varies from 68% to 133%. At a low salt concentration (0.44 M salt and 6.33 M KOH), all salts boost the discharge capacity, and at a high salt concentration (1.77 M salt and 5 M KOH), only KF and KBr additives enhance the discharge capacity. Figures 7 and 8 indicate that the concentration of salt additive in the electrolyte has a significant influence on the electrochemical performance. Figure 9 shows an apparent correlation between anion radii and the normalized degradation of low salt concentration electrolytes (0.44 M salt and 6.33 M KOH). This finding can also be explained by the energies of H–F, H–Cl, H–Br, and H–I bonds. In the traditional KOH electrolyte system, the OH group is an effective carrier for proton transfer between electrodes. In the salt containing electrolytes, both the OH groups and salt anions act as proton carriers. As F has a smaller radius than Cl, Br and I, its bond energy of H–F at 135 kcal·mol^{-1} is higher than that of O–H at 111 kcal·mol^{-1}, H–Cl at 103 kcal·mol^{-1}, H–Br at 87.5 kcal·mol^{-1}, and H–I at 71 kcal·mol^{-1}. The H–F bond is more stable than other hydrogen–halogen bonds during the charge/discharge process. Therefore, KF shows a decreased degradation compared to other potassium halogen salts. The bond energies of H–Cl, H–Br, and H–I are less than O–H, which implies that in KCl, KBr, and KI containing electrolytes, the OH⁻ group is preferred as the proton accepter instead of Cl⁻, Br⁻ and I⁻. Thus, KCl, KBr, and KI containing electrolytes (with a low salt concentration) show a higher degradation than the traditional KOH electrolyte.

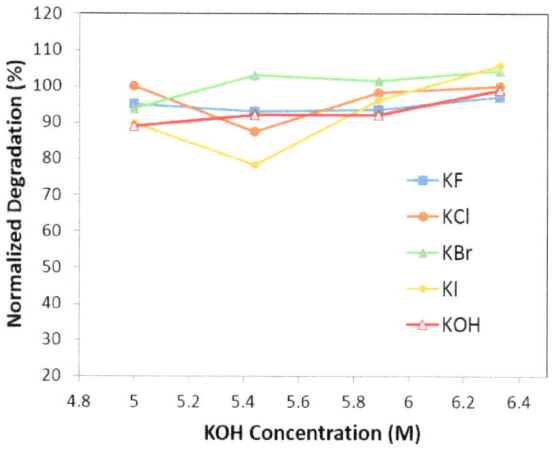

Figure 7. Degradation of electrolyte with 6.33 M KOH (and 0.44 M salts), 5.88 M KOH (and 0.88 M salts), 5.44 M KOH (and 1.33 M salts), 5.00 M KOH (and 1.77 M salts). Salts are KF, KCl, KBr, and KI, respectively.

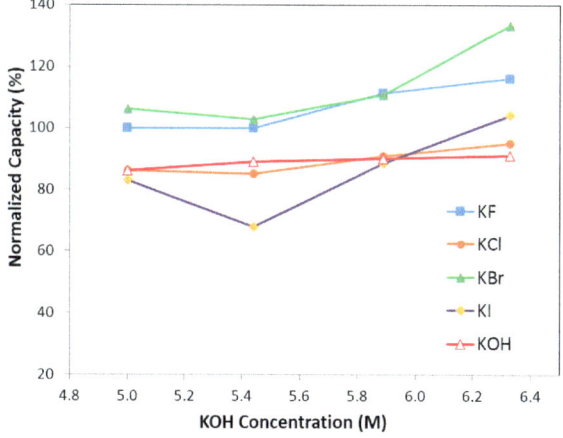

Figure 8. Capacity of electrolyte with 6.33 M KOH (and 0.44 M salts), 5.88 M KOH (and 0.88 M salts), 5.44 M KOH (and 1.33 M salts), 5.00 M KOH (and 1.77 M salts). Salts are KF, KCl, KBr, and KI, respectively.

415

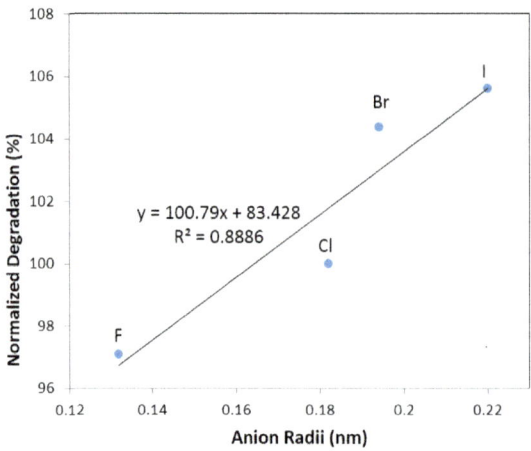

Figure 9. Linear fit of degradation with halogen anion radii (F^-, Cl^-, Br^-, and I^-).

3.3.5. Performance of Oxyacid Containing Electrolytes

Effects of low oxyacid salt concentration (and 6.33 M KOH) electrolytes on normalized discharge capacity and degradation are shown in Figure 10. Unlike halogen salt additives and alkaline salt additives, all oxyacid salt additives satisfy the desired properties of degradation <100%, and distribute only in Areas 1 and 4. The electrolytes in Area 1 (*i.e.*, normalized degradation <100% and normalized discharge capacity >100%) include: $NaC_2H_3O_2$ (0.44 M additives and 6.33 M KOH), $KC_2H_3O_2$ (0.44 M additives and 6.33 M KOH), K_2CO_3 (0.29 M additives and 6.33 M KOH), Rb_2CO_3 (0.29 M additives and 6.33 M KOH), Cs_2CO_3 (0.29 M additives and 6.33 M KOH), K_3PO_4 (0.22 M additives and 6.33 M KOH), Na_2WO_4 (0.29 M additives and 6.33 M KOH), Rb_2SO_4 (0.005 M additives and 6.33 M KOH), Cs_2SO_4 (0.005 M additives and 6.33 M KOH). The electrolytes in Area 4 (*i.e.*, normalized degradation <100% and normalized discharge capacity <100%) include: Rb_2SO_4 (0.005 M additives and 6.33 M KOH), Li_2SO_4 (0.005 M additives and 6.33 M KOH), Rb_2SO_4 (0.005 M additives and 6.33 M KOH), $CsCHO_2$ (0.44 M additives and 6.33 M KOH), KNO_3 (0.44 M additives and 6.33 M KOH), $RbNO_3$ (0.44 M additives and 6.33 M KOH), $LiNO_3$ (0.44 M additives and 6.33 M KOH), $NaNO_3$ (0.44 M additives and 6.33 M KOH), K_2CO_3 (1.77 M additives and 5 M KOH), $KCHO_2$ (0.44 M additives and 6.33 M KOH), $LiCHO_2$ (0.44 M additives and 6.33 M KOH), K_2SO_4 (0.005 M additives and 6.33 M KOH), Na_2SO_4 (0.005 M additives and 6.33 M KOH), and Na_2CO_3 (0.44 M additives and 6.33 M KOH).

The trend of discharge capacity is $NO_3^- < IO_4^- < CHO_2^- < SO_4^{2-} < WO_4^{2-} < CO_3^{2-} < PO_4^{2-} < C_2H_3O_2^-$. As shown in references [20,21], all of the above oxyacid anions, except for NO_3^-, could react with Mg or Ni ions on the surface of the metal hydride to form a solid layer that covers the metal hydride particles,

416

which could protect them from electrolyte corrosion. The solid layer covering the metal hydride particles had a significant influence on discharge capacity and degradation. Figure 11 illustrates the charge voltage curves of the first cycle of cells with electrolytes containing K_2CO_3, K_3PO_4, $CsCO_3$, Rb_2CO_3, Na_2WO_4, $NaC_2H_3O_2$, $KC_2H_3O_2$, Rb_2SO_4, and Cs_2SO_4, which demonstrated normalized degradation < 100% and discharge capacity > 100%. The full charge voltage of these oxyacid salt electrolytes is slightly lower than that of traditional KOH electrolyte, which means the resistance of these additive containing electrolytes is lower than that of traditional KOH electrolytes. A lower voltage also implies less corrosion to the electrodes. The charge voltage curves of other salt additive containing electrolytes (in Areas 2, 3, and 4 in Figure 3) were also studied. The values of the first cycle full charge voltage were generally higher than that of KOH electrolyte (data not shown). Further research on the formation mechanism of solid layer and the effects of solid layer on corrosion resistance and electrochemical performance will be performed.

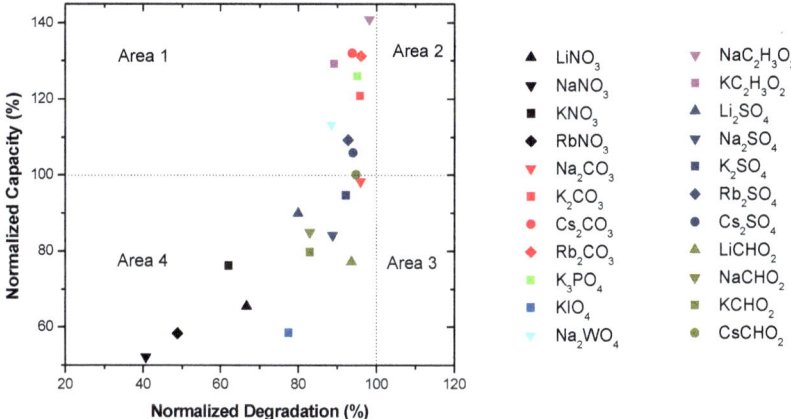

Figure 10. Discharge capacity and degradation for electrolytes containing oxyacid salts, including $NaC_2H_3O_2$, $KC_2H_3O_2$, K_2CO_3, Rb_2CO_3, Cs_2CO_3, K_3PO_4, Rb_2SO_4, Cs_2SO_4, Rb_2SO_4, Li_2SO_4, Rb_2SO_4, $CsCHO_2$, KNO_3, $RbNO_3$, $LiNO_3$, $NaNO_3$, K_2CO_3, $KCHO_2$, $LiCHO_2$, K_2SO_4, Na_2SO_4, and Na_2CO_3.

Figure 12 shows the effects of K_2CO_3 and K_3PO_4 containing electrolytes on the normalized discharge capacity and degradation. It was found that the concentration of oxyacid salts, similar to the concentration of halogen salts and alkaline salts, have a significant influence on electrochemical performance. There was a maximum value in discharge capacity for both 0.44 M K_2CO_3 and K_3PO_4 containing KOH electrolytes. On the other hand, the normalized degradation seemed to be stable for the concentration range of K_2CO_3 and K_3PO_4 studied, except for 0.89 M K_2CO_3 containing electrolytes which shows a lower normalized degradation.

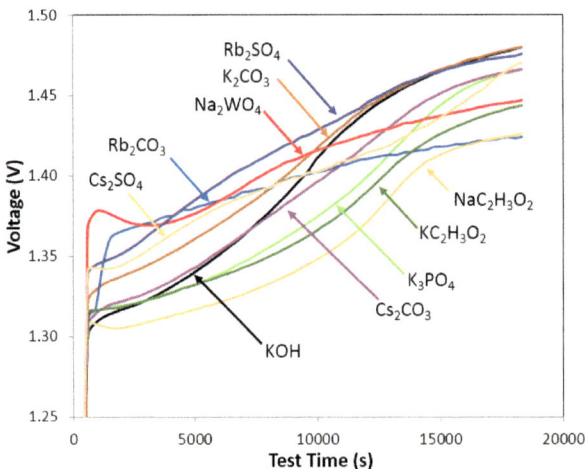

Figure 11. Charge voltage curves for first circle of electrolytes containing 0.29 M K_2CO_3 and 6.33 M KOH, 0.29 M Cs_2CO_3 and 6.33 M KOH, 0.29 M Rb_2CO_3 and 6.33 M KOH, 0.29 M Na_2WO_4 and 6.33 M KOH, 0.44 M $NaC_2H_3O_2$ and 6.33 M KOH, 0.44 M $KC_2H_3O_2$ and 6.33 M KOH, 0.005 M Rb_2SO_4 and 6.33 M KOH, 0.005 M Cs_2SO_4 and 6.33 M KOH, 0.22 M K_3PO_4, and 6.33 M KOH. The final charge voltage of these oxyacid salt containing electrolytes is lower than traditional KOH electrolyte.

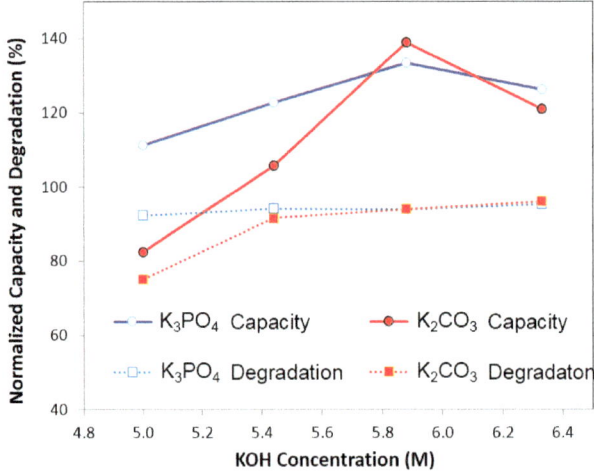

Figure 12. Degradation and capacity of the electrolytes containing K_2CO_3 and K_3PO_4 separately. The solutions are: 6.33 M KOH and 0.29 M K_2CO_3, 5.88 M KOH and 0.59 M K_2CO_3, 5.44 M KOH and 0.89 M K_2CO_3, 5.00 M KOH and 1.18 M K_2CO_3; 6.33 M KOH and 0.22 M K_3PO_4, 5.88 M KOH and 0.44 M K_3PO_4, 5.44 M KOH and 0.66 M K_3PO_4, 5.00 M KOH, and 0.88 M K_3PO_4.

418

Since the electrolyte with 0.44 M Cs_2CO_3 and 6.33 M KOH shows good performance on MgNi metal alloy, it was selected to test two kinds of BCC/C14 type metal alloy ($V_{44.0}Ti_{15.6}Ni_{18.5}Cr_{11.6}Mn_{6.9}Hf_{2.1}Co_{1.4}Al_{0.4}$ and $V_{44.0}Ti_{15.6}Ni_{18.5}Cr_{11.6}Mn_{6.9}Zr_{2.1}Co_{1.4}Al_{0.4}$). The BCC/C14 metal alloy has high hydrogen storage ability and high theoretical discharge capacity, and is also considered as a strong candidate material for metal alloy electrode in NiMH batteries as the MgNi metal alloy [7,8]. The theoretical discharge capacity of MgNi alloy ($Mg_{52}Ni_{39}Co_3Mn_6$) was reported as 999 mAh$\cdot$g^{-1} [7,10]. However, Figure 13 shows that the maximum value of practical discharge capacity in traditional 6.77 M KOH electrolyte over $Mg_{52}Ni_{39}Co_3Mn_6$ is only 212 mAh$\cdot$g^{-1}. When $Mg_{52}Ni_{39}Co_3Mn_6$ alloy electrode is tested in Cs_2CO_3 containing electrolyte, the maximum discharge capacity is 300 mAh$\cdot$g^{-1}, which suggests less corrosive in Cs_2CO_3 containing electrolyte. The theoretical discharge capacity of these two BCC/C14 alloys ($V_{44.0}Ti_{15.6}Ni_{18.5}Cr_{11.6}Mn_{6.9}Hf_{2.1}Co_{1.4}Al_{0.4}$ and $V_{44.0}Ti_{15.6}Ni_{18.5}Cr_{11.6}Mn_{6.9}Zr_{2.1}Co_{1.4}Al_{0.4}$) was reported as 767 mAh$\cdot$g^{-1} [13]. In traditional 6.77 M KOH, the maximum capacity for $V_{44.0}Ti_{15.6}Ni_{18.5}Cr_{11.6}$ $Mn_{6.9}$ $Hf_{2.1}Co_{1.4}Al_{0.4}$ is only 348 mAh$\cdot$g^{-1} and capacity decay after 86 cycles' running is 78% and for $V_{44.0}Ti_{15.6}Ni_{18.5}Cr_{11.6}Mn_{6.9}Zr_{2.1}Co_{1.4}Al_{0.4}$, the maximum is 366 mAh$\cdot$g^{-1} and capacity decay is 52%. In electrolyte of 0.44 M Cs_2CO_3 and 6.33 M KOH, the maximum capacity of $V_{44.0}Ti_{15.6}Ni_{18.5}Cr_{11.6}Mn_{6.9}Hf_{2.1}Co_{1.4}Al_{0.4}$ and $V_{44.0}Ti_{15.6}Ni_{18.5}Cr_{11.6}Mn_{6.9}Zr_{2.1}Co_{1.4}Al_{0.4}$ is 378 mAh$\cdot$g^{-1}. In addition, capacity decay of 100 cycles' running for $V_{44.0}Ti_{15.6}Ni_{18.5}Cr_{11.6}Mn_{6.9}Hf_{2.1}Co_{1.4}Al_{0.4}$ is 33%, and for $V_{44.0}Ti_{15.6}Ni_{18.5}Cr_{11.6}Mn_{6.9}Zr_{2.1}Co_{1.4}Al_{0.4}$ is 30%. As stated above, traditional 6.77 M KOH is quite corrosive to metal alloys and leads to a very low practical discharge capacity. The Cs_2CO_3 salt additive slightly improves discharge capacity and tremendously decreases capacity decay on electrodes of $Mg_{52}Ni_{39}Co_3Mn_6$, $V_{44.0}Ti_{15.6}Ni_{18.5}Cr_{11.6}Mn_{6.9}Hf_{2.1}Co_{1.4}Al_{0.4}$ and $V_{44.0}Ti_{15.6}Ni_{18.5}Cr_{11.6}Mn_{6.9}Zr_{2.1}Co_{1.4}Al_{0.4}$ as shown in Figure 13.

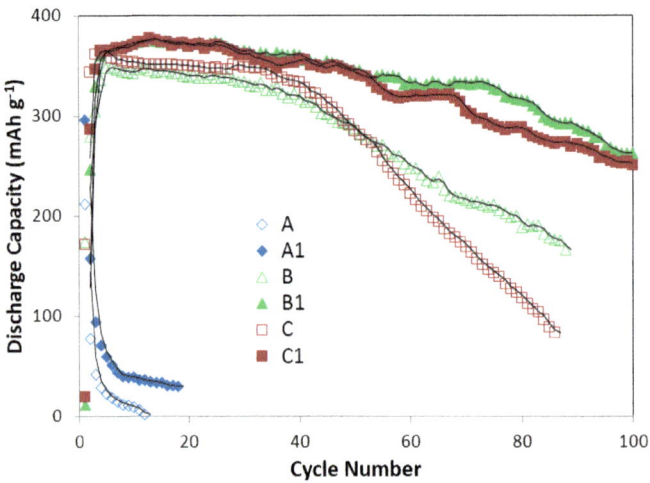

Figure 13. Discharge capacity of $Mg_{52}Ni_{39}Co_3Mn_6$, $V_{44.0}Ti_{15.6}Ni_{18.5}Cr_{11.6}Mn_{6.9}$ $Hf_{2.1}Co_{1.4}Al_{0.4}$ and $V_{44.0}Ti_{15.6}Ni_{18.5}Cr_{11.6}Mn_{6.9}Zr_{2.1}Co_{1.4}Al_{0.4}$ in electrolyte of 6.77 M KOH and electrolyte of 0.44 M Cs_2CO_3 and 6.33 M KOH. A: $Mg_{52}Ni_{39}$ Co_3Mn_6 in 6.77 M KOH; B: $V_{44.0}Ti_{15.6}Ni_{18.5}Cr_{11.6}Mn_{6.9}Hf_{2.1}Co_{1.4}Al_{0.4}$ in 6.77 M KOH; C: $V_{44.0}Ti_{15.6}Ni_{18.5}Cr_{11.6}Mn_{6.9}Zr_{2.1}Co_{1.4}Al_{0.4}$ in 6.77 M KOH; A1: $Mg_{52}Ni_{39}Co_3Mn_6$ in 0.44 M Cs_2CO_3 and 6.33 M KOH; B1: $V_{44.0}Ti_{15.6}Ni_{18.5}Cr_{11.6}$ $Mn_{6.9}Hf_{2.1}Co_{1.4}Al_{0.4}$ in 0.44 M Cs_2CO_3 and 6.33 M KOH; C1: $V_{44.0}Ti_{15.6}Ni_{18.5}$ $Cr_{11.6}Mn_{6.9}Zr_{2.1}Co_{1.4}Al_{0.4}$ in 0.44 M Cs_2CO_3 and 6.33 M KOH. It shows Cs_2CO_3 salt additive slightly improves discharge capacity and tremendously decreases capacity decay.

3.4. Mechanistic Analysis for Using Additive Containing Electrolytes in Novel Ni/MH Battery Systems

3.4.1. Synergistic Effect between KOH and Salt Additives

Comparing additive containing electrolytes to the pure KOH and pure salt electrolytes, significant changes in degradation and discharge capacity were observed. For salts in Area 1 of Figure 3 (*i.e.*, degradation <100% and capacity >100%), a synergistic effect between the salt additives and KOH electrolyte is presumed to explain the improvement in electro-performance. The synergistic effect can be explained by the following inferences: (1) a solid film forms on the surface of both electrodes. As shown in Figure 14, all the anions in Area 1 were able to react with Mg^{2+} and Ni^{2+} on the surface of MgNi metal hydride particles and Ni^{2+} on the surface of β-Ni(OH)$_2$ particles to form a solid deposit [20,21]. In this research, discharge capacity is closely related to the amount of MgNi metal hydride in negative electrodes. This solid film over the MgNi metal hydride particles could physically protect the electrode from electrolyte corrosion; (2) H-transfer is promoted in aqueous electrolyte. The salt additives' effects were considered from the aspects of the anions and cations. The anions of salts in Area 1 could provide more active sites for H adsorption and transfer [22,23]. The cations, together with their anions, could change the chemical environment of the OH group in the electrolyte, and promote H-transfer [22]. Lower degradation is closely related to the improvement in the amount of H carriers and the H-transfer process [7].

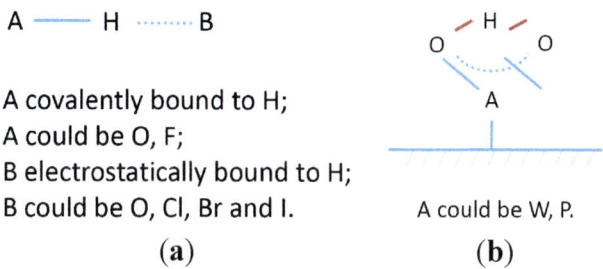

A covalently bound to H;
A could be O, F;
B electrostatically bound to H;
B could be O, Cl, Br and I.

(a)

A could be W, P.

(b)

Figure 14. (**a**) Hydrogen bond types in additive containing H bond type 1; and (**b**) H bond type 2.

3.4.2. Active Sites for Proton Transfer and H Bond Types

For traditional KOH electrolytes, the oxygen atom in the OH^- group acts as the basic site to accept H from the metal hydride and transfers H between electrodes [22,24]. As Basak [22] stated, H is covalently bonded to O with bond energy around 111 kcal·mol^{-1} and simultaneously has a weaker electrostatic interaction with another O with bond energy around 6–10 kcal·mol^{-1}. This H bond type in KOH electrolytes is called H bond type 1 (shown in Figure 14a).

In this research, the anions of the additive salts, together with OH^- from KOH, act as proton carriers [22–25]. For halogen containing salts, F^-, Cl^-, Br^- and I^- are basic sites for H adoption [24,25]. H bond type in halogen containing electrolytes can be presumed by analyzing the bond energy [24]. The H–F bond energy is 135 kcal·mol^{-1}, which is higher than O–H at 111 kcal·mol^{-1}. Thus, F should be covalently bonded to H (as the A atom in Figure 14a), and OH should be electrostatically bonded to H (as the B atom in Figure 14b). The bond energies for H–Cl, H–Br, and H–I are 103, 87.5, and 71 kcal·mol^{-1}, respectively, which are lower than the O–H bond energy. Thus, Cl, Br, or I should be electrostatically bonded to H (as the B atom in Figure 14a), and OH covalently bonded to H (as the A atom in Figure 14a) [24].

For most oxyacid salts, O is the basic site for H adsorption [22,24]. Analysis of the bonds containing O is important for understanding the H bond type in oxyacid salts. For CO_3^{2-}, $C_2H_3O_2^-$, and CHO_2^- salts, there are two types of O containing bonds: C–O, and C=O. H is covalently bonded to C–O, and electrostatically bonded to C=O [24]. WO_4^{2-} and PO_4^{3-} are special cases regarding H bond type. There are two identical W–O bonds and two identical W=O bonds in WO_4^{2-} and three identical P–O bonds in PO_4^{3-}. As Basak [22] stated, because of the identicality of the bonds, H is equally shared between two adjacent O by covalent interaction. In this paper, it is called H bond type 2, and its structure is as shown in Figure 14b.

3.4.3. Proton Transfer Mechanisms: Vehicle Mechanism and Structure-Diffusion Mechanism

Generally speaking, there are two kinds of proton transfer mechanisms in organic/inorganic solution systems: vehicle mechanism and structure-diffusion mechanism [22–25]. Vehicle mechanism is reported in traditional KOH electrolytes and is described as the transport of protonated species as whole molecules (for example: H_2O, H_3O^+) through aqueous media [24]. In this research, proton transfer in liquid electrolyte is considered as vehicle mechanism. In addition, proton transfer through the surface solid of the electrodes, is considered as a structure-diffusion mechanism [22]. The proton hops from site to site along a hydrogen-bonded network over the surface. Diffusion of protonic charges or protonic defects occurs through the formation and breaking of the hydrogen bonds. Both H bond types could contribute to structure-diffusion mechanism.

The H-transfer process in additive containing electrolytes during discharge is briefly illustrated in Figure 15. The solid film over the MgNi alloy and $Ni(OH)_2$ particles is formed during the charge process or the first few seconds of the discharge process. In the solid film covering the MgNi alloy, H is covalently or electrostatically bonded to other heteroatoms through bond type 1 or 2. H delivered through the solid film to the interface of the liquid electrolyte is through site-to-site hopping way (*i.e.*, structure-diffusion mechanism). The H is the delivered in the electrolyte through vehicle mechanism with H bond type 1. The H bond type and H-transfer mechanism for the solid film covering the $Ni(OH)_2$ are similar to that of the MgNi alloy. Figure 15 also illustrates the trend of the H concentration differential from the MgNi alloy to $Ni(OH)_2$ during the discharge process. The H concentration is the highest at the metal hydride-solid film interface and the H concentration is the lowest at the $Ni(OH)_2$-solid film interface.

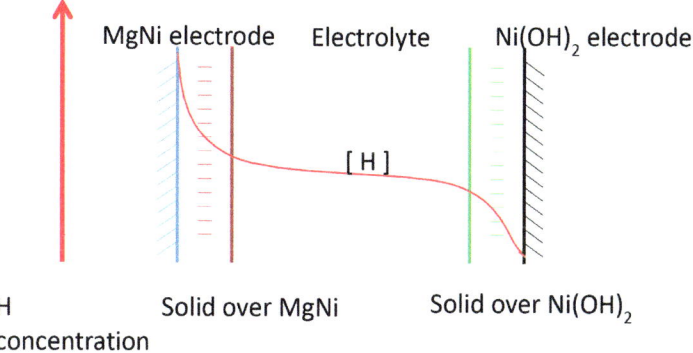

Figure 15. Model of H-transfer from the surface of MgNi electrode to Ni(OH)$_2$ electrode through the novel additive containing electrolyte.

3.4.4. Factors that Influence Electrochemical Performance of the Novel Additive-Containing Electrolytes

As stated before, the synergy between the additives and KOH is influenced by the nature of the salt additives. For additive containing electrolytes, both hydroxide ions and salt anions are basic sites and act as proton acceptors and carriers [22]. Thus, the parameters that can influence the formation of hydrogen bonds can influence the electro-performance. The parameters are summarized as follows:

(1) The nature of the basic sites. The stronger the basicity, the better the performance [23,24]. In this research, CO_3^{2-}, PO_4^{3-}, and F^- are strong Lewis bases and do well in decreasing degradation and boosting discharge capacity. NO_3^- and I^- are weak Lewis bases and have poor hydrogen ion adsorption. In Figure 3, NO_3^- and I^- containing electrolytes show poor performance.

(2) The amount of basic sites. Comparing CHO_2^- and $C_2H_3O_2^-$ containing electrolytes, it was found that $C_2H_3O_2^-$ shows better performance than CHO_2^-. For CHO_2^-, the Lewis basic site that acts as the proton carrier is the oxygen in the C–O bond. For $C_2H_3O_2^-$, the Lewis basic sites include the oxygen in the C–O bond and the C from the CH_3^- group. $C_2H_3O_2^-$ contains more basic sites than CHO_2^- [22].

(3) The stability of the salt additives. During the charge/discharge processes, the salt additives are in highly reductive/oxidative environments. The additives with low stability, such as KIO_4, perform poorly.

(4) The salt solubility in KOH solution. It has been found that all the SO_4^{2-} containing salts have low solubility in KOH solutions [20,21]. The maximum solubility of K_2SO_4 in KOH is only around 0.093% (0.005 M). In this research,

the SO_4^{2-} containing electrolyte performs poorly compared to other oxyacid salt electrolytes, and similarly to pure KOH solutions.

(5) The ionic charge of the additive ions. Anions with high ionic charge, such as CO_3^{2-}, PO_4^{3-}, WO_4^{2-}, perform well.

4. Conclusions

A systematic study of 32 salt additives in a traditional KOH electrolyte was performed on an MgNi-based Ni/MH battery systems. 12 salt additives were discovered to have efficiently decreased the corrosion of traditional KOH electrolyte: $NaC_2H_3O_2$, $KC_2H_3O_2$, K_2CO_3, Rb_2CO_3, Cs_2CO_3, K_3PO_4, Na_2WO_4, Rb_2SO_4, Cs_2SO_4, NaF, KF, and KBr.

The effects of the cations (Li^+, Na^+, K^+, Rb^+, and Cs^+) and anions (F^-, Cl^-, Br^-, I^-, CO_3^{2-}, NO_3^-, PO_4^{3-}, SO_4^{2-}, IO_4^-, WO_4^{2-}, $C_2H_3O_2^-$, and CHO_2^-) of the additives on charge/discharge performance were discussed. It was found that the oxidation potential of alkaline ions and the radii of halogen ions are linearly correlated with electrode degradation. For oxyacid salt electrolytes, a synergistic effect between the salt and KOH was observed: (1) a solid film forms on the MgNi particle surface, which physically prevents corrosion from KOH electrolyte and boosts discharge capacity; (2) H-transfer is promoted in aqueous electrolyte, which decreases degradation.

A brief discuss for H-transfer in the additive containing electrolyte system was included based on the screen test data. H bond type in the novel electrolyte system is not similar to traditional KOH electrolyte. Both type 1 and 2 hydrogen bonds exist in the additives containing electrolytes. Because of the formation of the solid film on the MgNi particles, the H-delivery mechanism in the novel electrolyte is also different from traditional KOH electrolytes. Both vehicle mechanism and structure-diffusion mechanism occurred in the novel electrolyte system. As the synergistic effect is determined by the nature of the salt additive, the factors that could influence the electro-performance were discussed based on the salt additive's chemical properties including the salt's anion's basicity, amount of base sites, stability during charging/discharging, solubility in KOH solution, and ionic charge.

Acknowledgments: This work is financially supported by Advanced Research Projects Agency-Energy (ARPA-E) under the robust affordable next generation EV-storage (RANGE) program (DE-AR0000386).

Conflicts of Interest: The authors declare no conflict of interest.

References

1. Linden, D.; Reddy, T.B. *Handbook of Batteries*; McGraw-Hill: New York, NY, USA, 2002.

2. Fetcenko, M.A.; Ovshinsky, S.R.; Reichman, B.; Young, K.; Fierro, C.; Koch, J.; Zallen, A.; Mays, W.; Ouchi, T. Recent advances in NiMH battery technology. *J. Power Sources* **2007**, *165*, 544–551.

3. Iwakura, C.; Fukuda, K.; Senoh, H.; Inoue, H.; Matsuoka, M.; Yamamoto, Y. Electrochemical characterization of $MmNi_{4.0-x}Mn_{0.75}Al_{0.25}Co_x$ electrodes as a function of cobalt content. *Electrochim. Acta* **1998**, *43*, 2041–2046.

4. Geng, M.; Feng, F.; Sebastian, P.J.; Matchett, A.J.; Northwood, D.O. Charge transfer and mass transfer reactions in the metal hydride electrode. *Int. J. Hydrog. Energy* **2001**, *26*, 165–169.

5. Niessen, R.A.H.; Notten, R.H.L. Electrochemical hydrogen storage characteristics of thin film MgX (X = Sc, Ti, V, Cr) compounds. *Electrochem. Solid State Lett.* **2005**, *8*, 534–538.

6. Notten, P.H.L.; Ouwerkerk, M.; Hal, H.V.; Beelen, D.; Keur, W.; Zhou, J.; Fei, H. High energy density strategies: From hydride-forming materials research to battery integration. *J. Power Sources* **2004**, *129*, 45–54.

7. Young, K.; Nei, J. The current status of hydrogen storage alloy development for electrochemical applications. *Materials* **2013**, *6*, 4574–4608.

8. Anik, M.; Özdemir, G.; Küçükdeveci, N. Electrochemical hydrogen storage characteristics of Mg–Pd–Ni ternary alloys. *Int. J. Hydrog. Energy* **2011**, *36*, 6744–6750.

9. Rongeat, C.; Grosjean, M.H.; Ruggeri, S.; Dehmas, M.; Bourlot, S.; Marcotte, S.; Roué, L. Evaluation of different approaches for improving the cycle life of MgNi-based electrodes for Ni-MH batteries. *J. Power Sources* **2006**, *158*, 747–753.

10. Liu, J. Effect of Ti-La substitution on electrochemical properties of amorphous MgNi-based secondary hydride electrodes. *J. Tianjin Norm. Univ.* **2011**, *313*, 67–70.

11. Anik, M.; Özdemir, G.; Küçükdeveci, N.; Baksan, B. Effect of Al, B, Ti and Zr additive elements on the electrochemical hydrogen storage performance of MgNi alloy. *Int. J. Hydrog. Energy* **2011**, *36*, 1568–1577.

12. Kalinichenka, S.; Röntzsch, L.; Riedl, T.; Gemming, T.; Weißgärber, T.; Kieback, B. Microstructure and hydrogen storage properties of melt-spun Mg–Cu–Ni–Y alloys. *Int. J. Hydrog. Energy* **2011**, *36*, 1592–1600.

13. Ma, F.; Liu, X.; Yan, S.; Ao, D.; Ren, Y. Research progress of influence factors on electrochemical performance of Mg-base hydrogen storage alloys. *Met. Func. Mater.* **2011**, *34*, 63–64.

14. Anik, M. Effect of titanium additive element on the discharging behavior of MgNi alloy electrode. *Int. J. Hydrog. Energy* **2011**, *36*, 15075–15080.

15. Danczuk, M.; Nunes, C.V., Jr.; Araki, K.; Anaissi, F.J.J. Influence of Alkaline Cation on the Electrochemical Behavior of Stabilized Alpha Nickel Hydroxide. *Solid State Electron.* **2014**, *18*, 2279–2287.

16. Karwowska, M.; Jaron, T.; Fijalkowski, K.J.; Leszczynski, P.J.; Rogulski, Z.; Czerwinski, A. Influence of electrolyte composition and temperature on behaviour of AB_5 hydrogen storage alloy used as negative electrode in Ni–MH batteries. *J. Power Sources* **2014**, *263*, 304–309.

17. Shangguan, E.; Li, J.; Chang, Z.; Tang, H.; Li, B.; Yuan, X.; Wang, H. Sodium tungstate as electrolyte additive to improve high-temperature performance of nickel–metal hydride batteries. *Int. J. Hydrog. Energy* **2013**, *38*, 5133–5138.

18. Vaidyanathan, H.; Robbins, K.; Rao, G.M. Effect of KOH concentration and anions on the performance of an Ni-H_2 battery positive plate. *J. Power Sources* **1996**, *63*, 7–13.

19. Ovshinsky, S.R.; Fetcenko, M.A.; Reichman, B.; Young, K.; Chao, B.; Im, J. Electrochemical Hydrogen Storage Alloys and Batteries Fabricated from Mg containing Base Alloys. U.S. Patent 5,616,432, 1 April 1997.

20. Schlapbach, L.; Züttel, A. Hydrogen-storage materials for mobile applications. *Nature* **2001**, *414*, 353–358.

21. Zaluska, A.; Zaluski, L.; Ström-Olsen, J.O. Structure, catalysis and atomic reactions on the nano-scale: A systematic approach to metal hydrides for hydrogen storage. *Appl. Phys. A* **2001**, *72*, 157–165.

22. Basak, D. *Proton Transfer in Organic Scaffolds*; University of Massachusetts Amherst: Amherst, MA, USA, 2012.

23. Gilli, P.; Bertolasi, V.; Ferretti, V.; Gilli, G. Covalent nature of the strong homonuclear hydrogen bond—Study of the O–H—O system by crystal structure correlation methods. *J. Am. Chem. Soc.* **1994**, *116*, 909–915.

24. Grabowski, J.S. What is the covalency of hydrogen bonding? *Chem. Rev.* **2011**, *111*, 2597–2625.

25. Wang, B.C.; Chang, J.C.; Jiang, J.C.; Lin, S.H. Ab initio study of the ammoniated ammonium ions $NH_4^+(NH_3)_{0-6}$. *Chem. Phys.* **2002**, *276*, 93–106.

Chapter 4:
Analytic Methodology

Microstructures of the Activated Si-Containing AB₂ Metal Hydride Alloy Surface by Transmission Electron Microscope

Kwo-hsiung Young, Benjamin Chao and Jean Nei

Abstract: The surface microstructure of an activated Si-containing AB₂ metal hydride (MH) alloy was investigated by transmission electron microscopy (TEM) and X-ray energy dispersive spectroscopy (EDS). Regions of the main AB₂ and the secondary TiNi (B2 structure) phases directly underneath the surface Zr oxide/hydroxide layers are considered electrochemically inactive. The surface of AB₂ is covered, on the atomic scale, by sheets of Ni_2O_3 with direct access to electrolyte and voids, without the buffer oxide commonly seen in Si-free AB₂ alloys. This clean oxide/bulk metal alloy interface is believed to be the main source of the improvements in the low-temperature performance of Si-containing AB₂ alloys. Sporadic metallic-Ni clusters can be found in the surface Ni_2O_3 region. However, the density of these clusters is much lower than the Ni-inclusions found in most typical metal hydride surface oxides. A high density of nano-sized metallic Ni-inclusions (1–3 nm) is found in regions associated with the TiNi secondary phase, *i.e.*, in the surface oxide layer and in the grain boundary, which can also contribute to enhancement of the electrochemical performance.

Reprinted from *Batteries*. Cite as: Young, K.-h.; Chao, B.; Nei, J. Microstructures of the Activated Si-Containing AB₂ Metal Hydride Alloy Surface by Transmission Electron Microscope. *Batteries* **2016**, *2*, 4.

1. Introduction

Sales of hybrid electrical vehicles (HEV), which are mainly based on the nickel/metal hydride (Ni/MH) battery technology, passed 10 million in July 2015 [1]. With a fast charge/discharge capability [2,3], a long cycle stability [4], high efficiency under a large charge current [5], superb abuse tolerance [6,7], and a wide operation temperature range [8,9], Ni/MH batteries are expected to power HEV for another decade. Therefore, efforts to improve the performance of Ni/MH batteries have never stopped. Ni/MH batteries have a catalytic oxide layer on the anode surface that facilitates the electrochemical reaction with the electrolyte, which is similar to the solid-electrolyte-interface (SEI) formation on the anode surface of a Li-ion battery [10–16]. In a conventional AB₂ metal hydride (MH) alloy, the catalytic surface oxide is composed of (1) a buffer oxide layer (approximately 100 nm and

50 nm for AB_2 and AB_5, respectively) on top of the metal bulk, followed by (2) a porous oxide layer (approximately 200 nm and 100 nm for AB_2 and AB_5, respectively), and finally (3) a surface oxide layer that is electrochemical inactive [17]. The porous surface oxide and metallic inclusions (mainly Ni) of several common AB_2 and AB_5 MH alloys have been studied by transmission electron microscopy (TEM) [13–26], electron energy loss spectroscopy (EELS) [17,27], X-ray diffraction (XRD) [28], X-ray photoemission spectroscopy (XPS) [29], and magnetic susceptibility measurements [26,30–33]. The general conclusion is that the metallic Ni inclusions embedded in the supportive oxides are the key catalyst for the electrochemical reaction in Ni/MH batteries.

Si-addition in a typical Laves phase based MH alloy demonstrated a large decrease in charge-transfer resistance at both room temperature and $-40\ °C$ [34], which was attributed to increased catalytic ability in the main C14 phase. Although both the abundances of the C15 and TiNi (B2 structure) secondary phases are higher in the Si-containing AB_2 alloy, their content remains relatively small ($\leqslant 5$ wt%). AC impedance results showed that both the surface area and surface catalytic ability increases with the incorporation of Si. The increase in the former was attributed to the higher solubility of Si oxide, whereas the origin of the increase in the latter remains unclear [34]. A similar enhancement in electrochemical reactivity with Si and AB_2 MH alloys was also previously reported [35], but the mechanism of the improvement is also not available. TEM has been used extensively to study the multi-phase nature of AB_2 MH alloys [36–40], as well as the surface oxides of conventional AB_2 and AB_5 MH alloys [17,26]. Therefore, in this study, we applied TEM techniques to reveal the detailed microstructure of the activated Si-containing AB_2 MH alloy surface.

2. Experimental Section

Arc melting of the designed alloy samples was performed with a non-consumable tungsten electrode and water-cooled copper tray, under a continuous argon flow. A piece of sacrificial titanium underwent several melting-cooling cycles before each run to reduce residual systemic oxygen levels. Each 12-g ingot sample was re-melted and turned over several times to ensure a uniform chemical composition. A JEOL-JSM6320F scanning electron microscope (SEM, JEOL, Tokyo, Japan) with X-ray energy dispersive spectroscopy (EDS) capabilities was used to study the phase distribution and composition of the as-cast ingots [34]. Ingots were hydrided, de-hydrided, crushed, and ground to powder and then sieved through a 200-mesh sieve. The final powder was etched in $100\ °C$ 30% KOH for 4 h (h) to emulate a lightly cycled electrode [17,26,30]. TEM examination was performed with a CM200/FEG microscope (Philips, Amsterdam, The Netherlands). The operating voltage was 200 keV. Magnetic susceptibility was measured using a Digital Measurement Systems Model 880 vibrating sample magnetometer (MicroSense, Lowell, MA, USA).

TEM sample preparation included the following steps:

(1) Mixing of the powders with M-Bond 610 glue, followed by placement of the mixture into a 3 mm diameter copper pipe. This was then cured for 8 h at 150 °C;

(2) Slicing the copper pipe containing the glued powders into 0.5 mm thick disks using a low-speed diamond saw;

(3) Polishing of the 3-mm-diameter disk down to approximately 100 μm using SiC media from 320 grit down to 1200 grit, then dimpling grinding;

(4) Finishing of thinning with 5 kV ion milling (on a liquid N_2 cooled stage);

(5) Finally, insertion of the sample into the microscope to examine for electron beam transparency areas. If none are detected, additional ion milling time, followed by TEM examination, is required. This step was repeated as many times as necessary to provide high quality thin areas for detailed TEM studies.

3. Results and Discussion

3.1. Summary of Alloy Properties

While the surface microstructure of the Si-free AB_2 MH alloy has been fully investigated in our previous paper [17], a Si4 alloy with a designed composition of $Ti_{12}Zr_{21.5}V_{10}Cr_{7.5}Mn_{8.1}Co_{8.0}Ni_{28.2}Si_{4.0}Sn_{0.3}Al_{0.4}$ from a previous report on the effects of Si-incorporation in the Laves phase based AB_2 MH alloy [34] was selected for this TEM/EDS study. Based on the full-phase analysis of the XRD data, this alloy (Si4) contained a major C14 phase (92.0 wt%), a C15 secondary phase (5.0 wt%), and a TiNi—like B2 secondary phase (3.0 wt%). EDS analysis showed that the Si-contents in C14, C15, and B2 (TiNi) phase were 3.2–5.5 at%, 1.6–1.9 at%, and 0.2 at%, respectively. Compared to the Si-free base alloy Si0, the Si4 alloy has a larger unit cell, higher C15 and B2 secondary phase abundances, a lower pressure-concentration-temperature (PCT) plateau pressure, a larger PCT hysteresis, a smaller gaseous phase and electrochemical storage capacities, a higher hydrogen diffusion coefficient, a higher surface exchange current, a lower charge transfer resistance and a higher surface reactive area as measured at −40 °C, and a higher surface catalytic ability [34].

3.2. Scanning Electron Microscope and X-Ray Energy Dispersive Spectroscopy Analyses on the Activated Alloy Surface

Both the pristine and activated alloy powder surfaces were examined by SEM and the resulting secondary electron images are shown in Figure 1a,b, respectively. Some fine particles seen on the surface of the pristine alloy are believed to be dust from hydrogenation and grinding operations. Interestingly, additional fine needle structures are observed on the activated alloy surface (Figure 1b). TEM studies indicated that these needles are mainly Zr oxide/hydroxide. The composition of the activated alloy surface was measured by EDS, although only semi-quantitative, and is in agreement with that of design (Table 1). The similar EDS results between design and etched samples indicate that the alloy surface affected by the activation procedure is much shallower than the X-ray escape depth of the EDS technique (typically 1–2 μm).

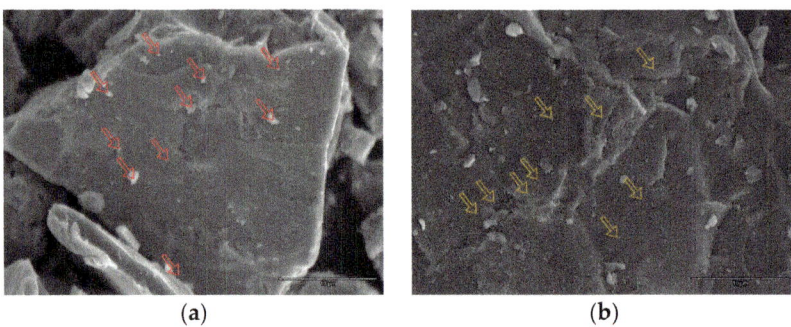

(a)   (b)

Figure 1. Scanning electron microscope (SEM) micrographs showing the surfaces of (**a**) pristine and (**b**) activated Si4 alloy. While dust particles on the pristine surface (a few examples indicated by red arrows) are from the pulverization/grinding operation, smaller sized precipitations of Zr oxide/hydroxide can be seen on the activated surface (representative examples indicated by yellow arrow).

Table 1. Design and surface composition of activated alloy Si4 determined by SEM-X-ray energy dispersive spectroscopy (EDS). All numbers are in at%.

Origin	Al	Si	Ti	V	Cr	Mn	Co	Ni	Zr	Sn
Design	0.4	4.0	12.0	10.0	7.5	8.1	8.0	28.2	21.5	0.3
Surface	0.7	4.2	11.2	10.1	7.5	8.4	8.2	29.3	20.3	0.1

3.3. Regions under the Zr Oxide/Hydroxide Layer

Figure 2a is a TEM bright field (BF) micrograph from a region containing two typical AB_2 crystals under the Zr oxide/hydroxide layer. Because of variations in sample thicknesses and weak EDS spectrum signal, resulting from a very small volume, composition calculations based on the TEM/EDS are much more challenging (if even possible) than those in SEM/EDS. Therefore, in this paper, semi-quantitative analyses were only carried out on several representative EDS spectra and most of the EDS spectra were qualitatively compared without delving into inaccessible quantitative details. EDS spectra recorded from the five numbered red box regions in Figure 2a are displayed in Figure 2b–f. The EDS spectra of Regions 1 (Figure 2b) and 3 (Figure 2d), which exhibit all expected elements of design with no detectable O signal, are assigned to the metallic AB_2 bulk phase. The EDS spectrum of Region 2 showed a similar metal content to the AB_2 bulk (Figure 2b,d) with an additional miniscule amount of oxygen signal. The observed Ca signal is due to artifacts from contamination during sample preparation. Figure 2e represents an EDS spectrum recorded from Region 4, the surface of the AB_2 phase, specifically heavily oxidized Zr, Ti, and Ni, some K from the etching solution, and very small amounts of V, Cr, and Mn. The spectrum is qualitatively similar to the buffer oxide in the previously reported activated Si-free AB_2 MH alloy surface [17]. Metallic Ni-inclusions were not visible in this region. The EDS spectrum (Figure 2f) of the top most surface represents mostly Zr and O, with smaller amounts of Ti, K, and Ni. The O/Zr atomic ratio in this region was estimated to be approximately 2.7, which is between the ratios of ZrO_2 (2.0) and $ZrO_2 \cdot 2H_2O$ or $Zr(OH)_4$ (4.0), and is closer to the former. This is the reason why this region was designated as a Zr oxide/hydroxide. With a surface oxide/hydroxide covering the entire surface and no metallic Ni embedded in the supporting oxide, the AB_2 region directly beneath the surface oxide is considered electrochemically inactive and cannot be the source of the improved low-temperature performance.

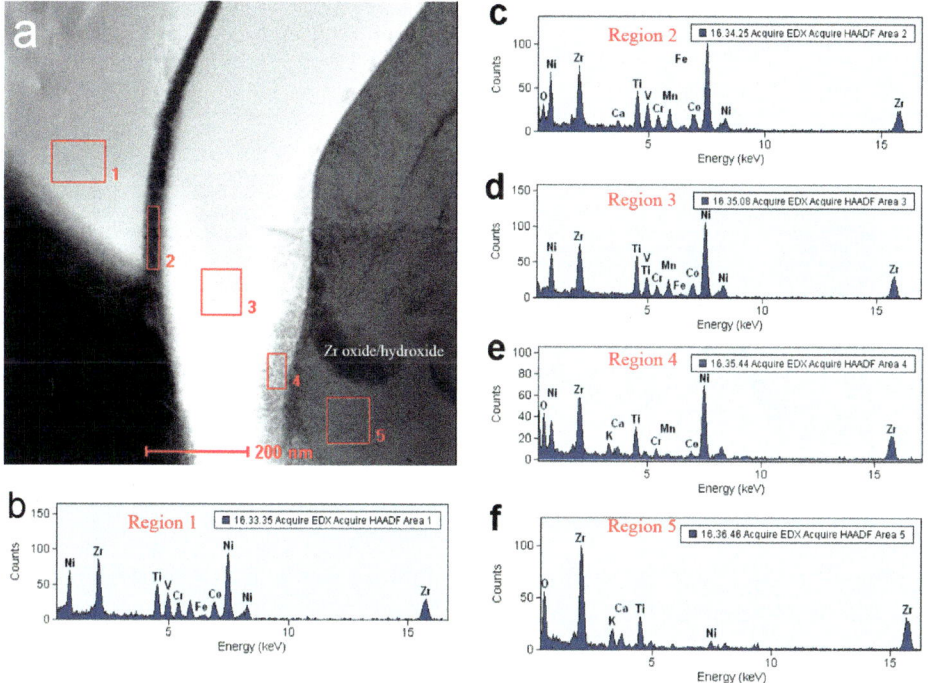

Figure 2. (**a**) A transmission electron microscopy (TEM) bright field (BF) micrograph showing the region representative of the AB_2 phase underneath the surface Zr oxide/hydroxide and (**b**)–(**f**) the corresponding EDS spectra from Areas 1–5.

A TEM BF micrograph taken from a different region shows three bright contrast areas in Figure 3a. Two are marked with A and one is marked with B. In addition to a dark contrast area on the left side of B, a crack clearly separates Areas A and B. Small cracks and fissures are normally generated during the hydrogenation and grinding processes. The EDS spectra from Areas A and B are shown in Figure 3b,c, respectively. Both spectra show no trace of O and a full compositional analysis (in at%) are presented in Table 2. According to the composition measurements, Areas A and B were identified as AB_2 and B2 phases, respectively. In summary: the AB_2 phase contains all the elements in the composition; the B2 phase consists mainly of Ti, Zr, Ni, and Co. Low solubilities of V, Cr, and Mn in the B2 phase are in agreement with a previous report [33] and almost all the Si resides in the AB_2 phase.

Table 2. EDS results from two phases identified by TEM BF micrography in Figure 3a. All numbers are in at%. ND denotes not detectable.

Location	Al	Si	Ti	V	Cr	Mn	Co	Ni	Zr
Area A (AB_2 phase)	1.25	5.23	9.77	13.06	12.47	9.86	8.60	23.59	16.12
Area B (B2 phase)	ND	0.19	24.08	1.08	ND	0.78	6.93	47.40	19.52

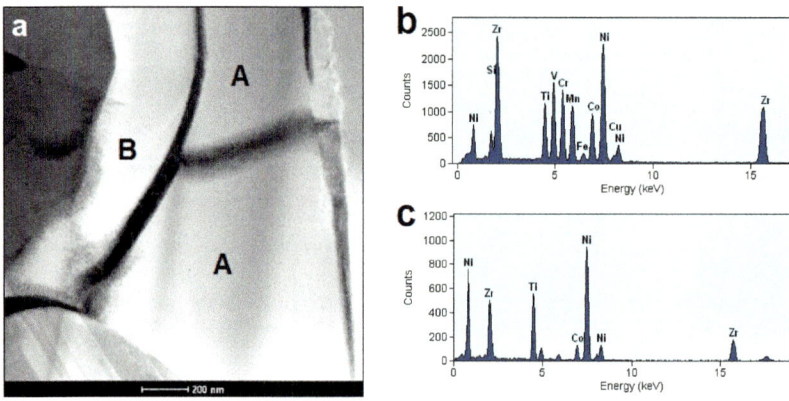

Figure 3. (a) A TEM BF micrograph and the corresponding EDS spectra from (b) Area A (AB_2 phase) and (c) Area B (B2 phase).

Further EDS spectra from the B2 phase and the dark contrast area will be discussed below in Figure 4. Figure 4a is a high-magnification TEM BF micrograph focusing on the bottom left corner of the B2 crystal and its surrounding areas. EDS examination was carried out in the following six areas: Region 1 represents the surface oxide/hydroxide, Region 2 shows the B2 bulk crystal (the same B2 crystal shown in Figure 3a), and Regions 3–6 are in the regions between B2 and surface oxide/hydroxide and the boundary areas surrounding the B2 phase. The corresponding EDS spectra are shown in Figure 4b–g. The EDS spectrum of Region 1 (Figure 4b) is very similar to that shown in Figure 2f representing a typical Zr oxide/hydroxide feature. The O/Zr ratios in the Zr oxide/hydroxide surface for both the AB_2 and B2 phases are very similar. Meanwhile, the EDS spectrum of Region 2—very similar to the one displayed in Figure 3c—is considered to represent the bulk of metallic B2 phase, and those from Regions 3–5 are hypothesized to be buffer oxide with high Ni-content. Region 6, a region next to an activated grain boundary, also shows a similar spectrum, but with an even higher Ni-content. No metallic Ni-inclusions were found in the region of the B2 phase under the surface oxide. This is similar to the AB_2 phase under the Zr oxide/hydroxide. Judging from the high packing density of the surface Zr oxide/hydroxide, the region of the B2 phase under the same oxide is not electrochemically active.

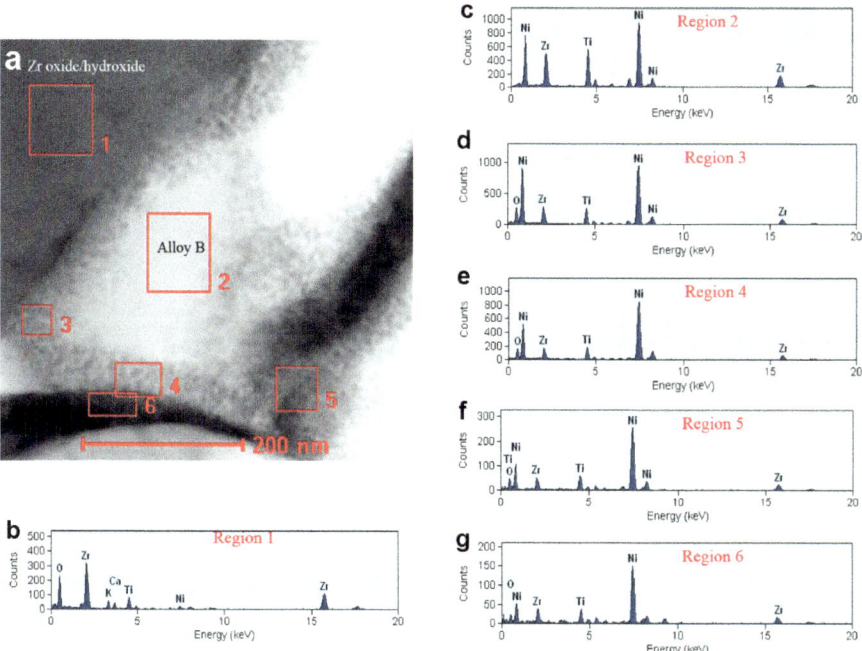

Figure 4. (**a**) A higher magnification TEM BF micrograph of Figure 3a, showing the B2 phase underneath the surface Zr oxide/hydroxide and (**b**)–(**g**) the corresponding EDS spectra from Areas 1–6.

3.4. Electrochemically Active Surfaces on the AB_2 Phase

Figure 5a is a TEM BF micrograph showing open space areas filled with fibrous-like microstructures near the middle portion of the view. A red rectangular region marked at the bottom edge of an open area (surface oxide) was closely examined by TEM BF micrography in higher magnification, shown in Figure 5b. The EDS spectra recorded from Areas 1–3 are present in Figure 5c–e, respectively. Area 1 contains mainly Ni oxide/hydroxide with small amounts of Cr, Zr, Ti, Mn, Si, and K (from the etching solution). Both Areas 2 and 3 are composed of partial oxides from all the metallic components and the oxygen signal intensity in Area 2 is higher than that in Area 3. In both areas, while the overall compositions are similar to that of the AB_2 phase (Figure 2b), the Ni-contents are significantly higher. This area can be considered as the surface oxide region of AB_2 phase, under the Ni oxide/hydroxide layer.

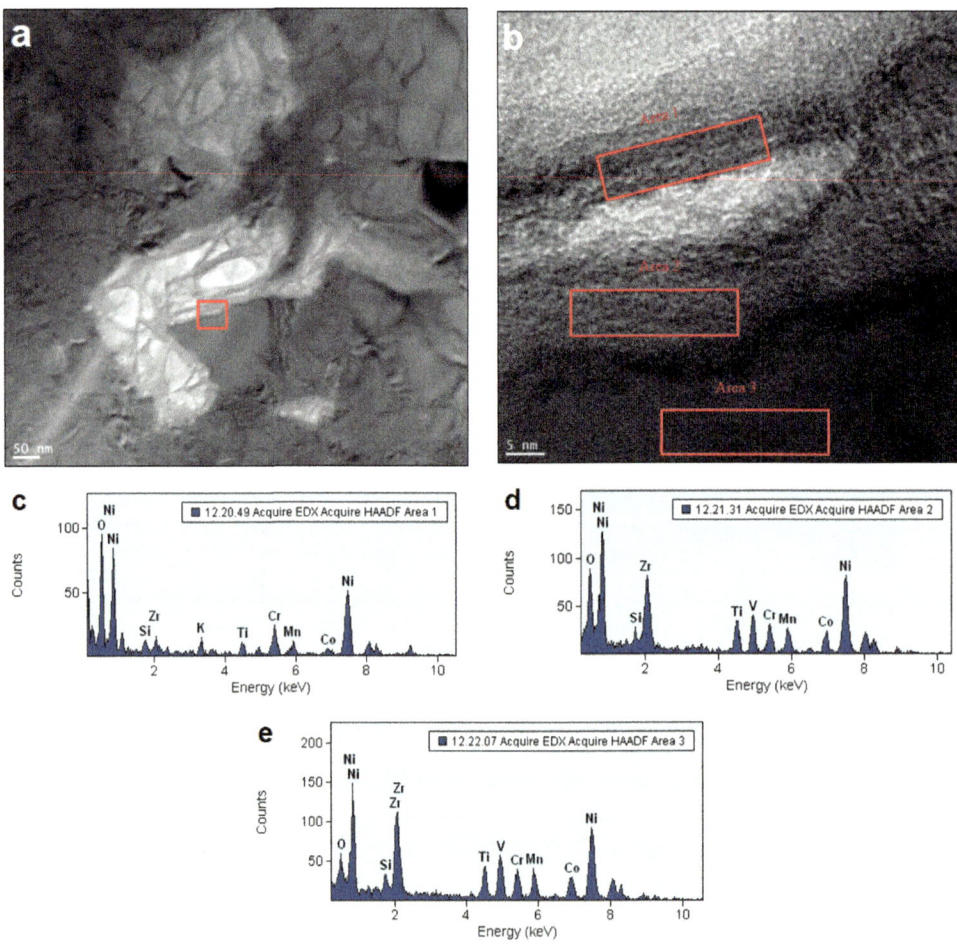

Figure 5. (**a**) A TEM BF micrograph showing the interface between oxide/AB_2 phases; (**b**) a higher magnification image from the red box region shown in (**a**); and (**c**)–(**e**) the associative EDS spectra from areas identified in (**b**).

436

The fine microstructure of the Ni oxide/hydroxide-AB$_2$ phase interface was further investigated with high resolution (HR) TEM. A representative HR-TEM micrograph in Figure 6a shows an oxide region sandwiched between two AB$_2$ metal alloy grains. The central portion is filled with a high density of fibrous-like features, similar to the oxide formation found in La-addition AB$_2$ metal alloys [26]. A higher magnification micrograph of the same region clearly shows layered structures associated with oxide formation (Figure 6b). Finally, the atomic fringes of the AB$_2$ crystal are clearly visible in Figure 6c. These atomic fringes have an average inter-planar distance of 0.24 nm, which is close to the (200) plane of a C14 structure. The interface between the oxide and AB$_2$ metal alloy bulk crystal (green line) is very clean and free of any amorphous phase, as is the buffer oxide layer reported earlier [17]. A selective area electron diffraction pattern (SAD) in the oxide region is shown in Figure 6f. The lattice spacing, calibrated with a Pd foil, is 0.325 nm, which is very close to the (101) inter-planar distance of 0.323 nm for a Ni$_2$O$_3$ crystal [41]. The compositions of the oxide and AB$_2$ phases were studied by EDS and the resulting spectra are shown in Figure 6e,f. The oxide area has an O/Ni ratio slightly higher than the stoichiometric 1.5 in Ni$_2$O$_3$, contains some Cr, possibly traces of Co, Ti, and no detectable Si. Therefore, this region is mainly Ni oxide with a small amount of Ni(OH)$_2$. The composition of the AB$_2$ alloy crystal is very close to the one found in the bulk (Figure 3b) with a small amount of oxygen, which suggests that a small portion of the AB$_2$ phase was oxidized. In summary, the region of AB$_2$ directly underneath the Ni$_2$O$_3$ sheets, which contain voids, is slightly oxidized and has direct access to the electrolyte and, therefore, is considered electrochemically active.

The surface Ni oxide region examined more thoroughly by TEM is shown in a regular TEM BF micrograph (Figure 7a), where some bright particles, marked by red arrows 1, 2, and 4, are present in the surface Ni oxide/hydroxide region. The composition of the four red arrow points shown in Figure 7a were examined by EDS and the resulting spectra are displayed in Figure 7b–e. Point 1 indicates a mixture of Ni and AB$_2$ phase with some degree of oxidation (Figure 7b). Points 2 and 4 are pure metallic Ni clusters. Point 3 exhibits a typical AB$_2$ composition. The Ni metallic inclusions are known to facilitate electrochemical reactions when they are imbedded in the surface porous oxide [30]. We believe the metallic Ni-clusters (10–20 nm) found here may not play an important role in the electrochemical reaction because of their relatively low density, when compared to the typical Ni inclusion in the supportive oxide on the conventional MH alloys.

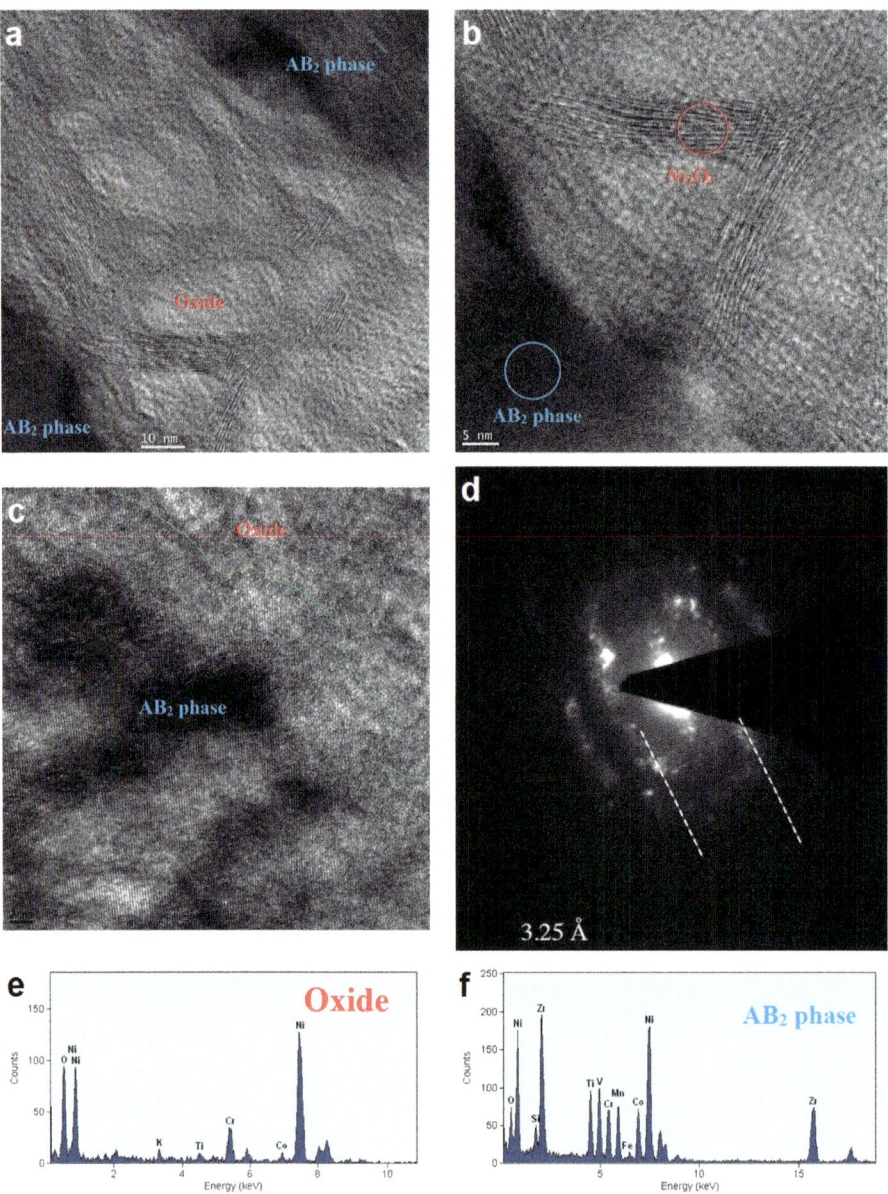

Figure 6. (**a**) A TEM BF micrograph showing the oxide between two AB_2 crystals; (**b**) a higher magnification image of one interface; (**c**) a clean interface image showing lattice fringes of 0.24 nm inter-planar spacing for the AB_2 crystal (200) plane and the oxide; (**d**) a SAD pattern from the oxide region showing an inter-planar distance of 0.325 nm and EDS spectra recorded from (**e**) the corresponding oxide and (**f**) AB_2 phase areas.

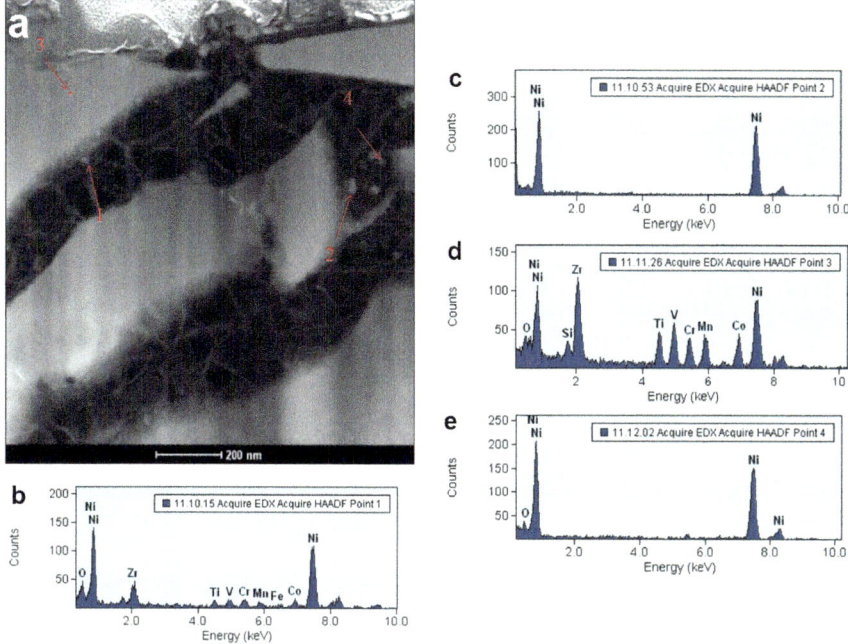

Figure 7. (**a**) A TEM BF micrograph showing surface oxide and imbedded Ni-clusters and (**b**)–(**e**) the corresponding EDS spectra from the areas identified in (**a**).

The Ni-inclusions were studied further using TEM micrography combined with EDS mappings. There are three clusters, each with a diameter of approximately 10 nm, denoted by red arrows in Figure 8a. The corresponding Ni- and O-mappings in the same area indicate that these three particles are made of Ni without any O and are therefore metallic Ni. Besides the Ni-inclusions in the region, the Ni-mapping coincides with O-mapping very well in the remaining areas, which signify that the surface of AB_2 phase in this region is covered mostly by Ni oxide/hydroxide.

Magnetic susceptibility measurements were used to quantify the total amount of metallic Ni on the surface, using the saturated magnetic susceptibility (M_S) and the average magnetic domain size, which is reversely proportional to the magnetic field strength at one-half of the M_S value ($H_{1/2}$). Details of the measurement protocol can be found in our earlier publications [30,33]. According to the results shown in Table 3, the Si-containing Si5 AB_2 MH alloy has a slightly lower total amount of metallic Ni, with a similar average Ni cluster size to those in the Si-free Si0 alloy. Therefore, the enhancement in the low-temperature performance with the addition of Si in the formula cannot be linked to metallic Ni-clusters. When compared to other AB_5 and A_2B_7 in Table 3, the AB_2 alloys have an insufficient amount of surface metallic Ni-that too small in size, which, therefore, increases in the surface catalytic ability by other means is extremely critical for improving the electrochemical performance of AB$_2$ MH alloys.

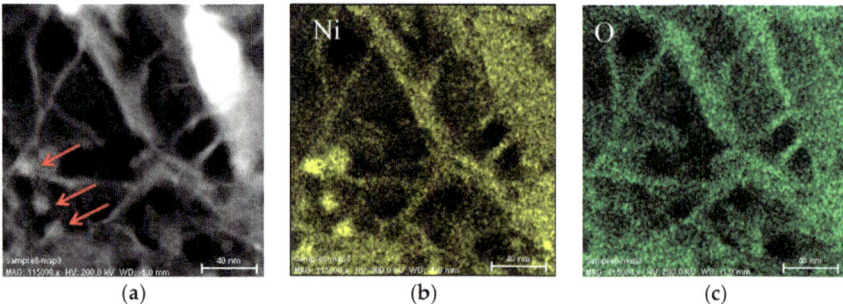

(a)   (b)   (c)

Figure 8. (a) A TEM BF micrograph showing three metallic Ni-clusters, indicated by arrows, in the oxide on the AB$_2$ phase surface, and EDS mappings of (b) Ni and (c) O.

Table 3. Comparison of saturated magnetic susceptibility (M_S) and applied magnetic field, corresponding to half of M_S ($H_{1/2}$), for Si-free Si0 and Si-containing Si5 AB$_2$, AB$_5$ and some A$_2$B$_7$ MH alloys. Mm stands for Ce-free misch metal.

Properties	Si5-AB$_2$	Si0-AB$_2$	AB$_5$	La-A$_2$B$_7$	Nd-A$_2$B$_7$	Mm-A$_2$B$_7$
M_S in emu·g^{-1}	0.0324	0.0372	0.434	0.369	0.679	0.314
$H_{1/2}$ in kOe	0.451	0.493	0.173	0.125	0.102	0.128
References	This work	[33]	[30]	[30]	Nd-AB5	[31]

3.5. Electrochemically Active Surfaces on the B2 Phase

The B2 phase, only 3.0 wt% of Si4, can also be an electrochemical active component in the AB$_2$ MH alloy [42,43]. A TEM BF micrograph in Figure 9a shows that the same surface of B2 phase was not entirely covered by the thick Zr oxide/hydroxide, as addressed in Section 3.3. The EDS spectra taken from the areas marked as 1, 2, 3, and 4 are shown in Figure 9b–e, respectively. The spectrum of Area 1 shows a typical element-distribution of a B2 phase (refer to Figure 3c). Area 2 contains mainly Ni and O with an O/Ni ratio of roughly 0.7, which suggest a mixture of metallic Ni and Ni oxide. The composition in Area 3 approaches a fully oxidized Ni oxide, while Area 4 may be a mix of metal (Zr, Ti, and Ni) oxides. From the structure, Region 2 is proposed to be the catalyst for the electrochemical reaction.

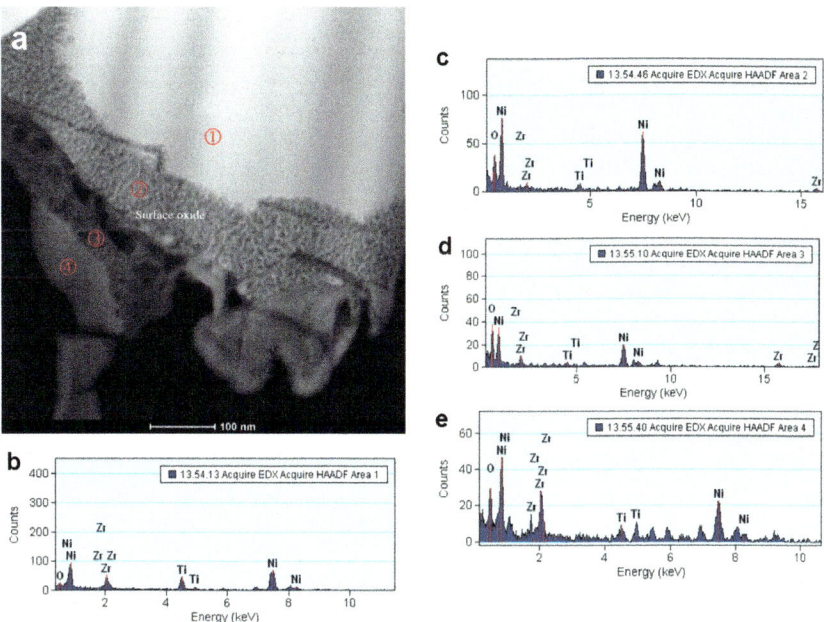

Figure 9. (**a**) A TEM BF micrograph showing the oxide on the B2 phase surface and (**b**)–(**e**) the corresponding EDS spectra from the areas identified in (**a**).

A catalytic mixture of Ni metal/oxide can also be found in the activated (etched) grain boundary between two B2 phase grains, as shown in the TEM BF micrograph in Figure 10a. In this case, the grain boundary region reacts rapidly with the KOH electrolyte, particularly with the connected voids, to remove the oxidation products and replace them with fresh electrolyte. The three square numbered areas were analyzed by EDS and the resulting spectra are shown in Figure 10b–d. Both grains (Areas 1 and 2) are assigned to the B2 phase, but with different compositions; while Square 1 shows a typical bulk B2 composition (Figure 3c), Square 2 has higher Zr and Sn content and a smaller Ti content. The dark spot in the grain boundary between two B2 crystals shows a high-Ni level and some degree of oxidation of Zr and Ti (Figure 10d). An HR-TEM BF image with an inserted SAD pattern is shown in Figure 10e. The micrograph is filled by the nano-scaled crystallites (~2 nm in diameter), attributing to the continuous halo-like rings in the SAD pattern. The ring patterns represent the face-centered cubic structure of the material. Among them, the most intense pattern corresponds to a lattice spacing of 0.21 nm, which results from the metallic Ni (111) plane. Figure 10f is a slightly lower magnification TEM dark field (DF) micrograph from the same area. The observed bright spots are the Ni nano-crystallites that are the dominating component in this area.

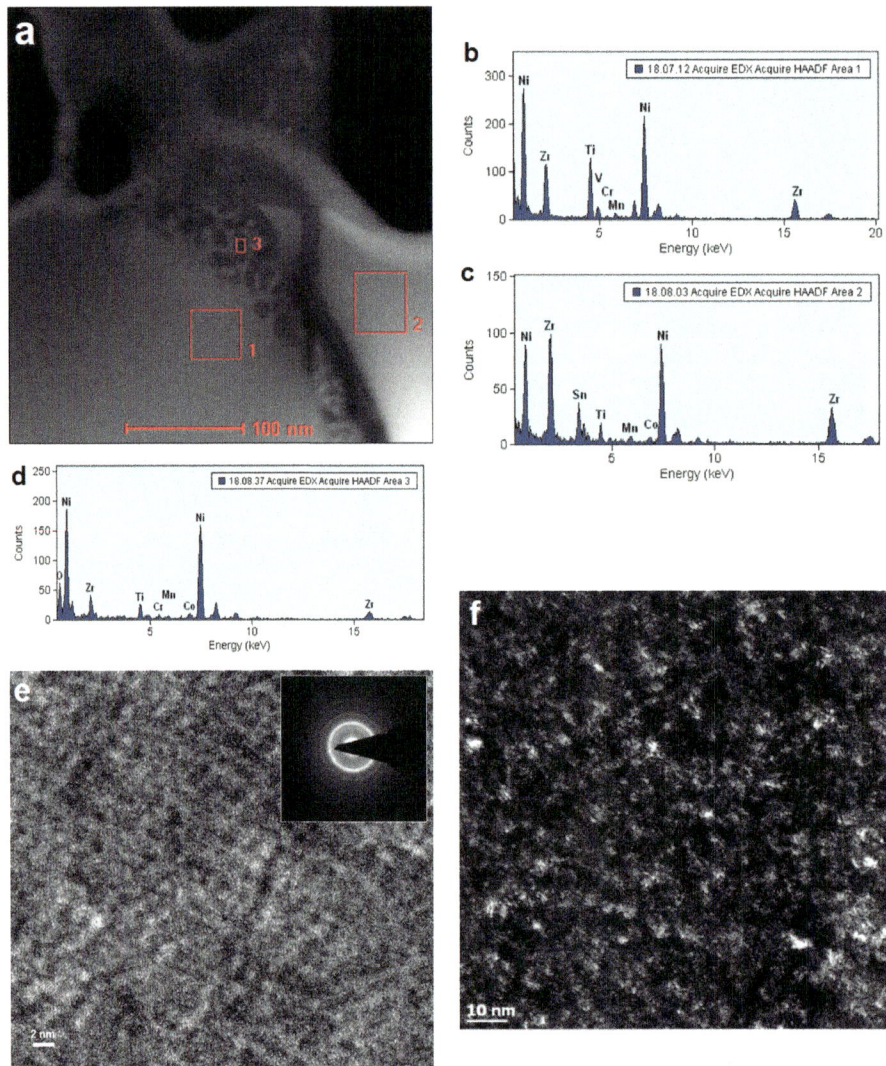

Figure 10. (**a**) A TEM BF micrograph showing the oxide between two B2 phases; (**b**)–(**d**) the corresponding EDS spectra from the areas identified in (**a**); (**e**) a high magnification TEM BF image with a SAD in the insert (**e**); and (**f**) a TEM DF micrograph from the same region as (**e**) showing the dominant Ni nano-crystallites.

Another HR-TEM BF micrograph demonstrating the presence of oxide in the B2 phase grain boundary is shown in Figure 11a with the corresponding EDS mappings of Ni (Figure 11b) and O (Figure 11c). Different from what is shown in Figure 8, the Ni-mapping in Figure 11a–c complements the O-mapping, suggesting a high density of metallic-Ni on the B2 phase surface.

442

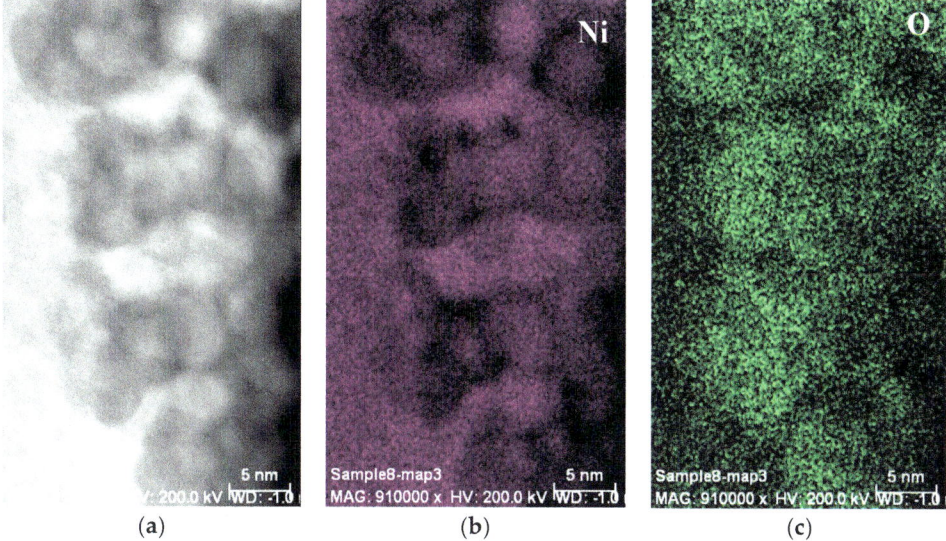

(a)   (b)   (c)

Figure 11. (**a**) A TEM BF micrograph showing oxide on the B2 phase surface, and EDS mappings of (**b**) Ni and (**c**) O. Different from Figure 8, the Ni-mapping here is complementary to the O-mapping, suggesting a high density of metallic-Ni on the B2 phase surface.

4. Conclusions

With a detailed TEM/EDS study, the surfaces of the main AB_2 phases not covered by Zr oxide/hydroxide are identified as the main source of improvement in low-temperature electrochemical performance. The electrolyte interface is a porous open space filled with fibrous Ni oxide sheets. The metal/oxide interface is free from amorphous buffer oxide, which can substantially reduce charge-transfer resistance. The surface of the B2 phase not directly under the Zr oxide/hydroxide is also electrochemically active, but the significance to the improvement is considered of secondary order because of the small percentage (3.0 wt%) and lack of Si, which is very common in Si-free AB_2 MH alloys. In contrast with the conventional MH alloys, the surface embedded metallic Ni nano-particles in the AB_2 phase in the Si-containing alloys, with a relatively low density, do not contribute directly to the improvement in the low-temperature performance of the MH alloys.

Acknowledgments: The authors would like to thank the following individuals from BASF-Ovonic for their help: Benjamin Reichman, Benjamin Chao, Baoquan Huang, Diana F. Wong, Taihei Ouchi, David Pawlik, Allen Chan, Ryan J. Blankenship, and Su Cronogue.

Author Contributions: Kwo-hsiung Young designed and conduct the experiment. Benjamin Chao and Jean Nei helped in the data interpretation and manuscript preparation.

Conflicts of Interest: The authors declare no conflict of interest.

References

1. Hybrid Electric Vehicle. Available online: https://en.wikipedia.org/wiki/Hybrid_electric_vehicle (accessed on 30 August 2015).

2. Takasaki, T.; Nishimura, K.; Saito, M.; Fukunaga, H.; Iwaki, T.; Sakai, T. Cobalt-free nickel–metal hydride battery for industrial applications. *J. Alloys Compd.* **2013**, *580*, S378–S381.

3. Nishimura, K.; Takasaki, T.; Sakai, T. Introduction of large-sized nickel–metal hydride battery GIGACELL® for industrial applications. *J. Alloys Compd.* **2013**, *580*, S353–S358.

4. Kai, T.; Ishida, J.; Yasuoka, S.; Takeno, K. The effect of nickel-metal hydride battery's characteristics with structure of the alloy. In Proceedings of the 54th Battery Symposium in Japan, Osaka, Japan, 6–9 October 2013; p. 210.

5. Kang, J.; Yan, F.; Zhang, P.; Du, C. Comparison of comprehensive properties of Ni-MH (nickel-metal hydride) and Li-ion (lithium-ion) batteries in terms of energy efficiency. *Energy* **2014**, *70*, 618–625.

6. Ovshisnky, S.R.; Fetcenko, M.A.; Ross, J. A nickel metal hydride battery for electric vehicles. *Science* **1993**, *260*, 176–181.

7. Dhar, S.K.; Ovshinsky, S.R.; Gifford, P.R.; Corrigan, D.A.; Fetcenko, M.A.; Venkatesan, S. Nickel/metal hydride technology for consumer and electric vehicle batteries—A review and up-date. *J. Power Sources* **1997**, *65*, 1–7.

8. Fierro, C.; Zallen, A.; Koch, K.; Fetcenko, M.A. The influence of nickel-hydroxide composition and microstructure on the high-temperature performance of nickel metal hydride batteries. *J. Electrochem. Soc.* **2006**, *153*, A492–A496.

9. Fetcenko, M.A.; Ovshinsky, S.R.; Reichman, B.; Young, K.; Fierro, C.; Koch, J.; Zallen, A.; Mays, W.; Ouchi, T. Recent advances in NiMH battery technology. *J. Power Sources* **2007**, *165*, 544–551.

10. Kong, F.; Kostecki, R.; Nadeau, G.; Song, X.; Zaghib, K.; Kinoshita, K.; McLarnon, F. *In situ* studies of SEI formation. *J. Power Sources* **2001**, *97–98*, 58–66.

11. Vetter, J.; Novák, P.; Wagner, M.R.; Veit, C.; Möller, K.-C.; Besenhard, J.O.; Winter, M.; Wohlfahrt-Mehrens, M.; Vogler, C.; Hammouche, A. Ageing mechanisms in lithium-ion batteries. *J. Power Sources* **2005**, *147*, 269–281.

12. Zhang, S.S. A review on electrolyte additives for lithium-ion batteries. *J. Power Sources* **2006**, *162*, 1379–1394.

13. Schranzhofer, H.; Bugajski, J.; Santner, H.J.; Korepp, C.; Möller, K.-C.; Besenhard, J.O.; Winter, M.; Sitte, W. Electrochemical impedance spectroscopy study of the SEI formation on graphite and metal electrodes. *J. Power Sources* **2006**, *153*, 391–395.

14. Bryngelsson, H.; Stjerndahl, M.; Gustafsson, T.; Edström, K. How dynamic is the SEI? *J. Power Sources* **2007**, *174*, 970–975.

15. Kim, S.; van Duin, A.C.T.; Shenoy, V.B. Effect of electrolytes on the structure and evolution of the solid electrolyte interphase (SEI) in Li-ion batteries: A molecular dynamics study. *J. Power Sources* **2011**, *196*, 8590–8597.

16. Colclasure, A.M.; Smith, K.A.; Kee, R.J. Modeling detailed chemistry and transport for solid-electrolyte-interface (SEI) films in Li–ion batteries. *Electrochim. Acta* **2011**, *58*, 33–43.

17. Young, K.; Chao, B.; Liu, Y.; Nei, J. Microstructures of the oxides on the activated AB_2 and AB_5 metal hydride alloys surface. *J. Alloys Compd.* **2014**, *606*, 97–104.

18. Schlapbach, L.; Stucki, F.; Seiler, A.; Siegmann, H.C. The formation of superparamagnetic metallic Ni and Fe particles at the surface of intermetallics by surface segregation. *Surf. Sci.* **1981**, *106*, 157–159.

19. Stucki, F.; Schlapbach, L. Magnetic properties of $LaNi_5$, FeTi, Mg_2Ni and their hydrides. *J. Less Comm. Metal.* **1980**, *74*, 143–151.

20. Stucki, F. Surface analysis by magnetization measurements on FeTi and $Fe_{0.85}Mn_{0.15}Ti$. *J. Appl. Phys.* **1982**, *53*, 2643–2644.

21. Kim, G.; Chun, C.; Lee, S.; Lee, J. A study on the microstructural change of surface of the intermetallic compound $LaNi_5$ by hydrogen absorption. *Scr. Metall. Mater.* **1993**, *29*, 485–490.

22. Broom, D.P.; Kemali, M.; Ross, D.K. Magnetic properties of commercial metal hydride battery materials. *J. Alloys Compd.* **1999**, *293–295*, 255–259.

23. Tai, L.T.; Hang, B.T.; Thuy, N.P.; Hieh, T.D. Magnetic properties of $LaNi_5$-based compounds. *J. Magn. Magn. Mater.* **2003**, *262*, 485–489.

24. Termsuksawad, P.; Niyomsoan, S.; Goldfarb, R.B.; Kaydanov, V.I.; Olson, D.L.; Mishra, B.; Gavra, Z. Measurement of hydrogen in alloys by magnetic and electronic techniques. *J. Alloys Compd.* **2004**, *373*, 86–95.

25. Li, W.K.; Ikeda, K.; Nakamori, Y.; Orimo, S.; Yakushiji, K.; Takanashi, K.; Ohyama, H.; Nakatsuji, K.; Dansui, Y. Size distribution of precipitated Ni clusters on the surface of an alkaline-treated $LaNi_5$-based alloy. *Acta Mater.* **2007**, *55*, 481–485.

26. Young, K.; Chao, B.; Pawlik, D.; Shen, H. Transmission electron microscope studies in the surface oxide on the La-containing AB_2 metal hydride alloy. *J. Alloys Compd.* **2016**.

27. Fetcenko, M.A.; Ovshinsky, S.R.; Young, K.; Reichman, B.; Fierro, C.; Koch, J.; Mays, W.; Ouchi, T.; Sommers, B.; Zallen, A. High catalytic activity disordered VTiZrNiCrCoMnAlSn hydrogen storage alloys for nickel–metal hydride batteries. *J. Alloys Compd.* **2002**, *330–332*, 752–759.

28. Maurel, F.; Knosp, B.; Backhaus-Ricoult, M. Characterization of corrosion products of AB_5-type hydrogen storage alloys for nickel-metal hydride batteries. *J. Electrochem. Soc.* **2000**, *147*, 78–86.

29. Song, D.; Gao, X.; Zhang, Y.; Lin, D.; Zhou, Z.; Wang, G.; Shen, P. Surface analysis of a TiNiB hydrogen storage electrode. *J. Alloys Compd.* **1993**, *199*, 161–163.

30. Young, K.; Huang, B.; Regmi, R.K.; Lawes, G.; Liu, Y. Comparisons of metallic clusters imbedded in the surface oxide of AB_2, AB_5, and A_2B_7 alloys. *J. Alloys Compd.* **2010**, *506*, 831–840.

31. Young, K.; Nei, J. The Current Status of Hydrogen Storage Alloy Development for Electrochemical Applications. *Materials* **2013**, *6*, 4574–4608.

32. Young, K.; Nei, J.; Wong, D.F.; Wang, L. Structural, hydrogen storage, and electrochemical properties of Laves phase-related body-centered-cubic solid solution metal hydride alloys. *Int. J. Hydrog. Energy* **2014**, *39*, 21489–21499.

33. Wong, D.F.; Young, K.; Nei, J.; Wang, L.; Ng, K.Y.S. Effects of Nd-addition on the structural, hydrogen storage, and electrochemical properties of C14 metal hydride alloys. *J. Alloys Compd.* **2015**, *647*, 507–518.

34. Young, K.; Ouchi, T.; Huang, B.; Reichman, B.; Blankenship, R. Improvement in −40 °C electrochemical properties of AB_2 metal hydride alloy by silicon incorporation. *J. Alloys Compd.* **2013**, *575*, 65–72.

35. Chen, J.; Dou, S.X.; Bradhurst, D.; Liu, H.K. Nickel Hydroxide as an Active Material for the Positive Electrode in Rechargeable Alkaline Batteries. In Proceedings of the Twelfth Annual Battery Conference on Applications and Advances, Long Beach, CA, USA, 14–17 January 1997; IEEE: Piscataway, NJ, USA, 1997; pp. 313–316.

36. Song, X.; Zhang, Z.; Zhang, X.B.; Lei, Y.Q.; Wang, Q.D. Effect of Ti Substitution on the microstructure and properties of Zr–Mn–V–Ni AB_2 type hydride electrode alloys. *J. Mater. Res.* **1999**, *14*, 1279–1285.

37. Shi, Z.; Chumbley, S.; Laabs, F.C. Electron diffraction analysis of an AB_2-type Laves phase for hydrogen battery applications. *J. Alloys Compd.* **2000**, *312*, 41–52.

38. Song, X.; Chen, Y.; Sequeira, C.; Zhang, Z. Microstructural evolution of body-centered cubic structure related Ti–Zr–Ni phases in non-stoichiometric Zr-based Zr–Ti–Mn–V–Ni hydride electrode alloys. *J. Mater. Res.* **2003**, *18*, 37–44.

39. Boettinger, W.J.; Newbury, D.E.; Wang, K.; Bendersky, L.A.; Chiu, C.; Kattner, U.R.; Young, K.; Chao, B. Examination of Multiphase (Zr,Ti)(V,Cr,Mn,Ni)$_2$ Ni-MH Electrode Alloys: Part I. Dendritic Solidification Structure. *Metall. Mater. Trans. A* **2010**, *41*, 2033–2047.

40. Bendersky, L.A.; Wang, K.; Boettinger, W.J.; Newbury, D.E.; Young, K.; Chao, B. Examination of Multiphase (Zr,Ti)(V,Cr,Mn,Ni)$_2$ Ni-MH Electrode Alloys: Part II. Solid-State Transformation of the Interdendritic B2 Phase. *Metall. Mater. Trans. A* **2010**, *41*, 1891–1906.

41. *Powder Diffraction File (PDF) Database*; MSDS No. 00-014-0481; International Centre for Diffraction Data: Newtown Square, PA, USA, 2011.

42. Young, K.; Reichman, B.; Fetcenko, M.A. Electrochemical performance of AB_2 metal hydride alloys measured at −40 °C. *J. Alloys Compd.* **2013**, *580*, S349–S352.

43. Young, K.; Wong, D.F.; Nei, J.; Reichman, B. Electrochemical properties of hypo-stoichiometric Y-doped AB_2 metal hydride alloys at ultra-low temperature. *J. Alloys Compd.* **2015**, *643*, 17–27.

Microstructure Investigation on Metal Hydride Alloys by Electron Backscatter Diffraction Technique

Yi Liu and Kwo-Hsiung Young

Abstract: The microstructures of two metal hydride (MH) alloys, a Zr_7Ni_{10} based $Ti_{15}Zr_{26}Ni_{59}$ and a C14 Laves phase based $Ti_{12}Zr_{21.5}V_{10}Ni_{36.2}Cr_{4.5}Mn_{13.6}Sn_{0.3}Co_{2.0}Al_{0.4}$, were studied using the electron backscatter diffraction (EBSD) technique. The first alloy was found to be composed of completely aligned Zr_7Ni_{10} grains with a ZrO_2 secondary phase randomly scattered throughout and a C15 secondary phase precipitated along the grain boundary. Two sets of orientation alignments were found between the Zr_7Ni_{10} grains and the C15 phase: $(001)_{Zr7Ni10A}//(110)_{C15}$ and $[100]_{Zr7Ni10A}//[0\bar{1}1]_{C15}$, and $(01\bar{1})_{Zr7Ni10B}//(\bar{1}00)_{C15}$ and $[100]_{Zr7Ni10B}//[313]_{C15}$. The grain growth direction is close to $[313]_{Zr7Ni10B}//[\bar{1}11]_{C15}$. The second alloy is predominated by a C14 phase, as observed from X-ray diffraction analysis. Both the matrix and dendrite seen through a scanning electron microscope arise from the same C14 structure with a similar chemical composition, but different orientations, as the matrix with the secondary phases in the form of intervening $Zr_7Ni_{10}/Zr_9Ni_{11}/(Zr,Ni)Ti$ needle-like phase coated with a thin layer of C15 phase. The crystallographic orientation of the C15 phase is in alignment with the neighboring C14 phase, with the following relationships: $(111)_{C15}//(0001)_{C14}$ and $[1\bar{1}0]_{C15}//[11\bar{2}0]_{C14}$. The alignments in crystallographic orientations among the phases in these two multi-phase MH alloys confirm the cleanliness of the interface (free of amorphous region), which is necessary for the hydrogen-storage synergetic effects in both gaseous phase reaction and electrochemistry.

Reprinted from *Batteries*. Cite as: Liu, Y.; Young, K.-H. Microstructure Investigation on Metal Hydride Alloys by Electron Backscatter Diffraction Technique. *Batteries* **2016**, *2*, 26.

1. Introduction

Nickel/metal hydride (Ni/MH) batteries have been the choice of energy storage medium for powering hybrid electric vehicles (HEVs) over the past fifteen years due to their unmatched safety record and durable cycle life. Facing challenges from other emerging battery technologies, further improvement in Ni/MH battery performance is needed, especially with regard to energy density. While potential improvement in the positive electrode (nickel hydroxide) is limited, the hope of improving the energy density falls to the negative electrode. Laves phase based metal hydride

(MH) AB_2 alloys have a great potential in gravimetric specific energy, compared to currently used AB_5 alloys [1–5]. A Mn-rich AB_2 MH alloy ($Ti_{0.9}Zr_{0.1}Mn_{1.6}Ni_{0.4}$) with the electrochemical capacity of 438 mAh·g^{-1} after 150 formation cycles has been reported [6]. Another Ni-rich AB_2 MH alloy, $Ti_{0.62}Zr_{0.38}V_{0.41}Cr_{0.30}Mn_{0.36}Ni_{0.89}$, exhibited a discharge capacity of 424 mAh·g^{-1} after activation in a 30% KOH bath at 110 °C for 4.5 h [7]. These capacity values are 40% higher than the 320 mAh·g^{-1} found in commercially available AB_5 alloys. Although higher in energy density, Laves phase based AB_2 MH alloys have lower nickel content in their chemical compositions due to the lower B/A ratio of 2, compared to 5 for AB_5 alloys. This deficiency in nickel makes AB_2 less catalytic in electrochemical reactions [8]. One way to improve the electrochemical performance in AB_2 alloys is through the introduction of minor secondary phases. Through synergetic effects between neighboring phases, the electrochemical properties of the multi-phase AB_2 alloys can be much improved [9–12]. Therefore, investigation of the primary/secondary phase interactions can provide very important insight in this area. We have previously reported results from studies of the synergetic effects in gaseous phase hydrogen storage [13], electrochemical charge/discharge [14], and a comparison between the two [15]. In both gaseous phase and electrochemistry, a "coherent" interface (free from amorphous, highly defective, and other interruptive region in between two different phases) was proposed to diffuse hydrogen and protons, and transfer stress from the hydrogenated side to the un-hydrogenated side (Figure 1). A few analytical works have been used to study the microstructure of the interface between the matrix and minor phases. Akiba and Iba [16,17] reported a microstructure between C14 and body-centered-cubic (bcc) phases by scanning electron microscopy (SEM). Chen and her coworkers reported the alignment in crystallographic orientations between the C15 and bcc phases [18]. Song and his coworkers [19–21] reported the microstructures among the C14, C15, (Zr,Ti)Ni, and Zr_9Ni_{11} phases. Shi and his coworkers [22] reported the crystallographic orientation alignment between the C14 and C15 phases. In addition, C14, face-centered-cubic (fcc), and bcc phases were also examined by Shibuya et al. [23]. The microstructures of the secondary phases in AB_2 MH alloys, including Zr_7Ni_{10}, Zr_9Ni_{11}, and TiNi, were studied by SEM and transmission electron microscopy (TEM) by Boettinger, Bendersky, and their coworkers [24,25].

Electron backscatter diffraction (EBSD), also known as backscatter Kikuchi diffraction or orientation imaging microscopy (OIM), is a microstructural-crystallographic technique that examines the crystallographic orientation of the constituent phases in a polycrystalline material. Combined with information of the chemical composition from X-ray energy dispersive spectroscopy (EDS), the capabilities of both techniques can be enhanced, including insight into the microstructure study of grains with mixed compositions and orientations [28]. In the past, we had employed EBSD in a study of Zr_7Ni_{10} based MH alloys [29], but the details regarding the crystallographic orientations were not discussed. We have also employed this technique to show the clean grain boundary in the C14/bcc multi-phase MH alloys [30]. In this paper, we further explore the strength of EBSD to study the crystallographic connection between the main and secondary phases of two MH alloys.

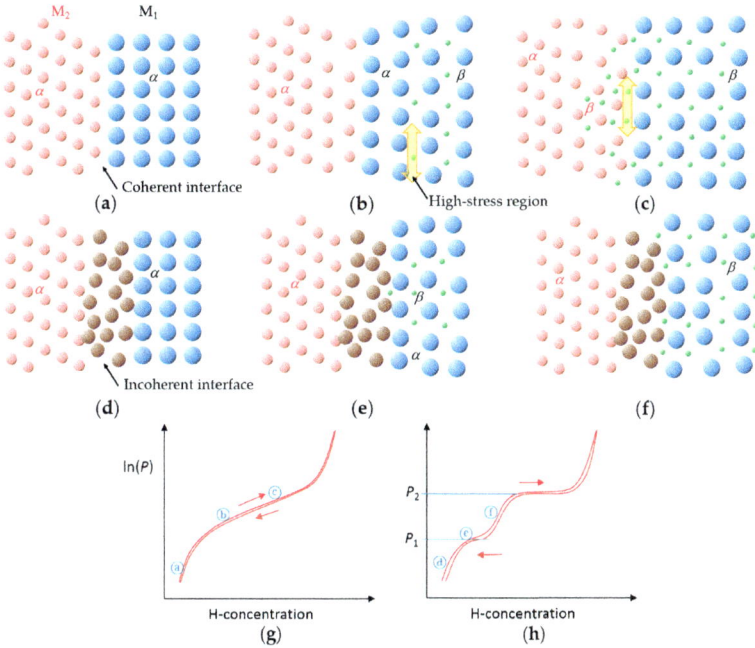

Figure 1. Schematic diagrams showing a coherent interface between the M_1 (blue balls) and M_2 (red balls) phases at the stages of (**a**) no hydrogen; (**b**) M_1 partially hydrided and the metal (α)–metal hydride (MH, β) phase boundary carrying the high-stress; and (**c**) M_1 completely hydrided and the M_2 lattice expanded due to the stress from the M_1 side, which facilitates the entrance of hydrogen (green dot) to enter the M_2 side, an incoherent interface at the stage of (**d**) no hydrogen; (**e**) M1 with mixed α and β phases; and (**f**) M_1 in the β phase and M_2 in the α phase, and pressure-concentration-temperature (PCT) isotherms corresponding to a multi-phase MH alloy with (**g**) a coherent interface (**a–c**) and (**h**) an incoherent interface (**d–f**). P_1 and P_2 are the equilibrium pressures for the M_1 and M_2 phases, respectively. Forward and backward arrows represent absorption and desorption isotherms, respectively. While (**g**) PCT isotherm can be found in most of the AB_2 multi-phase MH alloys with strong synergetic effects from secondary phases ([26], for instance), (**h**) isotherm can be found in a discrete system such as Mg_2Ni and Ni mixture [27].

449

2. Experimental Setup

Ingot samples were prepared by induction melting under an argon atmosphere in a 2 kg induction melting furnace using an $MgAl_2O_4$ crucible, an alumina tundish, and a steel pancake-shaped mold. A Philips X'Pert Pro X-ray diffractometer (XRD, Philips, Amsterdam, The Netherlands) was used to study the constituent phases in the samples. A piece of the ingot was cut off and went through a series of mechanical polishes. Final polishing was conducted by soaking the sample in a 0.05 μm silica colloidal suspension for several hours. It has been demonstrated that samples mechanically polished in this way show high quality Kikuchi patterns obtained by EBSD. The sample was observed using a backscattered electron (BSE) detector to show the contrast of different phases under a Hitachi S-2400 SEM (Hitachi High-Technologies Corp., Tokyo, Japan) equipped with EDS (EDAX Inc., Mahwah, NJ, USA) and EBSD/OIM systems. Diffraction patterns (Kikuchi patterns) were obtained by the EBSD system, which was attached to the SEM. The EBSD system was made by HKL Technology (Hobro, Denmark, now merged with Oxford Instruments, Inc., Abingdon, UK). The C14 phase exhibited an $MgZn_2$ type crystal structure with space group $P6_3/mmc$ (#194), while C15 phase has the $MgCu_2$ type crystal structure with space group $Fd3m$ (#227). Acquired using the EBSD pattern acquisition Software Flamenco 5.0 (Oxford Instruments Inc., Abingdon, UK), the computer simulated EBSD patterns for the C14 and C15 crystal structures are illustrated in Figures 2 and 3, respectively. The main differences between the two sets of EBSD patterns are: a four-fold symmetry exists in the (100) pole of the C15 (cubic) structure (Figure 3a) but not in the C14 (hexagonal) structure; $[1\bar{1}0]$ bands near the (110) pole of the C15 structure (yellow lines connecting poles (110) and (111) in Figure 3b) but not in the equivalent $(1\bar{1}00)$ plane in the C14 structure (Figure 2b); and $[3\bar{1}3]$ bands near the (111) pole of the C15 structure (pink lines connecting poles (231) and (332) in Figure 3c) but not in the equivalent (0001) projection in the C14 structure (Figure 2a). The operating voltage was 25 kV. Local chemical compositions were analyzed by EDS. Statistical analysis on the microstructure was performed by using "Scion Image", which is the PC version of the free software "NIH Image" (Scion Corp., Frederick, MD, USA) [31].

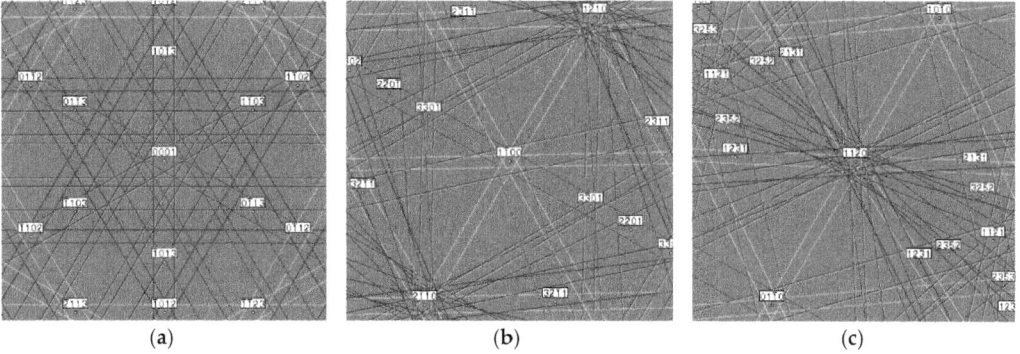

Figure 2. Computer generated electron backscatter diffraction (EBSD) patterns for: (**a**) (0001); (**b**) ($1\bar{1}00$); and (**c**) ($11\bar{2}0$) surfaces of a C14 (hexagonal) crystal structure.

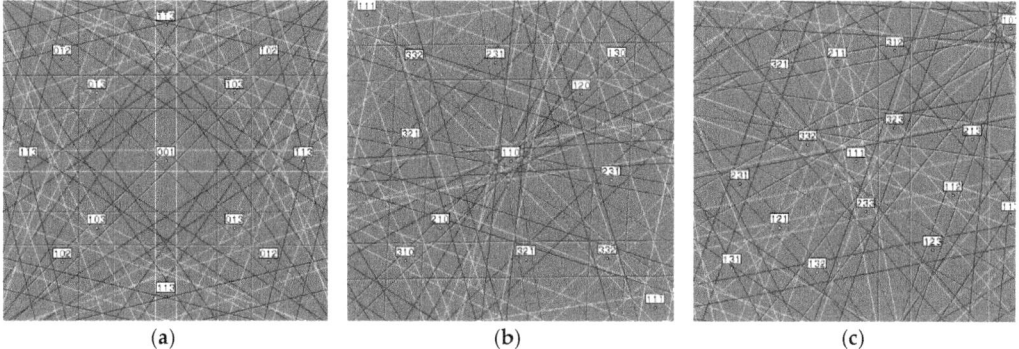

Figure 3. Computer generated EBSD patterns for: (**a**) (001); (**b**) (110); and (**c**) (111) surfaces of a C15 (cubic) crystal structure.

3. Results

Two multi-phase MH alloys were chosen for this study. Their structures and hydrogen storage properties are summarized in Table 1 based on previous studies [29,32]. The first alloy (ZN06) is from a Zr_7Ni_{10} based alloy family targeted to increase storage capacities [29]. The second alloy (AB2#2) is from a group of 33 high-performance AB_2 alloys prepared for composition optimization [32]. Although these alloys are not the most optimized, their simpler structures make them better candidates for the EBSD studies.

Table 1. Chemical compositions and hydrogen storage properties of two alloys used in this study. Compositions are in at%. High-rate dischargeability is defined as the ratio of the discharge capacities obtained from C/7 and C/70 rates in a half-cell configuration.

Alloy	Major Phase	Minor Phase(s)	Gaseous Phase Storage @30 °C (wt%)	Electrochemical Capacity (mAh·g^{-1})	High-Rate Dischargeability
ZN06	Zr_7Ni_{10}	C15	1.12	186	0.74
AB2#2	C14	C15, Zr_7Ni_{10}, Zr_9Ni_{11}, ZrNi	1.66	384	0.98

3.1. Zr_7Ni_{10}-Based Alloy

The ZN06 ($Ti_{15}Zr_{26}Ni_{59}$) alloy from a previous study on $Ti_xZr_{7-x}Ni_{10}$ MH alloys [29] was chosen for further structural studies using the EBSD technique. XRD analysis indicates an orthorhombic Zr_7Ni_{10} main phase and a C15 secondary phase (Figure 1 in [29]) in ZN06. A representative SEM-BSE image from the ZN06 alloy is shown in Figure 4. The main matrix showing different gray scale contrast is the Z_7Ni_{10} phase. The dark areas at the grain boundary and black boulders are C15 and ZrO_2 phases, respectively. The chemical compositions in the numbered areas (main matrix) in Figure 4 were analyzed by EDS and the results are listed in Table 2. The EDS technique is generally considered as only a semi-quantitative analytic tool and the results have to be interpreted carefully. The B/A ratios obtained using EDS (1.22–1.24) are well lower than the ideal Zr_7Ni_{10} stoichiometry (B/A = 1.43). The composition range of Zr_7Ni_{10} in the Zr-Ni binary phase diagram includes Zr content from 0.41 to 0.45, which corresponds to a B/A ratio of 1.22–1.44. The B/A ratios found in this study lie near the low end of the allowable solubility of the Zr_7Ni_{10} phase. Small variations in the compositions listed in Table 2 can be observed. The standard deviation in % for Ti, Zr, and Ni-contents are 2.1, 1.0, and 0.24, respectively. Most of the deviations are in the A-site atoms, especially for Ti. The variation in contrasts, as seen from the BSE image in Figure 4, is electron channeling contrast (ECC) due to the different orientations of the grains. The difference in orientation of the grain is sometimes very little, but, due to diffraction, can cause very different contrast, even due to mis-orientation between sub-grains. In order to confirm the source of the contrast, EBSD patterns were taken from an area with laminated layers exhibiting various BSE contrasts, as shown in Figure 5. The crystallographic orientations of these areas align very well, which includes the possibility of the contrast originating from different orientations. A typical EBSD pattern and a computer generated index from a Zr_7Ni_{10} phase are present in Figure 6 for reference.

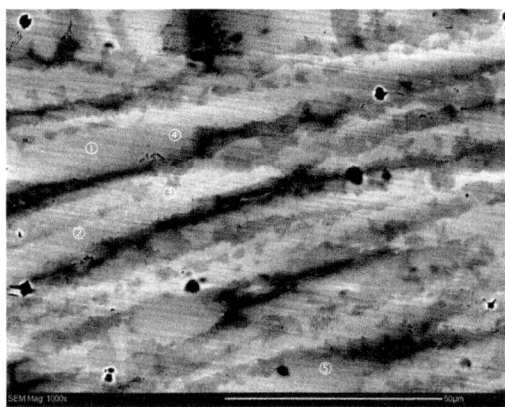

Figure 4. A scanning electron microscopy (SEM)-backscattered electron (BSE) micrograph from the ZN06 alloy. The main matrix, black phase in the grain boundary, and black boulders correspond to the Zr_7Ni_{10}, C15, and ZrO_2 phases. The chemical compositions in the numbered areas were analyzed by X-ray energy dispersive spectroscopy (EDS) and the results are listed in Table 2.

Table 2. Chemical compositions as measured by EDS in different areas shown in Figure 4 (in at%).

Location	Ti	Zr	Ni	B/A Ratio
1	12.65	32.30	55.05	1.224
2	12.83	31.76	55.40	1.242
3	12.75	31.88	55.37	1.241
4	12.44	32.43	55.13	1.229
5	12.08	32.65	55.26	1.235
Standard deviation (%)	2.1	1.0	0.24	0.55

The SEM micrograph, together with the four EBSD patterns from two Zr_7Ni_{10} phases, one C15 phase, and one ZrO_2 phase are shown in Figure 5 in [29]. Eight more EBSD patterns from the Zr_7Ni_{10} phase in the same area are presented in Figure 7. The Zr_7Ni_{10} phase has an orthorhombic structure with space group Cmca (#64). There are three distinctive orientations (A, B, and C) for the Zr_7Ni_{10} phase as shown in Figure 7. The orientations A and B deviate by only approximately $5°$ and the crystallographic orientation alignment between A and C is $(29\bar{3})_{\text{Orientation A}}//(01\bar{1})_{\text{Orientation C}}$ and $[313]_{\text{Orientation A}}//[111]_{\text{Orientation C}}$. The alignments in the crystallographic orientation of Zr_7Ni_{10} and C15 phases in Figure 5 in [29] are $(01\bar{1})_{\text{Zr7Ni10B}}//(\bar{1}00)_{\text{C15}}$ and $[100]_{\text{Zr7Ni10B}}//[313]_{\text{C15}}$, and $(001)_{\text{Zr7Ni10C}}//(110)_{\text{C15}}$ and $[100]_{\text{Zr7Ni10C}}//[0\bar{1}1]_{\text{C15}}$. The grain growth direction projected on the plane is close to: $[313]_{\text{Zr7Ni10B}}//[\bar{1}11]_{\text{C15}}$.

453

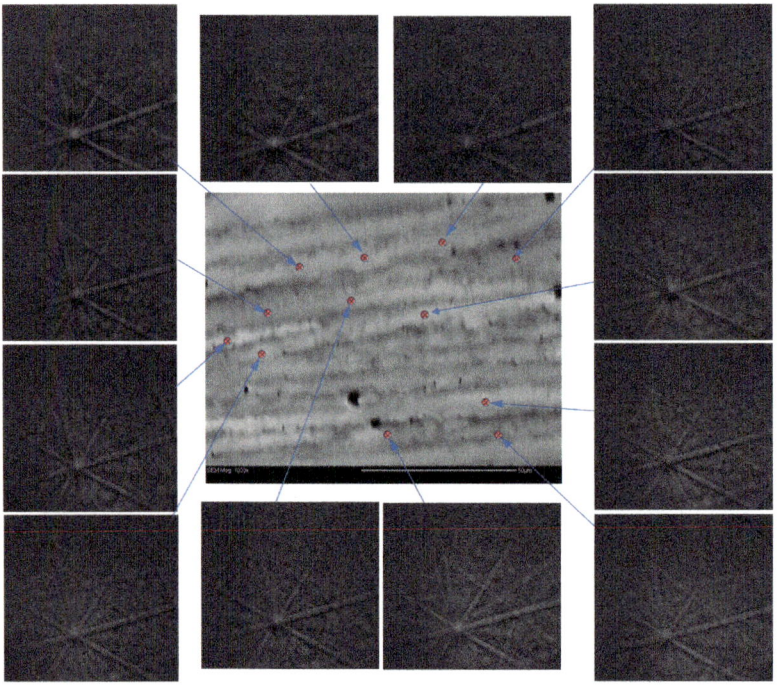

Figure 5. EBSD patterns of 12 different spots on a ZN06 alloy sample showing the crystallographic orientations of all the studied Zr_7Ni_{10} grains are well aligned.

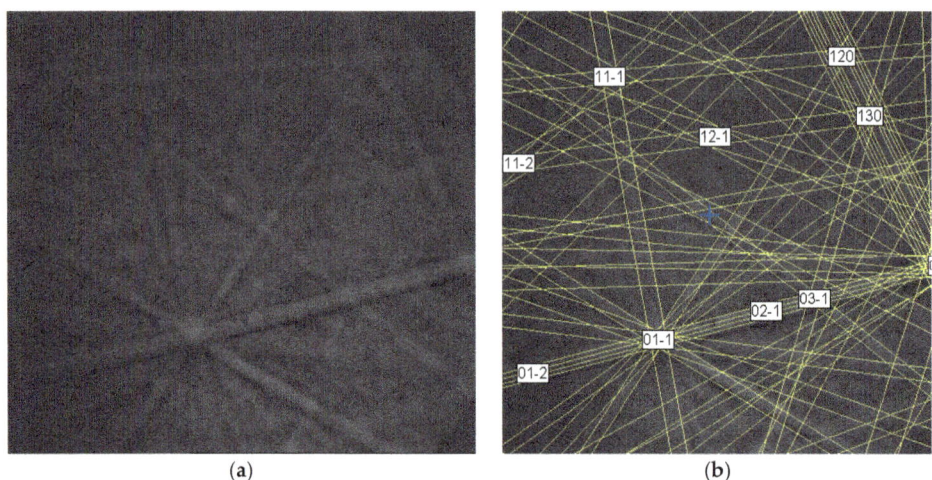

(a)   (b)

Figure 6. (**a**) A typical EBSD pattern; and (**b**) a computer generated index from a Zr_7Ni_{10} phase.

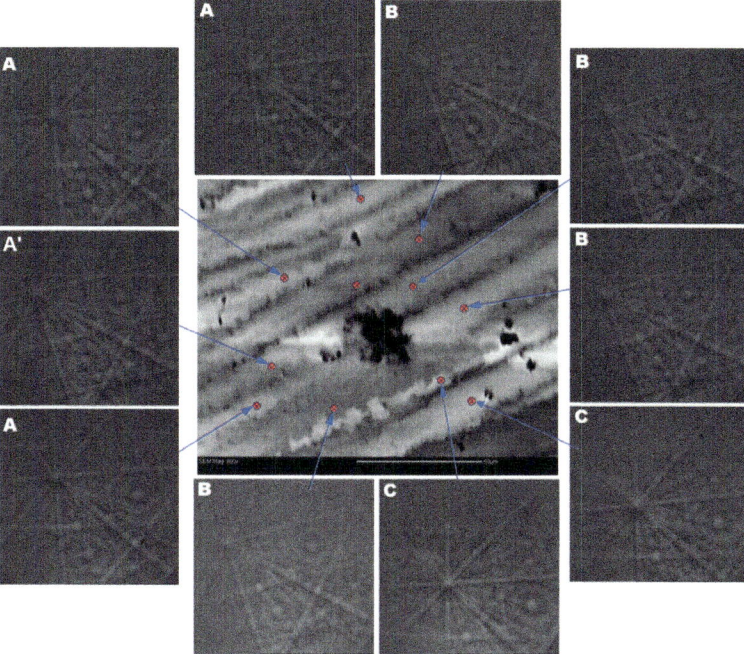

Figure 7. EBSD patterns at 10 different spots on a ZN06 alloy sample with three major crystallographic orientations (**A**, **B**, and **C**) and one minor crystallographic orientation with a small angle grain boundary (**A'**) present in Zr_7Ni_{10} grains.

From the crystallographic orientation alignments among Zr_7Ni_{10} phases and between the Zr_7Ni_{10} and C15 phases, the following solidification sequence is thought to occur during cooling. A Zr_7Ni_{10} phase with the average composition in the liquid is solidified congruently first, and then the Ni-rich C15 phase (average composition of $Ti_{17}Zr_{20}Ni_{63}$ [29]) is formed at the grain boundary as a solid reaction, resulting in the Zr_7Ni_{10} phase being pushed to be hypo-stoichiometric (average composition of $Ti_{15}Zr_{28}Ni_{57}$ [29]). This is different from the C14-C15-B2 solidification sequence found in the Ti-Zr-Ni alloy, which is closer to B/A ~ 2.0 [24].

3.2. Laves Phase Based AB_2 Alloy

A Laves phase based AB_2 alloy with the composition $Ti_{12}Zr_{21.5}V_{10}Ni_{36.2}Cr_{4.5}$ $Mn_{13.6}Sn_{0.3}Co_{2.0}Al_{0.4}$ from a previous study (Alloy #2 in [32]) was chosen for the EBSD study. The B/A in this alloy is 1.99. The XRD pattern of this alloy (shown in Figure 1b in [32]) shows it is C14-dominant. SEM-BSE images of the same alloy at different magnifications, shown in Figure 8, also reveal the same dendritic microstructure. Through careful and long-time mechanical polishing with 0.05 μm silica colloidal, the sample surface is almost free of lattice distortion caused by deformations from mechanical polishing. In addition to the Z-contrast shown in [32], ECC becomes available. This is because the BSE signals due to ECC are from the top surface of the sample. The different contrasts shown in Figure 8 are channeling contrasts representing the mis-orientation of different dendritic crystals. Therefore, the additional contrast observed in Figure 8b is from the crystallographic mis-orientation. In addition to the contrast from dendritic microstructures in Figure 8b, a different type of area can be found, as indicated by the arrows (Figure 8b). Figure 8c shows a magnified image of a similar area, where needle-like phases can be seen. The average chemical compositions of these areas, as measured by EDS are listed in Table 3. The composition of the grey and dark areas are very similar. For example, the difference in Ni content is 1.9%, which is lower than the 4.5% in Ni content between the phases identified in [32]. From the ratio of A site atoms (Ti and Zr) and B site atoms (V, Cr, Mn, Co, and Ni), both the grey and dark areas approach an AB_2 stoichiometry, while the needle-like secondary phase is closer to A_7B_9. There is no known intermetallic phase among the constituent elements matching an A_7B_9 stoichiometry, and, therefore, it can be considered as a mixture of a few possible intermetallic alloys. From a separate TEM study, it was found that this needle-like secondary phase was first solidified into a B2 structure and then decomposed into Zr_7Ni_{10}, Zr_9Ni_{11}, TiNi, and ZrNi phases through a solid-state chemical reaction [25]. By statistical analysis, the abundance of this mixture area is about 4.3% by volume fraction.

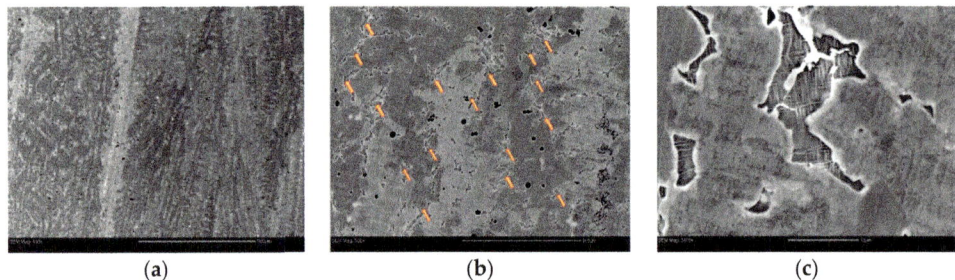

(a) (b) (c)

Figure 8. SEM-BSE micrographs from Alloy #2 at different magnifications. The scale bars represent: (**a**) 500 μm; (**b**) 100 μm; and (**c**) 10 μm.

Table 3. Chemical compositions measured by EDS in different areas in Figure 8c (in at%).

Region	Ti	Zr	V	Cr	Mn	Co	Ni	B/A Ratio
Grey	10.2	22.0	11.5	6.6	15.5	2.2	32.0	2.11
Dark	11.1	21.8	10.7	5.6	14.6	2.2	34.1	2.03
Second Phase	18.5	21.2	3.8	1.7	7.1	1.6	46.1	1.29

A dendrite microstructure usually forms due to a high cooling rate for alloy solidification, during which the nucleated grains may extend from the interface between the solid and liquid, and continue growing along the specific crystal orientations and along the temperature gradient, until the remaining area is solidified [33]. For this reason, the grains in the dendrite microstructure usually are sub-grains with close orientations that are different from the orientations of the matrix. This is the reason that the BSE images in Figure 8 show ECC between the dendritic area and the matrix, which was confirmed by obtaining EBSD patterns from different locations in the sample, as shown in Figure 9. The C14 phase has a hexagonal crystal structure and belongs to the $P6_3/mmc$ space group [34], while the C15 phase has an fcc crystal structure and belongs to the $Fd3m$ space group [35]. They are both AB_2 type Laves phases, which are very common in intermetallic alloys. The C14/C15 ratio can be influenced by the A/B atomic radii ratio [36], difference in electronegativity [37], electron concentration [38], stress [39,40], and process conditions [41,42]. In Figure 9, it can also be seen that most of the area is occupied by the C14 Laves phase, while only small isolated areas contain the C15 Laves phase. Locations 1, 2, and 3 in Figure 9 were indexed as C14 Laves phase, having different crystallographic orientations, while Locations 4, 5, and 6 were assigned to the C15 Laves phase. It is interesting to discover that Locations 1, 2, and 3, all having the C14 structure with very similar chemical compositions but different orientations, gave such a large contrast in the BSE micrograph. The EBSD patterns from Locations 2, 4, 5, and 6 aligned very well with each other. The C15 and neighboring C14 phases are aligned, with the following relationships: $(111)_{C15}//(0001)_{C14}$ and $[1\bar{1}0]_{C15}//[11\bar{2}0]_{C14}$, which was reported before from TEM study [22,40]. The contrasts in Locations 4, 5, and 6 are similar to each other, but differ significantly from Location 2 (C14). Therefore, we conclude that both the structure and orientation contribute to the contrast observed in the BSE micrographs.

Figure 10 shows another example of the microstructure at the interface between the C14 and C15 phases. In the middle of the micrograph is the needle-like secondary phase that has an average composition of A_7B_9. Locations 1 and 2 were indexed as C14 Laves phase with different crystallographic orientations. Locations 3, 4, and 5 were indexed as C15 Laves phase. The EBSD from Location 3 (C15) overlaps with that from Location 2 (C14), with the same relationships seen in Figure 8. The EBSD from Location 4 matches with that from Location 5 with the needle structure in between. This suggests that, during solidification, the C14 phase solidifies first, the C15 crystals grow secondarily, and the needle-like structure grows last inside the C15 phase as a peritectic phase. Comparing the EBSD patterns from Locations 2 and 4, we obtained the following relationships: $(101)_{loc.2}//(101)_{loc.4}$ and $[\bar{1}21]_{loc.2}//[\bar{1}21]_{loc.4}$. This is the micro-twinning of a cubic C15 structure, which is well-known in the field [43,44]. Some models have been proposed to explain the cause of the micro-twinning, which include stacking faults [20,45], dislocation movement [46], and phase growth [18]. This further proves that, at one stage of solidification, Locations 3 and 4 are connected but with micro-twinning in between.

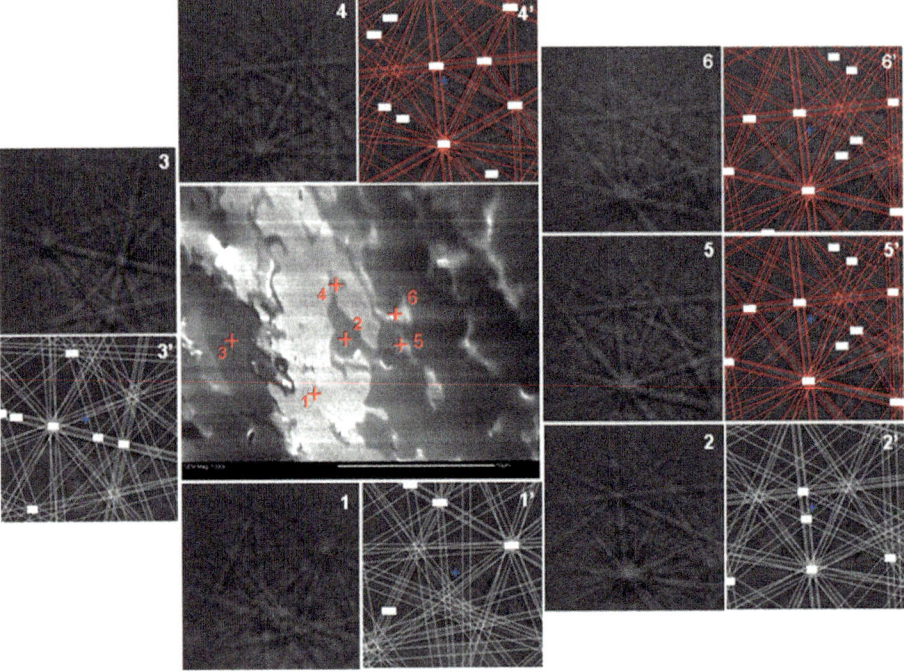

Figure 9. An example of EBSD point analysis on possible phases in the as-prepared Alloy #2. Points 1, 2, and 3 were indexed as C14 Laves phase, while Points 4, 5, and 6 were indexed as C15 Laves phase.

A compromise was made between high probe current mode and high resolution mode in the SEM experiments. In order to obtain high quality EBSD patterns, the SEM was adjusted to a high probe current mode, sacrificing resolution of the image, and thus the image shown in Figure 10 is not very clear. However, compared with Figure 11, it is apparent that the C15 Laves phase usually exists near the needle-like areas.

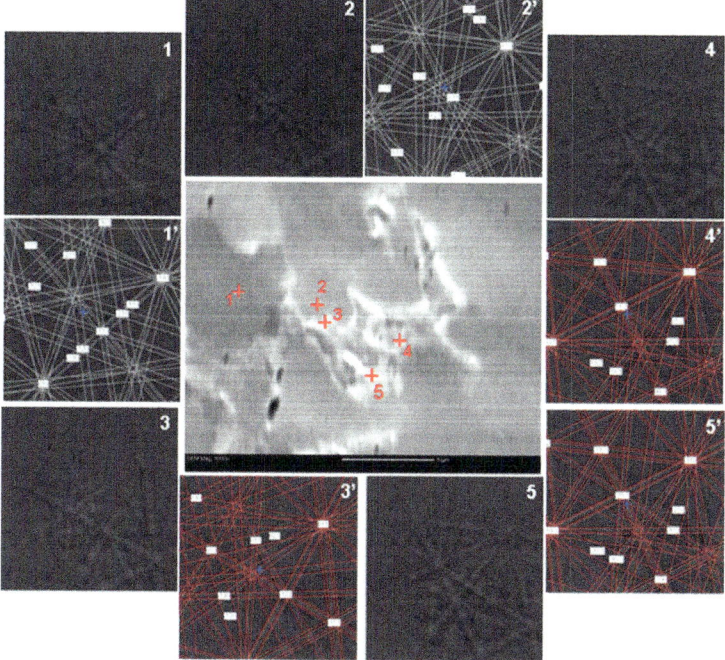

Figure 10. A further EBSD analysis around the needle-like phase area from Alloy #2. Points 1 and 2 are taken from the main C14 phase. The C15 Laves (Points 3, 4, and 5) phase is found to be along the boundary of needle-like phase area.

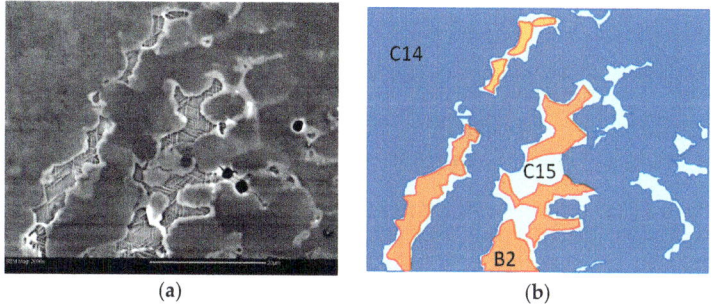

Figure 11. (**a**) A BSE image; and (**b**) a schematic showing the phases in different shades for a typical cross-section of Alloy #2. It was the alignment of the orientation for the fine needle structures observed in the SEM micrograph that initiated this series of microstructure studies.

From the binary alloy phase diagrams of Ti-Cr and Zr-Cr, it can be demonstrated that the C14 phase formed at a higher temperature and later transformed into a C15 structure under a slow cooling condition [47]. Therefore, the solidification path during cooling is when the C14 phase forms the dendrite structure with the same crystallographic orientation, and later a second C14 phase with a similar chemical composition forms the majority of the matrix. Finally, a needle-like secondary phase is formed with most of the vanadium and chromium segregated into the outside C15 layer and the A_7B_9 mixture forming the needles inside. Therefore, the cooling sequence of C14-C15-B2 in the AB_2 MH alloy proposed by Boettinger et al. [24] is verified. Similar microstructures observed by SEM in other AB2 multi-phase MH alloys are reported before [48,49].

The secondary phases are very important to the electrochemical performance of the multi-phase MH alloys. Eliminating (reducing) secondary phases through annealing results in severe degradations in capacity and high-rate dischargeability [9,12,50–54]. The secondary phases benefit the electrochemical reaction through the synergetic effects which require a clean, non-interrupted, and coherent interface adjacent to the main phase [14]. The highly aligned crystallographic orientations from various phases found in this study provide a strong support for the synergetic effect. Similar study with the same conclusion by EBSD and TEM has been performed on another multi-phase MH alloy family with a Laves phase related bcc structure [30].

4. Conclusions

EBSD techniques have been successfully employed to identify the structure and crystallographic orientation dependence of the phases from two Ti-Zr-Ni based MH alloys. This information contributed to the investigation of the cooling sequence of various phases. While the Zr_7Ni_{10}-based alloy (B/A = 1.43) was found to form a laminar-type solid, followed by solid-state phase segregation of Ni into the grain boundary and formation of the C15 secondary phase, the phase formation sequence is C14-C15-B2 in the AB_2-based alloy (B/A = 1.99). The alignment observed in the crystallographic orientation strongly indicates the presence of a clean and coherent interface for the synergetic effects to occur in both the gaseous phase and electrochemical hydrogen storage.

Acknowledgments: This work was financially supported by the Michigan Initiative for Innovation & Entrepreneurship (2009).

Author Contributions: Yi Liu performed the microscope work and prepared part of the manuscript, and Kwo-Hsiung Young organized the results and prepared the manuscript.

Conflicts of Interest: The authors declare no conflict of interest.

Abbreviations

MH	Metal hydride
bcc	body-centered-cubic
fcc	face-centered-cubic
PCT	Pressure-concentration-temperature
SEM	Scanning electron microscopy
TEM	Transmission electron microscopy
EBSD	Electron backscatter diffraction
OIM	Orientation imaging microscopy
EDS	X-ray energy dispersive spectroscopy
XRD	X-ray diffraction
BSE	Backscattered electron
ECC	Electron channeling contrast

References

1. Ovshinsky, S.R.; Fetcenko, M.A.; Ross, J. A nickel metal hydride battery for electric vehicles. *Science* **1993**, *260*, 176–181.
2. Ovshinsky, S.R.; Fetcenko, M.A. Development of high catalytic activity disordered hydrogen-storage alloys for electrochemical application in nickel-metal hydride batteries. *Appl. Phys. A* **2001**, *72*, 239–244.
3. Bowman, R.C., Jr.; Fultz, B. Metalic hydrides I: Hydrogen storage and other gas-phase application. *MRS Bull.* **2002**, *27*, 688–693.
4. Joubert, J.M.; Latroche, M.; Percheron-Guegan, A. Metallic hydrides II: Materials for electrochemical storage. *MRS Bull.* **2002**, *27*, 694–698.
5. Fetcenko, M.A.; Ovshinsky, S.R.; Reichman, B.; Young, K.; Fierro, C.; Koch, J.; Zallen, A.; Mays, W.; Ouchi, T. Recent advances in NiMH battery technology. *J. Power Sources* **2007**, *165*, 544–551.
6. Fetcenko, M.A.; Ovshinsky, S.R.; Young, K.; Reichman, B.; Fierro, C.; Koch, J.; Martin, F.; Mays, W.; Ouchi, T.; Sommers, B.; et al. High catalytic activity disordered VTiZrNiCrCoMnAlSn hydrogen storage alloys for nickel-metal hydride batteries. *J. Alloys Compd.* **2002**, *330–332*, 752–759.
7. Young, K.; Ouchi, T.; Koch, J.; Fetcenko, M.A. The role of Mn in C14 Laves phase multi-component alloys for NiMH battery application. *J. Alloys Compd.* **2009**, *477*, 749–758.
8. Young, K.; Fetcenko, M.A.; Regmi, R.K.; Lawes, G.; Liu, Y. Comparisons of metallic clusters imbedded in the surface oxide of AB_2, AB_5, and A_2B_7 alloys. *J. Alloys Compd.* **2010**, *506*, 831–840.
9. Zhang, W.K.; Ma, C.A.; Yang, X.G.; Lei, Y.Q.; Wang, Q.D.; Lu, G.L. Influences of annealing heat treatment on phase structure and electrochemical properties of $Zr(MnVNi)_2$ hydrogen storage alloys. *J. Alloys Compd.* **1999**, *293–295*, 691–697.

10. Zhang, Q.A.; Lei, Y.Q.; Yang, X.G.; Du, Y.L.; Wang, Q.D. Effects of annealing treatment on phase structure, hydrogen absorption-desorption characteristics and electrochemical properties of a $V_3TiNi_{0.56}Hf_{0.24}Mn_{0.15}Cr_{0.1}$ alloy. *J. Alloys Compd.* **2000**, *305*, 125–129.

11. Bououdina, M.; Soubeyroux, J.L.; Fruchart, D. Study of the hydrogenation/dehydrogenation processes of $ZrCr_{0.7}Ni_{1.3}$, a Laves phase-rich multi-component system, by in-situ neutron diffraction under hydrogen gas pressure. *J. Alloys Compd.* **2001**, *327*, 185–194.

12. Visintin, A.; Peretti, H.A.; Ruiz, F.; Corso, H.L.; Triaca, W.E. Effect of additional catalytic phases imposed by sintering on the hydrogen absorption behavior of AB_2 type Zr-based alloys. *J. Alloys Compd.* **2007**, *428*, 244–251.

13. Wong, D.F.; Young, K.; Nei, J.; Wang, L.; Ng, K.Y.S. Effects of Nd-addition on the structural, hydrogen storage, and electrochemical properties of C14 metal hydride alloys. *J. Alloys Compd.* **2015**, *647*, 507–518.

14. Young, K.; Ouchi, T.; Meng, T.; Wong, D.F. Studies on the synergetic effects in multi-phase metal hydride alloys. *Batteries* **2016**, *2*.

15. Mosavati, N.; Young, K.; Meng, T.; Ng, K.Y.S. Electrochemical open-circuit voltage and pressure- concentration-temperature isotherm comparison for metal hydride alloys. *Batteries* **2016**, *2*.

16. Iba, H.; Akiba, E. The relation between microstructure and hydrogen absorbing property in Laves phase-solid solution multiphase alloys. *J. Alloys Compd.* **1995**, *231*, 508–512.

17. Akiba, E.; Iba, H. Hydrogen absorption by Laves phase related BCC solid solution. *Intermetallics* **1998**, *6*, 461–470.

18. Chen, K.C.; Allen, S.M.; Livingston, J.D. Microstructures of two-phase Ti-Cr alloys containing the $TiCr_2$ Laves phase intermetallic. *J. Mater. Res.* **1997**, *12*, 1472–1480.

19. Song, X.; Zhang, X.; Lei, Y.; Zhang, Z.; Wang, Q. Effect of microstructure on the properties of Zr-Mn-V-Ni AB_2 type hydride electrode alloys. *Int. J. Hydrog. Energy* **1999**, *24*, 455–459.

20. Song, X.; Zhang, Z.; Zhang, X.; Lei, Y.; Wang, Q. Effect of Ti substitution on the microstructure and properties of Zr-Mn-V-Ni AB_2 type hydride electrode alloys. *J. Mater. Res.* **1999**, *14*, 1279–1285.

21. Song, X.Y.; Chen, Y.; Zhang, Z.; Lei, Y.Q.; Zhang, X.B.; Wang, Q.D. Microstructure and electrochemical properties of Ti-containing AB_2 type hydrogen storage electrode alloy. *Int. J. Hydrog. Energy* **2000**, *25*, 649–656.

22. Shi, Z.; Chumbley, S.; Laabs, F.C. Electron diffraction analysis of an AB_2-type Laves phase for hydrogen battery applications. *J. Alloys Compd.* **2000**, *312*, 41–52.

23. Shibuya, M.; Nakamura, J.; Enoki, H.; Akiba, E. High-pressure hydrogenation properties of Ti-V-Mn alloy for hybrid hydrogen storage vessel. *J. Alloys Compd.* **2009**, *475*, 543–545.

24. Boettinger, W.J.; Newbury, D.E.; Wang, K.; Bendersky, L.A.; Chiu, C.; Kattner, U.R.; Young, K.; Chao, B. Examination of multiphase $(Zr,Ti)(V,Cr,Mn,Ni)_2$ Ni-MH electrode alloys: Part I. Dendritic solidification structure. *Metall. Mater. Trans.* **2010**, *41*, 2033–2047.

25. Bendersky, L.A.; Wang, K.; Boettinger, W.J.; Newbury, D.E.; Young, K.; Chao, B. Examination of multiphase $(Zr,Ti)(V,Cr,Mn,Ni)_2$ Ni-MH electrode alloys: Part II. Solid-state transformation of the interdendritic B_2 phase. *Metall. Mater. Trans.* **2010**, *41*, 1891–1906.

26. Park, J.; Lee, J. Thermodynamic properties of the $Zr_{0.8}Ti_{0.2}(Mn_xCr_{1-x})$ system. *J. Less Common Met.* **1991**, *167*, 245–253.

27. Lv, J.; Zhang, B.; Wu, Y. Effect of Ni content on microstructural evolution and hydrogen storage properties of Mg_xNi_3La (x = 5, 10, 15, 20 at.%) alloys. *J. Alloys Compd.* **2015**, *641*, 176–180.

28. Matiland, T.; Sitzman, S. Electron Backscatter Diffraction (EBSD) Technique and Materials Characterizations Examples. In *Scanning Microscopy for Nanotechnology Techniques and Applications*; Zhou, W., Wang, Z.L., Eds.; Springer: New York, NY, USA, 2007.

29. Young, K.; Ouchi, T.; Liu, Y.; Reichman, B.; Mays, W.; Fetcenko, M.A. Structural and electrochemical properties of $Ti_xZr_{7-x}Ni_{10}$. *J. Alloys Compd.* **2009**, *480*, 521–528.

30. Shen, H.; Young, K.; Meng, T.; Bendersky, L.A. Clean grain boundary found in C14/body-center-cubic multi-phase metal hydride alloys. *Batteries* **2016**, *2*.

31. NIH Image Home Page. Available online: http://rsb.info.nih.gov/nih-image/ (accessed on 21 April 2006).

32. Young, K.; Ouchi, T.; Fetcenko, M.A. Roles of Ni, Cr, Mn, Sn, Co, and Al in C14 Laves phase alloys for NiMH battery application. *J. Alloys Compd.* **2009**, *476*, 774–781.

33. Reed-Hill, R.E. *Physical Metallurgy Principles*, 2nd ed.; Van Nostrand Company: Princeton, NJ, USA, 1973.

34. Friauf, J.B. The crystal structure of magnesium di-zincide. *Phys. Rev.* **1927**, *29*, 34–40.

35. Friauf, J.B. The crystal structures of two intermetallic compounds. *J. Am. Chem. Soc.* **1927**, *49*, 3107–3114.

36. Nakano, H.; Wakao, S.; Shimizu, T. Correlation between crystal structure and electrochemical properties of C14 Laves-phase alloys. *J. Alloys Compd.* **1997**, *253–254*, 609–612.

37. Zhu, J.H.; Liu, C.T.; Pike, L.M.; Liaw, P.K. Enthalpies of formation of binary Laves phases. *Intermetallics* **2002**, *10*, 579–595.

38. Zhu, J.H.; Liaw, P.K.; Liu, C.T. Effect of electron concentration on the phase stability of $NbCr_2$-based Laves phase alloys. *Mater. Sci. Eng.* **1997**, *239–240*, 260–264.

39. Stein, F.; Palm, M.; Sauthoff, G. Structure and stability of Laves phases. Part 1. Critical assessment of factors controlling Laves phase stability. *Intermetallics* **2004**, *12*, 713–720.

40. Hu, Y.; Vasiliev, A.; Zhang, L.; Song, K.; Aindow, M. Polymorphism in the Laves-phase precipitates of a quinternary Nb-Mo-Cr-Al-Si alloy. *Scr. Mater.* **2009**, *60*, 72–75.

41. Zhang, Q.A.; Lei, Y.Q.; Yang, X.G.; Ren, K.; Wang, Q.D. Annealing treatment of AB_2-type hydrogen storage alloys: I. crystal structure. *J. Alloys Compd.* **1999**, *292*, 236–240.

42. Shu, K.; Lei, Y.; Yang, X.; Liu, G.; Wang, Q.; Lu, G.; Chen, L. Effect of rapid solidification process on the alloy structure and electrode performance of $Zr(Ni_{0.55}V_{0.1}Mn_{0.3}Cr_{0.55})_{2.1}$. *J. Alloys Compd.* **1999**, *293–295*, 756–761.

463

43. Takasugi, T.; Yoshida, M. The effect of ternary addition on structure and stability of NbCr$_2$ Laves phases. *J. Mater. Res.* **1998**, *13*, 2505–2513.

44. Kumar, K.S.; Hazzledine, P.M. Polytypic transformation in Laves phases. *Intermetallics* **2004**, *12*, 763–770.

45. Yoshida, M.; Takasugi, T. TEM observation for deformation microstructures of two C15 NbCr$_2$ intermetallic compounds. *Intermetallics* **2002**, *10*, 85–93.

46. Zhou, O.; Yao, Q.; Sun, J.; Smith, D.J. Effect of V addition on the structure of ZrCr$_2$ Laves phase: A high-resolution transmission electron microscope study. *Philos. Mag. Lett.* **2006**, *86*, 347–354.

47. Massalski, T.B. *Binary Alloy Phase Diagrams*; ASM International: Materials Park, OH, USA, 1990.

48. Chang, S.; Young, K.; Ouchi, T.; Meng, T.; Nei, J.; Wu, X. Studies on incorporation of Mg in Zr-based AB$_2$ metal hydride alloys. *Batteries* **2016**, *2*.

49. Young, K.; Ouchi, T.; Nei, J.; Moghe, D. The importance of rare-earth additions in Zr-based AB$_2$ metal hydride alloys. *Batteries* **2016**, *2*.

50. Zhang, Q.A.; Lei, Y.Q.; Yang, X.G.; Ren, K.; Wang, Q.D. Annealing treatment of AB$_2$-type hydrogen storage alloys: II. Electrochemical properties. *J. Alloys Compd.* **1999**, *292*, 241–246.

51. Bououdina, M.; Lenain, C.; Aymard, L.; Soubeyroux, J.L.; Fruchart, D. The effects of heat treatments on the microstructure and electrochemical properties of the ZrCr$_{0.7}$Ni$_{1.3}$ multiphase alloy. *J. Alloys Compd.* **2001**, *327*, 178–184.

52. Young, K.; Ouchi, T.; Huang, B.; Chao, B.; Fetcenko, M.A.; Bendersky, L.A.; Wang, K.; Chiu, C. The correlation of C14/C15 phase abundance and electrochemical properties in the AB$_2$ alloys. *J. Alloys Compd.* **2010**, *506*, 841–848.

53. Nei, J.; Young, K.; Salley, S.O.; Ng, K.Y.S. Effects of annealing on Zr$_8$Ni$_{19}$X$_2$ (X = Ni, Mg, Al, Sc, V, Mn, Co, Sn, La, and Hf): Hydrogen storage and electrochemical properties. *Int. J. Hydrog. Energy* **2012**, *37*, 8418–8427.

54. Young, K.; Chao, B.; Huang, B.; Nei, J. Studies on the hydrogen storage characteristic of La$_{1-x}$Ce$_x$(NiCoMnAlCuSiZr)$_{5.7}$ with a B2 secondary phase. *J. Alloys Compd.* **2014**, *585*, 760–770.

Clean Grain Boundary Found in C14/Body-Center-Cubic Multi-Phase Metal Hydride Alloys

Hao-Ting Shen, Kwo-Hsiung Young, Tiejun Meng and Leonid A. Bendersky

Abstract: The grain boundaries of three Laves phase-related body-center-cubic (bcc) solid-solution, metal hydride (MH) alloys with different phase abundances were closely examined by scanning electron microscopy (SEM), transmission electron microscopy (TEM), and more importantly, electron backscatter diffraction (EBSD) techniques. By using EBSD, we were able to identify the alignment of the crystallographic orientations of the three major phases in the alloys (C14, bcc, and B2 structures). This finding confirms the presence of crystallographically sharp interfaces between neighboring phases, which is a basic assumption for synergetic effects in a multi-phase MH system.

Reprinted from *Batteries*. Cite as: Shen, H.-T.; Young, K.-H.; Meng, T.; Bendersky, L.A. Clean Grain Boundary Found in C14/Body-Center-Cubic Multi-Phase Metal Hydride Alloys. *Batteries* **2016**, *2*, 22.

1. Introduction

Metal hydride (MH) alloys are often used as the negative electrode active material in nickel/metal hydride (Ni/MH) batteries, which are currently dominating the hybrid electric vehicle application. To compete with Li-ion batteries in the battery-powered electric vehicles, the gravimetric energy density of Ni/MH batteries needs to be enhanced. Through an Advanced Research Projects Agency-Energy sponsored robust affordable next generation electric vehicles (RANGE) storage-BASF program, the capacities of both the positive and negative electrode active materials were improved [1]. The former was accomplished by using core-shell-structured β-/α-Ni(OH)$_2$, and the latter was developed from a Laves phase–related body-center-cubic (bcc) solid-solution alloy that demonstrated a 30% increase in storage capacity over conventional misch-metal-based AB$_5$ MH alloys. This group of materials is mainly composed of two interlaced three-dimensional frameworks consisting of both Laves and bcc phases (Figure 4 in [2]). The bcc phase can store a large amount of hydrogen [3,4], while the Laves phase acts as both a catalytic agent for electrochemical reactions [5,6] and a hydride activation facilitator due to its brittleness [7–9]. In these alloys, at some compositions and annealing conditions, TiNi-, Ti$_2$Ni-, and VNi-based secondary phases can also be observed [10]. The bcc phase, which has a higher V content, also has a higher

465

melting temperature and solidifies before the Laves phase (C14, for example) does. According to previous transmission electron microscopy (TEM) studies on C14-predominant MH alloys, the C14 phase solidified before the C15 phase, and the Ti(Zr)-Ni phase was the last to precipitate and it underwent a series of solid-state phase transformations [11–13]. Combining the high-capacity bcc storage phase with catalytic phases (C14, TiNi, Ti_2Ni, and/or VNi), the electrochemical performance of the multi-phase MH alloys can be dramatically improved through synergetic effects-beneficial effects observed in the presence of micro-segregated secondary phases that occurred in the multi-phase MH alloys [14,15]. The bcc storage phase has limited hydrogen absorption/desorption kinetics, though its capacity is high. When it comes in contact with the catalytic phases, hydrogen can move in and out of the bcc phase more freely through the catalytic phases functioning as a funnel. In this way, both the hydrogen storage capacity and the absorption/desorption kinetics can be maximized. For the synergetic effects to occur, it is critical to have a clean interface between the bcc storage phase and the catalytic phase, where the electron and proton transfers are not hindered and a strong bonding is maintained to allow plastic deformation during hydride formation from one side with the lower equilibrium pressure [16,17]. Although the microstructure in each phase has been previously well studied, there has been no definitive study on the microstructure at the interface/grain boundary in these multi-phase MH alloys. Therefore, we present a structural study of such interfaces using scanning electron microscopy (SEM), TEM, and electron backscatter diffraction (EBSD) techniques.

In this work, we focus on the interface of a relatively simple multi-phase alloy system which comprises mainly bcc and C14 phases. Alloys P1, P3 and P7 were selected from a group of Laves phase–related bcc solid-solution alloys with varying bcc/C14 contents [2] for the interface microstructure study. The x values in Equation (1) are 0.7, 1.3 and 2.4 for Alloys P1, P3 and P7, respectively. The compositions in at% and the inductively coupled plasma (ICP) results are summarized in Table 1.

$$Ti_{0.4+x/6}Zr_{0.6-x/6}Mn_{0.44}Ni_{1.0}Al_{0.02}Co_{0.09}(VCr_{0.3}Fe_{0.063})_x \tag{1}$$

With increasing x values, the bcc phase abundance increases in the following manner according to the X-ray diffraction (XRD) analysis: 8.4 wt% (P1), 27.9 wt% (P3), to 54.2 wt% (P7) [2]. In the pressure-concentration-temperature (PCT) measurement after a 2 h thermal cycle between 300 °C and room temperature at 2.5 MPa H_2 pressure, the maximum hydrogen storage capacities in Alloys P1, P3 and P7 increase from 0.61 wt%, 1.00 wt% to 1.12 wt% (Figure 1) [2]. Due to the high plateau pressures, the electrochemical discharge capacities measured in the open-air half-cell configuration at the second cycle are only 162.0 (P1), 160.7 (P3), and

170.0 (P7) mAh·g^{-1} [2]. Further cycling did not improve the discharge capacities significantly. Details of the electrochemical properties of the alloys can be found in [2]. Refinements in composition, fabrication, and activation methods, achieved through a series of studies conducted at BASF-Ovonic [10,18–21], increased the capacity of this family of alloys to 414 mAh·g^{-1} (P37) [1]. However, Alloy P37 is more disordered in phase structure and there are more catalytic phases (C14, TiNi, Ti_2Ni) involved. PCT curves in Figure 1 show that Alloy P7 exhibits absorption/desorption plateaus like P37 that are not observed in P1 and P3. Thus, comparison of the interfacial micro-structure of the relatively simple bcc/C14 alloys is of great significance in helping us understand part of the performance improvement of the more disordered P37 and shed light on the design of the MH alloy with better performance.

Table 1. Design composition and inductively coupled plasma (ICP) results for Alloys P1, P3 and P7 in at%.

Alloy	Ti	Zr	V	Cr	Mn	Fe	Co	Ni	Al
P1 (design)	14.7	13.8	20.0	6.0	12.6	1.2	2.6	28.5	0.6
P1 (ICP)	15.1	13.9	20.2	3.8	12.8	1.3	2.7	29.5	0.7
P3 (design)	14.3	8.9	30.1	9.0	10.2	1.9	2.1	23.1	0.5
P3 (ICP)	14.7	8.8	31.0	7.5	9.8	1.9	2.1	23.7	0.5
P7 (design)	13.7	3.1	42.0	12.6	7.4	2.6	1.5	16.8	0.3
P7 (ICP)	14.2	3.0	41.3	12.1	7.0	2.7	1.6	17.6	0.5
P37 (design)	14.5	1.7	46.6	11.9	6.5	-	1.5	16.9	0.4

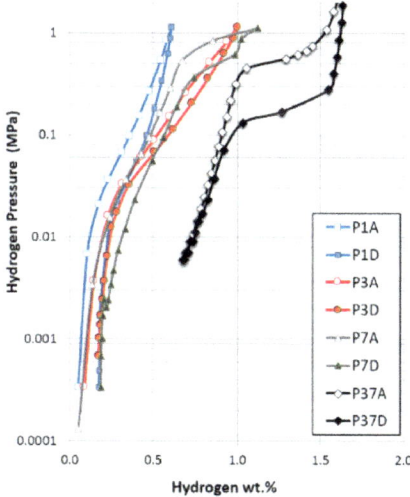

Figure 1. Pressure-concentration-temperature (PCT) isotherms measured at 30 °C for Alloys P1, P3, P7 and P37.

EBSD, also known as backscatter Kikuchi diffraction (BKD) and orientation imaging microscopy (OIM), is a microstructural crystallographic technique that reveals the crystal orientation of the constituent phases in a polycrystal structure using SEM. With an incident electron beam on a point of the sample, an EBSD pattern is collected and the crystal orientation can be obtained from the Kikuchi band positions. Based on the stereographic projections (Figure 2), a simulated pattern is generated and overlaid on the original diffraction pattern to verify if the grain orientation and crystal structure are the best matches. For example, three EBSD patterns collected from various grains in Alloy P3, as shown in Figure 3, are in good agreement with the simulated patterns and have been identified to originate from the same bcc structure but with different orientations. Together with X-ray energy dispersive spectroscopy (EDS), EBSD can provide insight into the microstructure of grains with mixed compositions and orientations [22]. In the research field of MH alloys, the combination of these techniques has been used to study Zr_7Ni_{10}-based [23] and C14-based MH alloys [13]. In this paper, we present results achieved by using EBSD to determine the crystallographic orientation relationship among bcc, C14, and B2 phases in Laves phase–related bcc solid solution alloys. Judging from the alignment of orientations observed among various phases, the cleanliness (being free from amorphous, highly defective, and other disruptive regions) of the interface is verified to provide the synergetic effect in the electrochemical environment.

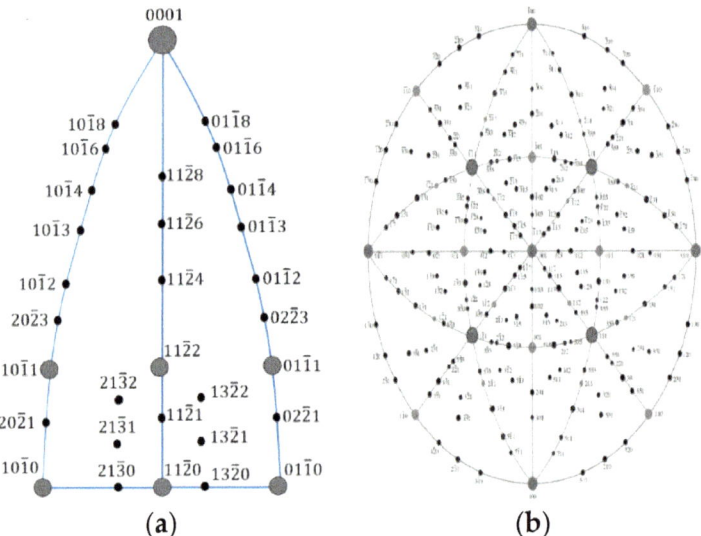

Figure 2. Stereographic projection crystallography for (**a**) hexagonal and (**b**) cubic systems.

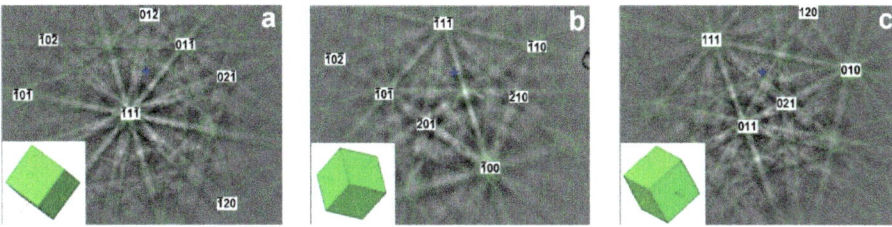

Figure 3. (**a–c**) Electron backscatter diffraction (EBSD) patterns at three different locations of Alloy P3 overlaid with simulated patterns (green lines). All three points are identified to have the same body-center-cubic (bcc) phase structures but different crystal orientations.

2. Experimental Setup

Arc melting under argon flow with tungsten electrode and water cooling was performed to prepare the ingot samples. Before each run, several heating-cooling cycles were performed with sacrificial titanium in the chamber to reduce the residual oxygen concentration. Next, 12 g ingots were re-melted and turned over several times to ensure uniformity with regard to chemical composition. The chemical composition of the ingot was analyzed using a Varian Liberty 100 ICP optical emission spectrometer (ICP-OES, Agilent Technologies, Santa Clara, CA, USA). The gaseous phase PCT analysis was performed with a Suzuki-Shokan Multi-Channel PCT (Tokyo, Japan) system. A JEOL JSM7100 field emission SEM with EDS capability (Tokyo, Japan) was used to study the microstructure of the samples. EDS mapping from the surfaces of polished samples was used to collect compositional distribution information. An FEI Titan 80–300 (scanning) transmission electron microscope (TEM/STEM, Hillsboro, OR, USA) was employed to study the microstructure of the alloy samples. For TEM sample preparation, mechanical polishing was used to thin samples and followed by ion milling.

3. Results and Discussion

3.1. Scanning Electron Microscopy and Energy Dispersive Spectroscopy Results

First, the microstructures and phase profiles of Alloys P1, P3 and P7 were identified using SEM and EDS mapping. The alloy samples are from the arc-melted alloy ingots and did not undergo any annealing process. The backscattered electron (BSE) image of Alloy P1 shows only one predominant phase (Figure 4a), and the EDS mapping shows that the elemental distribution is uniform in most regions, except occasional V and Zr inclusions (Figure 4b). In accordance with the ICP, EDS, and XRD results, this major phase is believed to be a hexagonal C14 (one of the AB_2 Laves phases) [2]. In Alloy P3, the V content increases to approximately 30 at% and the dark regions of Phase I are present in a significant amount and form a fishbone-like network, as shown in Figure 4c. The EDS results (Table 2 and Figure 4d) reveal that this fishbone-like phase is V-rich. The composition of the light regions (Phase II) in Alloy P3 is, on average, close to that of the C14 phase in Alloy P1, although the Ti and Ni contents slightly increase and the Zr and Mn contents slightly decrease, which can be attributed to the phase equilibrium for the V-rich phase (Phase I) in this multi-component alloy. The SEM image in a $100 \times 100 \ \mu m^2$ area also shows oriented alignments in the fishbone-like network in Alloy P3. The size of the regions where the frames are aligned is typically 50–100 μm. For Alloy P7, with the V content reaching 40 at%, the V-rich bcc phase volume fraction expands and, thus, the bcc becomes the major phase, as shown in Figure 4e. As revealed by the EDS mapping shown in Figure 4f, three phases with different elemental compositions are present in Alloy P7. The results of quantitative EDS analysis are listed in Table 2. Compared to Alloy P3, we notice that the compositions of Phase I and Phase II in Alloy P7 are similar to Phase I (green in Figure 4f) and Phase II (between green and red region in Figure 4f) in Alloy P2, which are expected to be bcc and C14 structures, respectively, based on their compositions. Phase III (red in Figure 4f) in Alloy P7 mainly consists of Ni and Ti (Figure 4b and Table 2), a composition commonly seen in the secondary phase of C14-predominating alloys [24,25] exhibiting a B2 crystal structure. This B2 phase is the precursor for the solid phase transformations to Zr_7Ni_{10}, Zr_9Ni_{11}, ZrNi, and TiNi in the C14-predominated MH alloys [11].

Table 2. EDS results (in at%) from Alloys P3 and P7 showing phase segregation.

Phase	Structure	Ti	Zr	V	Cr	Mn	Ni
P3 Phase I	bcc	4	<1	58	16	11	4
P3 Phase II	C14	19	12	19	4	10	31
P7 Phase I	bcc	6	<1	60	19	8	6
P7 Phase II	C14	22	8	20	5	6	32
P7 Phase III	B2	37	3	8	<1	3	43

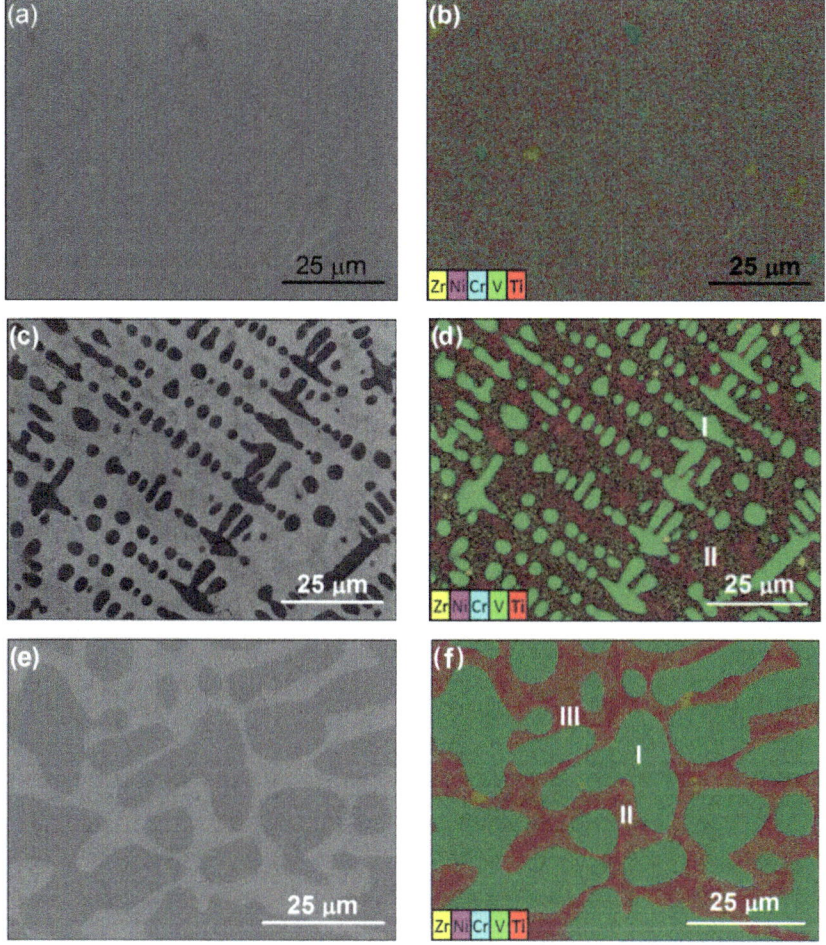

Figure 4. (**a,c,e**) Scanning electron microscopy (SEM) backscattered electron (BSE) images and (**b,d,f**) energy dispersive spectroscopy (EDS) mapping of Alloys (**a,b**) P1, (**c,d**) P3, and (**e,f**) P7, respectively. Phases I–III for P3 and P7 are shown in Table 2.

3.2. Electron Backscatter Diffraction Results

To verify the presence of different crystal structures, EBSD was performed at multiple spots in Alloy P3 (indicated by various symbols in Figure 5a). For the dark regions (Phase I), the analysis shows that the EBSD patterns are the same within each zone (Figure 3a–c corresponding to EBSD patterns from Zones A–C) and the patterns from different zones belong to the same structure and are oriented in different directions. Thus, this alignment of EBSD patterns within the same zone suggests that the dark grains belong to branches of the same dendritic tree that formed the primary phase during solidification. The observed lattice distortion can be explained by non-uniform stress. Analysis of the EBSD patterns from the regions shown in Figure 5a demonstrated that the dark region (Phase I) is consistent with the simulated EBSD of a bcc phase with a lattice constant of approximately 3 Å. Meanwhile, the light regions (Phase II) surrounding the bcc dendrites are believed to solidify later and originate from a liquid of modified composition after the network of bcc dendrites is established. The numerous diffraction patterns obtained by EBSD from these light regions (see Figure 5a and representative EBSD pattern in Figure 5e) show that the crystal structure is consistent with a C14 phase with lattice constants a = 4.9 Å and c = 8 Å. The EBSD data also reveals that, in certain regions (typically on the order of 100 µm), the Phase II structure has the same apparent orientation, similar to the Phase I structure.

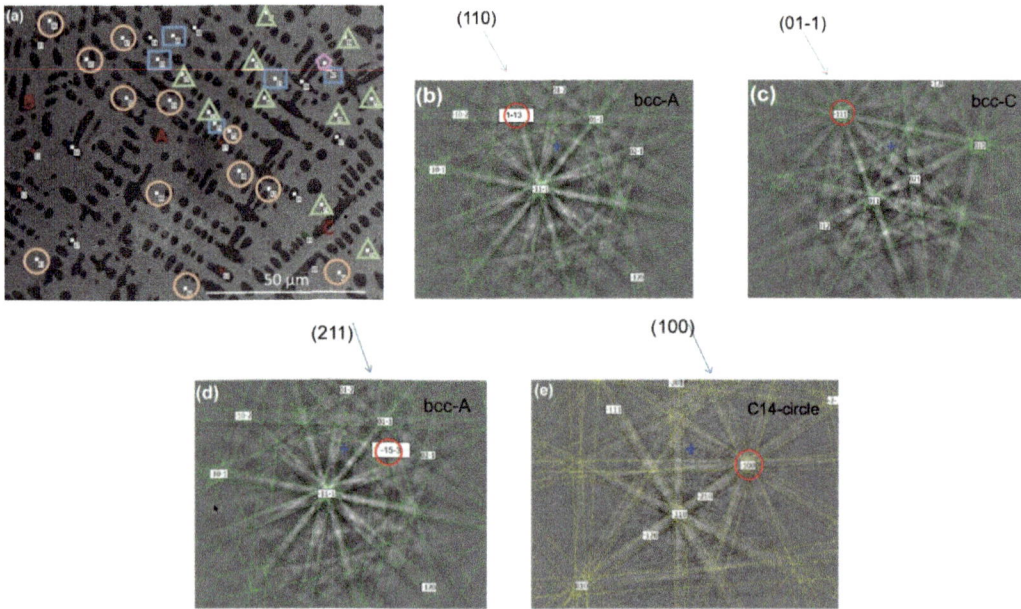

Figure 5. (a) The dark region, denoted as Phase I in the text, can be divided to Zones A–C (indicated by the red letters) based on the EBSD diffraction patterns. The light region, Phase II in the text, can be divided into four grains labeled by circles, triangles, rectangles, and pentagons (the numbers here are spot numbers in the measurement). Comparison between EBSD from (b) A and (c) C zones of the bcc phase indicates the following crystallographic orientation alignment: $(110)_A//(110)_C$ and $<311>_A//<111>_C$. Comparison between EBSDs from (d) Zone A of the bcc phase and (e) the circle grain of the C14 phase indicates the following crystallographic orientation alignment: $(211)_A//(100)_o$ and $<153>_A//<100>_o$. The circles in (b–e) highlight the aligned orientations.

472

It is also worth noting that the crystal orientation of neighboring grains is related. For example, for the EBSD pattern from Alloy P3 (Figure 5a), the (110) bcc crystal plane in bcc Zone A is aligned with the (0$\bar{1}$1) bcc plane in bcc Zone C, as demonstrated in Figure 5b,c. Such an orientation relationship is not only found between Phase I grains that share the same structure, but also between Phases I and II grains with very different crystal structures (bcc and C14). As shown in Figure 5d,e, the (211) bcc plane in bcc Zone A is found to be aligned with the (100) C14 plane in the C14 phase with a circled EBSD pattern (Figure 5a).

3.3. Transmission Electron Microscopy Results

It is challenging to analyze the crystal orientation of the grains and their relationships in Alloy P7 using only EBSD for the following reasons: (1) operation in the EBSD characterization mode requires the SEM system to be in the secondary electron imaging mode, during which the contrast between Phases II and III is weak; (2) the size of the Phase III grains approaches the rather limited space resolution of EBSD (a few microns), and causes difficulty in determining the point of interest. Thus, TEM was employed as an additional method for studying the crystal structures. Samples of Alloy P7 for the TEM study were prepared by mechanical polishing and followed by ion milling. Before loading the thinned samples for TEM study, compositional mapping, which serves as a guide map during TEM observation, was obtained from the sample using SEM/EDS, as shown in Figure 6a. While Phase I (bcc) is mainly composed of V, Cr, and Mn, Phase III (B2) is predominately Ti and Ni.

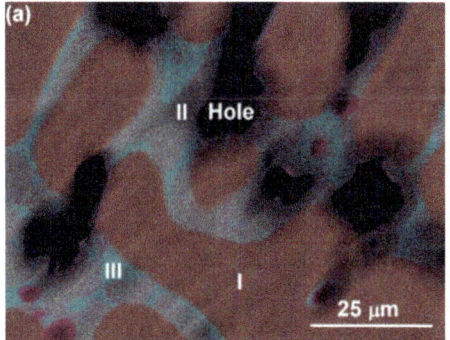 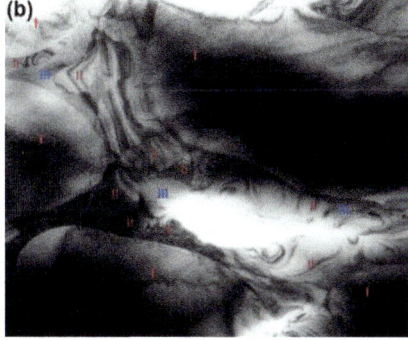

Figure 6. (a) SEM image for TEM touring and (b) low-magnification bright field TEM image from Alloy P7, with the phases labeled. Phases I–III are bcc, C14, and B2, respectively.

On the bright field image at lower magnification (Figure 6b), the dendritic phase (Phase I) can be recognized by the presence of interfaces with positive curvature. Accordingly, the dendritic phase is marked as "I" in Figure 6b, and a bcc phase with a lattice constant of about 3 Å is identified, according to selected area electron diffraction (SAED). Bright field images at higher magnification and SAED patterns from different phases (according to the compositional map) are shown in Figure 7. The interdendritic region with a small volume fraction was targeted to locate the interfaces that separate two (or more) phases. The two phases can be distinguished by the nature of the bright field contrasts. The phase marked as II shows dense diffraction bend contours. The contours presumably result from the TEM film bending to release internal stress during sample thinning. Such behavior is typical for large unit cell intermetallic phases with high elastic modules and no plastic accommodations. The phase marked as III shows a more uniform contrast, which also demonstrates the presence of fine precipitates. Although Phases I and III seem to have similar bcc structures with different compositions, tilting of the TEM samples proved that Phase III had a B2 (ordered bcc) structure. Figure 8 shows SAED taken from Phase III in two orientations, one with a [111] bcc zone axis and another obtained by tilting around the $(1\bar{1}0)^*$ to [110] bcc zone axis, which clearly reveals (100) order reflections, and thus the B2 structural ordering is confirmed. Dark field imaging with the (100) reflection did not show antiphase boundaries, which suggests that the phase directly crystallized as B2. The B2 structure is also consistent with the composition, which consists mainly of Ti and Ni in an atomic ratio of approximately 1:1 (Table 3) [26]. The fine coherent precipitates observed in the phase (and not contributing additionally to the B2 reflections) suggest that the B2 → B2' + bcc transformation, which occurs during cooling, forms into a two-phase field to adjust the composition of the B2 phase at lower temperatures. SAED patterns were also obtained from the Phase II intermetallic phase. The SAED pattern in Figure 9a is indexed and it belongs to the C14 structure in the [100] zone axis with lattice constants $a = 5$ Å and $c = 8$ Å. However, in Figure 9b, the pattern shows a face-centered cubic (fcc) structure in the [110] zone axis, which is neither a hexagonal C14 ($a = 4.9$Å, $c = 7.98$Å for suggested C14 of Alloy P7) nor a fcc C15 (e.g., $a = 7.45$Å for ZrV_2). This pattern indicates the presence of another intermetallic phase in the interdendritic region that can be indexed as an fcc with $a \approx 11$ Å in the [110] zone axis. According to the EBSD and TEM observations, the fcc phase is minor compared to the C14 phase. A similar (or the same) fcc phase (Fd-3m, $a = 11.319$ Å) has been observed in a number of V-Ti-Ni alloys [27].

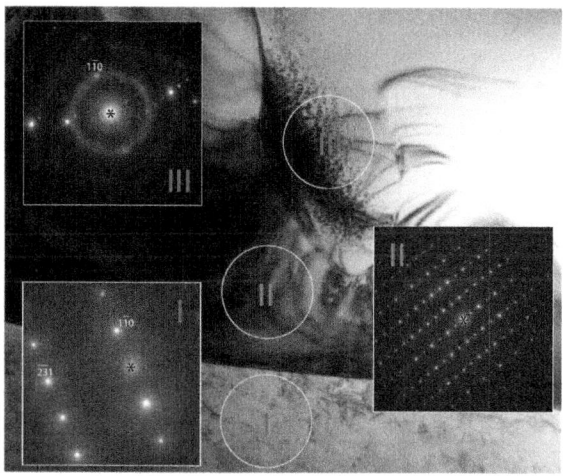

Figure 7. Selected area electron diffraction (SAED) patterns from Phases I–III in Alloy P7.

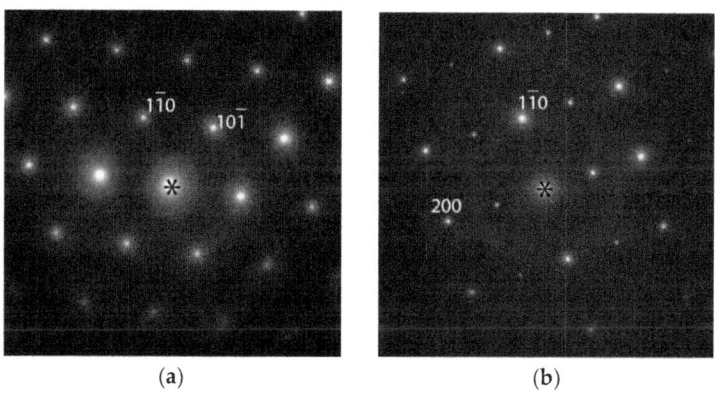

(a)   (b)

Figure 8. SAED taken from Phase III of Alloy P7 in two orientations: (**a**) with the [111] bcc zone axis; and (**b**) obtained by tilting around the $(1\bar{1}0)^*$ to the [110] bcc zone axis.

Table 3. EDS results (in at%) from selective spots in Figure 6a. B/A is the ratio between the sum of at% from other elements vs. Ti and Zr (bcc and C14) and Ti, Zr, and V (B2).

Area	Ti	Zr	V	Cr	Mn	Fe	Co	Ni	Al	B/A	Phase
I	5.8	0.0	59.4	18.3	7.4	2.8	1.1	5.0	0.5	16	bcc
II	23.1	8.3	20.0	4.3	7.3	3.6	2.8	30.0	0.9	2.2	C14
III	35.5	3.3	9.0	1.2	3.2	1.8	2.8	42.3	0.7	1.1	B2

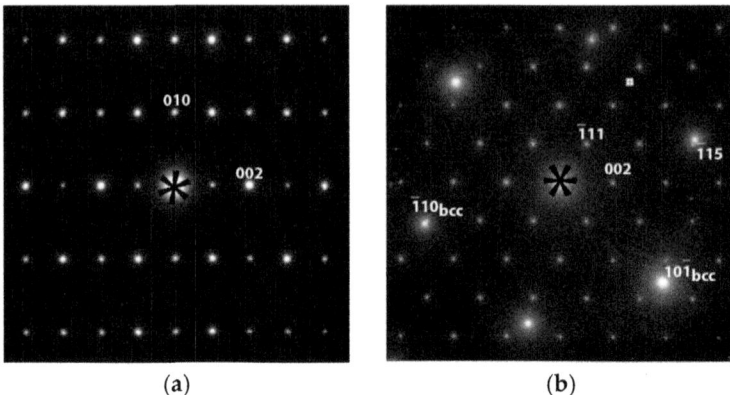

(a)   (b)

Figure 9. SAED patterns from Phase II of Alloy P7 that fit (**a**) a C14 structure in the [100] zone axis and (**b**) a face-centered cubic (fcc) structure with $a \approx 11$ Å in the [110] zone.

Based on the SAED patterns, Phase I (bcc) and Phase III (B2) were very close in orientation. Such alignments were present in more than one region, demonstrating a clear relationship in the crystal orientations between Phase I and Phase III. Based on the morphology of phases in Alloy P7, formation of the phases can be understood. Initially there is the dendritic growth of Phase I (bcc) and at the end of this growth, Phase I is surrounded with a liquid of modified composition. The composition of the liquid allows both Phase II (C14) and Phase III (B2) (and occasionally the fcc phase) to solidify from the liquid. In the case of the Phase III formation, it can be expected that the phase will grow coherently into similar structures, though they are slightly different in lattice parameters. However, if Phase II (C14) is formed first on Phase I (bcc), with a certain orientation determined by their structural relationship, it is expected that the formation of Phase III on Phase II will maintain the same structural relationship and orientation as Phase I/Phase II. Thus, Phase I and III will have very similar orientations. Further study is required to identify which mechanism or both play a role. However, in either case, the boundaries between the contacting phases are crystallographically clean and free of amorphous regions, which are easy to target during the initial formation in alkaline electrolyte, due to preferential etching in the grain boundary [28]. This type of clean interface can certainly facilitate the transportation of protons in the bulk of the alloy, which can be inferred from the relatively high diffusion coefficient (D) of Alloys P1–P8 (1.0–1.5×10^{-10} cm$^2 \cdot$s^{-1}) compared to a typical AB$_2$ multi-phase MH alloy ($<1 \times 10^{-10}$ cm$^2 \cdot$s^{-1}) [29]. The D value can be further increased to 2.13×10^{-10} cm$^2 \cdot$s^{-1} with an annealing at 900 °C for 5 h [10]. As elaborated in a recent paper devoted to the discussion of the synergetic effect in electrochemistry [17], a clean interface is necessary not only to transport protons freely, but also to have excellent electron conductivity. At the end of discharge, there is still some MH phase left on the side with a larger work function (M1 in Figure 14b in [17]), which will act as a nucleation center for the next hydrogenation process. With an elastically connected grain boundary, the stress from MH on the M1 side will enlarge the lattice constant of the M2 side (hydrogenated) and pave the road to the hydrogenation of the M2 phase (M1 in Figure 14b in [17]). In this case, M2 can be hydrogenated before the end of hydrogenation in M1 (Figure 14b in [17]). This elastic binding between M1 and M2 at the interface is a prerequisite for such synergetic effects, more specifically to lower the "equivalent" plateau pressure as in the case of Zr_2Ni_7 [30] and $ZrNi_{4.5}$ [31] multi-phase MH systems.

4. Conclusions

The microstructure of a group of Laves phase–related bcc solid-solution MH alloys with varying bcc/C14 content was studied using SEM/EBSD and TEM/SAED techniques. With the increase of V content in the alloy, the MH alloys develop from a C14-predominated phase for P1 to a bcc/C14 and bcc/C14/B2 multi-phase structure for P3 and P7, respectively. For the multi-phase P3 and P7, a high degree of alignment was found among the bcc grains, between bcc and C14, and bcc and B2 (P7), which suggests a very clean interface at the grain boundaries. These clean interfaces can facilitate the transport of protons between the bcc main storage phase and the catalytic phases (C14, B2) and ensure good electron conductivity, which supports the proposed synergetic effects found in the electrochemical testing of multi-phase MH alloys. Based on the works of P1–P7, a more complicated alloy, P37, with a higher V content, no Fe and additional Ti_2Ni catalytic phase shows further improvement in the electrochemical performance.

Acknowledgments: This work is financially supported by ARPA-E under the RANGE program (DE-AR0000386). The authors would like to thank the following for technical assistance from BASF-Ovonic: Jean Nei, Baoquan Huang, Taihei Ouchi, and Su Cronogue.

Author Contributions: Hao-Ting Shen designed the experiments and performed the SEM and TEM analysis. Kwo-Hsiung Young, Tiejun Meng and Leonid A. Bendersky analyzed and interpreted the data.

Conflicts of Interest: The authors declare no conflict of interest.

Abbreviations

bcc	Body-centered-cubic
EBSD	Electron backscattering diffraction
MH	Metal hydride
Ni/MH	Nickel/metal hydride
RANGE	Robust affordable next generation electric vehicles

TEM	Transmission electron microscopy
SEM	Scanning electron microscopy
ICP	Inductively coupled plasma
PCT	Pressure-concentration-temperature
BKD	Backscatter Kikuchi diffraction
OIM	Orientation imaging microscopy
EDS	Energy dispersive spectroscopy
OES	Optical emission spectrometer
BSE	Backscattering electron
XRD	X-ray diffraction
SAED	Selected area electron diffraction
fcc	Face-centered cubic
D	Diffusion constant

References

1. Young, K.; Ng, K.Y.S.; Bendersky, L. A Technical Report of the Robust Affordable Next Generation Energy Storage System-BASF Program. *Batteries* **2016**, *2*.
2. Young, K.; Nei, J.; Wong, D.F.; Wang, L. Structural, hydrogen storage, and electrochemical properties of Laves phase-related body-centered-cubic solid solution metal hydride alloys. *Int. J. Hydrog. Energy* **2014**, *39*, 21489–21499.
3. Inoue, H.; Arai, S.; Iwakura, C. Crystallographic and electrochemical characterization of TiV$_{4-x}$Ni$_x$ alloys for nickel-metal hydride batteries. *Electrochim. Acta* **1996**, *41*, 937–939.
4. Yu, X.B.; Wu, Z.; Xia, B.J.; Xu, N.X. A Ti-V-based bcc phase alloy for use as metal hydride electrode with high discharge capacity. *J. Chem. Phys.* **2004**, *121*, 987–990. PubMed]
5. Chen, N.; Li, R.; Zhu, Y.; Liu, Y.; Pan, H. Electrochemical hydrogenation and dehydrogenation mechanisms of the Ti-V base multiphase hydrogen storage electrode alloy. *Acta Metal. Sin.* **2004**, *40*, 1200–1204.
6. Qiu, S.; Chu, H.; Zhang, Y.; Sun, D.; Song, X.; Sun, L. Electrochemical kinetics and its temperature dependence behaviors of Ti$_{0.17}$Zr$_{0.08}$V$_{0.35}$Cr$_{0.10}$Ni$_{0.30}$ alloy electrode. *J. Alloy. Compd.* **2009**, *471*, 453–456.
7. Iba, H.; Akiba, E. The relation between microstructure and hydrogen absorbing property in Laves phase-solid solution multiphase alloys. *J. Alloy. Compd.* **1995**, *231*, 508–512.
8. Rönnebro, E.; Noréus, D.; Sakai, T.; Tsukahara, M. Structural studies of a new Laves phase alloy (Hf,Ti)(Ni,V)$_2$ and its very stable hydride. *J. Alloy. Compd.* **1995**, *231*, 90–94.
9. Tsukahara, M.; Takahashi, K.; Mishima, T.; Isomura, A.; Sakai, T. V-based solid solution alloys with Laves phase network: Hydrogen absorption properties and microstructure. *J. Alloy. Compd.* **1996**, *236*, 151–155.
10. Young, K.; Ouchi, T.; Nei, J.; Wang, L. Annealing effects on Laves phase-related body-centered-cubic solid solution metal hydride alloys. *J. Alloy. Compd.* **2016**, *654*, 216–225.

11. Boettinger, W.J.; Newbury, D.E.; Wang, K.; Bendersky, L.A.; Chiu, C.; Kattner, U.R.; Young, K.; Chao, B. Examination of multiphase (Zr, Ti)(V, Cr, Mn, Ni)$_2$ Ni-MH electrode alloys: Part I. Dendritic solidification structure. *Metall. Mater. Trans.* **2010**, *41A*, 2033–2047.

12. Bendersky, L.A.; Wang, K.; Boettinger, W.J.; Newbury, D.E.; Young, K.; Chao, B. Examination of multiphase (Zr, Ti)(V, Cr, Mn, Ni)2 Ni-MH electrode alloys: Part II. Solid-state transformation of the interdendric B$_2$ phase. *Metall. Mater. Trans.* **2010**, *41A*, 1891–1906.

13. Liu, Y.; Young, K. Microstructure investigation on metal hydride alloys by electron backscatter diffraction technique. *Batteries* **2016**, *2*.

14. Visintin, A.; Peretti, H.A.; Fruiz, F.; Corso, H.L.; Triaca, W.E. Effect of additional catalytic phases imposed by sintering on the hydrogen absorption behavior of AB$_2$ type Zr-based alloys. *J. Alloy. Compd.* **2007**, *428*, 244–251.

15. Wong, D.F.; Young, K.; Nei, J.; Wang, L.; Ng, K.Y.S. Effects of Nd-addition on the structural, hydrogen storage, and electrochemical properties of C14 metal hydride alloys. *J. Alloy. Compd.* **2015**, *647*, 507–518.

16. Mosavati, N.; Young, K.; Meng, T.; Ng, K.Y.S. Electrochemical open-circuit voltage and pressure-concentration-temperature isotherm comparison for metal hydride alloys. *Batteries* **2016**, *2*.

17. Young, K.; Ouchi, T.; Meng, T.; Wong, D.F. Studies on the synergetic effects in multi-phase metal hydride alloys. *Batteries* **2016**, *2*.

18. Young, K.; Wong, D.F.; Wang, L. Effect of Ti/Cr content on the microstructures and hydrogen storage properties of Laves phase-related body-centered-cubic solid solution metal hydride alloys. *J. Alloys Compd.* **2015**, *622*, 885–893.

19. Young, K.; Ouchi, T.; Huang, B.; Nei, J. Structure, hydrogen storage, and electrochemical properties of body-centered-cubic Ti$_{40}$V$_{30}$Cr$_{15}$Mn$_{13}$X$_2$ alloys (X = B, Si, Mn, Ni, Zr, Nb, Mo, and La). *Battereis* **2005**, *1*, 74–90.

20. Young, K.; Ouchi, T.; Huang, B.; Nei, J. Effects of Cr, Zr, V, Mn, Fe, and Co to the hydride properties of Laves phase-related body-centered-cubic solid solution alloys. *J. Power Sources* **2015**, *281*, 165–172.

21. Young, K.; Wong, D.F.; Nei, J. Effects of vanadium/nickel contents in Laves phase-related body-centered-cubic solid solution metal hydride alloys. *Batteries* **2005**, *1*, 34–53.

22. Liu, S.; Chu, H.; Zhang, J.; Zhang, Y.; Sun, L.; Xu, F. Effect of La partial substitution for Zr on the structural and electrochemical properties of Ti$_{0.17}$Zr$_{0.08-x}$La$_x$V$_{0.35}$Cr$_{0.1}$Ni$_{0.3}$ (x = 0–0.04) electrode alloys. *Int. J. Hydrog. Energy* **2009**, *34*, 7246–7252.

23. Young, K.; Ouchi, T.; Liu, Y.; Reichman, B.; Mays, W.; Fetcenko, M.A. Structural and electrochemical properties of Ti$_x$Zr$_{7-x}$Ni$_{10}$. *J. Alloy. Compd.* **2009**, *480*, 521–528.

24. Young, K.; Nei, J.; Ouchi, T.; Fetcenko, M.A. Phase abundances in AB$_2$ metal hydride alloys and their correlations to various properties. *J. Alloy. Compd.* **2011**, *509*, 2277–2284.

25. Young, K.; Ouchi, T.; Huang, B.; Chao, B.; Fetcenko, M.A.; Bendersky, L.A.; Wang, K.; Chiu, C. The correlation of C14/C15 phase abundance and electrochemical properties in the AB$_2$ alloys. *J. Alloy. Compd.* **2010**, *506*, 841–848.

26. Michal, G.M.; Singlair, R. The structure of TiNi martensite. *Acta Crystallogr.* **1981**, *B37*, 1803–1807.

27. Song, G.; Dolan, M.D.; Kellam, M.E.; Liang, D.; Zambelli, S. V-Ni-Ti multi-phase alloy membranes for hydrogen purification. *J. Alloy. Compd.* **2011**, *509*, 9322–9328.

28. Young, K.; Chao, B.; Nei, J. Microstructures of the activated Si-containing ab$_2$ metal hydride alloy surface by transmission electron microscope. *Batteries* **2016**, *2*.

29. Young, K.; Huang, B.; Regmi, R.K.; Lawes, G.; Liu, Y. Comparisons of metallic clusters imbedded in the surface oxide of AB$_2$, AB$_5$, and A$_2$B$_7$ alloys. *J. Alloy. Compd.* **2010**, *506*, 831–840.

30. Young, M.; Chang, S.; Young, K.; Nei, J. Hydrogen storage properties of ZrV$_x$Ni$_{3.5-x}$ (x = 0.0–0.9) metal hydride alloys. *J. Alloy. Compd.* **2013**, *580*, S171–S174.

31. Young, K.; Young, M.; Chang, S.; Huang, B. Synergetic effects in electrochemical properties of ZrV$_x$Ni$_{4.5-x}$ (x = 0.0, 0.1, 0.2, 0.3, 0.4, and 0.5) metal hydride alloys. *J. Alloy. Compd.* **2013**, *560*, 33–41.

Electrochemical Open-Circuit Voltage and Pressure-Concentration-Temperature Isotherm Comparison for Metal Hydride Alloys

Negar Mosavati, Kwo-Hsiung Young, Tiejun Meng and K. Y. Simon Ng

Abstract: In this study we compared the electrochemical pressure-concentration-temperature (EPCT) method with the gaseous phase pressure-concentration-temperature (PCT) method and demonstrated the differences between the two. Experimentally, this was done by electrochemically charging/discharging the electrodes of four different metal hydride (MH) alloys. The results indicate that in the PCT curve is flatter with a smaller hysteresis and a higher storage capacity compared to the EPCT curve. Moreover, while the PCT curves (up to around one third of the hydrogen storage capacity) reside in between the charge and discharge EPCT curves, the rest of the PCT curves are below the EPCT curves. Finally, we demonstrated a new calibration method based on the inflection points observed in the EPCT isotherms of a physical mixture of more than one alloy. This turning point can be used to find a preset calibration point to determine the state-of-charge.

Reprinted from *Batteries*. Cite as: Mosavati, N.; Young, K.-H.; Meng, T.; Ng, K.Y.S. Electrochemical Open-Circuit Voltage and Pressure-Concentration-Temperature Isotherm Comparison for Metal Hydride Alloys. *Batteries* **2016**, *2*, 6.

1. Introduction

Nickel/metal hydride (Ni/MH) batteries are an important energy storage medium for portable electronic devices, electric vehicles, and alternative energy. The active material used in the negative electrode (anode) of these batteries is a special kind of metal called hydrogen storage alloy (or metal hydride (MH) alloy) that is capable of storing hydrogen at ambient temperatures [1]. MH alloy can absorb hydrogen in two different environments: through an electrochemical charging process in wet chemistry or gaseous absorption in a dry container (Figure 1). While the former process requires splitting water molecules into protons and hydroxide ions, the latter depends on the splitting of hydrogen molecules into hydrogen atoms. Other differences between the electrochemical charging and gaseous phase hydrogen storage can be found in Table 1. Since the electrochemical environment is more complicated for hydrogen absorption compared to the gaseous phase, the pressure-concentration-temperature (PCT) measurement is commonly used to study the thermodynamics of gaseous phase hydrogen absorption of an MH

481

alloy [1]. The information obtained from PCT, including hydrogen storage capacity, metal-hydrogen bond strength (from the equilibrium pressure), hysteresis, and heat of hydride formation, are all essential for the design of MH alloys used for gaseous hydrogen storage and electrochemical applications. In wet chemistry, the hydrogen storage performance may be heavily influenced by the surface oxide [2], surface catalyst phase [3], proximity effect among phases [4], and activated grain boundary [5], which are difficult to isolate. Therefore, it is interesting and important to compare the electrochemical and gaseous phase hydrogen storages side by side through comparing the stabilized open-circuit voltage (V_{oc}) *versus* state of charge (SOC) curves in the electrochemical PCT (EPCT) and gaseous phase PCT curves.

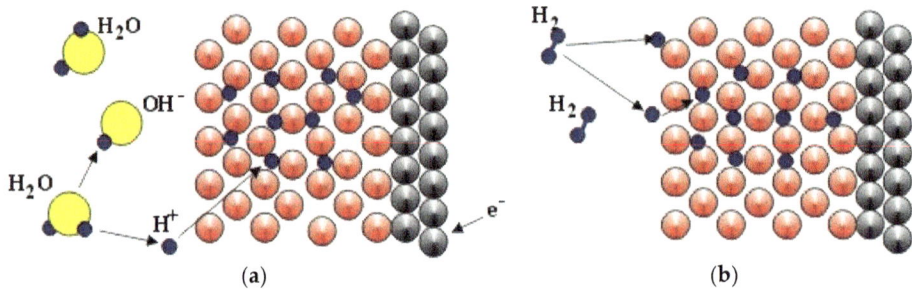

Figure 1. Schematic of hydrogen absorption through (**a**) an electrochemical reaction and (**b**) a gaseous phase chemical reaction.

Table 1. The main differences between electrochemical charging and gaseous phase hydrogen storage.

Difference in	Electrochemical Charging	Gaseous Phase Hydrogen Storage
Source of H	Splitting H_2O molecule at the electrode/electrolyte interface	H_2 dissociation at the surface
Environment	Alkaline oxidizing environment (KOH electrolyte)	H_2 gas, very susceptible to oxygen poisoning
Kinetics	Hydrogen storage/release at room temperature	Hydrogen storage/release at temperature range of 20–130 °C
Thermodynamics	ΔH from -30 kJ·mol·H_2^{-1} to -35 kJ·mol·H_2^{-1}	ΔH from -20 kJ·mol·H_2^{-1} to -55 kJ·mol·H_2^{-1}
Thermal conductivity	Not crucial	Very important
Electrical conductivity	Very important	Not crucial
Chemical reaction	$M + H_2O + e^- \leftrightarrows MH + OH^-$	H_2 (g) $\leftrightarrows$ 2H
Surface requirement	A thin and porous oxide allowing electrolyte penetration	Free from oxide and other contaminations
Common catalyst	Metallic Ni embedded in surface oxide [6]	Noble metals like Pd and Pt [7]

The EPCT method is not a new idea and had been applied in a number of MH alloy studies, including studies on AB_5 [8–11], AB_2-based multiphase alloys [12–14], Ce-Mn-Al-Ni [15], Ce-Mg-Ni-Ti-F [16], carbon nanofiber [17], and Mg_2Ni [18]. In EPCT, the stabilized V_{oc} is converted to equilibrium pressure via the Nernst Equation (1) [19]:

$$E_{MH,eq} \ (vs. \ Hg/HgO) \ = \ -0.934 - 0.029 \log P(H_2) \tag{1}$$

where $E_{MH,eq}$ is the equilibrium potential of hydrogen storage electrode *versus* the Hg/HgO reference electrode and $P(H_2)$ is the equilibrium hydrogen gas pressure, and the electrochemical capacity is converted to hydrogen storage capacity by using the conversion of $268 \ mAh \cdot g^{-1} = 1$ wt%. It should be noted that for the Ni/MH half-cell configuration used in this study, a partially pre-charged positive electrode was used and has a constant potential of 0.36 V *versus* Hg/HgO [20], and therefore $E_{MH,eq}$ (*vs.* Hg/HgO) $\approx V_{oc} - 0.36$ (V). Ideally, the stabilized V_{oc} obtained experimentally is close to the battery equilibrium potential.

The advantages of EPCT include simplicity, ease of operation, and ability to cover a wider voltage (pressure) range since a decade of pressure change is equivalent to a voltage difference of 29 mV at room temperature. It is also closer to the real life environment that batteries experience. Although many studies using EPCT to characterize MH alloys have been reported, a direct comparison between EPCT and the gaseous phase PCT had never been made and is the focus of this paper.

2. Experimental Section

Induction melting with an $MgAl_2O_4$ crucible, an alumina tundish, and a pancake-shaped steel mold in a 2-kg furnace under argon atmosphere were performed to prepare the ingot samples. AB_5 MH alloys were annealed at 960 °C for 8 h under vacuum. Gaseous phase hydrogen storage was measured using a Suzuki-Shokan multi-channel PCT system (Suzuki Shokan, Tokyo, Japan). Each ingot piece (about 2 g) was first activated by a 2 h thermal cycle between room temperature and 300 °C under 2.5 MPa H_2 pressure followed by PCT measurements at 30, 60, and 90 °C. To fabricate the electrodes for electrochemical tests, the ingot was hydrided and pulverized to -200 mesh powder and pressed onto an expanded Ni substrate without any binder. For EPCT analysis, a flooded half-cell configuration was used with the MH alloy as negative electrode, partially pre-charged and oversized $Ni_{0.9}Co_{0.1}(OH)_2$ as counter electrode, and 30 wt% KOH as electrolyte. Before the EPCT measurements, the electrode samples in half-cells were activated with 10 charge/discharge cycles at $25 \ mA \cdot g^{-1}$ and room temperature using a CTE MCL2 Mini cell test system (Chen Tech Electric MFG. Co., Ltd., New Taipei, Taiwan). Each half-cell was then cycled with several charge/discharge steps, each is 15 min long, at a current density of

$25 \text{ mA} \cdot \text{g}^{-1}$ and with a 30 min rest between each step. The electrochemical V_{oc} was recorded at the end of the 30 min rest after each charge/discharge step.

3. Results and Discussion

EPCTs of four MH alloys are included in this study, and their compositions and some gaseous phase properties are listed in Table 2. Fe1 is a C14-predominated AB_2 MH alloy developed in a previous Fe-addition study [21]. B37 and B65 are misch metal-based AB_5 MH alloys with very different equilibrium plateau pressures (Figure 2). The last alloy, YC#1, is a transition metal-based AB_5 MH alloy developed in a study of a series of $ZrV_xNi_{4.5-x}$ alloys showing a large discrepancy between gaseous phase and electrochemical hydrogen storage capacities [22]. EPCTs of these alloys will be reported in the following sections.

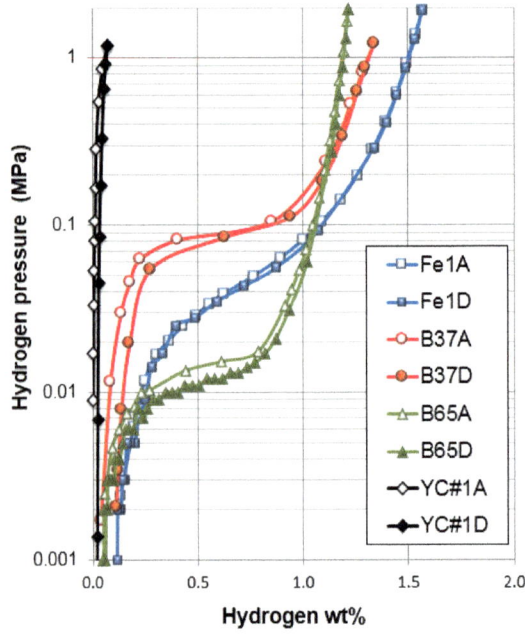

Figure 2. PCT isotherms of alloys in this study. Fe1 is a Fe-containing AB_2 metal hydride (MH) alloy. B37 and B65 are misch metal-based AB_5 MH alloys. YC#1 is a $ZrNi_5$-based MH alloy. Open and filled symbols stand for absorption and desorption isotherms, respectively.

Table 2. Design composition and hydrogen storage properties (at 30 °C) of the alloys in this study. PCT: pressure-concentration-temperature.

Alloy	Family	Composition (in at%)	Maximum H-storage Capacity	Desorption Plateau Pressure	PCT Hysteresis
Fe1	AB_2	$Ti_{12}Zr_{21.5}V_{10}Cr_{7.5}Mn_{8.1}Fe_{1.0}Co_{7.0}Ni_{32.2}Sn_{0.3}Al_{0.4}$	1.51 wt%	0.046 MPa	0.05
B37	AB_5	$La_{9.6}Ce_{4.0}Pr_{0.5}Nd_{1.3}Ni_{65.8}Co_{4.6}Mn_{4.2}Al_{5.5}Cu_{4.6}$	1.34 wt%	0.088 MPa	0.06
B65	AB_5	$La_{10.5}Ce_{4.3}Pr_{0.5}Nd_{1.4}Ni_{62.3}Co_{5.0}Mn_{4.6}Al_{6.0}Cu_{3.2}Zr_{0.2}Mo_{2.0}$	1.21 wt%	0.012 MPa	0.28
YC#1	AB_5	$Zr_{18.2}Ni_{81.8}$	0.075 wt%	>1 MPa	1.76

3.1. Electrochemical Pressure-Concentration-Temperature of AB_2 Metal Hydride Alloy

Room EPCT without re-calibration from Fe1 is shown in Figure 3a. Due to hydrogen gas evolution in the open-cell configuration, the starting point of the charge curve does not match the ending point of the discharge curve. After performing a self-discharge calibration, we found the capacity loss due to hydrogen gas evolution is linear with regard to SOC (from 0% to 100%), and the capacity loss for Fe1 is about $(0.3 + 1.2 \times SOC)$ mAh·g^{-1}·h^{-1}. The higher SOC of the electrode promotes increased hydrogen gas escaping from the electrode surface. After this calibration, the EPCT charge and discharge curves for Fe1 form a closed loop (Figure 3b). Compared to the PCT isotherm in the same graph, four observations can be made:

(1) The EPCT curve is much more slanted than the PCT curve.
(2) The hysteresis of EPCT is much larger than that of PCT, which has been demonstrated previously by Wójcik and his coworkers [13].
(3) The maximum capacity measured from EPCT is smaller than that from PCT, which has also been previously shown by Wójcik and his coworkers [13].
(4) The PCT isotherm is closer to the EPCT charge curve when the hydrogen storage content is small, but as the hydrogen storage content increases, it then moves to the center between the EPCT charge and discharge curves and finally flattens out.

The higher hysteresis (asymmetry) in EPCT compared to that in PCT (Observation 2) is related to the difference in equilibrium state. In the gaseous phase equilibrium state, hydrogen concentration in the alloy bulk is uniform. However, in the electrochemical open-circuit condition, distribution of protons is uneven if there is a voltage across the electrode. Therefore, V_{oc} during the charge state (high concentration of OH^- in the pores of surface oxide layer) and discharge state (lower concentration of OH^- in the pores of surface oxide layer due to water recombination) are different. In the open-circuit condition, the levels of hydrogen storage content in various parts of the electrode are different depending on the distance to surface, which results in a composite of hydrogen storage phases (components) with different metal-hydrogen bond strengths. This phenomenon increases the degree of disorder and lowers the critical temperature to where the PCT plateau disappears [23,24] and thus makes the EPCT curve more slanted (Observation 1).

As for the lower electrochemical capacity seen in EPCT compared to that in PCT (Observation 3), it is a very common observation in MH alloys [25,26]. Surface oxidation during the activation step in the electrochemical environment is one of the causes. In addition, some inter-grain phases acting as catalysts in the gaseous phase [27] can be etched away in the electrochemical environment [5]. In the open-cell configuration, the electrode cannot be fully charged due to hydrogen evolution, which can also contribute to the lower capacity measured by EPCT.

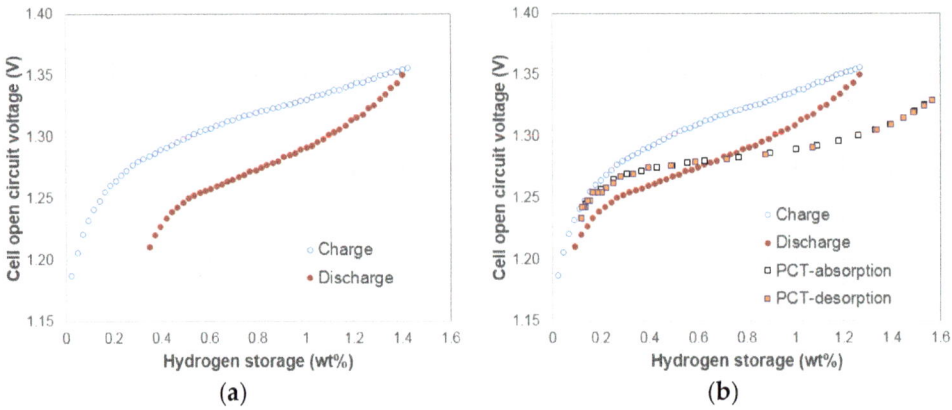

Figure 3. Room temperature electrochemical pressure-concentration-temperature (EPCTs) for Fe1: (**a**) before and (**b**) after taking self-discharge into consideration.

Observation 4 implies that the assumption of two-electron transfer with H_2 in Equation (1) is only valid at lower SOCs. As the hydrogen storage content increases, the electrochemical reaction becomes closer to the interaction with single H atoms, which will raise the voltage as dictated by the Nernst equation with single electron transfer Equation (2):

$$E_{\text{MH,eq}} \ (vs. \ \text{Hg/HgO}) \ = \ -0.934 - 0.059 \log P(\text{H}_2) \qquad (2)$$

3.2. Electrochemical Pressure-Concentration-Temperatures of Two AB₅ Metal Hydride Alloys

Room EPCTs of two AB_5 alloys with a large difference in plateau pressure, B37 and B65, were measured, and the charge and discharge curves after self-discharge calibration are shown in Figure 4. All four observations made in the previous section of the AB_2 MH alloy are also present for the AB_5 MH alloys. Therefore, these observations are likely to be universal for most MH alloys. Comparing these two EPCTs, the alloy with a lower PCT plateau pressure (B65) has a higher electrochemical capacity than the alloy with a higher PCT plateau pressure (B37), and hence the proposed cause for lower EPCT capacity discussed in the last section (incomplete charging due to hydrogen gas evolution) is validated.

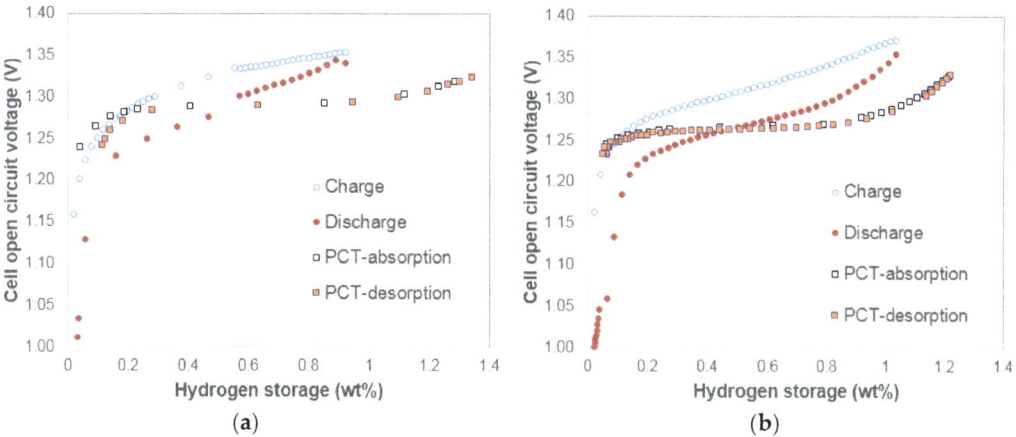

Figure 4.   Room temperature EPCTs for (**a**) B37 and (**b**) B65 after taking self-discharge into consideration.

3.3. Electrochemical Pressure-Concentration-Temperatures of Mixtures of Two AB$_5$ Metal Hydride Alloys

PCTs of materials with two phases (components) were classified into three categories: funneling mode, synergetic mode, and independent mode [4,28]. If two MH alloys were physically mixed together without any alloying, the resulting PCT isotherm will be a composite showing two individual plateaus (Figure 7c in [4]). It is interesting to see whether the discontinuity in the PCT isotherm of two-part alloy is present in the corresponding EPCT isotherm. If the discontinuity exists in the EPCT isotherm, it will contribute tremendously to SOC determination, which relies heavily on V_{oc} and impedance (Z) in today's battery management systems [29–35]. However, V_{oc} and Z depend not only on SOC but also on the state-of-health and environmental temperature, making a precise calculation of SOC impossible.

Room temperature EPCTs of B37/B65 mixtures (10:90, 30:70 and 50:50 by weight) are shown in Figure 5a–c, respectively. Discontinuities in slope can easily be found in both the charge and discharge curves in Figure 5a,b. In the case of the 50:50 mixture, the derivative of V_{oc} *versus* hydrogen storage content is plotted in Figure 5d to show the locations of the inflection points (when the derivative reaches the local maximum). The existence of these inflection points confirms the possibility of using V_{oc} to find a preset calibration point for SOC determination.

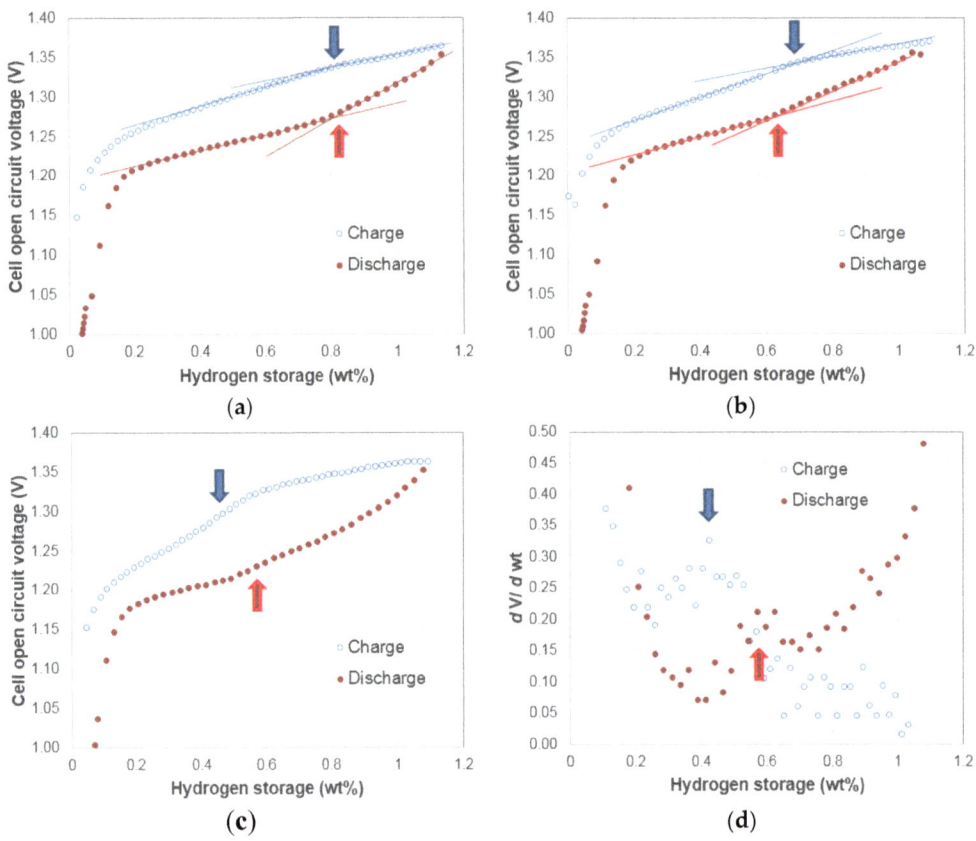

Figure 5. Room temperature EPCTs for (**a**) 10% B37 + 90% B65; (**b**) 30% B37 + 70% B65; (**c**) 50% B37 + 50% B65 mixtures and (**d**) derivative of Figure 5c. Arrows indicate positions of inflection points.

In order to investigate whether the existence of inflection points is sensitive to the environment temperature, EPCT of the B37/B65 50:50 mixture was measured at 10 °C. The resulting charge and discharge curves are shown in Figure 6a and clearly show the inflection points at approximately 50% SOC. Moreover, Figure 6b shows the derivative of V_{oc} *versus* hydrogen storage content, which can be used to better locate the inflection points.

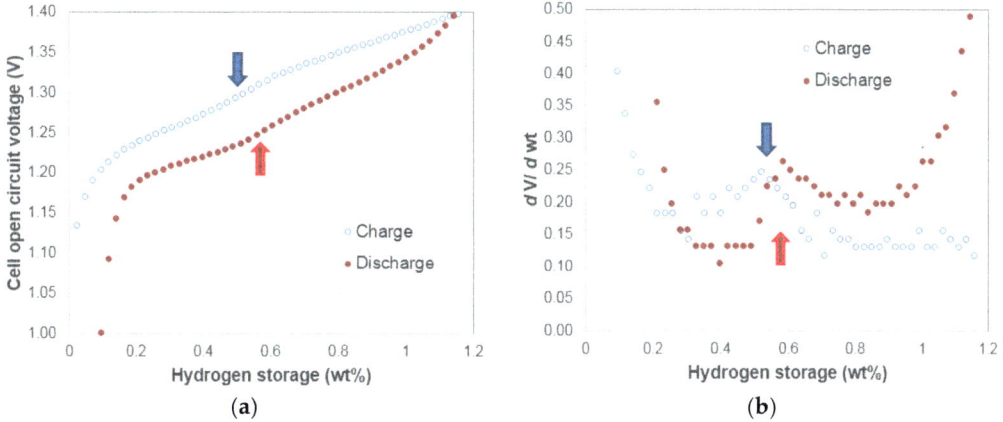

Figure 6. 10 °C: (**a**) EPCT and (**b**) derivative for 50% B37 + 50% B65 mixture. Arrows indicate positions of inflection points.

3.4. Electrochemical Pressure-Concentration-Temperature of ZrNi$_{4.5}$ Metal Hydride Alloy

In studies of multiphase Zr-Ni-V MH alloys, we found that the electrochemical capacities of these alloys were unusually high compared to the gaseous phase hydrogen storage capacities [22,36], which had been attributed to the decrease in PCT plateau pressure through synergetic effects in the electrochemical environment [22,36]. We now compare EPCT and PCT of one of those alloys, YC#1, in Figure 7. It is obvious that the charge and discharge curves are below the PCT curves with the assumption that the same correlation from AB$_2$ (Figure 3b) and AB$_5$ (Figure 4a,b) can be applied in YC#1. Therefore, we confirm that our previous statement: lowering the equilibrium pressure in the electrochemical environment is the main cause of increased capacity.

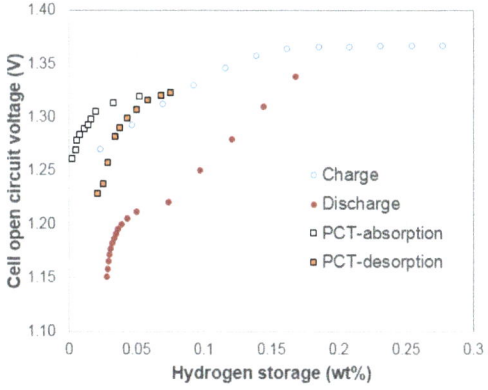

Figure 7. Room temperature EPCT for YC #1.

3.5. Discussion

The difference between PCT and EPCT results can be explained by the different natures of hydrogen storage processes in a single component hydrogen storage alloy (or metal) system. For hydrogen absorption in the gaseous phase, hydrogen gas molecules adsorb and split into two hydrogen atoms at the clean surface of metal (free from oxide), and then hydrogen atoms move into the bulk of alloy by diffusion. The protons occupy the interstitial sites among the host atoms while the accompanied electrons join the conduction band of host metal and increase the Fermi level of conduction electrons. In the electrochemical environment, the applied voltage forces electrons to flow into the negative electrode and also splits the water molecules at the electrode surface. While the protons enter the bulk of alloy driven by the electric field from the potential difference and are neutralized by the electrons from the current collector, the hydroxide ions are released into the electrolyte. Moreover, the equilibrium pressure in the gaseous phase is only dependent on the concentration of hydrogen in the metal, but what affects V_{oc} in the electrochemical environment is far more complicated and involves the dipole from the electrolyte-solid interface, chemical products from the activation process, and potential between the bulk and altered surfaces with different chemical compositions from the bulk. The discharge processes for the gaseous phase and electrochemistry are also different. In the gaseous phase, the movement of protons is intrigued by diffusion, and equilibrium is reached when the same amount of hydrogen is leaving and entering the metal. However, it is more complicated in the electrochemical environment. During discharge, electrons move away from the hydrided alloy into the current collector, forcing the protons to move in the opposite direction and reach the electrode surface. Unless the surface recombination of either proton-hydroxide or proton-oxygen is very slow (not likely) and a large amount of protons accumulation presents at the electrode-electrolyte interface, the protons should continue to flow towards the interface in order to reach charge neutrality until a depletion of protons.

Hystereses found in EPCT of the AB$_2$ and AB$_5$ MH alloys are much larger than the corresponding hysterese in PCT. While the hysteresis observed in PCT can be simplified as the energy barrier of elastic deformation at the α-β interface during hydrogenation [37], many other factors can influence the hysteresis in EPCT, such as the potential between the activated surface region and bulk, potential across the surface oxide region and oxide in the activated grain boundaries [38], remaining charge trapped within the defects and grain boundaries, residual protons trapped in the grain boundaries, and distribution of hydroxide ions in the surface pores and grain boundaries that are connected to the electrolyte by narrow channels.

Hydrogen storage in a multiphase alloy system is by far more complicated due to the presence of synergetic effects. Synergetic effects in multiphase MH alloys refer to the beneficial effects provided by the presence of micro-segregated secondary phases occurred in the melted alloys [39], which were reviewed, explained, and compared between gaseous phase and electrochemistry in one of our recent publications [40] and can be summarized as follows: the contact potential at the boundary between two neighboring phases with different work functions resulted from the increase in the Fermi level of the side with a lower Fermi level by the addition of electrons into the conduction band from the substituted hydrogen. This remaining hydrogen at the very end of dehydrogenation will act as the nucleation center for the β-phase (MH). It will also facilitate the hydrogenation process with a much reduced PCT hysteresis as seen from Figure 2—two single component AB_5 MH alloys (large hysteresis) and one multiphase AB_2 MH alloy (small hysteresis).

The above description, in general, works for both gaseous phase and electrochemistry. However, the synergetic effects in electrochemistry are more profound than those in the gaseous phase. With additional potential difference present across the alloy, the grain boundary acts as a liaison between two phases (changing from charged side to the uncharged side, resulting in two partially charged sides). The sharp transition in gaseous phase equilibrium pressure will not be seen in the electrochemical environment. In a separate article, a group of MH alloys composed of BCC and non-BCC (TiNi, C14, and Ti_2Ni) phases show two distinctive pressure plateaus in PCT but no noticeable jump or change in the slope of electrochemical charge/discharge curves even though x-ray diffraction analysis indicates the transformation from the completion of hydrogenation of the non-BCC phases (lower equilibrium pressure) to the start of hydrogenation of the BCC phase (higher equilibrium pressure) [40]. In the current study, the equivalent equilibrium pressure in the electrochemical environment of the phase with a higher plateau pressure ($ZrNi_5$) is lowered by the synergetic effects from the phase with a lower plateau pressure (Zr_2Ni_7).

According to our experience, the comparison between PCT and EPCT measurements is summarized in Table 3. Advantages of EPCT over PCT are its lower cost, ease of operation, ability to cover a wider voltage (pressure) range, and similarity to the real battery environment. The disadvantages of EPCT are that it is difficult to perform phase identification (catalyst *versus* storage) and separate the contributions from the surface and alloy bulk; it is also insensitive to several important MH alloy characteristics (metal-hydrogen bond strength, homogeneity, degree of disorder in both alloy and hydride, *etc.*).

Table 3. The comparison between PCT and EPCT techniques.

Aspect	PCT	EPCT
Main application	Gaseous phase hydrogen storage	Electrochemistry
Set-up	Valve, pressure gauge, vacuum pump, and heater	Basic electrochemical station
Equipment cost	~$20,000 USD per channel	~$1000 USD per channel
Sample size	0.5 g to 2 g	100 mg to 100 g
Sample shape	Ingot	-200 mesh powder
Pre-condition	Fresh cleavage or acid etched	Hot alkaline bath or electrochemical cycling
Main safety concern	Hydrogen gas safety; powder after PCT measurement can be highly pyrophoric	Cycled electrode after drying can be pyrophoric
Temperature range	$-20\,°C$ to $200\,°C$	$-40\,°C$ to $50\,°C$
Equivalent pressure range	0.0001 MPa to 10 MPa	Wider than 1 Pa to 1000 MPa
Synergetic effects in multiphase alloy	Determined mainly by the difference in metal-hydrogen bond strength	Determined by the difference in work function
Properties obtained	Storage capacity, metal-hydrogen bond strength, thermodynamic, hydride/dehydride hysteresis, and interaction among phases	Storage capacity up to 1 atm, contribution from the activated surface
Not suitable for	Alloys with very high or very low plateau pressure, alloys with poor surface catalytic abilities, alloys in the fine particle form	Alloys with low corrosion resistance in alkaline solution, alloys with high plateau pressure (>1 atm), alloys with very large hardness or ductility

4. Conclusions

Electrochemical PCT isotherms were obtained for several MH alloys. The results after self-discharge calibration show differences between gaseous phase PCT and EPCT. EPCT isotherm measurements of the AB_2 and AB_5 MH alloys prove that the PCT curves are much flatter than the curves for EPCT. PCT also has a smaller hysteresis and higher capacity. Moreover, when the hydrogen storage content is small, the PCT isotherm is closer to the EPCT charge curve. However, as the hydrogen storage content increases, the PCT isotherm then moves to the center between the EPCT charge and discharge curves and finally flattens out. In addition, previous and current studies on the multiphase Zr-Ni-V MH alloy (YC#1) demonstrate that the plateau pressure reduction in the electrochemical environment due to synergetic effects is the main cause of the increase in electrochemical discharge capacity compared to the gaseous phase measurements. Furthermore, it was found that the inflection points in the EPCT isotherm of a physical mixture of more than one alloy can be used to find a preset calibration point for measurement of SOC.

Acknowledgments: The authors would like to thank the following for their assistance: from BASF-Ovonic: Lixin Wang, Taihei Ouchi, Baoquan Huang, Jean Nei, Diana F. Wong, David Pawlik, Allen Chan, Ryan J. Blankenship, and Su Cronogue.

Author Contributions: Kwo-Hsiung Young and Tiejun Meng conceived and designed the experiments; Negar Mosavati performed the experiments; Negar Mosavati and Kwo-Hsiung Young analyzed the data; Kwo-Hsiung Young and K. Y. Simon Ng wrote the paper.

Conflicts of Interest: The authors declare no conflict of interest.

References

1. Young, K. Metal Hydrides. In *Reference Module in Chemistry, Molecular Sciences and Chemical Engineering*; Reedijk, J., Ed.; Elsevier: Waltham, MA, USA, 2012.
2. Young, K.; Chao, B.; Liu, Y.; Nei, J. Microstructures of the oxides on the activated AB_2 and AB_5 metal hydride alloys surface. *J. Alloy. Compd.* **2014**, *606*, 97–104.
3. Young, K.; Wong, D.F.; Nei, J.; Reichman, B. Electrochemical properties of hypo-stoichiometric Y-doped AB_2 metal hydride alloys at ultra-low temperature. *J. Alloy. Compd.* **2015**, *643*, 17–27.
4. Wong, D.F.; Young, K.; Nei, J.; Wang, L.; Ng, K.Y.S. Effects of Nd-addition on the structural, hydrogen storage, and electrochemical properties of C14 metal hydride alloys. *J. Alloy. Compd.* **2015**, *647*, 507–518.
5. Young, K.; Chang, S.; Chao, B.; Nei, J. Microstructures of the activated Si-containing AB_2 metal hydride alloy surface by transmission electron microscope. *Batteries* **2016**, *2*.
6. Young, K.; Huang, B.; Regmi, R.K.; Lawes, G.; Liu, Y. Comparisons of metallic clusters imbedded in the surface of AB_2, AB_5, and A_2B_7 alloys. *J. Alloy. Compd.* **2010**, *506*, 831–840.
7. Song, J.; Meng, B.; Tan, X.; Liu, S. Surface-modified proton conducting perovskite hollow fibre membranes by Pd-coating for enhanced hydrogen permeation. *Int. J. Hydrog. Energy* **2015**, *40*, 6118–6127.
8. Bliznakov, S.; Lefterova, E.; Bozukov, L.; Popov, A.; Andreev, P. Techniques for Characterization of Hydrogen Absorption/Desorption in Metal Hydride Alloys. In Proceedings of the International Workshop Advanced Techniques for Energy Source Investigation and Testing, Sofia, Bulgaria, 4–9 September 2004; pp. 19-1–19-6.
9. Raju, M.; Ananth, M.V.; Vijayaraghaven, L. Influence of temperature on the electrochemical characteristics of $MmNi_{3.03}Si_{0.85}Co_{0.60}Mn_{0.31}Al_{0.08}$ hydrogen storage alloys. *J. Power Sources* **2008**, *180*, 830–835.
10. Feng, F.; Ping, X.; Zhou, Z.; Geng, M.; Han, J.; Northwood, D.O. The relationship between equilibrium potential during discharge and hydrogen concentration in a metal hydride electrode. *Int. J. Hydrog. Energy* **1998**, *23*, 599–602.

11. Bliznakov, S.; Lefterova, E.; Dimitrov, N. Electrochemical PCT isotherm study of hydrogen absorption/desorption in AB_5 type intermetallic compounds. *Int. J. Hydrog. Energy* **2008**, *33*, 5789–5794.

12. Kopczyk, M.; Wójcik, G.; Mlynarek, G.; Sierczyńska, A.; Bełtowska-Brzezinska, M. Electrochemical absoption-desoprtion of hydrogen on multicomponent Zr-Ti-V-Ni-Cr-Fe alloys in alkaline solution. *J. Appl. Electrochem.* **1996**, *26*, 639–645.

13. Wójcik, G.; Kopczyk, M.; Mlynarek, G.; Majchrzyckia, W.; Bełtowska-Brzezinska, M. Electrochemical behavior of multicomponent Zr-Ti-V-Ni-Cr-Fe alloys in alkaline solution. *J. Power Sources* **1996**, *58*, 73–78.

14. Zhu, Y.F.; Pan, H.G.; Gao, M.X.; Ma, J.X.; Li, S.Q.; Wang, Q.D. The effect of Zr substitution for Ti on the microstructures and electrochemical properties of electrode alloys $Ti_{1-x}Zr_xV_{1.6}Mn_{0.32}Cr_{0.48}Ni_{0.6}$. *Int. J. Hydrog. Energy* **2015**, *40*, 6118–6127.

15. Hou, C.; Zhao, M.; Li, J.; Huang, L.; Wang, Y.; Yue, M. Enthalpy change (ΔH°) and entropy (ΔS°) measurement of $CeMn_{1-x}Al_{1-x}Ni_{2x}$ (x = 0.00, 0.25, 0.50 and 0.75) hydrides by electrochemical P-C-T curve. *Int. J. Hydrog. Energy* **2008**, *33*, 3762–3766.

16. Hu, F.; Zhang, Y.; Zhang, Y.; Hou, Z.; Dong, Z.; Deng, L. Microstructure and electrochemical hydrogen storage characteristics of $CeMg_{12} + 100wt\%Ni + Ywt\%TiF_3$ (Y = 0, 3, 5) alloys prepared by ball milling (in Chinese). *J. Inorg. Mater.* **2013**, *28*, 217–223.

17. Kudo, H.; Unno, K.; Kamegawa, A.; Takamura, H.; Okada, M. Synthesis and protium absorption properties of vapor growth carbon nano-fibers grown by Fe-based catalyst. *Mater. Trans.* **2002**, *42*, 1127–1132.

18. Yang, H.; Zhang, H.; Mo, W.; Zhou, Z. $Mg_{1.8}La_{0.2}$Ni-xNi nanocomposites for electrochemical hydrogen storage. *J. Phys. Chem. B* **2006**, *110*, 25769–25774. PubMed]

19. Kleperis, J.; Wójcik, G.; Czerwinski, A.; Showronski, J.; Kopczyk, M.; Bełtowska-Brzezinska, M. Electrochemical behavior of metal hydride. *J. Solid State Electrochem.* **2001**, *5*, 229–249.

20. Watanabe, K.; Koseki, M.; Kumagai, N. Effect of cobalt addition to nickel hydroxide as a positive material for rechargeable alkaline batteries. *J. Power Sources* **1996**, *58*, 23–28.

21. Young, K.; Ouchi, T.; Huang, B.; Reichman, B.; Fetcenko, M.A. The structure, hydrogen storage, and electrochemical properties of Fe-doped C14-predominating AB_2 metal hydride alloys. *Int. J. Hydrog. Energy* **2011**, *36*, 12296–12304.

22. Young, K.; Young, M.; Chang, S.; Huang, B. Synergetic effects in electrochemical properties of $ZrV_xNi_{4.5-x}$ (x = 0.0, 0.1, 0.2, 0.3, 0.4, and 0.5) metal hydride alloys. *J. Alloy. Compd.* **2013**, *560*, 33–41.

23. Young, K.; Huang, B.; Ouchi, T. Studies of Co, Al, and Mn substitutions in $NdNi_5$ metal hydride alloys. *J. Alloy. Compd.* **2012**, *543*, 90–98.

24. Sun, D.; Jiang, J.; Lei, Y.; Liu, W.; Wu, J.; Wang, Q. Effects of measurement factor on electrochemical capacity of some hydrogen storage alloys. *Mater. Sci. Eng. B* **1995**, *30*, 19–22.

25. Chang, S.; Young, K.; Ouchi, T.; Meng, T.; Nei, J.; Wu, X. Studies on incorporation of Mg in Ti-based AB_2 metal hydride alloys. *Batteries* **2016**, *2*. submitted.

26. Young, K.; Fetcenko, M.A.; Li, F.; Ouchi, T. Structural, thermodynamic, and electrochemical properties of $Ti_xZr_{1-x}(VNiCrMnCoAl)_2$ C14 Laves phase alloys. *J. Alloy. Compd.* **2008**, *464*, 238–247.

27. Iba, H.; Akiba, E. Hydrogen absorption and modulated structure in Ti-V-Mn alloys. *J. Alloy. Compd.* **1997**, *253–254*, 21–24.

28. Young, K.; Ouchi, T.; Huang, B.; Reichman, B.; Blankenship, R. Improvement in $-40\,^{\circ}C$ electrochemical properties of AB_2 metal hydride alloy by silicon incorporation. *J. Alloy. Compd.* **2013**, *575*, 65–72.

29. Bundy, K.; Karlsson, M.; Lindbergh, G.; Landqvist, A. An electrochemical impedance spectroscopy method for prediction of the state of charge of a nickel-metal hydride battery at open circuit and during discharge. *J. Power Sources* **1998**, *72*, 118–125.

30. Slepski, P.; Darowicki, K.; Janicka, E.; Sierczynska, A. Application of electrochemical impedance spectroscopy to monitoring discharging process of nickel/metal hydride battery. *J. Power Sources* **2013**, *241*, 121–126.

31. Pop, V.; Bergveld, H.J.; Notten, P.H.L.; Reftien, P.P.L. State-of-the-art of battery state-of-charge determination. *Meas. Sci. Technol.* **2005**, *16*, R93–R110.

32. Williams, M.; Mackie, C.; Barbour, P.W. Method and Apparatus for Determining the State of Charge of a Battery. U.S. Patent 4,958,127, 18 September 1990.

33. Galbraith, R.E.; Gisi, J.M.; Norgaard, S.P.; Reetz, D.D.; Ziebarth, D.J. Method and Apparatus for Estimating the Service Life of a Battery. U.S. Patent 6,191,556 B1, 20 February 2001.

34. Lim, J.H.; Ji, S.; Han, S. Method for Measuring SOC of a Battery Management System and the Apparatus Thereof. U.S. Patent 8,548,761, 1 October 2013.

35. Zhang, Y.; Du, X.; Salman, M.A.; Huang, S. Battery SOC Estimation with Automatic Correction. U.S. Patent 9,108,524, 18 August 2015.

36. Young, M.; Chang, S.; Young, K.; Nei, J. Hydrogen storage properties of $ZrV_xNi_{3.5-x}$ ($x = 0.0$–0.9) metal hydride alloys. *J. Alloy. Compd.* **2013**, *580*, S171–S174.

37. Young, K.; Ouchi, T.; Fetcenko, M.A. Pressure-composition-temperature hysteresis in C14 Laves phase alloys: Part 1. Simple ternary alloys. *J. Alloy. Compd.* **2009**, *480*, 428–433.

38. Young, K.; Chao, B.; Pawlik, D.; Shen, H.T. Transmission electron microscope studies in the surface oxide on the La-containing AB_2 metal hydride alloy. *J. Alloy. Compd.* **2016**.

39. Visintin, A.; Peretti, A.A.; Fruiz, F.; Corso, H.L.; Triaca, W.E. Effect of additional catalytic phases imposed by sintering on the hydrogen absorption behavior of AB_2 type Zr-based alloys. *J. Alloy. Compd.* **2007**, *428*, 244–251.

40. Young, K.; Ouchi, T.; Meng, T. Studies in synergetic effects in multi-phase metal hydride alloys. *Batteries* **2016**, *2*. submitted.

MDPI AG

St. Alban-Anlage 66

4052 Basel, Switzerland

Tel. +41 61 683 77 34

Fax +41 61 302 89 18

http://www.mdpi.com

Batteries Editorial Office

E-mail: batteries@mdpi.com

http://www.mdpi.com/journal/batteries

9 783038 423027